GENETICS IN THE COURTS

Henry M. Butzel

Studies in Health and Human Services
Volume 9

The Edwin Mellen Press
Lewiston/Queenston

Library of Congress Cataloging-in-Publication Data

Butzel, Henry M.
    Genetics in the courts.

    (Studies in health and human services ; v. 9)
    Bibliography: p.
    Includes indexes.
    1. Evidence, Expert--United States. 2. Forensic.
genetics. I. Title. II. Series.
KF8964.B88 1986   347.73'67   86-21725
            347.3-067
ISBN 0-88946-134-1 (alk. paper)

This is volume 9 in the continuing series
Studies in Health and Human Services
Volume 9 ISBN 0-88946-134-1
SHHS Series ISBN 0-88946-126-0

The Edwin Mellen Press   The Edwin Mellen Press
Box 450   Box 67
Lewiston, New York   Queenston, Ontario
USA 14092   LOS 1LO CANADA

Printed in the United States of America

To my father,
the late Justice Henry M. Butzel, Sr.
who ardently believed that justice and law
are synonymous.

TABLE OF CONTENTS

PREFACE

I am *obviously* not a lawyer, nor has this book been
written specifically for lawyers. Rather it is an attempt to
present, in one volume, all of those facets of law and of
genetics which intertwine in court decisions. I also do not
pretend to be a scholar of the law. Rather my point of view
will be that of a person trained in the field of genetics and
with a keen interest as an observer of the law. Thus a
certain amount of personal interpretations and feelings will
be presented in the writing of this book, and the reader
should be reminded that what is presented here is not
intended to be a definitive interpretation of the law as it
might be made by a judge or other arbiter, but rather a study
of how the field of genetics has influenced legal thinking
and judicial decisions. Another major theme will be the
problems of the law dealing with advances in the science of
genetics, and how the courts come to accept, or reject, new
findings in that science.

While there are many law reviews dealing with the various
fields presented here, in no instance have the multiple
aspects of the interaction of genetics and the courts been
comprehensively brought together in one volume. In the course
of study for this work, I have read each case cited herein in
its original form, and there will be an attempt to extract
not only the legal and genetic precedents set, but equally to
extract the flavor of the decision by allowing the courts to
speak for themselves.

Although my main purpose is to demonstrate and
discuss the use of genetic evidence in the courts, it is
essential that some cases which deal with non-genetic

matters be presented also so that the background for the use
of later genetic discoveries is established. For example, the
first cases of "wrongful life" did not arise from genetically
deformed children suing because of their birth, but from such
matters as illegitimacy or birth defects arising from the
teratological effects of such events as rubella or other
prenatal insults during early pregnancy. The precedents set
by these early decisions become the foundation for the
genetic cases and must be analysed first to set the
historical perspective.

When one first thinks of genetic evidence in the courts,
such obvious examples as paternity cases, analysis of body
fluids in criminal cases, or misdiagnoses of genetically
caused birth defects obviously come to mind; however, as will
become apparent, genetic evidence enters into almost all
fields of law. Such areas as contract law, patent law,
environmental law, adoption law, as well as cases involving
inheritance laws and antitrust laws all have some
interactions with genetic evidence. Indeed, if one interprets
the word genetics more broadly, then all cases in which the
person's sex or color are determined genetically, such as
racial or sexual discrimination, would also be considered as
cases involving genetic facts as well. From such cases as
these one is automatically drawn into constitutional law,
involving
both the Constitution itself and the Bill of Rights,
particularly the First, Fourth, Fifth, and Fourteenth
amendments. Such cases have arisen and will be frequently
referred to in this study.

It must also be pointed out that the monetary settlement
of each case is not the important point to be

made. What is important is to observe how the courts interpret what the law at any given point of time is assumed to be, particularly as new genetic and other scientific evidence is presented to the courts for evaluation and use in decision-making.

A note should be made concerning the selection of cases presented here. In many areas, such as paternity cases, there are probably hundreds, if not thousands, of such cases. I have chosen to use my discretion by selecting those which I feel are of importance in illustrating the use of genetic principles in the courts, and also those which seem to be the most interesting in terms of opinions and of the problems, both legal and ethical, which these cases bring to judges and juries.

Also, genetic evidence is not limited only to case law. All 50 states have statutes dealing with genetics in various ways, and there are several federal statutes dealing directly with genetic matters and many federal regulations which cite genetic principles; certainly these also must be considered in any study dealing with genetics and the legal process.

While there cannot help but be some overlap, the book is roughly divided into those cases affecting persons directly (chapters 1-6) and cases involving other issues such as liability and warranties where the main issues are financial rather than personal. Chapter 12 deals with a miscellaneous set of issues, some personal and some financial; either most of the issues here are unique or there are not enough cases for each to require a separate chapter. Chapter 13 is a brief summary of the present statutory laws dealing with genetics.

A note on the form of this text concerning footnotes and bibliography should also be made. Scientific writing abjures the use of footnotes while legal opinions and law reviews use them copiously. Similarly the form of bibliographic references also differs between the two styles. I have elected to compromise by not using footnotes, but by following the more simple form of legal citations, namely the volume, the title of the reference work cited, and the page number. I am, of course aware that in legal writing not only is the reference to the appropriate *Reporter* given, but also the exact citation to the original state reference. I have chosen to omit the latter as the reader may turn to the state records directly if that is desired. As the various citations are to the same case and opinion, it seems best in a general text to omit the state references and give only the more widely available citation from the various *Reporters*. In this way I hope that the reader's attention will not be diverted by the interruption of copious footnotes, but his or her ability to look up the original opinion or article cited will be made easier in any law library. In addition, the bibliography is in two forms, alphabetically for those interested in finding a particular case, and by year for those wishing to follow an historical approach.

Many people have helped to make this work possible. I would first like to thank Union College for the awarding of a Mellon Foundation Grant for Faculty Development which allowed me almost a year's paid leave. I would like also to express my thanks to Dean Richard Bartlett of the Albany Law School and to the Director and all of the staff of the Albany Law School Library

for their help and support and for their help in instructing
me in the use of the computerized legal search system. With
the use of that system what would have taken years was
reduced to months.

My wife, Miriam, has been most helpful in proof-reading
the drafts of this manuscript and in her criticism of the
work. In addition, Mr. Brian Quint, a former Union College
student, has done much of the work of checking the legal
citations for accuracy. Finally, I would like to thank the
students in my course in *Genetics & the Law* who unknowingly
served as guinea pigs for the experimental draft of this
manuscript. Their many helpful comments and corrections have
been fully incorporated into this text.

INTRODUCTION

Compared to the field of modern genetics, less than
100 years old, the "law" is ancient. In fact, in today's
parlance, the first case might well be named, *State v.
Adam, Eve, and an unnamed Serpent* (3 Genesis 7-24). In
this case the "State" served simultaneously as the law-
giver (the legislature today), the prosecuting attorney and
the judge, in the latter capacity as the Person who also
determined the sentence (exile and knowledge of sin for
the humans, pain in childbirth for the woman, and a
permanent apodal condition for the co-conspirator, the
serpent). The "trial" was swift (the basis for the right
to a speedy trial under our present Bill of Rights?),
and obviously there could be no appeal as the court of
original jurisdiction and the court of appeals were one
and the same. Perhaps the speed of the trial was due to
the fact that there was no other person to serve as
defense attorney!

"Laws" must have existed since the beginning of
civilization, either to insure the orderly working of
social intercourse or to define norms of conduct which,
if violated would lead to known consequences such as
exile from the tribe, loss of goods, or in extreme cases
death. These "laws" were originally not codified or set
down in writing, but were part of the culturally
inherited traditions. Some of these laws may have even
had a primitive, intuitive scientific basis, such as the
forbidding of the eating of pork by the Jews
(trichinosis), the almost universal incest taboos, with
the obvious exception of the Egyptian Pharaohs
(prevention of defective children) or the
reservation of the best seed for future planting or the

selection of the best animals for breeding by even
primitive societies.

Present United States Laws have an ancient
tradition; much of our law is based on English Common
Law, some of which in turn can be traced back to Roman Law
and even to Greek Law. (The laws in the State of
Louisiana of course are different, tracing their
origin to the Napoleonic Code of France.) But "laws" of
some form or other are basic and ancient inventions of
man.

On the other hand science is not ancient in the
sense that "law" is. While early man must have observed
some of the "laws of nature," such as the progression of
the seasons or the facts of sexual reproduction,
codification of science is relatively new. Modern
genetics is even more recent in its development,
encompassing little over a hundred years. While Mendel
established the basis of heredity in 1865, his discovery
went almost unnoticed for 35 years until the
simultaneous rediscovery of his work by three
independent investigators in 1900. The first examples of
heredity in man were forthcoming in the next two
decades, and modern molecular genetics is scarcely a
quarter of a century old. The original purpose of
genetics, in describing the laws of heredity and their
applications, was first to understand the passing of
traits from parents to offspring, and only recently has
the science turned to amelioration of human suffering
and to betterment of life for all. The first use of
human genetics, the so-called eugenics movement, arising
in the decades after World War I and with a demise
caused by the total misuse of the laws of heredity by
the Nazis, was designed to better mankind, but eventually led

to the racist views not only of the Germans, but also of much
of the racism and the racist policies applied to minority
groups in this country as well.

There is also a basic difference between the way
law operates and the methods of science and genetics.
The law must, because of its nature, deal with codified
procedural methods. Thus comparison to what has occurred
in the past is a part of most judicial opinions. The
principle of *stare decisis* (let the previous decision
stand), the application of earlier decisions to the case
at hand, arose so that some consistency in the law could
be observed. Thus the law exists in part so that
violators of known precepts may know the consequences of
their acts, and is constantly looking backward to
earlier cases. A major problem both for attorneys and
judges is to decide whether the facts which led to
previous decisions are sufficiently similar to the case
at hand as to render the earlier decision applicable. It
is always the task of one side of the case or the other
to argue that the facts of the present cases are
sufficiently different so that there is, or is not, an
equivalence between this case at hand and earlier
decisions.

Applications of "law" are based on the adversary
method; i.e., each side of a case tries to establish
that it represents the true facts and the correct law,
and the opposition either is factually inaccurate, or
does not apply the legal principles correctly. Trials
are based on this principle and it is left for either
the judge or the jury to determine whether one or the
other side has made the stronger case. In many instances
decisions by one tribunal can be appealed to another,
higher court, with the ultimate decision of American

higher court, with the ultimate decision of American laws being that of the United States Supreme Court. This final decision is made by humans, and is not subject to test by experimentation.

On the other hand, science does not operate in these fashions. While certainly using past discoveries as a basis for further work, science does not hold precedents as "sacred." Any investigator can, if proper data are found, overthrow single-handedly earlier stated conclusions, and science advances steadily both by looking forward and by "standing on the shoulders" of those who came before. Because the purpose of science is to make new discoveries to explain the laws of nature, it is constantly moving into new areas and affirming new concepts.

It is in this area, that of new concepts, that a natural tension between law and genetics arises. The present pace of molecular biology and genetics is simply too fast for the law to keep up with, and judges or juries are often presented with recent scientific discoveries which have not traditionally been used in court cases applying the laws to which these new discoveries are now relevant. For example, new methods of determining paternity with high probability require that the deciders of cases must deal with such complexities as statistics, modern methods of blood typing by such means as electrophoresis or HLA antigen tests, and eventually probably identification of individual chromosomes in an accused putative father and the offspring. Courts are simply not equipped, either by nature or training to deal with the flood of new scientific information, and are called upon constantly to establish which discoveries are valid and acceptable and which are only theoretical. The admission of scientific evidence is therefore an issue in many cases; is the genetic evidence correct in terms of science, and if so is it admissible in trials before either a judge or a judge and jury?

It is not just in paternity cases, but in many other fields of law, in which new genetic discoveries are of critical importance in reaching decisions. In all of the various areas to be discussed in this book, the issue either of validity of evidence or its relevance forms one of the central themes of discussion.

Science does not operate under the adversary principles. While many bitter arguments may arise between investigators, there is an ultimate test, further experimentation. Science must be replicable to be valid, and each individual experiment cannot simply use *stare decisis* to see whether past results match new data; the data must be analysed by themselves and the conclusion reached must be based on the most logical decision as to the meaning of the data.

Perhaps another philosophical view differentiates the law from science. This is the difference between objectivity and subjectivity. While all judges must attempt to bring an impersonal viewpoint to each case, this cannot help but lead to some subjective feelings on the part of the decider. It is for this reason, and the enactments of laws by various legislators, each of whom also cannot help but be motivated by personal past experiences or beliefs, that our state laws and state decisions vary so widely. Laws and decisions are made by humans with all of the best qualities and imperfections which we possess, and are not challengeable by experimental means. While, of course, science is a human

pursuit, it has certain different bases for its
procedures. A scientific experiment must be replicable,
it must be original in that it does not simply repeat
previous discoveries, and it has *predictability* as a
major premise. A scientific discovery is universal, and
is not true only in some places or at some times. It is
apparent that legal practitioners and scientists must
operate from different bases. Again, this leads to
conflicts between the two fields, although, as will be
shown these need not be irreconcilable. New scientific
discoveries will lead to changes in the laws, albeit
sometimes only after a much longer passage of time than
that found in science. Indeed as new scientific methods
are discovered, there is often a request from triers of
fact that legislatures act to bring state or federal
laws into line with the present state of science.
Unfortunately, as science is constantly making new
discoveries, legislation can only attempt to "catch up"
with science which has in many instances already moved
beyond such attempts into new areas which again make the
laws obsolete to some degree. Again, these are the types
of issues which will be dealt with in this book.

Another problem arises, as mentioned in the
epilogue, namely the almost incomprehensible numbers of
new cases and new applications of genetics to the law
which arise yearly. Almost as soon as this book is
finished some of the present status of the law will be
changed either by legislative or judicial acts. One can
only attempt to analyse the present cases at the time of
this writing, with the hope that this will form the
background to understand future decisions and future
changes. It is impossible to be completely up-to-date
when two fields, genetics and law, are changing almost

from day to day. However, the information presented here
is intended to be as complete and contemporary as
possible, and to allow an informed reader to grasp the
state of the art in both genetics and law and thereby be
prepared to recognize new discoveries and new legal
principles as they develop.

GENETICS IN THE COURTS

Henry M. Butzel

Chapter 1
WRONGFUL LIFE: THE EVOLUTION OF A TORT

THE ORIGINS

In 1963 Joseph Zepeda sued his father for having
caused him to be born an adulterine bastard, the first
case of "wrongful life." (An adulterine bastard, as
contrasted to an ordinary bastard, is a child born to
parents one or both of whom is married to another person
not a biological parent of the child. An ordinary
bastard is simply an illegitimate child born to two
unmarried parents.) The suit was basically for two
reasons: first it must be remembered that at that time
the social stigma of illegitimacy was much more damaging
to an individual than it might be today, and, second,
and more important, in most states in 1963 an
illegitimate child was nullius filius, literally the
"child of no one," with no rights of inheritance from
either parent. Thus Joseph was suing on the basis that,
given the conditions of his moral and legal status, he
would have been better off not to have been born at all.
The facts in the case were simple: the father had
assured his mistress that he was unmarried; however,
when she became pregnant he then admitted that he was
actually married with several legitimate children but
he also reassured her that he was in the process of
obtaining a divorce and that when the decree became
final he would then marry her. He had no such intention.
To complicate matters further, it should be remembered
that, at that time both adultery and fornication were
criminal offenses, and the elder Zepeda might well have
been subject to not only these charges, but possibly
also to fraud.

The Illinois Supreme Court in a split decision decided in a somewhat equivocal manner. First, they agreed that the tort of "wrongful life" was a genuine one, that is to say that the fact that no such tort had previously been recognized by any court was in itself no cause for banning such a claim from being legally brought and legally recognized. However, the court refused to recognize any claim for damages on the basis that, "...Be that as it may, the quintessence of his complaint is that he was born and that he is. Herein lies the intrinsic difficulty of this case, a difficulty which gives rise to this question: are there overriding legal, social, judicial or other considerations which should preclude recognition of a cause of action? ... Encouragement would extend to all others born into the world under conditions they might regard as adverse. One might seek damages for being born of a certain color, another because of race; one for being born with a hereditary disease, another for inheriting unfortunate family characteristics; one for being born into a large and destitute family, another because a parent has an unsavory reputation."

Also in the opinion, the court raised problems which at that time had not arisen, but which would later be a subject of much legal action. "Cases are appearing in the domestic relations field concerning children born as a result of artificial insemination...How long will it be before a child so reproduced sues in tort those responsible for its being? Will there be public sperm banks such as the blood banks we now have?....to protect the issue of the astronauts from mutations resulting from ionizing radiation in space, a technique has been developed for them to deposit their sperm in a sperm bank and to preserve it indefinitely through a refrigeration process. If there are public sperm banks in future years and if there are sperm injections like present day blood transfusions, with donors and donees unknown to each other, will there not be a basis for an action for wrongful life?"

Nevertheless, the court recognized that wrongful life was indeed a legal tort, even if no damages were to be allowed. In the final paragraph of the opinion the court acknowledged and established thereby this tort. "We have decided to affirm the dismissal of the complaint. We do this, despite our designation of the wrong committed herein as a tort, because of our belief that lawmaking, while inherent in the judicial process, should not be indulged in where the result could be so sweeping as here. The interest of society is so involved, the action needed to redress the tort could be so far-reaching, that the policy of the State should be declared by the representatives of the people."

Thus, the court recognized the principle that wrongful life was of itself a legal tort, but to award damages would require that legislative action be taken. More importantly, the establishment of this tort was an act of judicial recognition that there might be some conditions in which it might have been better for an individual not to have been born at all rather than live the type of life to which the conditions of his birth assigned him.

The second seminal case arose in New York State in 1966 when the Court of Appeals ruled upon a similar claim, although not so worded, in the case of *Williams v. State of New York*. In this instance suit was brought on behalf of an illegitimate child born as the result of an assault upon a mentally defective woman in a New York State mental institution. The particulars of the suit were the claims that due to negligence upon the part of the State the child had been "deprived of property rights; deprived of a normal childhood and home life; deprived of proper parental care, support and rearing; caused to bear the stigma of illegitimacy."

The court dismissed all claims by stating in part, "Impossibility of entertaining this suit comes not so much from difficulty in measuring the alleged 'damages' as from the absence from our legal concepts of any such idea as a 'wrong' to a later-born child caused by permitting a woman to be violated and to bear an out-of-wedlock infant. ...But the law knows no cure or compensation for it, and the policy and social reasons against providing such compensation are at least as strong as those which might be thought to favor it." While the court did cite *Zepeda*, it did not quote that part of the opinion which did acknowledge the tort of wrongful life. In a concurring opinion one judge added, "What troubles me about this case is what one commentator has described as the 'logical' difficulty of permitting recovery when the very act which caused the plaintiff's birth was the same one responsible for whatever damage she has suffered or will suffer ...The measure of damage which she is really seeking is based upon a comparison of the position she finds herself in now and the position she would have been in had she been born legitimately. Quite obviously, that is an unwarranted comparison here, for had the State acted responsibly, she would not have been born legitimately-- she would not have been born at all." Thus, the Court dismissed the claim and held essentially that there was no such tort as wrongful life, despite the Illinois decision in the *Zepeda* case.

These two cases, which will be cited in almost every subsequent such case, obviously disparate in the recognition of the tort of wrongful life, are the origins of such claims and will be of utmost importance in the cases which follow.

EARLY EVOLUTION: THE RUBELLA CASES

The precedent set by these two cases became more important as new, unanticipated causes of suit for wrongful life were brought. Again these were not due to genetic causes, *sensu strictu*, but were due to the birth of severely malformed children whose mothers had suffered from rubella early in their pregnancies. It is well known that this disease, if it occurs in early pregnancy, will almost invariably cause severe deformation. For this reason, many physicians advocate deliberate exposure to the rubella virus at an early age as a single exposure is thought to give lifelong immunity and thereby prevent the risks of the disease during pregnancy.

The causes and claims brought by plaintiffs in the rubella cases became more varied and more complex. Suits were brought claiming such damages as emotional suffering on the part of either or both parents of the defective child, for the physical pain and suffering during what would have been an unwanted pregnancy had the facts been known, for the special costs of medical treatment and raising of the deformed child, and for loss of consortium. All of these, in various combinations, were the bases for the three cases to be presented next.

The first, and by far the most cited, was *Gleitman
v. Cosgrove* in New Jersey in 1967. Here the parents of a
deformed child brought suit against Dr.Cosgrove for
allegedly not informing the mother of the risk to her
child due to her having had rubella during her first two
months of pregnancy. In fact, she claimed that the
physician had specifically denied any such risk several
times. Her son was born with substantial defects in
sight, hearing and speech. Dr. Cosgrove claimed that he
had informed her that some 20% of such children, exposed
to rubella in early pregnancy, developed some
malfunction but that he did not believe in aborting four
healthy babies because the fifth one might be malformed.
Further, at the time of this case, the Supreme Court
decisions regarding abortion had not been made and
abortion in New Jersey was a criminal offense unless
proven necessary to protect the health of the mother.
Mrs. Gleitman testified that had she known the real risk
she would have sought abortion elsewhere where it was
legal. The Gleitmans brought suit for wrongful life on
behalf of their son, and separately for their own
emotional suffering and also for the recovery of the
substantial costs of the rearing and care oftheir son.
The trial judge dismissed all three claims on the basis
that the action of the physician did not in itself cause
the malformations and that nothing he could have done
would have prevented the outcome of the pregnancy, and
the court further dismissed the last two claims because
of the illegality of abortion in New Jersey.

The New Jersey Supreme Court also denied any
redress to the Gleitmans. The opinion is of such legal
importance that it is necessary to quote from it at
length. "The infant plaintiff is therefore required to

say not that he should not have been born without defects but that he should not have been born at all. In the language of tort laws he says: but for the negligence of defendants he would not have been born to suffer with an impaired body. In other words, he claims that the conduct of defendants prevented his mother from obtaining an abortion which would have terminated his existence, and that his very life is 'wrongful.'...The infant plaintiff would have us measure the difference between his life with defects against the utter void of nonexistence, but it is impossible to make such a decision. This Court cannot weigh the value of life with impairments against the nonexistence of life itself. We hold that the first count on behalf of Jeffrey Gleitman is not actionable because the conduct complained of, even if true, does not give rise to damages cognizable at law." The Court cited the previous two cases, but chose only to consider the public policy part of the *Zepeda* decision and did not mention that the Illinois court had acknowledged that wrongful life was a legal tort.

The Court here went on to consider the claims of the Gleitmans' for emotional and financial redress and to deny them also, saying in part, "In order to determine their compensatory damages a court would have to evaluate the denial to them of the intangible, unmeasurable, and complex benefits of motherhood and fatherhood and weigh these against the alleged emotional and money injuries. Such a proposed weighing is similar to that which we have found impossible for the infant plaintiff. When the parents say their child should not have been born, they make it impossible for a court to measure their damages in being the mother and father.

The right to life is inalienable in our society. A court cannot say what defects should prevent an embryo from being allowed life such that the denial of the opportunity to terminate the existence of a defective child in embryo can support a cause for action.

In one concurring opinion one judge based his affirmation solely upon the illegality of abortion and described at some length the history of abortion laws. Much more important for later decisions, a dissenting opinion was also filed. In this dissent the judge, after pointing out that the physician's failure to inform Mrs. Gleitman of the risk to her child was both a moral and a legal wrong, continued and said, "While the law cannot remove the heartache or undo the harm, it can afford some reasonable measure of compensation towards alleviating the financial burdens. In failing to do so, it permits a wrong with serious consequential injury to go wholly unredressed. That provides no deterrent to professional irresponsibility and is neither just nor compatible with the expanding principles of liability in the field of torts. ...I find no substantial basis for the majority's opinion that it would be 'impossible' for a court to deal properly with the matter of compensatory damages." During the course of his dissent the judge cited many instances where damages for emotional suffering were awarded, although certainly not in this type of case.

A second dissenting opinion was also written, this one dealing with the issue of the abortion laws. This judge used his opinion to state his strong opposition on constitutional grounds to the New Jersey abortion laws and their applicability to the case at hand. He also took the opportunity to challenge the constitutionality of the New Jersey laws which forbade the open sale of contraceptives "without just cause." The dissents foreshadow much of what later courts will decide, but this, of course, was of no value to the Gleitmans who were left with the difficult task, both emotional and financial, of dealing with their birth-defective child.

In *Stewart v. Long Island College Hospital*, a New York Supreme Court (it must be remembered that in this state the Supreme Court is a lower court) dealt with much the same issue in 1968. In this case a pregnant woman who knew she had suffered from rubella applied to the hospital for an abortion. The hospital abortion committee of four doctors was divided evenly and her request was denied. However, it was also established that Mrs. Stewart was not informed of the committee's decision, and that she was explicitly told she should not seek an abortion elsewhere. It was also shown that such therapeutic abortions were routinely performed in this hospital when deemed necessary. Suit was therefore brought on behalf of the birth-defective child for being wrongfully born with physical and mental handicaps which she would have throughout her life. Suit was also brought by the parents. The Supreme Court denied the claims for wrongful life on the basis of the *Williams* decision previously mentioned; however it upheld, albeit reluctantly, the jury award of $10,000 to the mother and $1.00 to the father. Thus it was established that a mother <u>can</u> recover for damages either emotional or physical in this type of case.

The third basic case arose in Texas in 1975. Here the Supreme Court of that state (in this case the highest state court) was not dealing with a true case of wrongful life, but only with the issue of costs of rearing the birth-defective child. In *Jacobs v. Thiemer* the facts were quite similar to the above cases in that Dr. Thiemer was sued both for being negligent in failing to diagnose a case of rubella in Mrs. Jacobs's early pregnancy and for failing to advise the parents of the risk to a child from this disease during pregnancy. The parents sued to recover the costs of raising their child and both the trial court and the Court of Civil Appeals denied their claim. However, the Texas Supreme Court allowed the claims and remanded the case to the lower court for retrial. The basis of their decision was the fact that the economic burden was well established, and it denied the lower courts' decisions that the abortion laws precluded the rights of the parents to make their own decision concerning the abortion. They also expressly denied the claims of Dr. Thiemer who had stated that had he advised them to have an abortion he would have thereby committed a felony himself and been liable to a heavy fine or to being jailed. The court cited many precedents, some of which will be discussed later, in dealing with other aspects of wrongful life cases, saying in part, "The plaintiffs George and Dortha Jacobs have stated a cause of action against Dr. Thiemer. The suit for recovery of expenses reasonably necessary for the care and treatment of their child's physical impairment, due to Mrs. Jacobs having contracted rubella, is not barred by considerations of public policy." The damages being sued for, $21,472, could be readily ascertained from the records. The important precedent here is not the amount of damages, nor the discussion of the abortion laws at that time, but that the parents can recover costs from the claimed negligence of the physician.

The various legal problems raised by these cases
should now be made more clear in terms of tort law. In
order for an act to be considered a tort it must be
shown that the committer of the tort (tort-feasor) had
an obligation to the person claiming injury as well as
there being both negligence and foreseeable harm to the
person claiming damages. Damages may also be of two
kinds, compensatory damages and punitive damages. The
former is an award to the person damaged in order to
restore him to a condition which existed prior to the
alleged tort, and the latter are damages to penalize
the tort-feasor and, secondarily, to set a precedent so
that others will not commit the same tort.

This raises the legal issues involved in the
previous cases and most cases involving claims of
personal injury. In the rubella cases the physicians
claimed that no act of theirs had created the disease
and nothing they could have done would have prevented
the birth defects and there could therefore be no tort
committed upon the foetus by the physician. As a result,
no claim for wrongful life would be a valid one under
tort laws. The second defense claim was that there was
no way to determine actual damages as the child cannot
be "made whole" by any financial award. However, damages
to the parents for their suffering, both physical and
mental, as well as the extra costs of rearing a defective
child might be awarded as it is possible to estimate
these costs. In addition, the costs of rearing a normal
but unwanted child can also be readily determined.

Another legal issue is whether damages can be
inflicted upon an infant *in utero* (literally in the
uterus of the mother). Here for many years a conflict
between criminal law and tort law existed. In criminal

cases the child was considered *esse* (in existence) from the time of "quickening" while in tort law the child was not considered to be in existence until the time of birth. The interpretation under criminal law, of course, gave rise to the expression "the quick and the dead." A negligent act which resulted in a defective child could not therefore be considered a tort as the act, if proven, would have occurred prior to the existence of the infant; hence no tort suit can be brought. There can be, according to Prosser, no "negligence in the air" so to speak. These legal points must be kept in mind as we follow the attempts by the courts to deal with wrongful life, wrongful birth, and wrongful conception cases.

Another feature which appears in most of the following cases is the "speciation" of the wrongful life claim. Henceforth, wrongful life refers to cases where the infant himself, through a guardian, brings suit for having been born. Two new derivative claims arise from this. The first is for "wrongful conception," and the second for "wrongful birth." In the former the suit usually follows an unsuccessful vasectomy or tubal ligation, while the second arises when the physician either through ignorance or negligence fails to inform the parents of the possible malformation of their child in time for them to obtain a legal abortion during the early part of a pregnancy, a right established by the United States Supreme Court in 1973.

Further, some courts now distinguish between three types of abortion. First, a therapeutic abortion is an abortion performed to protect the health of the pregnant woman. Second, an eugenic abortion is performed when there are either genetic or other reasons leading to a high probability of a defective child being born. The third type, a so-called sociological abortion, may be sought when an undesired pregnancy occurs and the woman does not wish to carry to term. Cases where a couple already have a large number of children and for economic reasons do not want another child, or where pregnancy occurs to an unmarried woman, would fit in this rubric. There are, of course, cases where more than one reason may be involved in seeking abortion; for example a woman with a large number of children and at genetic risk for producing a defective child would fit both the socioeconomic and the genetic abortion classification. And cases in which both wrongful birth and wrongful pregnancy occur will arise. These cases are numerous, and in the remainder of this chapter an attempt will be made to select those which have the greatest impact on other cases as well as those which present a high degree of ethical problems.

With these considerations in mind, we can now examine other rubella cases, again to set the framework for the genetic cases which arise later. The first, *Sylvia v. Gobeille* was decided by the Supreme Court of Rhode Island in 1966. Suit was brought against Dr. Gobeille for allegedly failing to administer gamma globulin to a pregnant woman even though he was aware that she had been exposed to rubella early in her pregnancy (gamma globulin, a factor derived from blood, may, if given early enough, prevent the effects of rubella). The court dealt primarily with the legal issue whether a tort could be committed upon an unborn child and decided positively, stating in part, "...since 1946 in an almost unbroken line of decisions the courts have rejected the arguments once relied upon and have said that a cause of action accrues for prenatal injuries.

Where once they recognized a child's legal existence while *en ventre sa mere* [literally in its mother's belly] in the criminal law and with respects to his property rights and his right to inherit but denied it for purposes of allowing suit for prenatal injury, they now extend that recognition to the field of torts. They answer the suggestion that such an extension may encourage false claims by saying that the issue of causal connection between the injury to the unborn child and the negligence of another is one of proof which differs, if at all, only in degree from other medical questions which arise in negligence cases.

"While we could, as has sometimes been done elsewhere, justify our rejection of the viability concept on the medical fact that a fetus becomes a living human being from the moment of conception, we do so not on the authority of the biologist but because we are unable logically to conclude that a claim for an injury inflicted prior to viability is any less meritorious than one sustained after." In this case the parent was permitted to bring suit for damages and to bring the case again before the lower court to determine the amount of damages to be awarded. There was no claim of wrongful life in this case; today it would probably be considered as a case of wrongful birth.

In *Dumer v. St. Michael's Hospital* wrongful life was an issue. Here, in 1975, the Supreme Court of Wisconsin dealt with a suit brought against the Hospital, claiming a breach of duty based upon the fact that the emergency room nurses and the attending physician diagnosed her condition as being due only to allergy and not to rubella; at the time she was approximately one month pregnant. As a result of this misdiagnosis, her child was born severely malformed. The court held that there was no claim against the Hospital as the attending nurses had done their duty properly and called a physician, and there was no claim that the Hospital staff had failed to carry out an order by the doctor. Next, the Court, relying heavily on *Gleitman* ruled that there was no tort of wrongful life recognizable under law. However, the Court did rule that there could be a suit against the attending physician for negligence. The case was remanded to the lower court to ascertain whether such negligence was shown by the doctor's failure to diagnose the occurrence of rubella properly, and in the failure to inquire whether the patient was pregnant. In the event negligence could be shown the lower court could then determine what damages to assess. Again this could be a case of wrongful birth as the mother testified that had she been properly informed concerning the risk of a defective child she would have sought an abortion.

The Court of Appeals of Michigan dealt with a similar issue in 1981. In *Eisbrunner v.Stanley* the facts were basically the same as the above case, failure of the physician to diagnose rubella and to advise the pregnant woman of the risks involved to her child. Suit was brought for damages for mental distress of the parents and for the costs of the 5-year treatment of the infant who had died before the trial. Again, the *Gleitman* decision was cited and no case for wrongful life was allowed, but as in *Dumer* the rights of the parents to sue for their claims were upheld.

The U. S. Court of Appeals, Seventh Circuit recognized wrongful birth as a legal tort. The case, appealed from the U. S. District Court in Illinois, involved a federal issue as the alleged tort of failing to diagnose and to advise a pregnant woman who contracted rubella during her pregnancy occurred in a military hospital. In this case, *Robek v. United States*, the Court decided there was just cause for the suit. A complicating factor was that the case was originally brought as the result of an act committed in Alabama, and the Court had to decide whether wrongful birth was indeed a tort under the Alabama state laws. It was ruled that there was nothing to prevent such a claim, despite the defense's plea that at the time of the origin of the case, 1972, there was no cause for action "because of the strong public policy against abortion." The Court reasoned in part that the *Roe* decision had been made and the State of Alabama would now recognize such a claim as the mother could have sought a legal abortion. While other factors entered into the decision, the major point is that the recognition of wrongful birth is now recognized as a tort by the federal court system.

These cases have been described in some detail because, while they occurred contemporaneously with many of the genetic cases, they are essential as a background to establish the basis for suits brought solely for genetic reasons. Admittedly rubella is not an hereditary disease, but the legacy of the rubella decisions is the basis for much that follows.

As pointed out previously, a basic issue in cases involving genetics in the courts is the determination whether a tort can be committed upon an unborn infant, regardless of the stage of pregnancy of the mother. Again the courts have vacillated on this issue, and a review of some of this type of case is also necessary for understanding the genetic cases. It should also be

understood that many of the later genetic facts were not
known at the time of the decisions to be discussed next,
but these cases have direct applicability to the later
genetic issues.

One of the earliest cases, *Stemmer v. Klein*, dates
as far back as 1942. The Court of Errors and Appeals of
New Jersey wrote a lengthy opinion summarizing many
previous cases dealing with the time at which "legal"
life began. In this particular case a child sued for
damages caused to him before birth. He claimed to have
been born defective due to the misuse of x-rays during
his mother's pregnancy. It was also shown during the
presentation of evidence in the lower courts that the
physician had misdiagnosed the pregnancy as being a tumor and
that the massive dosage of x-rays administered to the mother
was the cause of the infant's prenatal injury leading to
birth defects. The majority opinion held there was no
possibility of a tort as the child was unborn and it upheld
the defendant physician. However, a dissenting opinion
strongly disagreed with the majority, citing English Common
Law and particularly Blackstone who said, "Life is an
immediate gift of God, a right inherent by nature in every
individual and it begins in *contemplation* of law as soon as
an infant is able to *stir* in the mother's womb."
The dissenting judge went on to ask, "If the civil rights
of an unborn begin with conception as to property and the
like, upon what reasonable ground may there be a distinction
against its rights to recover for serious personal injuries
done it through the negligence of a physician?

Apparently this dissenting opinion held
considerable influence upon the New Jersey Courts, for
in 1960, in the case, *Smith v. Brennan*, the Supreme

Court of that state overruled the previous decision and held that an unborn could recover for damages caused to him by injury to his mother in an automobile accident. There were other legal cases to cause the Court to overlook the principle of *stare decisis*, but the strong opinion stated, "And regardless of analogies to other areas of the law, justice requires that a child has a right to begin life with a sound mind and body. If the wrongful conduct of another interferes with that right, and it can be established by competent proof that there is a causal connection between the wrongful interference and the harm suffered by the child when born, damages for such harm should be recoverable by the child. ... Therefore the rule of *Stemmer v. Klein* is no longer the law of this state."

Some of the intervening cases referred to in this decision should also be examined. In 1951 the Court of Appeals in New York examined this question in *Woods v. Lancet* as well as reviewing other cases. They concluded that a child who was injured severely enough during the ninth month of his mother's pregnancy so as to be born with permanent defects could recover from the one who caused his injury. To the argument made by the defendant that there was no such law, the Court responded curtly, "...and if that were a valid objection the common law would now be what it was in the Plantagenet period. When these ghosts of the past stand in the path of justice clanking their mediaevel chains, the proper course for the judge is to pass through them undeterred." Despite a dissent stating that such changes in the law were best left to legislatures, the law allowing recovery for the unborn stands.

A case which illustrates both the acceptance of the

principle of prenatal injury, and the ignorance of genetics at the time was *Sinkler v. Kneale* decided by the Supreme Court of Pennsylvania in 1960. Here the court awarded damages to a Down Syndrome child based on the syndrome being caused by an automobile accident when the foetus was approximately one month of age. An interesting conclusion to the opinion was the statement, "It is not the fact that an unborn child is part of its mother, but rather in the unborn state it lived with the mother, we might say, and from conception on developed its own distinct, separate personality." Of course, with today's knowledge of the genetic basis of Down Syndrome (the presence of an extra 21st chromosome, known as trisomy 21) the case would not have been allowed, but the principle of the separate existence of the child from the time of conception is clear.

A U.S. District Court in 1956, in a case which foreshadows one to be discussed later, ruled in *Morgan v. United States* that a negligent use of the wrong type of blood so as to cause Rh incompatability and birth defects was a legitimate cause for action. However, in this case the issue was moot as the statute of limitations for bringing the suit had expired before the action began, even though there would have been a tortious action had the suit been brought in time.

The Appellate Division of the New York court system again in 1953 decided in *Kelly v. Gregory* that the courts would assume that prenatal injury could be inflicted any time after conception. They ruled that damages to a 3-month foetus caused by an auto accident were recoverable, and affirmed the lower court's decision in this regard. "It is to be noticed that no distinction was attempted to be drawn in determining the

point of vestiture of a legal right. Conception and
vestiture become coincidental in the full sense of the
word."

In *Hornbuckel v. Plantation Pipe Line* the Supreme
Court of Georgia came to a similar opinion. Here a
dissenting opinion captured the thinking of many
legalists at that time. "...I am unwilling now, in the
absence of legislation...to hold that the life of a
person, possessing or forming the subject of individual
personality, begins when the male and female elements of
procreation unite to form the seed of a person.
Assuredly, we could not call an acorn a tree. The
eternal riddle which came first, the egg or the
chicken?, can be solved by saying they are one and the
same...How can there ever be a definite time fixed when
the egg in the body of the mother is fertilized by the
father's spermatozoa? Does it take place in one minute,
or two hours, or two days, or two weeks after
copulation?" It might be added that the judge was
somewhat naive about biology if he thought that the
spermatazoa could survive two weeks, but otherwise the
dilemma of when legal life begins is clear.

Perhaps the most remarkable case of prenatal injury
occurred in Illinois in 1977. A divided Supreme Court in
that state in *Renslow v. Mennonite Hospital* dealt with
these facts. A mother brought suit on behalf of her
minor daughter for damages brought about by the
Hospital. When the mother was 13 years old she underwent
a blood transfusion at that Hospital receiving 500cc of
Rh positive blood. The girl was Rh negative. The
transfusion caused a sensitization so that several years
later when she conceived a child the expected Rh
incompatibility between her and the child occurred (her

husband must have been Rh positive) and the child was brought into the world with severe mental and physical damages. The lower court dismissed her claim on the basis that she was not "at the time of the alleged infliction of the injury conceived." The higher court reversed the lower court on various legal grounds. Many points were dealt with in deciding whether a tort could be committed upon a child not conceived until years after the time of the tort. The defense's claims that this could not be done were not recognized as valid, even though this is certainly a new interpretation of the tort law. The Supreme Court, quoting from Professor Corbin, said, "It is the function of our courts to keep the doctrines up to date with the *mores* by continual restatement and by giving them a continually new content. This is judicial legislation, and the judge legislates at his peril. Nevertheless, it is the necessity and duty of such legislation that gives to judicial office its highest honor: and no brave and honest judge shirks the duty or fears the peril. ... Obviously courts create law. If it were otherwise the common law would be as out of touch with life as is a corpse. Courts must take an active part in the development of the common law, although this may mean creativeness." Thus the court ignored the plea that only the legislature could rule on this issue, and despite the many years which passed between the alleged tort and the birth of the child, a suit for damages could be recognized. A further statement was made to the effect that "It was foreseeable that this 13-year old girl would grow up, marry and become pregnant." Thus the issues of duty and foreseeable injury, upon which tort law is based, were together present and damages could be recovered. It should be added that several strong dissents were entered by four of the judges, but the majority opinion stands, and it is now possible for recovery by an infant for damages done long before the child is conceived.

Other medical malpractice cases based on injury to the unborn have also been allowed. For example a 1978 case, *Bergstresser v. Mitchell*, decided by the U.S. Court of Appeals, allowed recovery for damages inflicted when a physician in performing a Caesarean section caused rupture of the mother's uterus ten weeks before expected delivery of the infant, and caused her to undergo a premature emergency Caesarean during which alleged damages (hypoxia or anoxia) to the infant resulted. Thus the federal courts also acknowledge tortious action prior to birth.

Surprisingly, and perhaps anachronistically, a 1981 N.Y. case did not take the previous cases into account. During the course of an abortion the mother's uterus was punctured and a subsequent pregnancy resulted in a defective child. In this case, *Albala v. City of New York*, the Appellate Court ruled that no injury to the latter could be claimed under law. While admitting this was the first case of this kind in the state, the court fell back upon *Williams* and seemingly ignored the previously cited cases. Although there are many intermediate cases cited in the opinion, the decision basically returned to *Zepeda* for its reasoning. Indeed the Court used some of the cases presented above as the basis for their decision although the precedents seem completely to be in the opposite direction.

These cases, dealing with whether a tort can be committed both on the unborn and the unconceived must be

considered as similar to many genetic cases, involving
such things as wrong advice from a genetic counselor or
from faulty laboratory tests for genetic defects which
may induce the mother to carry to term a pregnancy she
would otherwise have aborted. Damages caused to the
infant as a result of this type of action are obviously
upon the unborn infant, and in order to constitute a
tort, it must first be established that a tort of this
sort is legally recognized.

One final set of points which must be discussed
before turning to the actual genetic cases concerns the
awarding of damages in all these suits. As mentioned
earlier, damages for compensation can be brought in
several ways. The medical costs of an unwanted pregnancy
are a basis for suit as are the costs of the birth
itself. Also suits are brought for mental anguish and
suffering brought about by the pregnancy, and often by
the additional costs of rearing a birth-defect child.
Such costs can be brought in behalf of the mother and
father, and in wrongful life cases in behalf of the
infant itself. Other bases for costs may include loss
of wages by an employed woman who finds herself pregnant
unintentionally after a tubal ligation or a vasectomy
have failed. Suit has even been brought, although
usually not allowed, on behalf of the already existing
children who claim loss of a share of parental affection
and even loss of what now would become an extra share in
an estate should the parents die.

Further complications arise in many wrongful birth
or wrongful pregnancy cases dealing with awards to the
parents for negligence by the physician. This is due to
the problem of awarding financial costs for a *healthy*
but unwanted child as against the costs awarded for a

defective child under similar circumstances. For many years the law and the courts have held that the awards of parenthood, either emotional or financial, are so substantial that even an unwanted child is in itself sufficient cause for emotional and financial gain to the parents that no suit for loss based on these two reasons should even be awarded. Perhaps this arose during the times before the child labor laws were enacted and when children went to work at an early age both to earn their keep and to help increase the overall family earnings. Whatever the cause, these two obstacles to the recovery of damages must first be overcome before monetary awards for such types of cases can be established, and this section will deal primarily with this problem in the law.

In the wrongful life cases discussed earlier, it will be recalled that in *Zepeda*, *Williams*, and *Gleitman* no costs were allowed for rearing the child. In the first this was on a basis of public policy, while in the latter two no tort under existing laws was recognized. However, in *Thiemer* the judge found that costs could be awarded for the extra financial burden imposed upon the parents by the birth of their rubellaaffected child. Here the extra medical costs could be clearly determined and there were no claims of suit by the parents for their own emotional suffering, thereby making the issue a simple one of negligence.

Nowhere in the various fields of law does one find more differences of opinions than in the cases dealing with the rights of parents to recover for the costs of rearing a healthy, but unwanted child. While the earlier courts prior to the 1970's tended to disallow any such claims, in the later years more courts have tended to

regard these costs as perfectly sound claims due to the
negligence of the physicians who negligently perform
vasectomies, tubal ligations, abortions, or in some
cases, simply misdiagnose a pregnancy. The best way,
perhaps, of understanding the difficulties is again to
let the courts speak for themselves. While there are
many such cases which could be cited, the author has
again exercised the prerogative of selection of those
deemed most important. It should also be added that
almost all of the more recent decisions are in
themselves good legal reviews, and a reader wanting a
more complete bibliography of wrongful birth or wrongful
pregnancy should find it in almost any recent case.

In *Shaheen v. Knight*, a district court in
Pennsylvania brought the issue sharply into focus. They
simply stated in 1956, "To allow damages in a suit such
as this would mean that the physician would have to pay
for the fun, joy and affection which plaintiff Shaheen
will have in the rearing and education of this, the
defendant's fifth child.... He wants to have the child
and wants the physician to support it. In our opinion to
allow such damages would be against public policy." In
the same vein, a decision by the Circuit Court of
Appeals in California in 1976 may be cited. "Who can
place a price tag on a child's smile or the pride of a
child's achievement? Even if we consider only the
economic point of view, a child is some security for the
parents' old age....Every child's smile, every reason
for parental pride in a child's achievement, every
contribution by the child to the welfare and well being
of the family and parents is to remain with the mother
and father. For the most part these are intangible
benefits, but they are real." (*Sills v. Gratton*)

Similarly, the 1974 opinion of the Supreme Court of
Wisconsin in *Reick v. Medical Protection Company*
stated, "To permit the parents to keep their child and
shift the entire costs of its upbringing to a physician
who failed to determine or to inform them of the fact of
pregnancy would be to create a new category of surrogate
parent." Even as recently as 1982, the Supreme Court of
Arkansas in *Wilbur v. Kerr* stated, "Litigation cannot
answer every question; every question cannot be answered
in dollars and cents. We are also convinced that damages
to the child will be significant; that being an unwanted
or 'emotional bastard,' who will someday learn that its
parents did not want it, and, in fact, went to court to
force someone else to pay for its raising, will be
harmful to that child. It will undermine society's need
for a strong and healthy family relationship. We have
not become so sophisticated a society to dismiss that
emotional trauma as nonsense." In *Public Health
v. Brown*, a 1980 decision by a District Court in Florida
put it in a different light by stating, "On a more
practical level the validity of the principle may be
tested by simply asking any parent the purchase price
for that particular youngster. It is a rare but happy
instance in which a specific judicial decision can be
based solely upon a reflection of one of the most humane
ideals which form the foundation of our entire legal
system." It should be noted that a vigorous dissent was
entered in this case, attacking the majority decision
quite bitterly. "I firmly believe the result reached by
the majority in the name of humaneness is, unwittingly,
inhumane. I see nothing humane in denying a parent the
wherewithal which might save a child from deprivation
or, in many cases, abject poverty. I see nothing humane

in a rule of law that could enhance the already dire need of parents and existing siblings. I see nothing humane in a decision which effectively immunizes physicians from their negligence and victimizes a mother who sought to relieve herself and her family from the additional burden of an additional child."

On the other hand, not just a dissenting opinion, but the majority decision in many cases has upheld the rights of parents to collect damages for the upbringing of the unwanted child. Perhaps the most cited of all the cases allowing consideration of costs for rearing an unwanted child is *Custodio v. Bauer*, decided by the Court of Civil Appeal of the First District in California in 1967. Here the courts remanded for trial a case in which the lower court had refused to consider costs of raising an unwanted tenth child, born after an unsuccessful tubal ligation. After citing all of the above cases (as well as many which have been omitted in the interest of brevity) the court did decide that the issue should be faced. "With fears being echoed that Malthus was indeed right, there is some change in social ethics with respect to the family establishment. City, state and federal agencies have instituted programs for dispensing of contraceptive information with a view toward economic betterment of segments of the population. One cannot categorically say whether the tenth arrival in the Custodio family will be more emotionally upset if he arrives in an environment where each of the other members of the family must contribute to his support, or whether he will have a happier and more well-adjusted life if he brings with him the wherewithal to make it possible." This case thereby is one of the first which makes it possible that damages

from the negligent carrying out of sterilization may be
assessed for the extra cost of rearing an unwanted
child. In *Terrell v. Garcia*, the Court of Civil
Appeals in a 1973 Texas case stated, "The question is
whether a negligent doctor should be held responsible
for the consequences of his negligence. There is no
basis for the assumption that plaintiffs here will
derive any joy and satisfaction from the raising of the
unwanted child. Perhaps these parents, in deciding they
did not want to pay the price of the enjoyment and
pleasures which 'normal' parents would derive from the
birth of an unwanted child, were not acting as 'normal'
persons. But it is hornbook law that a tort-feasor must
take his victim as he finds him and has no right to
insist upon a 'normal' victim." Similarly, in *Schroeder
v. Perkel* the Supreme Court of New Jersey found in 1981
that, "A family is woven of the fibers of life; if one
strand is damaged the whole structure may suffer. The
filaments of family life, although individually spun,
create a web of interconnected legal interests. ...In
this context a wrong should not go unrequited and an
entire family left to suffer because of the dry
technicality that Thomas [the child] has the primary
obligation to pay for his own medical expenses. ...In
brief, the problem of wrongful conception and wrongful
birth involve an evaluation not only of law, but also of
morals, medicine, and society. Thus it is not surprising
that the same issue may elicit divergent judicial
responses." To which an observer of the law can only add
"Amen!" In a similar vein, in 1981 the Court of Appeals
of California in *Turpin v. Sortini* stated, "New and
nameless torts are being recognized constantly and the
progress of the common law is marked by many cases in

which the court has struck out boldly to create a new
cause of action where none has been recognized before.

"The law of torts is anything but static and the
limits of its development are never set. When it becomes
clear that the plaintiff's interests are entitled to
legal protection against the conduct of the defendant,
the mere fact that the claim is novel will not of itself
operate as a bar to remedy. ...There is good public
policy favoring the state leaving to its citizens the
question of when a couple should procreate. There is no
public policy favoring medical malpractice."
Unfortunately, it must be noted that this is only a
dissenting opinion, and the majority had ruled that the
suit was not valid.

Even earlier, the Supreme Court of Minnesota in
Sherlock v. Stillwater had stated emphatically,
"Ethical and religious views aside, it must be
recognized that such costs [rearing the unwanted child]
are a direct financial injury to the parents from the
wrongful conception and birth of the child. Although
public sentiment may recognize that to the vast majority
of parents the long-term and enduring benefits outweigh
the economic costs of rearing a healthy child, it would
seem myopic to declare today that those benefits exceed
the costs as a matter of law. ....Today it must be
recognized that the time-honored command to be 'fruitful
and multiply' has not only lost contemporary
significance to a growing number of potential parents
but is contrary to public policies embodied in the
statutes encouraging family planning. ...Compensatory
damages for the cost of rearing the child to an age of
majority would also, in our opinion, serve the useful
purpose of an added deterrent to negligent performance

of sterilization operations. ...We remain unconvinced
that a physician should be held harmless for the
economic costs of supporting an unplanned child. The
result we reach today is at best a mortal attempt to do
justice in an imperfect world. In this endeavor we are
not unmindful of the deep and often times painful
ethical problems that cases of this nature will continue
to pose for both courts and litigants." Again in two
similar cases, joined in Illinois as one, *Cockrum v.
Baumgartner*, The Appellate Court in 1981 ruled that
costs of rearing an unwanted child were actionable under
tort law. "The defendants do not dispute the legal
sufficiency of the allegation that their negligence was
the direct and proximate cause of the expenses which the
plaintiffs seek to recover. Instead they argue that for
reasons of public policy, damages should be limited to
pregnancy and birth-related costs. ...While we agree
that most parents hold the sentiment that the birth of a
healthy albeit unplanned child is always a benefit, we
are not inclined to raise this sentiment to the level of
public policy. The uniqueness of life is in no way
denigrated by a couple's choice not to have a child.
Neither the individual nor society as a whole is harmed
by the exercise of this choice. .... Regardless of
motivation, a couple has a right to determine whether
they will have a child. That right is legally protected
and need not be justified or explained. The allowance of
rearing costs is not an aspersion upon the value of the
child's life. It is instead a recognition of the
importance of the parent's fundamental right to control
their reproductivity. ...It has been suggested that the
parents who seek to recover the costs of raising and
educating an unplanned child should be required to

mitigate damages through abortion or adoption. ....These alternatives are uniquely personal choices which cannot be forced upon parents as a means of mitigating damages."

Yet, despite these cited cases, a Federal Court in 1981 found the opposite. In *White v. United States* the U.S. District Court, D. Kansas, denied the claim for expenses of rearing the unwanted child, saying in part, "In ruling these are not recoverable damages, we are following a growing number of courts that have refused to allow parents to recover costs to raise a child born after preventive measures have failed." The Court then cites many of the previous cases I have mentioned, but concludes that "Allowing the parents to recover for the costs of raising a child constitutes a windfall to the parents and an unreasonable burden upon physicians." Thus, even though many courts find no such "unreasonable burden," this court chose to follow its own path and to take a position which upheld the early decisions not to allow costs and disagreed strongly with the other decisions allowing recovery. The issue is still unsettled and probably will continue to be so as predicted in the 1981 *Schroeder* decision. In fact, a search of the various court decisions, of which there are scores, indicates not so much a final statement of the law, but more a series of individual decisions by various judges in various courts as to what the law is. The chance of recovery for negligence by a physician seems to be more dependent upon *where* the suit is brought, and in some cases to which court in a particular state the suit is made. It would seem to this observer that a lawyer either for the plaintiff or the defendant in these cases can pick and choose those cases upholding his client's position and then the court can do the same in making its own decision with good legal precedence for either side.

A different way of viewing these cases arose in Michigan in 1971 in *Troppi v. Scarf*. Here the facts are somewhat different from the usual failure of a physician to carry out his role in a sterilization procedure. Mrs. Troppi had already had seven children and during her eighth pregnancy suffered a miscarriage. She and her husband then decided to limit their family to the seven children. She consulted a physician who prescribed *Norinyl* and phoned this prescription to the druggist, Scarf. He mistakenly filled the prescription with *Nardyl*, a mild tranquilizer. As might be expected, Mrs. Troppi was soon pregnant again and eventually gave birth to a normal eighth child. The Troppis brought suit for various damages, including the economic costs of rearing this child. The Court of Appeals after first allowing the costs of the pregnancy and birth as well as loss of wages of the mother due to her pregnancy then dealt with the issue of the financial burden of raising the unplanned child. They invoked the "so-called benefit rule" to reach their decision. "Where the defendant's tortious conduct has caused harm to the plaintiff or to his property and in so doing has conferred upon the plaintiff a special benefit to the interest which was harmed, the value of the benefit conferred is considered in negation of damage where this is equitable. (Restatement, Torts 920, p.616.) "Consider, for example, the case of the unwed college student who becomes pregnant due to a pharmacist's failure to fill properly her prescription for oral contraceptives. Is it not likely that she has suffered far greater damage than

the young newlywed who, although her pregnancy arose
from the same sort of negligence, had planned the use of
contraceptives only temporarily, say while she and her
husband took an extended honeymoon trip? Without the
benefits rule, both plaintiffs would be entitled to
recover substantially the same damages. Application of
the benefits rule permits a trier of fact to find that
the birth of a child has materially benefitted the newlywed couple, notwithstanding the inconvenience of an
interrupted honeymoon and to reduce the net damage award
accordingly. Presumably the trier of fact would find
that the 'family interests' of the unmarried coed had
been enhanced very little." The court also ringingly
denied the defense claim that the mother could have had
an abortion or placed the child for adoption and thus
avoided the costs of raising the infant. "The defendant
does not have the right to insist that the victim of his
negligence have the emotional and mental makeup of a
woman who is willing to abort or place a child for
adoption. ...we are persuaded to rule, as a matter of
law, that no mother, wed or unwed, can reasonably be
required to abort (even if legal) or place her child for
adoption." The court then goes on to discuss the
uncertainty of damages and concludes that, while
difficult, it could be done. They cite an earlier
opinion of 1960 in which the same court had said, "But
where injury to some degree is found, we do not preclude
recovery for lack of precise proof. We do the best we
can with what we have. We do not, in the assessment of
damages, require a mathematical precision in situations
of injury where, from the very nature of the
circumstances, precision is unattainable." This Michigan
decision has been cited many times even in some of the

cases above. There has been no general agreement as to how to ascertain the benefits rule, but it has been left to individual lower courts or juries to attempt to apply it to cases of negligence resulting in either wrongful birth or wrongful pregnancy cases. This, along with the general lack of conformity of the courts in awarding costs of rearing a normal child after its unwanted birth due to inadequately performed vasectomies, tubal ligations, or even failed abortions, only adds to the difficulty of trying to predict how such cases will be decided in the future.

These cases have been described at some length. Of course, there are scores more which could have also been analysed. For example the following cases, dealing with the birth of a normal but unwanted child while not cited in detail also apply here: *Christensen v. Thornby* (Minn), 1934, no costs allowed; *Jackson v. Anderson* (Fla), 1970, remanded to determine costs; *Wilmington Medical Center v. Coleman* (Del) 1972, no costs; *Aronoff v. Snider* (Fla) 1974, no costs for other childrens' loss of share of love, affection or estate; *Cox v. Stretton* (NY) 1974, no costs for prior children; *Ziemba v. Sternberg*, 1974, costs allowed; *Coleman v. Garrison* (Del) 1974, no costs; *Stephens v. Spiwak* (Mich), 1975, no costs; *Betancourt v. Gaylor* (NJ) 1975, costs allowed; *Stills v. Gratton* (Ca) 1976, costs allowed on basis of *Troppi* above; *Anonymous v. Hospital* (Conn) 1976, costs allowed, also on basis of *Troppi*, *Ladies' Center of Clearwater,Inc. v. Reno* (Fla) 1977, cannot plead that failure of unwed father to use contraception removes negligence by physician in a failed tubal ligation; *Garwood v. Locke* (NY) 1977, no suit allowed for costs of raising child; *Sala v. Tomlinson* (NJ) 1979, no costs;

Rivera v. State of New York (NY) 1978, costs allowed;
Green v. Sudakin (Mich) 1978, costs allowed; *Wilczynski
v. Goodman* (Ill) 1979, no costs; *Clapham v. Yanga* (Mich)
1980, costs allowed; *Hartke v. McKelway* (US) 1981, no
costs; *Delaney v. Krafte*, (NY), 1984, no costs after a
failed abortion; *Jones v. Malinowski* (Md), costs after
failed tubal ligation; *Flowers v. District of Columbia*
(DC), 1984, no costs; *Jackson v. Baumgardner* (N.Car),
1984, costs allowed for failure to properly insert an
intrauterine device; *O'Toole v. Greenberg* (NY), 1985, no
costs. As can be seen, the viewpoint of the courts
varies widely from state to state and sometimes from
court to court within a state. These cases form, in part,
the basis for the genetic cases which will be discussed
next. But these, and the other cases already discussed
are the foundations upon which all of the genetic cases
are based, and without them an understanding of the
cases to be discussed next would be difficult. Although
the genetic cases almost always involve a *defective*
child, the judges will continually rely upon many of the
ones which involve normal children. Perhaps it is
obvious that if recovery for the birth of an unwanted,
but healthy, child is possible, then it should be much
easier to recover for the damages caused by the birth of
a genetically afflicted child, whether wanted or
unwanted.

The other issues in many cases, as pointed out
above, involve suits for damages for a variety of
causes: medical expenses of the unwanted pregnancy, loss
of earnings by the pregnant woman, emotional and
psychological suffering, and loss of services
(consortium). Table 1 shows a summary of a few such
cases. Another point which the above and following

cases raise is a most difficult one, in terms of the present controversy concerning the right to abortion. It must be noted that in many of these cases the courts have admitted that in tort law an infant is *in esse* from the time of conception, and in *Renslow* even prior to conception. Thus those advocates who insist upon the rights of the foetus as an individual can find many court decisions upholding their viewpoint. It must also be remembered that the extension of the legal rights of a foetus from criminal law to cases dealing with wrongful birth or conception is a relatively recent change in the interpretation of tort law. Nonetheless, it is somewhat surprising that the advocates of the so-called "Right to Life" movement have not used these legal decisions as a basis for their claims. The advocates of the right of a woman to opt for abortion might well be warned to be prepared to rebut these court cases which, despite *Roe*, seem to give their opponents a legal basis for their claims that life begins at conception.

With this background, we may now turn to those decisions in which genetic evidence is the main issue in the case. Of course, some of the cases cited earlier, particularly those dealing with Down Syndrome and with Rh factors have been essentially dealing with genetic evidence in establishing the principles of tort law; they have been included previously essentially as they established precedence for suits brought for preconceptual harm or for damages wrought upon an infant *in utero*. The cases which follow are those in which the main claim is made for genetic causes only. They are, however, firmly based on the previous cases already noted, and depend for the most part on those precedents.

The spread of the tort throughout the land.

Among the earliest cases in which genetics played a major role in the suit is *Naccaroto v. Grob*, decided by the Court of Appeals in Michigan in 1968. The facts are simple: an action was brought in behalf of an infant suffering from the genetic disease, phenylketonuria. This inherited disease is caused by the inability of the victim to metabolize one of the most common amino acids in foods, phenylalanine. Failure to diagnose the disease in early childhood results invariably in the infant's becoming feeble-minded. If the disease is diagnosed at very early infancy, a very restricted, and difficult to maintain, diet can be prescribed and the child will develop normal intelligence. In this case the Naccarato child was not diagnosed and the suit followed. The case is of interest not only for the introduction of genetic information, but also because it dealt mainly with the use of expert witnesses. The plaintiff brought two outside witnesses, one, Dr. Hsia from Chicago, who was a recognized authority in cases of phenylketonuria, and the other, Dr. Koch, who was equally renowned in this field. Despite the testimony of both, the court refused to be persuaded, as neither, it was felt, could testify on the state of medical care in Detroit, and evidence showed that at the time of the trial, treatment and diagnosis of the disease was not usual in this city. Although the court expressed sympathy for the infant, it felt constrained to deny the plea on the basis above. Thus the lack of knowledge of the two out-of-state doctors as to the prevailing standards of medical practice in Detroit was considered to be of more importance than their knowledge of the disease and the

case was decided in favor of the defendant physician.
(In order to establish malpractice it must be shown that
the accused physician violated the generally accepted
medical practice in the area where the supposed
malpractice occurred.)

Another early case, brought in federal court in
1973, is *Jorgensen v. Meade Johnson Laboratories*. Here
the U.S. Court of Appeals, 10th Circuit, ruled in favor
of the plaintiff in a case involving the claimed
teratrogenic effects of a birth control pill, Oracon.
The woman had been using this means of contraception for
several months and then, desiring to become pregnant,
ceased their use. Shortly afterwards pregnancy ensued,
and after a normal term she gave birth to twins.
Unfortunately, both children suffered from Down
Syndrome. The District Court dismissed the suit, but the
higher court vacated that decision and remanded the
case. One of the twins died a few years later, but the
other was surviving at the time of the suit. The claim
stated, "Specifically, the aforesaid birth control pills
altered the chromosome structure within the body of the
plaintiff's wife, Alta J. Jorgensen, and as a result
thereof, a Mongoloid deformity was created within the
viable fetus of the minor plaintiffs during the period
of development prior to birth." The court in deciding
that a tort was committed upon the children stated, "If
the view prevailed that tortious conduct occurring prior
to conception is not actionable in behalf of an infant
ultimately injured by the wrong, than an infant
suffering personal injury from a defective food product,
manufactured before his conception, would be without
remedy. Such reasoning runs counter to the various
principles of recovery which Oklahoma recognizes for

those ultimately suffering injuries proximately caused
by a defective product or instrumentality manufactured
and placed on the market by the defendant. ...For these
reasons we conclude that the complaint should not have
been dismissed. We are convinced that the Oklahoma
courts would afford an opportunity to present a case on
these allegations. It does not appear beyond doubt that
the plaintiff can prove no set of facts in support of
his claims which would entitle him to relief." In other
words, if the Jorgensens can present sufficient evidence
to the trier of facts that a causal relationship exists
between the use of the pills and the chromosomal
abnormality causing Down Syndrome, then they are
legally entitled to damages.

In the 1976 case in New York, *Howard v. Lecher*, the
courts were faced with the following issue: Could a
couple recover for the failure of a physician to advise
them of the risk of a child having Tay-Sachs disease?
This genetic defect is relatively common among Ashkenazi
Jews and relatively rare in the rest of the population.
Prenatal tests, particularly amniocentesis (the taking
of amniotic fluid by use of a hypodermic and culturing
cells from it), can detect the disease in the foetus.
The disease is particularly heart-rending as the newborn
infant appears normal, but gradually, over a two to three
year period, wastes away and eventually becomes a
"vegetable" and dies. Also it must be remembered that
there are adequate means of testing parents for their
being carriers of the trait. If both are carriers there
is then a one in four chance of any infant conceived by
them having the disease and amniocentesis should
probably be used routinely in these cases so that the
parents have the option, should the test reveal the

disease in the foetus, of having a eugenic abortion and attempting another pregnancy with the same safeguards. In this case, the parents claimed that the physician was negligent in not taking a case history, particularly as he knew the parents were both of South European Jewish ancestry. When their Tay-Sachs infant was born they brought suit on the basis that the necessary tests and advice were not given. The Supreme Court held in favor of the parents and an appeal was brought by the physician to the Appellate Division. This court, adhering rigidly to legal precedents, held that if a tort had been committed it was upon the infant and not the parents, and therefore they had no basis for suit. The Court split three to two, with a strong dissent saying in part, "Moreover, genetic counselling, which involves the communication process between physicians and patients in an attempt to deal with the occurrence or the risk of genetic disorders in a family, has become widespread in the last several years. Mrs. Howard, having conceived late in 1971, might therefore have expected her obstetrician, at the minimum, to ascertain the fact of her background and, once having done so, to conduct a serum assay (blood test) and the amniocentesis procedure." The dissent then goes on to cite many cases where damages to a foetus were recoverable by parents despite the majority opinion noted above, and in addition gives a detailed and thoughtful analysis of the disease, its genetic aspects, and concludes that the reasoning denying the claim given by the majority is wrong. "Inasmuch as it may be found that Mrs. Howard's injury was directly caused by the obstetrician's breach of duty, it follows that the damages are recoverable for her emotional and mental anguish. She was in the 'orbit of duty' for the breach of which the wrongdoer may be held liable. ...The very nature of the child's condition makes certain that the mother and father will suffer grave emotional harm." This dissent, however, obviously was not considered sufficient reason for the majority of the court to allow the parents to recover any damages they might have undergone.

Another genetic cause of disease of a newborn, polycystic kidney disease, also fatal, is the basis of the claims in *Park v.Chessin*. This case began in the Supreme Court in New York and found its way up through the Appellate Division and ultimately to the highest New York Court, the Court of Appeals, after being joined with a similar case, *Becker v. Schwartz*, in 1978. In the Park case, the facts are again undisputed. Dr. Chessin had been the physician in charge of Mrs. Park when she delivered a polycystic kidney child. Shortly thereafter she conceived again, and delivered a second birth-defect child with the same disease. Claims were brought here for wrongful life, for mental and physical damages on behalf of the deceased infant, for costs prior to its death, and for mental and physical anguish on behalf of the parents. The Supreme Court judge ruled that the physician was negligent in not properly advising the Parks, following the birth of the first child, of the risk of their next child falling victim to the same disease. Indeed, the parents claimed that Dr. Chessin had repeatedly told the Parks there was no risk involved. The lower court reviewed all of the earlier cases in a scholarly fashion and concluded that this case differed from Howard above as it was not really a wrongful life case at all, but a suit brought for the suffering of the infant prior to its death. "The infant decedent does not seek damages for being born, per se, but rather seeks damages for the pain suffered by her *after* her birth based upon the tort committed prior to conception."

The court then goes on to make a ringing declaration of the duty of the courts to render justice, "The law must ever move forward; it must not be allowed to become a motionless pool stagnated by mere lack of precedent or fall prey to the antiquated theory of 'public policy'...What statute or theory of law grants preferential treatment or immunity to the medical profession? The court is guardian of the rights of all the citizenry, not only of a chosen few. This court believes that the medical profession is not 'unreasonably burdened' if held liable for the injuries caused to those who depend upon it for their very lives and who are dependent upon that profession conducting itself within the legal standards set for it. Unlike other professions, the medical profession deals not with money or property, but with the continuance of life and avoidance of death. To use the worn-out, rejected cliche of 'public policy' is to grant preferential treatment to the medical profession over all other professions and enterprises where malpractice could result in payment of ensuing resultant damages. This was never truly contemplated by either the general public, the Legislature or the Court." The Appellate Division upheld most of the claims and modified the decision slightly, but added back the validity of the claims for mental anguish and for consortium. However, the Court of Appeals, consistent with the opinion in Howard above, dismissed the claim for wrongful life sustained by the Appellate Division, but the court "also held that a cause for 'wrongful life' by the parents in their own right does exist for the pecuniary damages suffered as a result of the birth as the damages are ascertainable; however, public policy reasons prevent parents from recovering damages for emotional harm as calculations for such an injury are too speculative" (Quotation from the summary). The court cites *Williams* supra and obviously rebutted the opinion by the Supreme Court judge cited previously.

As noted above, the courts consolidated two cases in hearing the appeals. While the claims were similar, the basis was slightly different. In *Park*, as noted, the birth defect was polycystic kidney disease. However, in *Becker*, the birth defect was Down Syndrome, a disease caused by the presence of an extra 21st chromosome, and one found in vastly increased numbers among the offspring of older mothers. In this case, Mrs. Becker was in the high risk age group, being 35 at the time. The Beckers' claims were similar to those of the Parks in that their claim for negligence was that the doctor did not monitor the pregnancy and warn them of the risks involved. As Down can be readily ascertained by amniocentesis, the Beckers' claimed that the failure of the physician in their case to recommend this procedure was the act of negligence for which suit was brought. Thus the courts in consolidating these two different genetic defects dealt with essentially the same issue, as did Howard, namely failure of proper genetic counseling.

Yet another genetically-caused birth defect, cri-du-chat, was the basis for the claims in *Johnson v. Yeshiva University* in 1977. This fatal disease is caused by a chromosomal defect which can also be detected by

amniocentesis. The Johnsons brought suit on the basis
that when they sought genetic counseling during the
course of pregnancy they were not advised to have
amniocentesis performed so that they could make an
informed decision regarding the possibility of aborting
the defective foetus. The lower court was again
sympathetic, but, as might be expected, the Court of
Appeals again remained firm in its denial of any claim
by the parents or the child on a somewhat different
ground, one which essentially begged the issue of
wrongful life. "But, however interesting those questions
may otherwise be, we cannot reach them here. For our
review of the record does not demonstrate that an issue
of fact was raised by the plaintiffs in the face of the
uncontroverted fact showing by the defendants that, on
the basis of the patient's medical history and the state
of medical knowledge regarding the use of the
amniocentesis test in 1969, the defendants' failure to
perform this test was no more than a permissable
exercise of medical judgement and not a departure from
the then accepted medical practice." Thus the issue of
wrongful life was not considered and the court, as in
Naccarato, based its decision solely upon the issue of
standard medical practice at that time.

Again in 1977 the Court of Appeals denied yet
another wrongful life suit. In *Karlsons v. Guerinot*, the
facts are somewhat similar to the previous ones. In this
case, despite the fact that the pregnant mother was 37
years old, had a thyroid condition and had previously
given birth to a deformed child, the defendants failed
to inform her of the risks and of the existence of the
amniocentesis test. Her child was born with Down
Syndrome and the Karlsons brought suit, citing no less

than 11 causes of action, including lack of informed
consent, breach of contract, negligence, malpractice,
wrongful life, damages for pain, suffering and mental
anguish, care and support of the child, loss of
potential income due to the care of the child, loss of
consortium, and for alleged development of cancer of the
breast related to emotional and physical stress. In
addition the father brought suit for many of the same
causes, and suit in behalf of the child for living an
impaired life was also involved. Although the court
denied most of these claims, an important new
development was reached when the court upheld the claims
for compensatory damages for the anguish of the parents
in rearing the child. "Additionally, the court grounded
its holding upon the conclusion that it is virtually
impossible to evaluate as compensatory damages the
anguish to the parents of rearing either a malformed
child or a child born with a fatal disease, measured
against the denial to them of the benefits of
parenthood. We reject this approach however and find the
cause of action here to be maintainable." Thus, the
failure of the physician to inform the parents or to
prescribe amniocentesis would appear in this instance to
be a tort. Surprisingly, although the court did allow
damages of this kind, there was no reference to the
"benefits" rule or to the Michigan case of *Troppi*.

A different approach was used in a federal suit in
1977. Here, in *Simon v. United States*, the malpractice
suit was brought against a government hospital for
allegedly causing an infant to be stillborn. The claim
here was that an unnecessary amniocentesis was performed
and also that improper foetal monitoring by not
administering an oxytocin challenge test were the causes

of the stillbirth. The United States District Court for
the Southern District of Florida dismissed the case for
wrongful *death* on the basis that the laws of the State
of Florida did not recognize such a claim for a
stillborn infant, but they upheld the claim for the
mother's physical pain and mental anguish and the
father's claim for loss of consortium. It should be
noted that no description of the facts was given in
detail in this opinion, and moreover that the risk of
damage to a foetus by amniocentesis is certainly not
more than 1%, thereby causing one to question whether
this was in any way a contributing factor in the case.

Another 1977 case, *Debora S. v. Sapega* concludes
the cases for the earlier years. Here the Supreme Court,
Appellate Division in New York, concluded that a
physician who did not properly diagnose a pregnancy of a
15-year old rape victim who probably would have
undergone an abortion but for the misdiagnosis was
liable for suit. The court carefully distinguished this
from Howard on the basis that that case dealt with a
wanted, but birth-defect child, while in the present
instant there was a healthy, but obviously unwanted
child. It seems somewhat peculiar that a birth-defect
infant whose existence could have been prevented had the
physician been more careful can not recover, while a
healthy infant born unwanted can.

The number of cases involving genetics increases
almost linearly in the years to come. It seems best to
deal with them in a more or less chronological order as
each sets some precedents for its successor. Also, as
will be seen, the cases are not limited to New York, but
become a national record of suits brought on genetic
bases.

In 1978 the Supreme Court of Alabama ruled on a
wrongful life case, *Elliot v. Brown*. In this case suit
was brought against the physician for a failed vasectomy
and a subsequent birth-defect child. The court, like so
many others, ruled there was no such tort and the case
brought for the infant was therefore not a matter for
the courts. The court did rule that separate suits on
behalf of the parents were justified. The usual
citations to *Gleitman* and others were made. In addition,
however, the court raised a new, and possibly major point
against such suits. "We are not unaware of the rapid
progress made in medical science in recent years. Many
mysteries of the how and why of human development have
succumbed to medical knowledge. However, we do not
understand that the state of the art in the medical
profession is such that it can be said that no child
need be born deformed. With deference to *Park v.
Chessin, supra,* we do not feel that it is possible to
say that technological, social and economic changes
merit the recognition of a cause of action on behalf of
the plaintiff for 'wrongful life.'" (It should be noted
that this decision was handed down prior to the reversal
by the New York Court of Appeals which denied a cause of
action in the case cited.)

"Upon what legal foundation is the court to
determine that it is better not to be born than to be
born with deformities? If the court permitted this type
of cause of action, then what criteria would be used to
determine the degree of deformity necessary to state a
claim for relief. We decline to pronounce judgement in
the imponderable area of nonexistence." Here the court
touches upon a real issue heretofore ignored. How
serious must a defect be before a cause of action may be

brought? For some such defects which can be readily
treated, with no subsequent ill effects, such as cleft
palate or extra digits, might seem an unbearable burden,
while for others these are the risks of childbirth which
must be faced by all. For some, a Down Syndrome child
would not be an overwhelming burden, while to others it
might lead to disastrous consequences. The very real
issue of what is a "defect" sooner or later must be
faced, and this court deserves credit for raising what
until now has been a hidden issue.

Also in 1978, in *Gildiner v. Thomas Jefferson
Memorial Hospital* a U.S. District Court again deals with
the issue of wrongful life in a case involving Tay-Sachs
disease and amniocentesis. In this case the diagnostic
test was carried out, but the results were wrong, and
the parents were informed that there was no chance of
their child having the disease. The child was born with
Tay-Sachs disease with a life expectancy estimated to be
5 years or less. Again the court referred to *Gleitman*
and other cases in refusing the claim of wrongful life
but admitted that the parents may have a case for
emotional damages. The defense raised the issue that
their claim was derivative from that of the child's and
that their alleged negligence did not *cause* the defect.
The court rejected both of these claims and further
stated, "Society has an interest in insuring that
genetic testing is properly performed and interpreted.
The failure to properly perform or interpret an
amniocentesis could cause either the abortion of a
healthy fetus, or the unwanted birth of a child
afflicted with Tay-Sachs disease. Either of these is
contrary to public policy of Pennsylvania. The
recognition of a cause of action for negligence in the

performance of genetic testing would encourage the
accurate performance of such testing by penalizing
physicians who fail to observe customary standards of
good medical practice. ...Our holding that the
plaintiffs have stated a cause for damages caused by
negligence in the performance and interpretation of an
amniocentesis involves the application of the doctrine
of negligence, established by the common law of the
State of Pennsylvania, to a recently developed medical
procedure. The determination of the scope of the common
law doctrine of negligence is within the province of the
judiciary." Thus the process of amniocentesis is now
fully recognized as a normal part of medical care, and
the earlier cases which referred to this test as being
outside of that standard would no longer hold if this
decision is to be used by other courts. In other words,
the scientific advance represented by amniocentesis is
now not only accepted, but in some instances may be a
necessary medical procedure.

Another New York case which found its way through
the series of courts there began in 1978, when suit was
brought for damages inflicted upon an infant *in utero* by
the ingestion of the drug, Delalutin, by the mother
during her pregnancy. The claim was that this drug
caused her child to be born without limbs and with other
serious and permanent injuries and defects. The Supreme
Court held that the claim for injuries to the parents
due to emotional damage, personality changes and extreme
mental anguish were properly brought and the suit was
valid. This case, *Vaccaro v. Squibb Corporation*, led the
lower court to declare, "Before a consideration of these
cases, the tragedy implicit in the facts of this case
set forth above compels this court to observe that it is

long overdue for our system of jurisprudence to give
rational effect to the realities of life and to the vast
changes which have occurred in modern society. Blind
adherence to the past and outmoded shibboleths do not
serve our society but only create distrust and cynicism
among our citizens in the viability of our system of
justice." Further, the judge stated, " In recent years,
advancements in medical science, and the development of
new drugs, have imposed additional duties and
responsibilities on physicians, to keep abreast of the
literature and the latest developments in the use of
such drugs. It is fair to say that some drugs are so
relatively new that we do not as yet know their long-
term effects on the human body, whatever may be their
claimed immediate therapeutic value." Earlier in the
opinion it had been stated, "Parenthetically, the drug
'Delalutin' is a progestational hormone known
generically as hydroxyprogesterone caproate (injection),
intended to prevent miscarriage. The record before this
court shows that the infant's mother, Inez Vaccaro, has
had three pregnancies, one which terminated in the birth
of a stillborn child at defendant hospital...another
which was terminated by a miscarriage, and the third
resulting in the birth of the infant plaintiff."

The Appellate Division essentially agreed with this
opinion, only modifying it slightly to deny some of the
many causes of action brought, specifically to deny the
claim against Squibb by the father and all other claims
against Squibb and the physician involved on his behalf.
Not in the least surprising, the Court of Appeals, which
is consistent, to say the least, overruled the lower
courts and ordered all suit for emotional and mental
harm dismissed. It is of interest to this observer that

in the cases mentioned to date in New York, the lower
courts seem to have delivered the more analytical and
certainly more humane decisions, while the highest court
seems to have been firmly conservative and to have
relied on *stare decisis* rather than to have considered
the facts and the human suffering involved. It might
seem that there are sufficient differences of opinion
between the lower and higher courts so that the latter
might have taken more cognizance of the attempts by the
lower court judges to recognize humane decisions which
they, at least, did not find to be in disaccord with
legal precedence. One is particularly referred to the
lower court's opinion in the first of these cases, *Park
v. Chessin*, in which a compassionate and scholarly judge
attempted to render both a strong legal and an equally
strong humanistic decision.

By far the most peculiar case arose in 1978, *Del
Zio v. The Presbyterian Hospital, Vande Wiele, and the
Trustees of Columbia University*. (The names of all the
defendants are given here as they play a role in the
factual description of this case.) The Del Zios were a
remarried couple, each of whom had had children during
their former marriages. However, it was found that Mrs.
Del Zio had subsequently developed lesions which totally
blocked her fallopian tubes and, despite the desire of
the couple for children of their own, she was
effectively sterile. She underwent surgery for the
correction of the blockage and shortly thereafter became
pregnant. Unfortunately, the pregnancy was terminated by
a miscarriage. It was also found that her tubes were
again blocked, and she underwent a second operation in
order to be able to become pregnant; in addition
to the operation, she was placed on fertility pills.

This was in vain as she failed to conceive during the next 14 months. She then learned of the possibility of an in *vitro fertilization* and of the experiments being carried out by Dr. Shettles, then at Columbia Presbyterian Hospital, and after consultation with her own physician decided to try this procedure as a very last resort. Accordingly the next 12 months were spent in preparation, continuing the fertility drug, and taking daily temperature records to determine the time of ovulation so that the procedure could be carried out at the proper time in her cycle. She entered the hospital where she underwent surgery. The operation was successful and an egg and fluids to maintain it were obtained. The egg was placed in a sterile tube, and her husband then took the tube to Dr. Shettles who obtained semen from Mr. Del Zio. The mixture of the egg and semen were then placed in an incubator in Dr. Shettle's laboratory to await the outcome, and it was planned, if fertilization occurred, to reimplant the egg in Mrs. Del Zio's uterus.

Sometime the next day, Dr. Shettles was summoned to Dr. Vande Wiele's office. To the shock of Dr. Shettles, the tube which he had carefully kept sterile was now on the desk of the defendant, Dr. Vande Wiele, and the stopper had been removed, rendering the culture useless as it was no longer sterile. Dr. Vande Wiele informed Dr. Shettles that he had personally ordered the removal of the culture and deliberately rendered it useless. He refused to return the culture to Dr. Shettles. He also informed the Del Zios' physician of his actions, based on the alleged proscription by the National Institutes of Health against such procedures.

The Del Zios then brought the suit against the

Hospital, Dr. Vande Wiele, and Columbia University for the destruction of their potential child. The defense claimed that no tort had been committed under New York law, and there should be no case, and further that Vande Wiele's actions were taken with complete justification under the Hospital and University procedures. In a short decision, a United States District Court ruled there was sufficient legal basis for the case to go to trial. A jury awarded a substantial amount to the Del Zios. The case, apparently, was not appealed. Indeed no record of the case appears in the New York or Federal Records, and the only facts available are in the form of the trial before the federal judge and the data reported by newspapers. Several interesting results of this case, however, will be of importance in the years to come. It should be remembered that at the time of these actions, there was absolutely no assurance of the success of the procedure as the pioneer work by Steptoe and his group in England using similar techniques had not yet been accomplished. But today when the number of *in vitro* fertilizations carried out successfully is rapidly rising, there may well be similar cases coming to the courts. One thinks of instances where a defective child might result, and of the problem of negligence by a hospital or physician should such occur. It will be interesting to see whether the Del Zio case will, like *Gleitman*, become the case upon which further suits are bottomed.

Another case not reported in the N.W. Reporter was that of *Werth v. Paroly* in 1979. Here a decision by the Wayne County Circuit Court for the plaintiff followed the presentation of the fact that a woman claimed the doctor did not properly inform her of the correct procedure for using an IUD. She had elected to use this method of birth control as she already had six children. Because of the alleged failure, she did not notice that the device was not functioning properly. In addition, the resulting pregnancy resulted in the birth of a Down Syndrome child. An additional claim was made that, despite the age of the plaintiff, 38, the physician assured her that amniocentesis was not necessary, and none was performed. A very large jury award was made to the Werths. The case was to have been appealed, but the parties decided to settle out of court and the Appellate Court thereby had no opportunity to rule on the suit. But the precedent set by the lower court in allowing trial for costs of raising the mentally retarded child as well as for the pain for mental anguish claimed by the parents was allowed. As there was no claim for wrongful life by the infant, this case is technically one of wrongful conception or wrongful birth, and one can only speculate upon what the higher courts might have decided. Apparently the defendant felt that it was wiser to settle rather than to go to trial again.

Also in 1979, the Supreme Court of New Jersey dealt with a wrongful life and wrongful birth case *Berman v. Allan*. In this malpractice suit it was alleged that the 38 year old mother was not given an amniocentesis, contrary to what the parents claimed was standard medical practice at this time. Their child was born with Down Syndrome. The lower court held that there was no basis for the Bermans' suit, nor for that of the infant. However, the Supreme Court modified that decision. Although they again refused the claim for wrongful life, using *Gleitman* and others as precedents, they decided to allow the suit for wrongful birth based upon the fact

that the mother had not been given the opportunity to
choose abortion had she known of the genetic defect of
the foetus. They also ruled that no costs for rearing
the defective infant would be awarded, but that the suit
for mental and emotional anguish the parents suffered
and will suffer in the future can be legally sought. In
denying the wrongful life claim the court said, in part,
"One of the most deeply held beliefs of our society is
that life--whether experienced with or without a major
physical handicap--is more precious than non-life.
Concrete manifestations of this belief are not hard to
discover. The documents which set forth the principles
upon which our society is founded are replete with
references to the sanctity of life. The federal
constitution characterizes life as one of three
fundamental rights of which no man can be deprived
without due process of law. Our own state constitution
proclaims that the 'enjoying and defending [of] life' is
a natural right. The Declaration of Independence states
that the primacy of man's 'inalienable' right to life is
a 'self-evident truth.' Nowhere in these documents is
there to be found that the lives of persons suffering
from physical handicaps are to be less cherished than
those of non-handicapped human beings." A second
opinion, concurring in part and dissenting in part,
states, "Without doubt, expectant parents, kept in
ignorance of severe and permanent defects affecting
their unborn child, suffer greatly when the awful truth
dawns upon them with the birth of the child. Human
experience has told each of us, personally or
vicariously, something of this anguish. Parents of such
a child experience a welter of negative feelings--
bewilderment, guilt, remorse and anguish--as well as

anger, depression and despair. ...A full perception of
the mental, emotional--and I add, moral --suffering of
parents in this situation reveals another aspect of
their loss. Mental, emotional and moral suffering can
involve diminished parental capacity. Such incapacity of
the mother and father *qua* parents is brought about by
the wrongful denial of a reasonable opportunity to learn
of and anticipate the birth of a child with permanent
defects, and to prepare for the heavy obligations
entailed in rearing so unfortunate an individual. Such
parents may experience great difficulty in adjusting to
their fate and accepting the child's impairment as
nature's verdict." The judge acknowledges the source of
his statement as being from Wolfensberger & Menolacino,
" *Theoretical Framework for the Management of Parents
of the Mentally Retarded.*" He goes on to quote other
authorities, but the essential reasoning is contained in
the citation given. In reality, the second opinion is
more than just that, but comes close to being a law
review article on cases such as this. Obviously, he is
more in sympathy, or at least expresses more sympathy,
than did the majority opinion.

One of the most complex and tragic cases also
arose in 1979, *Speck v. Finegold and Schwartz*, decided
by the Superior Court of Pennsylvania. The sad facts are
as follows: Frank Speck suffered from a genetic disease,
neurofibromatosis, a disease which is crippling as it
affects the nervous system. He and his wife had two
children, both affected with the disease, one of them
seriously. The Specks decided not to have more children
in view of the risk involved. Mr. Speck then underwent a
vasectomy under the care of Dr. Finegold who assured him
he was now sterile. Nevertheless, Mrs. Speck again

became pregnant; obviously the operation had failed.
Mrs. Speck, concerned that yet another defective child
would be born to the family, then consulted the second
defendant, Dr.Schwartz, a gynecologist. He was then
engaged to perform an abortion upon Mrs. Speck. The
operation was performed, and the doctor informed her that it
was successful. Nevertheless, the pregnancy continued,
despite the doctor's reassurance several times that this
was not possible. The pregnancy resulted in the birth of
a third child, afflicted with neurofibromatosis.

The Specks, in a sense triple losers, brought suit
against the physicians for a variety of good causes. The
lower court found no legal basis for suit and the case
was appealed to the Superior Court. This court,
following precedent, found no legal basis for the
wrongful life suit brought in behalf of the afflicted
infant, but found just cause for awarding the parents
damages for expenses incurred in care and treatment of
the child. However, no award was allowed for emotional
suffering on the part of the parents. In their denial of
the claim for wrongful life, they did express their
feelings for the problem, "Finally we hold that the
impossibility of this suit as to Francine [the child]
comes not so much from the difficulty in measuring the
alleged damages as from the fact, unfortunately, that
this is not an act cognizable in law." Two dissenting
opinions were written, one which would have denied all
claims, and another which would have allowed in addition
to the costs of rearing the infant those in behalf of
the parents for emotional suffering. Thus, in this case
one can choose any of three opinions for future use,
although, of course, the majority opinion allowing only
cost of care for the child stands. It is obvious the

judges were moved by the facts of the case, but that the majority and the two dissenting opinions differed sharply in the interpretation of the law.

Other 1980 cases show again a diversity of opinions by the courts. In *Stribling v. deQuevedo*, decided by the Superior Court of Pennsylvania, the court ruled on a suit which involved the birth of a child with dextrocardia, an assumed genetic defect. The basis of the malpractice suit was again a failed tubal ligation which Mrs. Stribling had undergone in order to limit her family. Suit was brought for mental and emotional pain, recovery of lost wages and earning capacity of the mother, costs of medical expenses and rearing of the child, costs of the medical expenses of the woman, and loss of consortium, all claimed by the father. The court ruled in favor of the Striblings on all claims except for emotional and mental distress. They also ruled against any claim by the infant for damages due to being born with a birth defect. The latter ruling stated, "First, there is no precedent in appellate judicial pronouncements that holds a child has a fundamental right to be born as a whole, functional human being. Whether it is better to never have been born at all rather than to have born with serious mental defects is a mystery more properly left to the philosophers and the theologians, a mystery which would lead us into the realm of metaphysics, beyond the realm of our understanding or ability to solve. The law cannot assert a knowledge which can resolve this inscrutable and enigmatic issue." The court then goes on to rule in favor of the parents on costs of rearing and consortium, citing their own opinion in the previous *Speck* case. Again, the same judge who dissented in that case dissents here and would allow damages for emotional suffering as he would have before.

In *Donadio v. Crouse-Irving Memorial Hospital*, the

issue was whether genetic counseling should have been given. The Appellate Division in New York in 1980 dealt with the following facts. The infant plaintiff was born 10 weeks prematurely and after treatment for a respiratory problem was rendered permanently blind. His mother had a record of 11 previous pregnancies, of which 8 resulted in miscarriages. The plaintiffs contended that because of this record the attending obstetricians should have advised genetic counseling and also taken extraordinary care during the pregnancy. The court ruled that these were true issues in law, stating, "The court recognized the tort of negligent counseling in *Becker v. Schwartz*. If plaintiff can prove her doctors negligently failed to provide or suggest adequate genetic counseling, they may recover for medical malpractice...

Accordingly, since we believe the mother's unusual medical history of miscarriages raises a triable issue of fact concerning whether her doctors should have provided more than routine prenatal care, we would reverse the order of summary judgement [in favor of dismissal of all counts against the physicians] and deny the motion." Once again the right to proper genetic counseling is considered a legal fact, and physicians dealing with difficult genetic issues are warned to seek competent consultation regarding genetic risks.

In the same year, 1980, the same New York Appellate Division dealt with the case, *Aquilio v. Nelson*. Here the issue dealt with was the failure of the physicians to deal properly with an inherited blood incompatibility which caused the death of the infant born to the Aquilios. The facts state that the attending physicians had been informed of the development by their first child of thrombocytopenia and that they knew that the

second child would be born with the same defect. The
claims were again multiple, but the ones concerning this
issue were the ones of emotional and psychological pain.
(The other issues involved were the claims that the
physicians, knowing of the previous medical history, did
not take adequate care to deal with the blood
incompatibility of the second child, and therefore also
of wrongful death of the infant.) The court again
concluded that damages for emotional and mental harm
were not valid claims for suit. A long dissenting
opinion declared that in this case the precedents set by
Becker, Park and *Howard* did not apply, stating, "Here
the claim is not that some omission or representation of
the physicians prompted plaintiffs to initiate the
pregnancy or to continue it to full term. The
circumstances are just the reverse. The Aquilios desired
a second child and had every reason to expect an infant
who, like their firstborn son, would thrive.
Interruption of the pregnancy was never a consideration.
As a result of the defendants' breach of duty to Daria
Aquilio, plaintiffs claim that the blood condition which
could and should have been controlled was not and that,
therefore, their son was born afflicted and died. Herein
lies the difference. In *Becker*, *Park* and *Howard*, the
plaintiffs had children when they would have had none.
In the case at bar, the Aquilios, who would have had a
robust youngster, through the defendants' malpractice
had one who developed thrombocytopenia, a fatal
condition." In other words, this is not a wrongful life
case, but a simple malpractice issue wherein, had the
physicians taken the normal precautions to use
transfusions at birth there would have been no death of
the infant. Once again, the majority opinion reflects

the essentially conservative opinion found throughout the
New York cases; however the dissents continue.

In three cases, entitled *Phillips v. United States*,
the difference between claims of wrongful life and
wrongful birth are made clear. The first case, in 1980,
before the United States District Court in South
Carolina gave the facts clearly. A Down Syndrome child
was born to the Phillipses while the husband was on active
duty with the Navy, and the child was born in a naval
hospital-thus the federal suit. Mrs. Phillips had an
early miscarriage with her first pregnancy. At that time
she had told the physicians that she was 22 years of
age, had no previous pregnancies, but that she did have
a mentally retarded sister with Down Syndrome. She was
given no genetic counseling nor was any genetic testing
carried out at the time of her second pregnancy which
resulted in the Down child. The District Court in the
first instance held that as the State of South Carolina
did not recognize the tort of wrongful life, for which
suit was brought, the court could not recognize this as
a basis for claims by the infant. The same court, in
1981, again faced the same facts, but this time the suit
was brought for wrongful birth, and the court decided
that suit for damages could be entertained and was a
legitimate basis for claims. The opinion is lengthy, and
cites most of the cases previously discussed here. A
difference, however, is that here the second decision
was founded upon the Federal Torts Claims Act, rather
than the South Carolina laws. Thus the second case was
not the same as the first, both in that its claim was
for wrongful birth, and also the suit was brought under
Federal law. The court hence could come to its
conclusion without denying the first opinion disallowing

wrongful life claims.

The third *Phillips* case was again before the U.S. District Court in South Carolina in 1983. Here the court decided to allow recovery for the cost of rearing, educating, and caring for the child until the age 40, presumably the life span of the infant. In addition they allowed damages for emotional suffering on behalf of the parents. However, the court applied the "benefits rule" and reduced the amount awarded for emotional suffering to 50% of the award for that claim.

A different approach was tried in New York in 1981 in *Feigelson v. Ryan*. The case again involved the birth of a Down child to a woman 36 years of age. In this instance, Mrs. Feigelson had wished to become pregnant and one of the defendant physicians had performed an artificial insemination upon her, resulting in her becoming pregnant. Some time after the birth of the child, a chromosome test was made and it was determined that the child carried the extra 21st chromosome found in all Down children. The parents claimed that the defendants had improperly failed to inform them of the relatively high risk of this occurring with older females and had also failed to advise amniocentesis. The plaintiffs also attempted to apply a different approach to the law, those cases dealing with an "extra object," the so-called Flanagan rule. Their claim was that a foetus *per se* was an extra object in the body of the woman and also, that the third 21st chromosome which caused the Down Syndrome was an extra object in their child. The court refused this extension of the Flanagan Rule on the basis that the rule stated explicitly, "...the term 'foreign object' shall not include a chemical compound, fixation device or prosthetic aid or device." The

importance of this attempt to use the Flanagan rule is
further underlined by the fact that this case was not
brought within the 3 year time limit of the statutes of
limitation, but under the Flanagan rule this could be
interpreted as only beginning when the foreign object
was first discovered. The court ruled that neither a
foetus nor an extra chromosome can be considered a
foreign object, and as a result the Flanagan rule did
not apply and no suit could be brought due to the
statute of limitations, and the plaintiffs were left
with no cognizable case.

Although no genetic defect was claimed in the birth
of a normal child in *Mason v. Western Pennsylvania
Hospital*, mentioned previously, there is a precedent set
in 1981 by the Superior Court of Pennsylvania. This
decision, coming after Speck above, stated in part,
citing their previous opinion, "The damages recoverable,
then, in a 'wrongful but healthy life case' may be
reduced due to the application of the benefits rule."
Again, there is not a claim of wrongful birth, but the
difference between wrongful life and wrongful birth seem
here to be less obvious; it is apparent in this and the
U.S. cases just noted that the sole difference now lies
in whether the claim is made by the infant or by the
parents. In the former instance no tort will be
admitted, while in the latter, suits for identical
damages will be considered under tort law.

Schroeder v. Perkel, also cited previously, brings
us back to genetic cases. Here cystic fibrosis is the
genetic defect involved. In this instance, in 1981, the
facts indicate that Mrs. Schroeder had a previous infant
whom the physicians negligently failed to diagnose
properly as suffering from this disease. Consequently

she was not given genetic counseling and was deprived of th
choice of not having additional children. The result was th
birth of a second child also afflicted by the disease. The
opinion of the Supreme Court of New Jersey reads in part
as if it were a textbook of medical genetics, and also
as if it were part of a similar book on medical
diagnosis and practice. According to claims of the
parents, the physicians failed to use standard
diagnostic practices to determine the presence of the
disease in the illness of their first child. The claims
were many: wrongful life, wrongful birth, claims for
medical costs for the second infant, and damages for pain
and suffering. The Superior Court reversed a lower court
decision and held that the case be remanded for trial
and that if the facts were proven the defendants could
be held liable for incremental medical costs of the
child born with cystic fibrosis. "The damages sought by
Mr. and Mrs. Schroeder are the medical, hospital and
pharmaceutical expenses needed for Thomas's [the second
child] survival. That kind of expense, regularly
measured by the courts, includes the prescription of
enzymes and antibiotics, checkups every one or two
months and hospitalization, which typically occurs in
emergencies or towards the end of the life of the
victim. Mr. and Mrs. Schroeder also intend to seek
recovery for the cost of daily therapy to Thomas. Until
now that therapy has been provided by them without any
outside assistance. At trial they must prove the
probability not only of the need and cost of the
therapy, but that it will be rendered by someone other
than themselves. They cannot recover for services that
they have rendered or will render personally to their
own child without incurring financial expense." The

court upheld the dismissal of the claim for wrongful life and other damages. Again a dissenting opinion pointed out the contradiction noted by the writer previously, saying, "The majority implicitly clings to the rule there can be no cause of action for 'wrongful life' but its reasoning completely undercuts the rule's *raison d'etre*. It accepts the thesis the infant has no cause for action based on whether he or she should not have been born. It accepts the public policy that the child's existence outweighs any discomfort the parents might endure in rearing and caring for the child. The consequence of adopting these propositions is that the parents can have no claim for medical expenses unless they are injured in their own right. Conversely, once the parents are held to have such a claim, it follows logically and conceptually that the child should be entitled to recover for its incapacity, and the parents for rearing the child." What the judge seems to be saying is that the fine line between wrongful birth and wrongful life is an artificial one, as once the latter is established the same rights adhere to the parents, and thus to the child, as would have been the case if wrongful life were admitted as a tort.

Another wrongful life-wrongful birth case arose in 1981 in *Moores v. Lucas*. The case, brought before the District Court of Appeals in Florida, concerned the inheritance of Larsen's Syndrome which causes physical and mental problems and high costs to maintain the victim suffering from the disease. The Moores had a previous child suffering from the disease and sought the advice of the physician concerning the chances of a subsequent child also suffering from the disease. They claim they were told the trait was not inheritable and as a result began and carried out a second pregnancy. The resulting infant had Larsen's Syndrome. The parents brought suit for the usual reasons, medical expenses for care and treatment of the second infant, past and future emotional suffering, physical pain and mental anguish of the pregnancy, as well as costs of the suit. The suit for wrongful life was dismissed as there was no such recognized tort in the State of Florida. The court upheld the validity of the other claims, "We hold there is a cause of action for a physician to diagnose and/or warn of an inheritable disease which results in the birth of a deformed child." Thus the court seems to recognize in essence the claim of wrongful birth, and allows suit for damages to be brought. Proper genetic counseling now appears to be a right of parents and a duty of physicians.

A very similar decision was reached by the Supreme Court of Virginia in the 1982 case, *Naccash v. Burger*. Here Tay-Sachs disease was again involved but with different factual background. The father, of Eastern Jewish ancestry, underwent a blood test to determine whether he was a carrier of the disease. The results of the test were negative and there was thus no need to test the mother who then proceeded with a pregnancy which resulted in a Tay-Sachs infant. Upon later retesting of both parents it was found that they were, of course, both carriers of the disease, and that the blood sample originally studied had been mistakenly confused with that of an entirely different person. Had the samples been properly identified and the fact that that the husband was a carrier been discovered, the wife would have also been tested, and upon finding her also to be a carrier, amniocentesis would have been performed and the defective foetus in all probability aborted. The negligence of the laboratory technician in mislabelling the blood samples is virtually conceded in this case. The court made a thorough review of most of the cases referred to previously, and decided that the Burgers were entitled to recover damages for the expenses for their afflicted child, and for emotional damages as well. The only claim which was disallowed was for the funeral expenses and a grave marker for the child.

Another point of law was raised by the defense, namely whether a "master-servant" relationship existed between the defendant physician and the laboratory technician who made the mistake in the blood tests. It is not clear just who was in charge of the hospital's cytogenetic laboratory. The lab was to have been closed by the hospital board of trustees, and doubtless would have been had not Dr. Naccash volunteered to run it without pay, and with full powers of hiring and supervision of personnel in the laboratory. After carefully reviewing the case, the court held that the above-mentioned relationship existed and that Dr. Naccash was therefore responsible for the mistake of his technician. Two dissents were filed, one disagreeing with the allowance of emotional damages, and the second disagreeing with the decision not to award funeral expenses. The latter said, "It is illogical for the majority to say that expenses of care and treatment due to the wrongful birth are recoverable and in the same breath to deny recovery of the burial expenses, when without the wrongful birth there would have been no death necessitating the expense of burial." Be that as it may, one can now add Virginia to the growing number of states with affirmative decisions as to the validity of the tort of wrongful birth.

Two other 1982 cases also should be examined. In *Nelson v. Krusen*, the Fifth Court of Appeals in Dallas, Texas, dealt with a case involving the genetic disease, Duchenne Muscular Dystrophy. The parents, after consulting with their physician, the defendant, decided not to terminate a pregnancy despite the fact that their previous child suffered from this disease. The physician had advised them that the wife was not a carrier of this sex-linked trait after performing the usual tests upon her. Nonetheless, when their second son was almost 3 years of age, he showed symptoms of the disease and more careful diagnosis by a specialist in pediatrics showed that indeed he had inherited the trait. The problem of the statute of limitations was also involved here, as the suit was not filed until more than three years after the alleged negligent genetic counseling by the defendants. The court ruled that, based upon earlier decisions, the statute of limitations did not apply, but the suit for wrongful life was denied on the basis that there was no such recognizable tort in Texas. "In summary, until a tort of wrongful life is recognized by the Supreme Court or the Legislature, summary judgement [by the lower court] denying relief to the child on such a claim was proper." The court made no attempt in this opinion to review the legal cases previous to this one with the exception of *Jacobs*, cited previously. The decision relied solely on this case, which it would seem did not deal with wrongful life, but only with damages awarded for what would now be considered wrongful birth or wrongful pregnancy.

The second 1982 opinion came from the Supreme Court of Connecticut. Here the genetic trait in *Ochs v.*

Borrelli was that of an orthopedic defect. The Oches had two previous children born with this mild defect as well as a gynecological history of miscarriages and ovarian surgery. The defendants performed, at the request of the female plaintiff, a tubal ligation which obviously failed as she became pregnant again soon thereafter. The third child, with the mild defect, was a result of this pregnancy. Suit was brought, a case of first impression, in Connecticut for the additional costs of rearing the child on the basis of the negligence of the physician in performing the sterilization operation, a case of wrongful pregnancy. The court decided in favor of the plaintiffs, stating in part, "In our view, the better rule is to allow parents to recover for the expenses of rearing an unplanned child to majority when the child's birth results from negligent medical care. The defendants ask us to carve out an exception grounded in public policy, to the normal duty of a tort-feasor to assume liability for all the damages that he has proximately caused. But such public policy cannot support an exception to the tort liability when the impact of such an exception would impair the exercise of a constitutionally protected right. It is now clearly established that parents have a constitutionally protected interest 'located within the zone of privacy created by several fundamental constitutional guarantees'... We may take judicial notice of the fact that raising a child from birth to maturity is a costly enterprise, and hence injurious, although it is an experience that redundantly recompenses most parents with intangible rewards. There can be no affront to public policy in our recognition of these costs and no inconsistency in our view that parental pleasure softens but does not eradicate economic reality." The opinion of the Connecticut Supreme Court was, for a change, unanimous.

THE REDISCOVERY OF THE TORT

Two recent California cases, as well as a case from the State of Washingon, and others based on these illustrate the rediscovery of the tort of wrongful birth. Like the rediscovery of the dawn redwood, long believed to be extinct but now available from most nurseries, the tort of wrongful life also was not extinct. The California cases demonstrate the differing views of lower courts and the need of a superior court to resolve them.

In July of 1980 the case of *Curlender v. Bio-Science Laboratories* was decided by the Court of Appeal, Second District, of California. The cases arose due to the following facts: the laboratory in question apparently misdiagnosed the parents as non-carriers of Tay-Sachs. As a result of this misinformation the pregnancy was undertaken, and after a normal term the infant was born with the disease. The court was obviously intrigued by the opportunity to deal with this issue. "The appeal presents an issue of first impression in California: What remedy, if any, is available to a severely impaired child--genetically defective--born as the result of the defendants' negligence in conducting certain genetic tests on the child's parents--tests which, if properly done, would have disclosed the high probability that the actual, catastrophic result would occur?" The court goes on to review once again the history of this tort, beginning with *Zepeda*, and covering most of the cases to date. The opinion refers in detail to the dissenting opinion in *Gleitman* and criticizes other courts for following the majority in that case. It also cites the dissenting opinion in *Berman*, cites *Speck* at length and also refers to many other cases as noted. "The decisional law of other jurisdictions, while not dispositive of Shauna's [the infant here] claim pursuant to California Law, is of considerable significance in defining the basic issues underlying the true 'wrongful life' action--one brought by the infant whose painful existence is a direct and proximate result of negligence by others. That decisional law demonstrates some measure of progression in our law. Confronted with the fact that births of these infants may be directly traced to the negligent conduct of others, and that the result of that negligence is palpable injury, involving not only pecuniary loss but untold anguish on the parts of all concerned, the courts in our sister states have progressed from a stance of barring all recovery to a recognition that, at least, the parents of such a child may state a cause of action founded on negligence.

"We note that there has been a gradual retreat from the position of accepting 'impossibility of measuring damages' as the sole ground for barring the infant's right of recovery, although the courts continue to express divergent views on how the parents' damages should be measured...

"The concept of public policy has played an important role in this developing field of law. Public policy, as perceived by most courts, has been utilized as the basis for denying recovery; in some fashion, a deeply held belief in the sanctity of life has compelled some courts to deny recovery to those among us who have

been born with serious impairments. But the dissents,
written along the way, demonstrate that there is not
universal acceptance of the notion that 'metaphysics' or
'religious beliefs' rather than law, should govern the
situation. The dissents have emphasized that
consideration of public policy should include regard for
social welfare as affected by careful genetic counseling
and medical procedures." The court also noted that the
U.S. Supreme Court decision in *Roe*, giving rights to
women to have an elective abortion, apply here and the
failure of genetic counseling to apprise them of the
risk of the disease in their infant essentially deprived
them of the use of amniocentesis and the right to make
an abortion decision. However, the main focus remains on
the wrongful life issue. "The circumstances that the
birth and injury have come hand in hand has caused other
courts to deal with the problem by barring recovery. The
reality of the 'wrongful life' concept is that such a
plaintiff both *exists and suffers*, due to the negligence
of others. It is neither necessary nor just to retreat
into meditation on the mysteries of life. We need not be
concerned with the fact that the plaintiff might not
have come into existence at all. The certainty of
genetic impairment is no longer a mystery. In addition,
a reverent appreciation of life compels recognition that
plaintiff, however impaired she may be, has come into
existence as a living person with certain rights." But
the court now went even further in their discussion of
this case. The defense had claimed that allowing such
suits as this might lead to suit being brought by the
defective infant against the parents! Said the court,
"If a case arose where, despite due care by the medical
profession in transmitting the necessary warnings,

parents made a conscious choice to proceed with a
pregnancy, with full knowledge that a seriously impaired
infant would be born, that conscious choice would
provide an intervening act of proximate cause to
preclude liability insofar as defendants other than the
parents were concerned. Under such circumstances, we see
no sound public policy which would protect those parents
from being answerable for the pain, suffering and misery
which they have wrought upon their offspring." The court
also stated as part of the basis of their opinion, "In
addition, we have long adhered to the principle that
there should be a remedy for every wrong committed.
Fundamental in our jurisprudence is the principle that
for every wrong there is a remedy and that an injured
party should be compensated for all damage proximately
caused by the wrongdoer.'" Thus the tort of wrongful
life is reestablished by this decision.

But is it really established? In that same year
another California court, the Court of Appeal, Fifth
District, thought not. The case, Turpin v. Sortini, also
dealt with incorrect diagnosis of existing children and
hence inadequate or wrong genetic counseling. The
Turpins had an older child with some learning problems.
Their second child was born totally deaf; the facts
established that the physician had examined the older
child and said her hearing was within normal range,
despite the fact that she was actually "stone deaf." The
parents were thereby misled and had no opportunity to
determine whether to produce another child with the risk
of its having the same handicap. A suit was brought by
the infant on the claim of the tort of wrongful life.
This court flatly refused the claim, saying, "After a
thorough review of these and other authorities we reject

Curlender as unsound under established principles of
public policy clearly within the competence of the
Legislature. The reasons for changing these principles
do not comport with the long-standing rules of
law...specifically, as will be pointed out, there has
been no dramatic shift, indeed no shift, in the weight
of authority against allowing recovery, and it cannot be
said that the reasons for the rule no longer exist." A
long, analytical dissent taking up and repeating most of
the reasoning given in *Curlender* was made.

Obviously there is now a dilemma, with one court
finding a tort of wrongful life and a court at an equal
level in the California judicial system finding the
opposite. Such a difference can only be settled by
appeal to a higher court. This appeal was made in 1982
by the Turpins. The Supreme Court of California was
confronted with the issue. Although it is not quite
clear what was exactly meant by their decision, they
reversed the lower court opinion and found in favor of
the Turpins. "In sum, we conclude that while a
plaintiff-child in a wrongful life case may not recover
general damages for being born impaired as opposed to
not being born at all, the child--like his or her
parents--may recover special damages for the
extraordinary expenses necessary to treat the hereditary
ailment." In other words, wrongful life claims for
medical and rearing costs apparently are a proper basis
for suit, but no general damages for being born may be
considered. This apparent contradiction was pointed out
in a dissenting opinion, "An order is internally
inconsistent which permits a child to recover for a so-
called wrongful life action, but denies all general
damages for the very same tort. While the modest

compassion of the majority may be commendable, they
suggest no principle of law that justifies so neatly
circumscribing the nature of damages suffered as a
result of a defendant's negligence."

This, then, was the status of these cases in
1985. What can one now conclude as to the nature of
wrongful life, wrongful birth, or wrongful pregnancy
cases? Perhaps it can be fairly stated that the latter
two claims are well established as genuine torts, and
cases of this kind are to be dealt with by the courts
as general negligence or malpractice cases. Yet, the
tort of wrongful life still remains a difficult issue.
The two California cases seem to point the way for a
more general acceptance in other states of this tort.
California has often led the way in establishing case
law, for example in *Custodio* the costs of rearing an
unwanted child were early on recognized and other states
eventually followed, although certainly not unanimously.
On the other hand, the Court of Appeals in New York has
been a bastion of defense against allowing any costs in
such cases. Perhaps as did the *Custodio* decision, these
two cases in California will spread to other states.
Certainly it will make it more easy to bring wrongful
life cases now that the California precedent, both in
Curlender and in the Supreme Court's decision in Turpin
have been made. It remains to be seen how widespread the
decisions in wrongful life cases will be in the future.

However, the tort of wrongful life has already
spread on the west coast. In January of 1983, the case
of *Harbeson v. Parke Davis* was heard before the Supreme
Court of Washington. The case had been sent to the
Supreme Court from a Federal Court to determine whether
the tort of wrongful life and the tort of wrongful birth

would be recognized by the State of Washington. The
facts were again not simple, but basically they involved
an epileptic mother (epilepsy is an inherited disease)
taking the drug Dilantin during two pregnancies in
order to avoid seizures. The drug has the effect of
causing a risk that children conceived and carried to
term during the use of the drug may be born with birth
defects. Such was the unfortunate case here; two
afflicted children were born to the Harbesons. As
plaintiffs they brought suit against the drug company
and against the United States both for wrongful birth,
and as *guardians ad litem* for the children for wrongful
life. The federal issue arose as Mrs. Harbeson's husband
was in the military and they had several times, at
different military hospitals, sought medical advice
concerning the use of the drug during pregnancy and had
been assured by military doctors that it was advisable
to continue its use during her pregnancy. The two
children were born with foetal hydantoin disease,
presumably induced by the use of Dilantin during
pregnancy.

Thus, the Harbesons had twice sought from military
physicians advice concerning the risks of defective
children while using Dilantin and they had been advised
that Mrs. Harbeson should continue the treatment. The
result was not one, but two pregnancies, producing the
type of defective child described.

The Washington Supreme Court, in a long and
detailed opinion, divided the issue and discussed both
wrongful birth and wrongful life. In the first instance,
they held that failure of the military hospital
physicians to inform the Harbesons of the possible risks
of birth-defective children and to allow them to have

the choice of an abortion or to undertake the risk of defective children was a legal tort and that wrongful birth was a recognized tort and would allow their suit to be upheld. But more importantly, they reviewed the entire history of cases, including the California *Turpin* decision, and ruled that the suit by the two children, brought through their *guardian ad litem* (in this case, their parents) for wrongful life should be upheld. The tort of wrongful life is in 1983 now recognized by these two states. This opinion is both lengthy and scholarly, reviewing the entire concepts of torts and of the rights of parents and children in cases of both wrongful birth and wrongful life. Most of the other cases previously discussed in this chapter are referred to. In concluding the Court's opinion it is stated, "The causation in a wrongful life claim is whether, but for the physician's negligence, the parents would have avoided conception, or aborted the pregnancy, and the child would not have existed... Some early cases advanced a proximate cause argument based on the fact that the negligence of the physician did not cause the defect from which the plaintiff suffered; rather, the negligence was in failing to disclose the existence of the defect. ...This argument does not convince us. It is clear in the case before us that, were it not for the negligence of the physicians, the minor plaintiffs would not have been born, and consequently not have suffered fetal hydantoin syndrome. More particularly, the plaintiffs would not have incurred the extraordinary expenses resulting from that condition. There appears to be no reason a finder of fact could not find that the physicians' negligence was a proximate cause of the plantiffs' injuries.

"For these reasons we hold that a claim for

wrongful life may be maintained in this state.

"Elizabeth and Christine Harbeson [the two children born with the defect] may maintain a wrongful life action...The minor plaintiffs suffer an actionable injury to the extent that they require special medical treatment and training beyond that required by children not afflicted with fetal hydantoin syndrome. They may recover damages to the extent of the cost of such treatment and training..."

Perhaps because of the *Harbeson* and the *Turpin* decisions, many more cases arose in 1984 and early 1985, all dealing with the validity of wrongful life as a legally recognized tort. The first, *Andalon v. Superior Court*, before the Court of Appeal, Third District, dealt with this issue along with a large number of legal points. The mother gave birth to a Down Syndrome child, and the parents claimed that the sudden shock of learning of the child's defect caused great mental distress. In addition suit was brought to cover the future earnings which would have accrued to the child had it not been defective. The opinion dealt at length with the issue of emotional shock and the wrongful life issue was in many ways not the main cause, as the parents had waived any claims of this kind. (In fact the concurring opinion points out that the long sections in this opinion dealing with wrongful life are "extraneous, interesting dissertations on a fascinating topic; however, they are totally lacking in necessity in the proceeding and would be better suited for exposition in a law review as an expression of the writer's opinion." In the majority opinion, the issue of further earnings was disposed of simply, "There is no loss of earning capacity caused by the doctor in negligently permitting the child to be born with a genetic defect that precludes earning a living. One cannot lose what one never had." The majority also seems to be siding with the earlier California case and perhaps would have sustained the claim for wrongful life had it been put forward in a straightforward manner.

The next case, also in 1984, was tried in the Pennsylvania Courts. Again in *Ellis v. Sherman*, the issue was whether wrongful life is a valid tort in that state. Here the claim was that negligent genetic counseling led to the birth of a child with Van Recklinhausen's disease (neurofibromatosis, a disease which in its mild state causes cafe-au-lait spots and, in its more severe form, slow-growing neurofibromas). Despite the references to both *Harbeson* and *Turpin*, the court held there is no recognizable tort of wrongful life in Pennsylvania.

The Pennsylvania Court was quite consistent in its opinion dealing with *Rubin v. Hamot Medical Center*, also in 1984. Here the case resulted from the birth of a child with Taye-Sachs disease. Suit was brought against both physicians and the Medical Center. The latter had performed tests showing that the parents were carriers of the disease but for some reason neither the Medical Center nor the physician informed the parents of these results. Despite what would appear to be obvious neglect, the court ruled on the sole claim before them, wrongful life; no such tort was allowed. The court also felt that the two reference cases, here *Turpin*, and also *Curlender*, were wrongly decided by the California Courts.

The most surprising of the 1984 decisions came in *Procanik by Procanik v. Cillo*. The affirmation that

wrongful life *can* be a recognized tort in this case came
from the Supreme Court of New Jersey, the state in which
the basic case, *Gleitman*, arose 17 years previously.
This was also a rubella case to make the parallel facts
more striking; however, in the present case, the alleged
tort occurred when the physician failed to carry out
tests for rubella in an adequate fashion. The opinion
considers most of the cases previously referred to. In
particular, they point out that abortion is now legal
and the mother has the right to decide whether to carry
an at risk pregnancy to term. There is considerable
reference to the dissenting opinion in *Gleitman* and the
court also states in part, "When a child requires
extraordinary medical care, the financial impact is felt
not just by the parents, but also by the injured child.
As a practical matter, the impact may extend beyond the
injured child to his brothers or sisters. Money that is
spent for the health care of one child is not available
for the clothes, food, or college education of another
child. ...Our decision is consistent with recent
decisions of the Supreme Courts of California and
Washington. The Supreme Court of California has ruled
that special damages related to the infant's birth
defects may be recovered in a wrongful life suit. ...
Following *Turpin*, the Supreme Court of Washington has
held that either the parents or the child may recover
special damages for medical and other extraordinary
expenses incurred during the infant's minority, and that
the child may recover for those costs during majority.
Two dissents were filed, one which would have allowed
none of the claims, and one which would have allowed
general as well as special damages. The important point
is that the former dissenting opinion in *Gleitman* has

now become the majority opinion and wrongful life is recognized as a tort in New Jersey.

Similarly, in *Azzolino v. Dingfelder*, before the Court of Appeals of North Carolina, again in 1984, the tort of wrongful life was affirmed. This court based its opinion squarely upon the New Jersey case just cited as well as the earlier authorities used by the New Jersey court. The case dealt here with the failure of the defendant to advise a 36-year-old mother of the possible need for amniocentesis, and in fact it was advised against as being unnecessary, and the physician's nurse, who had strongly religious beliefs against such a procedure, also had told the woman that it should not be done. Much of the opinion here is based on the standards of medical care at that time in this state, as malpractice could not be proven unless such standards had been violated. Expert witnesses testified that amniocentesis was a standard practice for women of this age, and the court accepted the fact that malpractice had occurred. The court went further in its opinion, after first discussing the earlier cases.

"It has been argued that even though the impaired child may experience a great deal of physical and emotional pain and suffering, he or she will be able 'to love and be loved and to experience happiness and pleasure--emotions which are truly the essence of life and which are far more valuable than the suffering she (or he) may endure.'" The citation is from *Berman*.

"While we agree this may be arguably true in some cases, we do not agree it is always or necessarily so. We are unwilling, and indeed unable, to say as a matter of law that life even with the most severe and debilitating of impairments is always preferable to nonexistence. We believe a child who is as severely impaired as Michael Azzolino has suffered a legally cognizable injury; therefore, Michael's action for wrongful life should not be dismissed for lack of actionable injury." The court then goes on to rule that special damages for the child's maintenance and medical care can be calculated. They take care to point out that the money which can be allotted should be vested in the child itself and there cannot be equal claims for damages by the parents. "Thus allowing the child to sue on his own behalf, through his guardians, will not only protect the best interests of the child, it will not in any way injure the interest of the parents." Subsidiary claims by the other Azzolino children for loss of parental consortium on the basis that Michael was legally a "nuisance" were quickly denied. "Furthermore we find the idea of classifying a child who is impaired, such as Michael Azzolino, as a nuisance for legal purposes to be morally repugnant." Nonetheless, North Carolina now joins the previous three states in recognizing wrongful life as a legal tort.

Lest it be thought that a tidal wave of acceptance of the wrongful life claim is sweeping through the nation, several cases in which the claim was denied will be mentioned. The Supreme Court of Texas in a second opinion in *Nelson v. Krusen*, also in 1984, stated at the onset, "We withdraw our opinion and judgement of November 16, 1983, and substitute the following." Although the original opinion was withdrawn, the Supreme Court of Texas did not change their ruling that wrongful life cannot be brought as a tort.

In addition New York again in 1984 denied recognition of a wrongful life claim. *Alquijay v. St.*

Lukes-Roosevelt Hospital also involved a mistake in analysis of amniotic fluid and the subsequent birth of a Down Syndrome child. The court in a brief opinion dismissed the suit on the basis of there being no such tort as wrongful life in the state of New York.

Illinois also has refused to admit wrongful life as a valid tort. The 1984 case, *Goldberg by and through Goldberg v. Ruskin*, dealt with Tay-Sachs disease and the failure of the defendant physician to either inform the parents of the simple test for the disease or to carry out the tests. While this court, the Supreme Court of Illinois, would not allow a wrongful life claim, they did however agree that the parents have a cause of action for wrongful birth. Again a dissent was entered stating in part that "Next the majority's conclusion that Jeffrey Goldberg did not suffer any injury as a result of defendant's negligence is plainly untenable. On this point the majority totally ignores reality and adopts instead an ontological precept that a child is better off enduring real pain and suffering than the child would have been had the child not been born. In reaching its conclusion, the majority overlooks the fact that judges must deal with reality and not metaphysical concepts. Metaphysics is a subject that is best left to the theologians and the philosophers and others similarly inclined. Here the sober reality with which we, as judges, must come to grips is that Jeffrey Goldberg endured immense pain and suffering he would not have had to endure had the defendant not been negligent. To hold that he was not injured as a result of defendant's negligence not only shunts reality but also imposes a cruel hoax upon the people involved in these tragic circumstances. I find the majority's conclusion indefensible." Again, as in *Gleitman*, it may seem as if the dissent were written for posterity.

The Supreme Court of Idaho early in 1985 also refused to recognize a wrongful life claim. In *Blake v. Cruz*, the issue here was once more the failure to diagnose rubella during pregnancy. The Supreme Court of Idaho held that no suit for wrongful life could be brought but that the parents may recover for the expenses of rearing and treating the child.

Perhaps the most peculiar twist on the wrongful life cases arose in 1984 in New York. Here, no child was involved in the case *Scott v. Brooklyn Hospital*. Instead, a woman with severe cancer received radium treatment from physicians at the hospital. While the treatment may have saved her life, she suffered severe side effects which will be life-long. Her claim is that the defendants' negligent treatment has caused her suffering. The negligence claim is based on the fact that had she gone to another facility the same type of treatment might have been given more carefully and she would not have the present pain and suffering. The defense claimed that this was in essence a wrongful life case and as such had no standing in New York Courts! The Supreme Court, Special Term, Kings County, was more than skeptical of the defense claim, to say the least, "Another distinguishing factor is the different analytical scheme-work involved in each case. In a case of 'wrongful life' or 'wrongful birth' there are only two possiblities for a court to consider. The plaintiff would either have an impaired life or no life at all. In the present case, there was a third possibility. Mrs. Scott could have gone through treatment and enjoyed a life without the impairments caused by the defendants'

negligence. The presence of this third possibility is
the basis of the law suit. Since this possibility is not
present in a cause for 'wrongful life' or for 'wrongful
birth,' the cases cited by defendants are not
determinative in this case. ...Just as life is preferred
over death, an unimpaired life is preferred over an
impaired one. A physician has a duty to act in such a
way as to minimize injury to the patient, even when
saving that patient's life. A breach of that duty which
results in unnecessary injury to the patient is
negligible and constitutes a valid cause of action in a
medical malpractice." Perhaps this illustrates best the
ingenuity and imagination of defense lawyers rather than
any great legal issue.

As more persons are aware of and seek genetic
counseling, the room for more errors in this field
expands. Genetic counseling is now a medical
subspecialty, licensed by the American Society of Human
Genetics, and as such certainly is no less free from
malpractice cases than any other medical subspecialty.
The establishment of wrongful life cases, if it becomes
widespread, will certainly introduce an additional need
for competent and able counseling, and act as a positive
inducement against errors in this field. One cannot
predict what decisions will be made by other courts,
but one can predict with reasonable certainty that
following the California, Washington, New Jersey and
North Carolina cases, the incidence of wrongful life
suits cannot help but increase.

The need for protection and support of the
defective child, not only during its minority but in
later years is the critical issue here. By allowing the
child to bring suit and awarding damages to him the

courts may be attempting to assure that he is taken care of for his life span. Obviously, the money will be given to either a guardian or a trust fund established solely for the purpose of supporting the unfortunate child. In this manner cost of the negligence by the tort-feasor will come as close to making the child "whole" as is humanly possible.

Table 1. Summary of cases involving costs of
raising unwanted child, medical expenses of pregnancy
and for child if necessary, awards for emotional and
psychological suffering (for either or both parents),
cost of rearing unwanted child, and awards for loss of
wife's services (consortium). The interpretations are by
the author.

Yes: award possible; no: no award granted by court;
---- not an issue, or not dealt with by decision; ? case
decided upon on other legal issues.

Note: Cases are referred to by name of plaintiff
only; for full citation see bibliography.

DATE	PLAINTIFF	MEDICAL	EMOTIONAL	REARING	CONSORTIUM
1934	Christ'sen	no	no	no	----
1942	Stemmer	no	no	----	----
1953	Kelly	yes	----	----	----
1956	Hornbuckel	yes	----	----	----
1956	Morgan	no	----	no	----
1957	Shaheen	----	----	no	----
1960	Sinkler	yes	----	----	----
1960	Smith	no	----	----	----
1963	Zepeda	----	no	----	----
1966	Sylvia	yes	----	----	----
1967	Custodio	yes	yes	yes	----
1967	Gleitman	no	no	no	no
1968	Stewart	yes	----	----	----
1970	Jackson	yes	----	no	----
1971	Troppi	yes	yes	yes	----
1972	Will'n Med	?	?	?	?
1973	Terrell	----	----	no	----
1974	Aronoff	----	----	no	----
1974	Coleman	yes	yes	no	yes
1974	Cox	no	no	no	----
1974	Reick	----	----	no	----
1974	Ziemba	yes	yes	yes	----
1975	Betancourt	yes	yes	yes	----
1975	Dumer	yes	----	----	----
1975	Jacobs	yes	----	yes	----
1975	Stevens	----	----	yes	----
1976	Anonymous	yes	yes	yes	----
1976	Stills	yes	yes	yes	----
1977	Garwood	yes	yes	----	----
1977	Renslow	yes	----	----	----

1977	Sherlock	yes	yes	yes	yes
1978	Bergstre'r	yes	----	----	----
1978	Green	----	yes	yes	----
1978	Rivera	yes	yes	yes	----
1978	Sala	yes	no	no	----
1979	Wilczynski	yes	----	no	----
1980	Pub.He.Tr.	yes	yes	no	----
1980	Yanga	----	----	yes	----
1981	Abala	no	----	no	----
1981	Cockrum	----	----	yes	----
1981	Eisbrenner	yes	yes	----	----
1981	Hartke	yes	yes	no	----
1981	P.& Husband	yes	yes	no	yes
1981	Robak	----	----	yes	----
1981	White	yes	yes	no	yes
1982	Wilbur	yes	----	no	----

ALPHABETICAL LIST OF CASES CITED

CASE	CITATION	YEAR
Albala v. City of New York	78 AD 2d 389	1981
Alquijay v. St.Lukes-Roosevelt Hospital	63 N.Y.2d 978	1984
Andalon v. Superior Court	208 Cal.Rptr. 899	1984
Anonymous v. Hospital	366 A.2d 204(Conn)	1976
Aquilio v. Nelson	78 AD2d 195	1980
Aronoff v. Snider	292 So.2d 418(Fla)	1974
Azzolino v. Dingfelder	322 S.E.2d 567 (N.C)	1984
Becker v. Schwartz	60 A.D.2d 587	1977
Becker v. Swartz	46 NY2d 401	1978
Bergstresser v. Mitchell	577 F.2d 22	1978
Berman v. Allan	404 A.2d 8(N.J.)	1979
Betancourt v. Gaylor	344 A.2d 336(N.J.)	1975
Blake v. Cruz	698 P.2d (Idaho)	1985
Bowman v. Davis	356 N.E.2d 496(Ohio)	1976
Christensen v. Thornby	255 N.W.2d 620(Minn)	1934
Clapham v. Yanga	300 N.W.2d 727(Mich)	1980
Clegg v. Chase	89 Misc 2d 510	1977
Cockrum v. Baumgartner	425 N.E.2d 968(Ill)	1981
Coleman v. Garrison	327 A.2d 757(Del)	1974
Cox v. Stretton	77 Misc 2d 155	1974
Curlender v. Bio-Science Laboratories	165 Cal. Rptr. 477	1980
Custodio v. Bauer	50 Cal.Rptr. 463	1967
Debora S. v. Sapega	392 N.Y.S.2d 79	1977
Del. Zio v. Presby. Hosp.et al	74 C.Div. 3588(N.Y.)	1978
Johnson v. Yeshiva University	42 N.Y.Rptrs.2d 818	1977
Jones v. Malinowski	473 A.2d 429 (Md)	1984
Jorgensen v.Meade Johnson Laboratories	483 F.2d 237	1973

Delaney v. Krafte   98 AD 2d 128   1984
DiNatale v.Lieberman;Allen v.
                    Col.Lab   409 S.2d 512(Fla)   1982
Doerr v.Villate   220 N.E.2d 767(Ill)   1966

Donadio v. Crouse-Irving
          Memorial Hosp.   75 AD2d 715   1980
Dumer v. St. Michael's Hosp.   233 N.W.2d 216(Wisc)1975
Eisbrenner v. Stanley   308 N.W.2d 209(Mich)1981
Elliot v.Brown   361 So.2d 546(Ala)   1978
Ellis v. Sherman   478 A.2d 1339 (Pa)   1984
Feigelson v. Ryan   108 Misc 2d 191   1981
Flowers v. D. Columbia   478 A.2d 1073 (D.C.)1984
Garwood v. Locke   552 S.W.2d 892(Tex)   1977
Gildiner v.Thomas Jefferson
                    Univ. Hosp.   451 F.Supp 692   1978
Gleitman v. Cosgrove   227 A.2d 689(NJ)   1967
Goldberg by & through Goldberg
                    v. Ruskin   471 N.E.2d 530 (Ill)1984
Green v. Sudakin   265 N.W.2d 411(Mich)1978
Harbeson v. Parke Davis, Inc   656 P.2d. 483(Wash) 1983
Hartke v. McKelway   526 F.Supp. 97   1981
Hays v. Hall   488 S.W.2d 412(Tex)   1973
Hornbuckle v. Plantation
                    Pipe Line   93 S.E.2d 727(Ga)   1956
Howard v. Lecher   386 N.Y.S.2d 460   1976
Jackson v. Anderson   230 So.2d 503(Fla)   1970
Jackson v. Bumgardner   N.C.No.8411SC6,Slip 1984
Jacobs v. Theimer   519 S.W.2d 846(Tex)1975
Karlsons v. Guerinot   57 AD2d 73   1977
Kelly v. Gregory   125 N.Y.S.2d 696   1953
LaPoint v. Shirley   409 F.Supp. 118   1976
Ladies Center of Clearwater,
               Inc.v. Reno   341 So.2d 543(Fla)   1977

Mason v. Western Pa. Hospital	428 A.2d 1366(Pa)	1981
Moores v. Lucas	405 So.2d 1022(Fla)	1981
Morgan v. United States	143 F.Supp. 580	1956
Naccarato v. Grob	162 N.W.2d 305(Mich)	1968
Naccash v. Burger	Slip Opinion, S.C.	1982
Nelson v. Krusen	21044 Tex. Slip Opn.	1982
Nelson v. Krusen	678 S.W.2d 918 (Tex)	1984
O'Toole v. Greenberg	64 NY 2d 427	1985
Ochs v. Borelli et al	187 Conn 253	1982
P. and Husband v. Portadin	432 A.2d 556(N.J.)	1981
Park v. Chessin	387 N.Y.S.2d 204	1976
Park v. Chessin	60 AD2d 80	1977
Park v. Chessin	88 Misc 2d 222	1976
Phillips v. United States	575 F,Supp. 1309	1983
Phillips v. United States (I)	508 F.Supp.537	1980
Phillips v. United States (II)	508 F.Supp. 544	1981
Phillips v. United States (III)	575 F.Supp. 1309	1983
Procanik v. Cillo	478 A.2d (N.J.)	1984
Public Health Trust v. Brown	388 So.2d 1084(Fla)	1980
Reick v. Med. Protective Co.	219 N.W.2d 242(Wisc)	1974
Renslow v. Mennonite Hospital	367 N.E.2d 1250(Ill)	1977
Rivera v. State of New York	94 Misc 2d 157	1978
Robak v. United States	658 F.2d 471	1981
Roe v. Wade	410 U.S. 113	1973
Roman v.City of New York	110 Misc 2d 799	1981
Rubin v. Hamot Medical Center	478 A.2d 869 (Pa)	1984
Sala v. Tomlinson	73 A.D.2d 724	1979
Schroeder v. Perkel	432 A.2d 834(N.J.)	1981
Scott v. Brooklyn Hospital	125 Misc 2d 765	1984
Shaheen v. Knight	11 D.&C.2d 41(Pa)	1957
Sherlock v. Stillwater Clinic	260 N.W.2d 169(Minn)	1977
Simon v. United States	438 F.Supp. 759	1977
Sills v. Gratton	127 Cal.Rptr. 652	1976
Sills v. Gratton	55 Cal.Rptr. 652	1976

Sinkler v. Kneale	164 A.2d 93(Pa)	1960
Slawek v. Stroh	215 N.W.2d 9(Wisc)	1974
Smith v. Brennan	157 A.2d 497(N.J.)	1960
Speck v. Finegold & Schwartz	408 A.2d 496(Pa)	1979
Stemmer v. Kline	26 A.2d 489(N.J.)	1942
Stephens v. Spiwak	233 N.W.2d 124(Mich)	1975
Stewart v. L. I. College Hosp.	59 Misc 2d 432	1968
Stribling v. deQuevedo	422 A 2d 505(Pa)	1980
Sylvia v. Gobeille	220 A.2d 222(R.I.)	1966
Terrell v, Garcia	496 S.W.2d 124(Tex)	1973
Troppi v. Scarf	187 N.W.2d 511(Mich)	1971
Turpin v. Sortini	174 Cal.Rptr. 128	1981
Turpin v. Sortini	182 Cal.Rptr. 337	1982
Vaccaro v. Squibb Corp.	52 N.Y.Rpts 2d 810	1980
Vaccaro v. Squibb Corp.	71 AD2d 271	1979
Vaccaro v. Squibb Corp.	97 Misc 2d 907	1978
Watson v. State of Florida	292 S.E.2d 418(Fla)	1974
Werth v. Paroly	Wayne Cty. 74025162	1979
White v. United States	510 F.Supp.146	1981
Wilbur v.Kerr	628 S.W.2d 568(Ark)	1982
Wilczynski v. Goodman	391 N.E.2d 479(Ill)	1979
Williams v. State of New York	18 N Y 2d 481	1966
Wilmington Medical Center.		
v. Coleman	298 A.2d 320	1972
Woods v. Lancet	303 N.Y. 349	1951
Zepeda v. Zepeda	190 N.E.2d 849	1963
Ziemba v. Sternberg	45 A D 2d 230	1974

CHRONOLOGICAL LIST OF CASES CITED

YEAR	CASE	CITATION
1934	Christensen v. Thornby	255 N.W.2d 620(Minn)
1942	Stemmer v. Kline	26 A.2d 489(N.J.)
1951	Woods v. Lancet	303 N.Y. 349
1953	Kelly v. Gregory	125 N.Y.S.2d 696
1956	Hornbuckle v. Plantation Pipe Line	93 S.E.2d 727(Ga)
1956	Morgan v. United States	143 F.Supp. 580
1957	Shaheen v. Knight	11 D.&C.2d 41(Pa)
1960	Sinkler v. Kneale	164 A.2d 93(Pa)
1960	Smith v. Brennan	157 A.2d 497(N.J.)
1963	Zepeda v. Zepeda	190 N.E.2d 849
1966	Doerr v.Villate	220 N.E.2d 767(Ill)
1966	Sylvia v. Gobeille	220 A.2d 222(R.I.)
1966	Williams v. State of New York	18 N Y 2d 481
1967	Custodio v. Bauer	50 Cal.Rptr. 463
1967	Gleitman v. Cosgrove	227 A.2d 689(NJ)
1968	Naccarato v. Grob	162 N.W.2d 305(Mich)
1968	Stewart v. Long Island College Hospital	59 Misc 2d 432
1970	Jackson v. Anderson	230 So.2d 503(Fla)
1971	Troppi v. Scarf	187 N.W.2d 511(Mich)
1972	Wilmington Medical Center v. Coleman	298 A.2d 320
1973	Hays v. Hall	488 S.W.2d 412(Tex)
1973	Jorgensen v.Meade Johnson Laboratories	483 F.2d 237
1973	Roe v. Wade	410 U.S. 113
1973	Terrell v, Garcia	496 S.W.2d 124(Tex)
1974	Aronoff v. Snider	292 So.2d 418(Fla)

1974	Coleman v. Garrison	327 A.2d 757(Del)
1974	Cox v. Stretton	77 Misc 2d 155
1974	Reick v. Medical Protective	
	Company	219 N.W.2d 242(Wisc)
1974	Slawek v. Stroh	215 N.W.2d 9(Wisc)
1974	Watson v. State of Florida	292 S.E.2d 418(Fla)
1974	Ziemba v. Sternberg	45 A D 2d 230
1975	Betancourt v. Gaylor	344 A.2d 336(N.J.)
1975	Dumer v. St. Michael's	
	Hospital	233 N.W.2d 216(Wisc)
1975	Jacobs v. Theimer	519 S.W.2d 846(Tex)
1975	Stephens v. Spiwak	233 N.W.2d 124(Mich)
1976	Anonymous v. Hospital	366 A.2d 204(Conn)
1976	Bowman v. Davis	356 N.E.2d 496(Ohio)
1976	Howard v. Lecher	386 N.Y.S.2d 460
1976	LaPoint v. Shirley	409 F.Supp. 118
1976	Park v. Chessin	387 N.Y.S.2d 204
1976	Park v. Chessin	88 Misc 2d 222
1976	Stills v. Gratton	127 Cal.Rptr. 652
1976	Stills v. Gratton	55 Cal.Rptr. 652
1977	Becker v. Schwartz	60 A.D.2d 587
1977	Clegg v. Chase	89 Misc 2d 510
1977	Debora S. v. Sapega	392 N.Y.S.2d 79
1977	Garwood v. Locke	552 S.W.2d 892(Tex)
1977	Johnson v. Yeshiva	
	University	42 N.Y.Rptrs.2d 818
1977	Karlsons v. Guerinot	57 AD2d 73
1977	Center of Clearwater,Inc.	
	v. Reno	341 So.2d 543(Fla)
1977	Park v. Chessin	60 AD2d 80
1977	Renslow v. Mennonite	
	Hospital	367 N.E.2d 1250(Ill)
1977	Sherlock v. Stillwater	

		Clinic	260 N.W.2d 169(Minn)
1977	Simon v. United States		438 F.Supp. 759
1978	Becker v. Swartz		46 NY2d 401
1978	Bergstresser v. Mitchell		577 F.2d 22
1978	Del. Zio v. Presbyterian		
		Hospital et al	74 C.Div. 3588(N.Y.)
1978	Elliot v.Brown		361 So.2d 546(Ala)
1978	Gildiner v.Thomas Jefferson		
		University Hospital	451 F.Supp 692
1978	Green v. Sudakin		265 N.W.2d 411(Mich)
1978	Rivera v. State of New York		94 Misc 2d 157
1978	Vaccaro v. Squibb Corp.		97 Misc 2d 907
1979	Berman v. Allan		404 A.2d 8(N.J.)
1979	Sala v. Tomlinson		73 A.D.2d 724
1979	Speck v. Finegold & Schwartz		408 A.2d 496(Pa)
1979	Vaccaro v. Squibb		
		Corporation	71 AD2d 271
1979	Werth v. Paroly		Wayne County,
			Mich. 74025162
1979	Wilczynski v. Goodman		391 N.E.2d 479(Ill)
1980	Aquilio v. Nelson		78 AD2d 195
1980	Clapham v. Yanga		300 N.W.2d 727(Mich)
1980	Curlender v. Bio-Science		
		Laboratories	165 Cal. Rptr. 477
1980	Donadio v. Crouse-Irving		
		Memorial Hospital	75 AD2d 715
1980	Phillips v United States (I)		508 F.Supp.537
1980	Public Health Trust v. Brown		388 So.2d 1084(Fla)
1980	Stribling v. deQuevedo		422 A 2d 505(Pa)
1980	Vaccaro v. Squibb Corp.		52 N.Y.Rpts 2d 810
1981	Albala v. City of New York		78 AD 2d 389
1981	Cockrum v. Baumgartner		425 N.E.2d 968(Ill)
1981	Eisbrenner v. Stanley		308 N.W.2d 209(Mich)

1981	Feigelson v. Ryan	108 Misc 2d 191
1981	Hartke v. McKelway	526 F.Supp. 97
1981	Mason v. Western Pennsylvania Hosp.	428 A.2d 1366(Pa)
1981	Moores v. Lucas	405 So.2d 1022(Fla)
1981	P. and Husband v. Portadin	432 A.2d 556(N.J.)
1981	Phillips v.U.S. (II)	508 F.Supp. 544
1981	Robak v. United States	658 F.2d 471
1981	Roman v.City of New York	110 Misc 2d 799
1981	Schroeder v. Perkel	432 A.2d 834(N.J.)
1981	Turpin v. Sortini	174 Cal.Rptr. 128
1981	White v. United States	510 F.Supp.146
1982	DiNatale v.Lieberman;Allen v.Col.Lab	409 S.2d 512(Fla)
1982	Naccash v. Burger	Slip Opinion, S.C.Va
1982	Nelson v. Krusen	21044 Tex. Slip Opn.
1982	Ochs v. Borelli et al	187 Conn 253
1982	Turpin v. Sortini	182 Cal.Rptr. 337
1982	Wilbur v.Kerr	628 S.W.2d 568(Ark)
1983	Harbeson v.Parke Davis,Inc.	656 P.2d. 483(Wash)
1983	Phillips v. U.S. (III)	575 F.Supp. 1309
1984	Alquijay v. St.Lukes-Roosevelt Hospital	63 N.Y.2d 978
1984	Andalon v. Superior Court	208 Cal.Rptr. 899
1984	Azzolino v. Dingfelder	322 S.E.2d 567 (N.C)
1984	Delaney v. Krafte	98 AD 2d 128
1984	Ellis v. Sherman	478 A.2d 1339 (Pa)
1984	Flowers v. District of Columbia	478 A.2d 1073 (D.C.)
1984	Goldberg by & through Goldberg v. Ruskin	471 N.E.2d 530 (Ill)
1984	Jackson v. Bumgardner	N.C.No.8411SC6,Slip
1984	Jones v. Malinowski	473 A.2d 429 (Md)

1984	Nelson v. Krusen	678 S.W.2d 918 (Tex)
1984	Procanik v. Cillo	478 A.2d (N.J.)
1984	Rubin v. Hamot Med.Center	478 A.2d 869 (Pa)
1984	Scott v. Brooklyn Hospital	125 Misc 2d 765
1985	Blake v. Cruz	698 P.2d (Idaho)
1985	O'Toole v. Greenberg	64 NY 2d 427

Chapter 2
IN MOST STATES YOU MUST BE PSYCHOTIC
TO FIND OUT WHO YOU ARE
ADOPTION LAWS

Adoption laws, per se, do not exist in common law;
all laws pertaining to adoption are therefore enacted by
legislatures and represent a departure from English
Common Law upon which so much of our legal system
depends. As each state is thereby free to enact any
adoption acts it chooses, an attempt was made, in 1953,
to develop a uniform law which could be accepted by, and
enacted by, all states.

In 1971, the Uniform Adoption Laws were again
reviewed and the 1953 proposals were revised, partly due
to the enactment of the so-called "right to know" acts.
The original uniform code had read, "...(2) All papers
and records pertaining to the adoption shall be kept as
a permanent record of the court and withheld from
inspection. No person shall have access to such records
except on order of the judge of the court in which the
decree of adoption was entered for good cause shown"
(emphasis added). The 1971 revision states "...all
papers and records pertaining to the adoption ...are
subject to inspection only upon the consent of the Court
and all interested persons; or in exceptional cases only
upon an order of the Court for good cause shown; and (3)
except as authorized in writing by the adoptive parent,
the adopted child, if [14] or more years of age, or upon
order of the court for good cause shown in exceptional
cases, no person is required to disclose the name or
identity of either an adoptive parent or an adopted
child."

An additional condition of legal adoption, designed
specifically for the benefit of the adoptee is the

issuing of a new birth certificate for the child. This
certificate now lists the names of the adoptive parents.
The original birth certificate which listed the name of
the natural mother (and the father if known) remains
part of the sealed record and is not destroyed. The
purpose of these acts, enacted by many, but not all,
states was intended to foster adoption by guaranteeing
to both the natural parents and the adoptive parents
that privacy would encloak the adoption proceedings and
that the adopted child would be protected as well. At
the time that this suggested law became enacted by many
states the full implications of its impact was not
clearly understood, certainly not in psychological
terms, nor in light of today's growing concern by many
to know their true ancestry.

To many adult adoptees this law has erected strong
barriers in their attempt to achieve self-identification
and self-fulfillment. As was claimed in *Society v.
Mellon* in 1978, "...As they put it, an adoptee is
someone upon whom the State has, by sealing his records,
imposed lifelong familial amnesia...injuring the adoptee
in regard to his personal identity when he was too young
to consent to, or even know, what was happening." Many
cases have now been brought by adult adoptees seeking a
variety of information concerning their background,
religious, economic, racial, and, more important to this
study, their genetic background.

In addition, cases involving the right to inherit
from natural as well as adoptive parents have raised
issues with which the courts must deal. Most of the
cases raise further constitutional issues including the
right of privacy, due process under the XIV Amendment,
and in some cases the right to practice one's proper

religion.

Among the first of many cases to test these actions was *In the matter of Katrina K. Maxtone-Graham*. This 1975 New York case involved a 30-year-old adopted woman who sought to learn her true genetic identity. At the time she was undergoing psychiatric treatment and her psychiatrist testified that the information she sought was essential to her successful treatment. The Surrogate Court agreed and directed the adopting agency to make a search for and, if possible, to obtain the consent of the natural mother to open the records. The search was unsuccessful, but the petitioner herself was somehow able to obtain the name of her mother who then gave consent for the information sought to be given to Maxtone-Graham. The Court, however, declined to release the names of the foster parents with whom she had been placed prior to the completion of the adoption procedures, holding that such information was not required for the treatment. The Court did state, however, "Copies of all remaining records of the respondent agencies regarding petitioner are to be made available to petitioner's psychiatrist without the names of the foster parents. By the consent of all parties, copies of these records shall constitute petitioner's property." This case was in many ways simplified as the mother's consent was freely given and the psychiatric need well established. The case apparently set sufficient precedent that in 1976 another Surrogate Court in New York acceded to a request from another person, also suffering from psychiatric problems, to have the records open. However, in this case, *In the Matter of the Application of Anonymous*, the court concluded that the natural parents be informed of their

legal rights in the case and actually only determined
that they must also be parties to the suit. They
reasoned that if the petitioner had the right to open
the records due to his incapacity, caused by his not
knowing who his parents were, the natural parents were
in the same class. "Surely the natural parents, whose
identity is secret and cannot presently come forward to
protect their rights, fit the statutory definition of
'incapacitated person.'" The court therefore determined
that a designated person with the status of *guardian ad
litem* be given the names of the natural parents and
should "determine the present whereabouts of the natural
parents; counsel them as to their rights and, in
general, represent and protect their interests."

In 1977, another court in New York, this time an
appellate court, came to a different conclusion in the
case of *In the matter of Joann N. Chattman*. In this case
a Surrogate Court had denied the request for the records
to be open. Chattman had requested the records for
genetic purposes, wishing to know whether she might be
subject to some genetic problem, or be a carrier of a
genetic trait which she in turn might pass on to any
future offspring.

The Appellate Court ruled that the names of the
natural parents could not be given but that "...the
petitioner shall be entitled to have access to the file
of her adoption to review and inspect her medical
records and those of her natural parents, and any other
material therein relating to genetic or hereditary
conditions. ...Petitioner is a married woman and is
thinking of starting her own family. Her concern or
preoccupation with the possibility of some genetic or
hereditary factor in her background which might foretell

a problem for any issue she might bear constitutes 'good cause' to allow her access to any medical reports or related matter contained in the records of her adoption." Thus, while not entitled to the names of her natural parents, she was entitled to any biological information concerning their genetic background.

In the Matter of "Anonymous," a N.Y. Surrogate Court in 1977 again granted right to an emotionally disturbed man to have the adoption records opened. The court ruled in part, "The legislative requirement is not prompted by any delusion that a man donning a black robe is automatically endowed with the omniscience of a Deity or the sublime wisdom of a Solomon; but merely the state's desire to insert an objective human being into what may be an emotionally supercharged situation, in the hope that a decision can be reached which is in the best interest of all concerned--the linchpin from which hangs the determination of whether 'good cause' has been shown. ...As alleged in his pleading, the petitioner is suffering from a severe psychological disorder caused by not knowing his true identity...Accordingly the *guardian ad litem* ....is directed to prepare a final report in which he shall state the names, addresses, and ethnic background of the petitioner's natural parents."

Despite these cases, the opening of records is not as general as these may make it seem to be. In 1981 the New York Court of Appeals, New York's highest level court, upheld the lower court's refusal to open an adoption record. In the case of *In the Matter of Linda F.M.*, the court held that despite her claims, Linda was not severely enough damaged psychologically to have good cause for the records to be opened to her despite testimony that she had suffered sufficient emotional

upset as to have had her marriage broken up and her
artistic and musical creativity impaired. She
petitioned, "to know who I am. I feel cut off from the
rest of humanity. I was given birth to the same way as
everyone else, but everyone else can send away $3 and
get a copy of their birth certificate. I want to know
who I am. The only person in the world who looks like me
is my son. I have no ancestry. Nothing." Despite this
pathetic plea, the court denied the request, stating
"...By its very nature good cause admits of no
universal, black-letter definition. Whether it exists,
and the extent of disclosure that is appropriate, must
remain for the courts to decide on the facts of each
case. Nevertheless, mere desire to learn the identity of
one's natural parents cannot alone constitute good cause
or the requirement of section 114 [of the New York Code]
would become a nullity." If this is to be the principle,
then claims of severe emotional agony are not enough;
one must be certified psychotic by the testimony of a
psychiatrist so that a proof of such a claim is provided
to a court in order for it to allow the adoptee to
learn his or her true identity.

A quirk in the earlier law was revealed in the
1969 case, *Anonymous v. Anonymous*. Here a natural mother
who had placed her illegitimate child for adoption
sought to find the names of the adoptive parents. The
lawyer for the latter refused to supply the names, and
the mother went to court to obtain them on the basis
that it should be determined whether the adoptive
parents were fit parents for caring for the infant. The
Supreme Court ordered the lawyer to furnish the court with
these names so that a full inquiry could be made. The
court pointed out, "The applicable statute contains

nothing which would preclude or prohibit the natural parent from learning the names of adoptive parents in any adoption proceeding which the adoptive parents might initiate. Section 114 of the Domestic Relations Law, which applies to agency adoptions, provides that the surname of the child must be concealed from the adoptive parents, but there is no requirement of concealment of the names of the adoptive parents from the natural parent. The court then pointed out that, given this section of the law, the names of the adoptive parents were required as the case was not really one concerning adoption, but one concerning the welfare of the adoptee. "The sole question is whether there should be an inquiry to determine whether the present custody is in the best interests of the child. There is no question that such an inquiry should be had, but such an inquiry cannot be had without the identification of the persons who now have custody of the child." It is obvious, from the unfairness of this law, that is there is no privacy for the adopting family, but only for the natural parent, and that some sort of change to protect the interests of all was necessary.

In another case where the mother also had placed a child for adoption, *People ex rel. Scarpetta v. Spence-Chapin Adoption Service*, a judgment to return a child placed for adoption, but not yet adopted, was made. The mother changed her mind within twenty-three days after placing the child for adoption and wished now to keep her out-of-wedlock child. The issue was whether the mother lacked sufficient "stability of mind and emotion" at the time of her having authorized the placement for adoption. The Court of Appeals in this 1971 case ruled that such stability did not exist at the time and, as the mother now had both the emotional stability and more than adequate financial resources to give the child a good home and upbringing, the child should be returned to her custody.

Other states which have adopted the uniform law have also consistently refused access to adoption records. In *Mills v.Atlantic City Department of Vital Statistics*, an opinion by the Superior Court of New Jersey in 1977, a constitutional challenge was brought against that state's law. The court here decided against opening records based on the Constitutional right to privacy of the First, Fourth, Fifth and Ninth Amendments, holding that these provided sufficient legal basis to uphold the New Jersey statute. The court stated, "In the present case the right to privacy asserted by plaintiffs is in direct conflict with the right to privacy of another party, the natural parent. ...the natural parent surrenders a child for adoption with not merely an expectation of confidentiality but with actual statutory assurance that his or her identity as the child's parent will be shielded from public disclosure. In reliance on these assurances the natural parent of an adult adoptee has now established new life relationships and perhaps a new family unit. It is highly likely that he or she has chosen not to reveal to his or her spouse, children or other relations, friends or associates the facts of an emotionally upsetting and potentially socially unacceptable occurrence 18 or more years ago. ...the statute herein does not totally deny plaintiffs' access to the information they seek. It only requires that they as members of a class in which there is an overwhelming state interest must demonstrate good cause in order to protect the countervailing privacy rights of the natural

parents." The court goes on to deny any violation of the
XIV Amendment which plaintiffs had claimed based on the
grounds that others could obtain a birth certificate
while they, as adult adoptees, were denied that right.
However, the court then continues, at the request of the
deputy attorney general, to lay new groundwork for
subsequent cases. First, there is to be no change in
those cases involving minor adoptees. More importantly,
the court further states, "When the child reaches
maturity, he or she may have a desire to search for and
find his or her natural parents. The need to search is
now becoming more publicized and researched than ever
before. With this publicity more and more adoptees daily
are beginning to realize that their desire to meet their
natural parents is not unusual nor does it have any
reflection on their relationship with their adoptive
parents whom they love dearly.

"In the case of adult adoptees who request access
to their birth records, the burden of proof should shift
to the state to demonstrate that good cause is not
present. In certain situations the request of the adult
adoptee for information should be granted as a matter of
course. Where the natural parent has placed on file a
consent to identification with the agency who placed the
child, access by the adult adoptee should be
automatically allowed." The court then suggested
procedures for a state-wide registry for this purpose.
The opinion also considers the psychological needs and
gives them great sympathy. The final problem dealt with
is whether the adoptee receiving such information might
then seek to confront his or her natural parents in
cases where the registry was not used. In such cases,
the court will refer the request to the agency that made

the initial placement and appoint it to act as an arm of
the court. If the natural parent consents the opening of
the record is to be automatic. If consent is refused,
the adoptee will have the right to appeal to the
Superior Court which will decide whether the state can
meet the burden of proof to keep the records sealed. As
yet, no case seems to have arisen involving this new
reversal of roles with the state having to establish
cause for refusing to open the records. It will be
interesting to see what occurs when such cases arise,
and to note whether a similar approach will now be taken
by courts, and possibly legislatures, in the future. In
fact, a bill containing most of the provisions from the
New Jersey decision was prepared for the New York
legislature in 1982, but did not reach the floor.
(However, see below for the recent enactment of this
bill and its consequences.)

Other states which have also enacted laws similar
to the Uniform Adoption Laws have also decided in a
manner similar to that ofhas New York. Thus in *Sherry H. v.
Probate Court* the Supreme Court of Connecticut set aside
the refusal by a lower court to open the adoption
records when it was found that the genetic mother
objected vigorously to this procedure. The case was
complicated by the fact that since the time of
adoption the laws of Connecticut had been revised to
give a Probate Court judge powers to decide whether
there was sufficient cause to open the sealed file. The
court here ruled that the change in the law was not
meant to be retroactive to the time of the original
adoptive procedures. It also was noted by the court that
the present law, as written, would seem to give an
absolute veto to the natural parents as no record could

be opened without their direct consent. The Supreme
Court ruled that the Probate Court had not, under the
law, made adequate inquiry into the effect the
examination of the adoptee's record would have upon the
welfare of the adopted person, the adoptive parents, and
the genetic parents, and must do so before refusing the
adult adoptee's request.

Similar cases of the courts refusing to open the
records without the adoptee showing "good cause" are
found in most states today. For example, a 1980 case in
Illinois, *In the Matter of Roger B.*. the appellate court
in that state denied the right of an adult adoptee to
see his records on the basis of Constitutional grounds.
The appellant had claimed that this violated his rights
under the First, Ninth and Fourteenth Amendments. His
claims that he was denied freedom of information, the
right to identity, and due process were all rejected and
his claim to see the records on the basis that he was
now an "adult" were held to be without merit. However a
dissenting opinion would have allowed his plea. "The
fact that we are dealing here with a fundamental right
is illustrated by the questions we all have about
ourselves. What are the physical characteristics to
which my children may be genetically prone? What is my
ancestral nationality or religious persuasion? What
achievements or feats can I point to with ancestral
pride? Can anyone seriously deny that one's liberty to
pursue these and similar questions is 'so rooted' in the
traditions and conscience of our people as to be ranked
as 'fundamental'? Thus the right to know one's
individually created identity must be considered a
fundamental right. For those persons that are adopted,
however, the only way that this fundamental right can be

meaningful is to include within it one's liberty to know the identity of his genetic parents. ...I agree that protecting the genetic parents from public disclosure of a traumatic emotional event may be a concern of the state. But I cannot agree that it is a compelling state interest. Laws are not generally enacted to protect persons from the consequences of their own acts on their private lives." The judge goes on to cite that some states, Alabama, Kansas, and Florida, do not endorse the Uniform Adoption Act, and also that several foreign countries, England, Finland, and Wales, have either changed their laws to allow access to adult adoptees or never enacted prohibitions against such access. (He might also have included Scotland and Israel which also allow access.) As will be seen in cases to be mentioned below, there is also a problem in the right to inheritance of these adoptees. There is no law prohibiting inheritance by illegitimate children in most states, yet the adoptee cannot inherit from parents whose identity is unknown. This opinion is quoted at some length, because as was seen previously, dissenting opinions in many instances become the basis for majority opinions in later cases and may allow a judge of a different case to find a basis for a different decision.

The inheritance cases mentioned above occurred in Louisiana. In *Massey v. Parker*, a case brought in 1979, the Louisiana Supreme Court handled the matter of whether the records should be opened by the appointment of a curator *ad hoc* to examine them to determine if the names of the parents or other relatives could be ascertained and if so to discover whether there were inheritance rights and to determine how these could be exercised in behalf of the adult adoptee without denying

protection to the natural parents. An additional
consideration, not found often, was the testimony by the
appellant that he would also like to help his biological
parents should *they* be in financial need. Again a
dissent was filed. "The remedy so authorized by the
state is chimerical...Leaving aside the questions of
diligence and thoroughness of representation inherent in
the often perfunctory appointment of a curator at a low
fee, the best that might be ascertained is that, *at a
given point in time*, the natural parent living in
Louisiana possesses no property. It does not negate that
the parent will acquire some thereafter. Must the
adopted child bring suit again and again to
reinvestigate the issue, in order to vindicate his
inheritance rights protected for him by our law when,
without his consent, he is detached from his family and
placed in a new family?" The judge continues his dissent
by pointing out that he would regard the adoptee's right
to inheritance to be a compelling one which should be
regarded as sufficient to open the records.

A quite similar case arose in 1980 in the same
state. In *Kirsh v. Parker* an adult woman sued to have
the records opened both for determination of her
inheritance rights and also to determine her
medical background which would then enable better
diagnosis and treatment of any ailment she might be
subject to as well as to alleviate her mental anguish
from flashbacks she thought might be caused by her early
childhood. The lower courts, in this case, ruled in
favor of Ms. Kirsh and the case was appealed by Parker,
the Registrar of the Bureau of Vital Statistics of New
Orleans. The Supreme Court, however, decided that the
previous *Massey* majority should hold as far as

determination of any inheritance rights and also that there was insufficient medical evidence concerning anguish presented in the trial. The case was remanded and the trial judge was required to hold another hearing and to appoint the curator *ad hoc* referred to previously.

Another case dealing with inheritance, but from a different approach was the 1945 case, *In re Tilliski's Estate*. Here the question was the apportionment of an estate after a child was adopted by a couple. This couple died and the adopted child inherited property from them. Subsequently the natural parent died and the child brought suit for a share in her estate as well. The Illinois Supreme Court determined that the fact that the child had inherited from the adoptive parents was no barrier and that there was no reason not to allow inheritance from the natural parent.

Similarly in *In re Estate of Anna Mae Cregar*, an Illinois appellate court was faced with the determination whether an adopted child could claim double shares of an estate because of dual relationship to the deceased by virtue of blood and adoption. The child was a relative of the adoptive parent. The natural niece and nephew of the deceased aunt who had adopted the the child opposed the adoptee claiming dual shares, first as a relative of the late aunt, and second by virtue of being her adopted child. The appellate court decided that there was no legal basis to deny such a claim and the claim for dual shares was upheld by the appellate court.

An unusual case involving conflict of state laws arose in 1981. This involved the determination whether the marriage of a half-brother to his half-sister, the latter having been adopted, was legal. Under ordinary

conditions, marriage of close relatives is a crime in
the state of Delaware where the case *State of Delaware
v. Sharon H. and Dennis H.* was decided by the Superior
Court. The adopted woman in question had found that she
had a half-brother who was raised as a ward of the
state. Both were children of the same mother, but of
different fathers. The young woman, upon maturity, found
the young man in a correctional institution, succeeded
in obtaining parole for him and sometime later they were
married. The state sought to prosecute them for
violation of the laws prohibiting marriage of close
relatives. Their legal claim was a novel one, and rather
ingenious. Sharon claimed that under the adoption laws
when she was legally adopted, she was legally no longer
the child of her natural mother, and therefore in the
eyes of the law should not be considered to be related
by blood to her husband! She pointed out that the
records were sealed and that there should not have been
access to them without good cause. A second point which
the court dealt with was that the Delaware laws
prohibited marriage between brother and sister, but
nowhere did the legislative act mention half-brothers or
half-sisters, and therefore the law by this omission did
not apply here. There was also an issue of unnecessary
delay but it too was not considered to be an important
point. As the two persons involved were in jail for
their offense, however, this may not have been in the
best interests of the parties. A further complication
was a charge of perjury brought against Sharon for
swearing under oath that she was not related to her
husband. The Superior Court judge examined the various
claims carefully. He decided that the failure of the
legislature to include this kind of relationship in

their ban against marriages of close relatives was not
pertinent as the purpose of the law was plain. The issue
of whether a legal relationship between Sharon and her
half-brother existed was also decided again on the basis
of what the judge believed the intent of the legislative
act dealing with adoption was--namely that although the
act stated clearly, "...the adopted child shall no
longer be considered the child of his natural parent or
parents and shall no longer be entitled to any of the
rights of privileges or subject to any of the duties or
obligations of a child with respect to the natural
parent or parents....," it was not the intent of the act
to permit cases such as these to be legal. Finally, the
judge determined that the wording of the adoption law
regarding the access of the state to these records
(without them there would be, of course, no proof of the
relationship of the couple) did not prevent their use in
this case. The apparent problem here, it would seem, and
a point raised in the case but not dealt with by the
opinion at any length, is the unfairness inherent in the
fact that there is no doubt that had Sharon herself
sought to have access to her adoption files it would
have been denied on the basis of no compelling need, yet
the State was able to open the files for use in a
criminal procedure.

A major Federal Court case is *The Alma Society v.
Mellon*, decided by the U.S. District Court, Southern
District, New York in 1978. The Alma Society is a group
of adult adoptees who together are fighting for their
rights as adoptees to have the adoption records open to
those who wish to see them. They brought action in the
Federal District court to have the New York (and
consequently all other states') laws dealing with sealed

records declared unconstitutional on the basis that such laws violated their rights, and that these records should be opened upon request with no need to show cause. Their claim was brought under the First, Fourth, Ninth, Thirteenth and Fourteenth Amendments. The court dismissed all of these complaints, and denied the Society's claims entirely. The court ruled there was no claim under the various amendments, as had the state courts in many of the cases mentioned previously (indeed, several of these had cited this decision). The novel claim here was that the XIII Amendment was pertinent. The Federal Court took pains to point out that while this Amendment abolished slavery, saying, "With respect to the Thirteenth Amendment, the gist of the plaintiffs' argument is that the requirement of good cause [to open the adoption records] is a badge or incident of slavery because it is the equivalent of the sale and separation from their parents of slave children too young to remember who their parents were." The same issues applied in the appeal to the U.S. Court of Appeals in 1979. The same essential arguments were presented and again the Court of Appeals reached the same decision, namely that there was no violation of federal rights by the statutes requiring compelling reasons for the records to be opened. It was held that the New York statutes, with the provision for limited access by the adult adoptee, were sufficiently fair so as not to infringe upon any constitutional rights of the adoptee. The court dealt at length with each of the issues raised and determined in each claim that there was sufficient state interest in protecting the records so as to render the appellants' case invalid.

In another opinion, handed down at the same time,

Rhodes v.Laurino the court also denied a request made by the plaintiff who had claimed that the attempt to have his records made available to him had been subjected to undue procrastination. The court ruled that the plaintiff, despite his impatience to have the Surrogate Court of New York move more quickly, had not exhausted his legal remedies in this court and therefore no federal claim existed under law. The appellant had requested information concerning the medical records of his natural parents as well on the basis that he might be subject to cardiac ailments which could be hereditary. The court ruled that although it sympathized with the appellant's claim there was no federal appeal as the case was still under the jurisdiction of the New York court.

A different issue arose in *Caban v. Mohammed* in 1979. Here the issue was the validity of New York"s law which permitted an unwed mother, but not an unwed father, to block the adoption of their child by simply withholding her consent. The U. S. Supreme Court held that such sexual discrimination violated the Equal Protection Clause of the Fourteenth Amendment "because it bears no substantial relation to any important state interest." The court was far from unanimous; the 5:4 majority opinion delivered by Justice Powell and joined by 4 others was vigorously opposed by two dissenting opinions. Thus the Court, as has been the case in many major opinions recently has not been able to speak with real authority and it will probably be necessary for other similar cases to be brought again before the court before the exact status of such cases can be determined.

Many similar cases to *Alma* are on record. For example in *Application of Maples* decided in 1978 by the

Supreme Court of Missouri the court upheld the
constitutionality of the adoption laws of that state
ruling there was no denial of the First or Fourteenth
Amendments. Here Ms. Maples had appealed lower court
decision denying her access to the names of her natural
parents although much information concerning her
background and the physical health of her real parents
had been furnished. A concurring opinion, however was
less severe. "...I do not believe an adopted adult
seeking the identity of his natural parent or the
natural parent seeking the identity of her natural child
should have the heavy burden indicated by the principle
opinion merely to get the juvenile court to inquire of
the other whether they were agreeable to the information
requested. The inquiry ought also to go to the adoptive
parents; but, after the child is emancipated, the
desires of the adoptive parent to continued
confidentiality would not necessarily be as weighty as
the agreement of the adopted person and the natural
parent that they know the identity of each other."
Another concurring opinion also stated, "I note briefly
the current fascination with the profound achievement of
author Alex Haley in his recorded search for
genealogical roots. These sensations of the
consciousness of personal history are ample testimonials
to the unique anxiety of Americans; for we are, with
rare exceptions, a nation of uprooted immigrants whose
family crests are little more than the remnants of
graffiti on the steerage decks of a generation of
vessels. All of us need to know our past, not only for a
sense of lineage and heritage, but also for a fundamental and
crucial sense of our very selves: our identity is
incomplete and our sense of self retarded without a real

personal historical connection....Adoptees often suffer from what has been termed 'geneological bewilderment.'"

A similar attempt to have the records open, but with a novel plea, is found in *Application of Gilbert* decided by the same court also in 1978. Here a 49-year-old man appealed to have the records opened on the basis that he was a member of of the Church of Jesus Christ of Latter Day Saints and that a "fundamental belief of that church is that in order to be saved and exalted after death every person must trace their ancestry and perform certain ordinances for their blood relatives." There was, however, conflicting testimony concerning the Church's position, and the case was remanded to the lower court, which had denied the request for the opening of the adoption records, in order to determine exactly what the positions of the appellant and the Church were.

The State of Washington also has declined to open the records, despite a plea that the adoptee wished to know whether his natural father might have died of a heart attack. In this case, *Application of Sage*, the Court of Appeals again denied the request on the basis of insufficiency of the plea. An additional ruling was made concerning his plea that he was entitled to the information on the basis of the freedom of information act mentioned earlier. The court also ruled that there was no compelling cause on that basis to allow access to the adoption records.

The other side of the coin was a case in Rhode Island in 1979. In *In re Christine*, the Supreme Court of that state declined the request of the natural mother who wished to determine the welfare of her child by discovering the names and circumstances of the adoptive parents. The Court denied good cause here and upheld the lower court's decision not to open the records.

The issue of proving mental need to open records arose again in *Bradley v. Childrens Bureau of South Carolina*. Here the Supreme Court of that state found in 1981 that, "The finding of 'good cause' was therefore unwarranted. Bradley feels some insecurity; he shows some distraction; he has a sincere desire to know his identity. He has not, however, required medical assistance. We find it significant, too, that Bradley has enjoyed steady employment and an apparently stable family life of his own." In other words, Bradley was apparently not sufficiently deranged to allow the plea of necessity for psychological reasons to allow access to his records.

In a series of cases, again in Missouri, running from 1981-1982, the Supreme Court dealt with a complicated issue in which genetic principles are definitely important. In these cases, all titled *Application of George*, the matter arose when an adult adoptee suffered from chronic myelocytic leukemia, a malignant disease of the bone marrow. This is a progressive disease and the only possibility of remission is by means of a bone marrow graft. This graft can only be successful if the proper genetic match between donor and recipient is possible. The ideal match would come from the use of an identical twin. The next best results come from full siblings, and here there is only about a one in four chance of finding a suitable donor. If such siblings are not available, then the chances of finding a suitable donor diminish with the degree of relationship. Neither parent can usually be the donor, as the recipient child has genes from both parents, and the use of only one parent's marrow is not possible. Thus in this case the appellant wished to open the records to determine if there were other siblings or half-siblings who might be found to be suitable donors. In the first decision, the lower court did not attempt to locate the father and the case was then remanded for such action to be taken. The lower court judge then, in a genuine act of humanity took it upon himself to act and contacted the natural father, which had not been done before (it should be noted that the natural mother was willing to cooperate and was willing to have both herself and her child by a later marriage tested). The judge approached the named natural father who strongly denied that he was the father despite the records, and stated that the natural mother had named him falsely. The judge, and his own wife, spent many hours with the man and his present wife in an attempt to persuade him to allow himself to be tested, but he continued to deny any relationship with the natural mother and the test was not carried out.

The final decision by the Supreme Court in its second opinion in 1982 on this case was to deny George access to the records. This was not due to a lack of humane feeling, but was based on the fact that medical testimony made it clear that, even if he were indeed the natural father, the odds of his being a suitable donor were so fantastically low that there was no real reason to allow his name to be given to the appellant. It is not clear what would have been done were there a favorable chance of the man's being a donor. The issue was thus solely decided upon the genetic and medical evidence that he could not serve as a donor and therefore no good reason for him to be named was present.

Lehr v. Robertson decided by the U.S. Supreme Court in 1983 exemplifies rigid adherence to the fine points of the law and this unwillingness to take into account the human values involved. Here the issue was simple; what rights does a father have to block adoption when he is desirous of establishing his paternity and a meaningful relationship with his child? The case rose through the New York Courts and involved that state's laws concerning establishment of paternity. Lehr was the acknowledged father of an illegitimate offspring. Later the mother married another man and the couple then legally adopted the child, thereby terminating all Lehr's rights as the father. New York State has established several methods of declaring paternity, none of which, unfortunately, Lehr had followed. Instead he had brought a filiation procedure, rather than using the formal methods which form the laws of this state. A Family court judge, despite complete knowledge of Lehr's pending suit, allowed the adoption to be accomplished, thereby ending Lehr's right to be recognized as the father. The case was appealed through the various levels of the State Courts and ultimately to the United States Supreme Court on several bases, mainly involving constitutional rights and the claimed discrimination against males who cannot block an adoption while the woman involved can do so.

The Supreme Court in a split decision denied Lehr any rights whatsoever. What seems astounding is that the facts of the case, as given by the dissenting opinion, were all but completely ignored by the majority. The majority simply stated that as two years had elapsed before the man brought the filiation procedure, and he had not used any of the previously mentioned methods to establish paternity he had shown no interest in the child. But the minority opinion went into much more detail, pointing out the following: 1) He had lived with the woman for 2 years before the child's birth; 2) The woman acknowledged him as the father and reported this to the New York State Department of Social Services; 3) He visited her and the infant while she was in confinement; 4) After she left the hospital she disappeared with the child while Lehr frantically tried, at times successfully, to find her and eventually hired a detective agency to track them down (when he finally located her she had already married); 5) She threatened to have him arrested despite the fact that he offered to set up a trust fund for the infant; 6) He then hired a lawyer to assert his rights of visitation whereupon she began the adoption procedure. Yet none of these acts satisfied the state law. Had Lehr spent 14 cents for a postage stamp, so as to be listed by the appropriate agency, he would have established his legal rights to make a paternity claim. Instead, he spent considerable sums hiring both detectives and lawyers to attempt to obtain his acknowledged infant, sums which were in the long run wasted. But the majority opinion, ignoring all of Lehr's actions, simply held that he failed to comply with the exact letter of the law; despite all of his efforts the court found that there had been no "bond" between him and the child, a bond he was desperately trying to establish only to be thwarted by the mother, and therefore he was not legally entitled to any rights of paternity.

Two Attorney General's Opinions in the State of Utah further define the problems involved with secrecy in adoption. The first dealt with the issue whether a court could release the birth name of a missing child to the State Bureau of Criminal Identification in order to facilitate a search for the missing youth. After careful search of the Utah Laws, the Attorney General ruled that such revelation was in accord with the statutes governing adoption in Utah and the name was presumably released despite the sealed court records.

The second opinion by the same office dealt with the release of the information concerning the birth parents of an adult adoptee by a private adopting agency. Here, the adoptee and the natural parents both agreed to the release of the records. However, again, quoting Utah Statutes, the Office of the Attorney General ruled that only a court could open the records and there was no right of the adoptee to have the birth parents' names released.

Such a wide discrepancy, it would seem, might be highly discriminatory. In the first instance, no permission from the adoptee of the natural parents was sought or given, and the records were opened. In the second case, the records remained sealed even when permission was given. As in the earlier case of the marriage of half-siblings, it would seem that the state has powers which the individuals do not. In the case just discussed, it seems that law and justice are opposed; where no objections were raised the records could not be opened, and where no consultation with individuals directly concerned was made, the records were opened.

A final case, *Humphers v. First Interstate Bank*, heard by the Supreme Court of Oregon, illustrates still another problem in the adoption laws. In this instance a physician had assisted an umarried woman in giving birth

to an unwanted child and the child was placed for
adoption. The physician had registered the woman in the
hospital under an assumed name, and the usual sealing of
the records concerning the adoption occurred. Some years
later, the adoptee, now an adult, wished to find her
natural mother, and could not, of course, have the
records opened. However, she did somehow find the
attending physician and he, apparently sympathetic to
her search, gave the adoptee a false medical record
stating that he had given the mother diethylstill-
besterol and that this, with its possible   )⊥.
effects upon the daughter seeking knowledge of her
birth, made it essential for her to have the hospital
medical records pertaining to that event. The hospital,
acting on this recommendation, allowed the adoptee to
see the medical records, and from this the adoptee was
able to locate her natural mother.

The mother was extremely displeased and brought
suit for a number of reasons, including "outrageous
conduct" among other claims. Meanwhile the physician had
died, and the suit was brought against the bank as
executor of the physician's estate. Further claims were
for invasion of privacy and for breach of confidence, as
well as for unprofessional conduct by the physician. The
court, in a thoughtful opinion, dealt with these issues,
first ruling that there was no such tort as "outrageous
conduct," then ruling that the long history of invasion
of privacy cases did not apparently apply here, but that
the breach of confidentiality was a recognizable tort in
this instance. The case was remanded to the lower court
to be tried again on that issue only.

All of these cases leave a feeling of
irresolution and of uncertainty in the mind of the

observer. It is obvious that protection of a young
woman, or man, who has had an illegitimate child at some
time long in their past is an important factor in their
lives. Equally, it is understandable that such persons,
having gone through the severe emotional trauma in
making the decision to place the child for adoption
do so not out of lack of love for the infant, but out of
the societal pressures upon them. Many of these persons
will go on to live lives where it would be of
considerable stigma to them for the facts of their
youthful indiscretions to be made public; indeed such
revelation might serve to destroy a marriage, reflect
harmfully upon legitimate children of such a union, and
perhaps bring severe economic consequences as well. As
long as the parent cannot be fully assured that the
facts concerning his or her early life will be concealed
against the intrusion of others, a perpetual dread of
revelation might indeed induce severe psychological
difficulties. On the other hand the best interests of
the child must also be a foremost consideration,
particularly as the adoptee had no choice in determining
the position in which he finds himself. The need to know
one's origins has been stressed in these cases, and is
certainly a real one. The obstacles placed in the path
of the adult adoptee are formidable; psychiatric
symptoms which perhaps are brought about by the person's
self-conceived sense of non-identity seem to be the
major, if not the only way in which the adoptee can
force the opening of the records. While some courts
recognize the force of an argument to open the records
for inheritance purposes, but only by a curator *ad hoc*,
this is not likely to be a universal action. Also it may
be argued that in some cases the prolonged sense of

guilt of the natural parent might be alleviated by
finally bringing the facts to light. But the moral
issues are conflicting; on the one hand the well-meaning
state laws seek by sealing the records to protect the
best interests of all and, on the other hand the need,
and perhaps the right, of the child upon reaching legal
maturity, to know his origins presents an almost
irresolvable conflict. Even more troublesome may be the
fact that the decision whether to allow access is in
each case left to an individual judge who, like all
others, has certain moral and ethical values of his or
her own, thereby leaving the dilemma even more
difficult. The suggestion of a registry, made by the New
Jersey court does not in reality solve the problem as
many parents may well refuse to give such permission.

Indeed, "open adoption" is being tried in Texas by
the Lutheran Social Service and several successful cases
where the biological parent and the adopting parents are
in communication have been reported by the press during
1983. The adoptive parents apparently are required to
attend seminars prior to meeting the natural mother; in
all cases, however, despite face to face meetings of the
birth mother and the adoptive parents, names are
apparently withheld. It will be interesting to see
whether this Texas agency's plan becomes widespread.

Further, in 1983 the New York Legislature did enact
the Lasher bill to provide all but identifying
information to adoptees who seek to learn their ethnic,
religious, or other knowledge of their identity. Of
course, unlike the Texas case, this does not satisfy the
natural desire of the adoptee to learn the true names of
the birth parents. The law would therefore seem to
accomplish little which is not already available to

adoptees seeking this type of information except that it will not take a court order to make such general information available.

However, an example of courts placing obstacles in the path of the search by an adoptee has occurred in New York. As mentioned above the Lasher bill, establishing a registry and allowing adult adoptees over the age of 21 to obtain medical, genetic, ethnic and other characteristics of the natural parents, did pass in 1983. Soon thereafter more than 1000 such requests were made to the State Health Bureau by adult adoptees. Due to the difficulty of obtaining the records, the Bureau recommended that court procedures might be quicker. Yet it is apparent that New York State judges are falling back on an older law and ignoring the Lasher Act on the basis that it is too vague. The state's chief administrative judge issued a memorandum to family and surrogate courts suggesting they follow the older law; while not an order, this certainly bears considerable weight. As a result, instead of turning over the information to the State Health Bureau, as the Lasher bill provides, they are insisting on the usual hearing and court order to unseal any records. For a time both Lasher and the Department of Health considered going to court "to challenge the denials by the judges." (New York Times, July 11, 1984, pg. B2). The need for adoptees to learn their identity is increasing, and the need for the law to recognize this fact is also a reality which as yet seems not to be understood.

Further emphasis of such a need to know one's origin is the recent attempt in Sweden to change their laws so that the identity of donors to artificial insemination clinics would be recorded and the name of the donor would become available to the child at the legal age of 18. People in the United States are strongly opposed to such an act, claiming that the number of sperm donors would drop precipitously if they were threatened with the revelation of their names to the A.I.D. children. It is also predicted that many women would also avoid this if they knew that the child, raised by them legally as their own, would someday be able to find out that its biological father was not the same as the person who had raised the child for its entire minority. It is estimated that approximately 8000 women annually use this procedure; in most countries the name of the donor is not given, and only a few relevant facts concerning the health and other non-identifying information is revealed. One leading physician, Dr. Joseph Kaufman of the UCLA Medical Center, has been quoted as saying "I can't imagine the Swedish medical profession permitting the passage of any such almost-guaranteed troublemaking legislation." (*Parade* 7/29/84, pg. 18.) The unsealing of the records for adoption to reveal the names of both biological parents might seem to point the way towards opening records where only one biological parent is sought. In a sense, A.I.D. is a type of semi-adoption, and there may be inconsistency in keeping one type of record sealed while the other is available in these cases. The right of the adoptee or the A.I.D. child to know its origins may be paramount, and this proposed change in Sweden may have a more profound legal basis than it would first appear.

As of the present time, with excellent arguments on behalf of both sides, there seems to be no clear cut way to resolve the issue. However, as more and more children are being born out of wedlock, perhaps society will eventually resolve the problem outside of the courts. If the bearing of illegitimate children becomes socially a norm for large parts of the population, there will be less need for concealment and the need for sealed records will no longer be felt. But as long as society treats both the illegitimate child and his or her parents as less than desirable, the courts will continually have to wrestle with the problem as to whose interests are best protected by sealed records, the child's, the natural or adoptive parents' or those of society at large. This is not to imply that all adopted children are illegitimate. In many instances married couples decide to place their offspring for adoption either for economic or other reasons. The case for sealing these records to the adult adoptee would seem much weaker in these instances, but the present laws do not differentiate between such legitimate adoptees and others.

Indeed, as more and more seek their origins, a book entitled *Search, A Handbook for Adoptees and Birthparents* was published in 1982 in which the authors list all the possible procedures by which an adult adoptee can search for his parents and what such a person's rights may be in each state. This handbook will almost certainly cause many more searches and raise many legal cases which must come before the courts.

ALPHABETICAL LIST OF CASES CITED

CASE	CITATION AND DATE	
Alma Society v. Mellon	459 F.Supp. 912	1978
Anonymous v. Anonymous	298 N.Y.S.2d 345	1969
Application of George	625 S.W.2d 151(Mo)	1981
Application of George	630 S.W.2d 614	1982
Application of George	630 S.W.2d 614(Mo)	1982
Application of Gilbert	563 S.W.2d 768(Mo)	1978
Application of Maples	563 S.W.2d 760(Mo)	1978
Application of Sage	586 P.2d 1201(Wash)	1979
Bradley v.Childrens Bureau of S.Carolina	274 S.E.2d 418(S.C.)	1981
Caban v. Mohammed et Ux	441 U.S. 380 1979	
Humphers v. First Interstate Bank	696 P.2d 527 (Ore)	1985
In re Christine	397 A.2d 511(R.I.)	1979
In re Estate of Cregar v.Pesikey	333 N.E.2d 540(Ill)	1975
In re Tilliski's Estate	61 N.E.2d 24(Ill)	1945
Kirsch v. Parker	383 So.2d 384(La)	1980
Lehr v. Robertson	---U.S---	1983
Massey v. Parker	369 So.2d 1310(La)	1979
Matter of Anonymous	92 Misc 2d 224	1977
Matter of Application of Anonymous	89 Misc 2d 132	1976
Matter of Joann Chattman	57 AD2d 618	1977
Matter of Katrina K. Maxtone-Graham	90 Misc 2d 107	1975
Matter of Linda F. M.	52 N Y 2d 1981	1981
Matter of Roger B.	407 N.E.2d 884(Ill)	1980
Mills v. Atlantic City Dept. Vital.Stat	372 A.2d 646(N.J.)	1977

Off. Atty Gen State of Utah   Slip No. 83-80   1983

Office of Atty Gen, State
              of Utah   Slip No.84-36   1985
Rhodes v. Laurino   601 F.2d 1239   1979
Scarpetta v.Spence-Chapin
         Adoptive Service   269 N.E.2d 787(N.Y.) 1971
Sherry H. v. Probate Court   411 A.2d 931(Conn)   1979
State Delaware v.Sharon H.
         and Dennis H.   429 A.2d 1321(Del)   1981
Uniform Adoption Law   1953
Uniform Adoption Law   1971

CHRONOLOGICAL LIST OF CASES CITED

YEAR   CASE   CITATION

1945 In re Tilliski's Estate   61 N.E.2d 24(Ill)
1953 Uniform Adoption Law
1969 Anonymous v. Anonymous   298 N.Y.S. 2d 345
1971 Scarpetta v.Spence-Chapin
            Adoption Service   269 N.E.2d 787 (NY)
1971 Uniform Adoption Law
1975 In re Estate of Cregar
            v.Pesikey   333 N.E.2d 540(Ill)
1975 Matter of Katrina K.
            Maxtone-Graham   90 Misc 2d 107
1976 Matter of Application
            of Anonymous   89 Misc 2d 132
1977 Matter of Anonymous   92 Misc 2d 224
1977 Matter of Joann Chattman   57 AD2d 618
1977 Mills v. Atlantic City
            Dept. Vital.Stat   372 A.2d 646(N.J.)
1978 Alma Society v. Mellon   459 F.Supp. 912

1978	Application of Gilbert	563 S.W.2d 768(Mo)
1978	Application of Maples	563 S.W.2d 760(Mo)
1979	Application of Sage	586 P.2d 1201(Wash)
1979	Caban v. Mohammed et Ux	441 U.S. 380
1979	In re Christine	397 A.2d 511(R.I.)
1979	Massey v. Parker	369 So.2d 1310(La)
1979	Rhodes v. Laurino	601 F.2d 1239
1979	Sherry H. v. Probate Court	411 A.2d 931(Conn)
1980	Kirsch v. Parker	383 So.2d 384(La)
1980	Matter of Roger B.	407 N.E.2d 884(Ill)
1981	Application of George	625 S.W.2d 151(Mo)
1981	Bradley v.Childrens Bureau of S.Carolina	274 S.E.2d 418(S.C.)
1981	Matter of Linda F. M.	52 N Y 2d 1981
1981	State Delaware v.Sharon H. and Dennis H.	429 A.2d 1321(Del)
1982	Application of George	630 S.W.2d 614
1982	Application of George	630 S.W.2d 614(Mo)
1983	Lehr v. Robertson	---U.S.---
1983	Off. Atty Gen State of Utah	Slip No. 83-80
1985	Humphers v. First Interstate	
1985	Office of Atty General, State of Utah	Slip No.84-36

Chapter 3
THE RIGHT TO LIVE, THE RIGHT TO PROCREATE
INFORMED CONSENT FOR THE GENETICALLY DEFECTIVE

The entire area of informed consent for those suffering from genetic defects, especially the mentally impaired, involves serious emotional stress as well as legal complexity. The rights to "Life, liberty, and the pursuit of happiness" are basic to our system of government. The right to procreate, as will be seen in the second part of this chapter, has been considered a fundamental liberty, as is the right to limit one's family. Exceptions have been made, as many early cases will show, but the basic right, unless it seriously interferes with society, has been maintained.

More difficult for the courts to handle are the cases which will be discussed first. These involve problems of life itself and all are uncomfortable cases to deal with from an emotional viewpoint. An observer cannot help but be moved by the plight of the persons involved, and the courts themselves have had to attempt to balance their own emotional responses against their strict legal duty. These cases involving organ transplants are, then, an example of such legal attempts.

Basically, the problem is a genetic one. In all cases of organ transplant there is a risk that the transplanted organ will invoke an immune response in the recipient, and the result will be the failure of the transplant. If closely genetically related persons are involved both as donor and recipient, the risk is much lower and most such transplants are now successful. The ideal transplant would be between identical twins who share an identical genetic constitution. (A case involving this has already been discussed in the chapter on adoption, i.e., *Application of George* where the needed transplant material was bone marrow.) Obviously transplants between "dead" persons and living persons do not present the type of problem dealt with here.

The real issue in these cases deals with the fundamental legal problem of who can speak for the mentally deficient and who can legally grant permission for an organ transplant. There is the added difficulty that without such *legal* permission a physician conducting the operation, or the hospital in which such a transplant operation is carried out, may be sued and without that permission they would almost certainly lose the case and be subject to high damages. The other problem involved is whether parents of defective children may give legal permission, or whether such permission may only be given by the courts. No general laws passed by legislatures exist, and each case must involve the court's decision as to what the law either is or should be.

All of the first cases involve kidney transplants. This technique is relatively sure, and causes only mild discomfort and only a short hospitalization for the donor. As, of course, the donor has two kidneys, only one of which is sufficient to maintain life and health, there is no problem of risk to the donor. In fact many perfectly healthy and normal people may have only one kidney from birth and would never have known this if medical examinations for other reasons had not revealed that information.

The 1969 case, *Strunk v. Strunk*, decided by the

Court of Appeals of Kentucky is among the earliest
cases. As it will be cited frequently by the others,
some detail is necessary. The suit was brought by a
guardian ad litem for a mentally defective 27-year-old
incompetent (Jerry) with a mental age of approximately
a 6 years old or less. His IQ was deemed to be
about 35; he was institutionalized and had severe speech
defects as well. The defendant in this case was his
mother who was acting as a committee representing her
son. The case arose as she sought permission for the
incompetent to serve as a kidney donor for his older
brother, Tommy, who was married, employed, and a part-
time college student. Unfortunately, he suffered from
chronic glomerulus nephritis, a fatal kidney disease. At
the time of the case he was being maintained by an
artificial kidney, but his physicians felt that this
support could not be continued much longer and only a
transplant would save him. The doctors felt that it
might be possible to use a kidney from a cadaver, but
these were not easy to come by, and they conducted blood
tests on all of the family to see whether one might be a
more suitable donor. Because of blood incompatibilities,
all were rejected as possible donors, including the
parents and a number of collateral relatives. As a last
resort, the incompetent brother was tested and found to
be a highly acceptable donor, presenting the legal
problem of what could be done to secure permission for
the transplant and leading to the formal issue brought
before the court.

Another important facet in this case was the
psychological dependence of Jerry upon Tommy. Testimony
by Jerry's attending psychiatrist showed that the death
of the older brother would be devastating to Jerry, and

therefore the operation, which would be life-saving for Tommy would benefit the younger brother. "Jerry Strunk, a mental defective, has emotions and reactions on a scale comparable to that of a normal person. He identifies with his brother Tom; Tom is his model, his tie with his family. Tom's life is vital to the continuity of Jerry's improvement at Frankfort State Hospital and School. ...We the Department of Mental Health must take all possible steps to prevent the occurrence of any guilt feelings Jerry would have if Tom were to die.

"The necessity of Tom's life to Jerry's treatment and eventual rehabilitation is clearer in view of the fact that Tom is his only living sibling and at the death of their parents, now in their fifties, Jerry will have no concerned intimate communication so necessary to his stability and optimal functioning." It was also pointed out that even if a cadaver-donor could be found Tommy would not be able to survive for a second transplant should the first fail. These then are the facts.

The court made a careful survey of the laws. They found that Jerry had been thoroughly represented by the *guardian ad litem* who questioned fully the right of the state to authorize the operation. It should be pointed out that this was not done in any inherently nasty way, but simply to be sure that the legal aspects were correctly analysed and that the operation, if performed, carried express legal consent by the courts. The court found it did have a right, acting as *parens patriae*, to interfere, based on English law and thus by inheritance from the English system to our laws. The doctrine of substituted judgment, whereby the courts could act as

interpreters of what the incompetent would wish were he
able to express his desires, was found broad enough for
the court in this case to lead to its approval of the
transplant. The court also gave a detailed history of
kidney transplant operations, and studied the medical
risk of a donor, concluding that there was little or
none. After this, the judge based his opinion upon the
fact that the operation would be to the benefit of
Jerry. "We are of the opinion that a chancery court does
have sufficient inherent power to authorize the
operation. The circuit court having found that the
operative procedures in this instance are to the best
interests of Jerry Strunk and this finding having been
based upon substantial evidence, we are of the opinion
the judgment should be confirmed..." It must be pointed
out, however, that the decision to grant permission was
far from unanimous, indeed the court split 4:3 on the
issue of whether the state had such power. The
dissenters quoted many other cases and made several
legal arguments based on those, including the citation
from one of them, "Parents may be free to become martyrs
themselves. But it does not follow that they are free,
in identical circumstances, to make martyrs of their
children before they have reached the age of full and
legal discretion when they can make that choice for
themselves. The ability to fully understand and consent
is a prerequisite to the donation of a part of the human
body..." If this were followed, as the legalistic
dissenters would have done, it is obvious that Jerry
could never have given consent as he would never reach
the legal capability to do so. The dissenters were also
not convinced of the benefit to Jerry and did not think
that benefit had been sufficiently established. The

minority opinion ended by saying, "I am unwilling to hold that the gates should be open to permit the removal of an organ from an incompetent for transplant, at least until such time as it is conclusively demonstrated that it will be of benefit to the incompetent. The evidence has not risen to that pinnacle. To hold that committees, guardians or courts have such awesome power even in the persuasive case before us, could establish legal precedents, the dire results of which we cannot fathom."

These legal details have been presented here because the two views of the legal aspects of the instant case are set forth forcibly and they will be part of the fundamental reference points upon which future cases will depend. It should also be noted that the minority opinion, while seeking to deny the transplant did not do so from a lack of humane feelings, but from the sense that the legal precedence was not clear, and from the fear that should the case be decided as it was there would soon be other cases, not as clear as this, which would lead the courts into an inexorable tangle of laws and actions.

Even if there is no genetic defect, both physicians and hospitals are reluctant, as pointed out before, to perform transplants when the donor is a minor. In the case *Hart v. Brown* the Superior Court of Connecticut faced this issue. In this case one of a pair of identical twins developed uremic syndrome at the age of 7; haemodialysis was undertaken and subsequently it was found that a more serious, and fatal, disease, malignant hypertension, had developed. A bilateral nephrectomy was performed, leaving the child with no kidneys and totally dependent upon dialysis for survival. At the time of the case, she was in the hospital awaiting a suitable donor for kidney transplant, her only hope of leading a normal life, or indeed of survival for any length of time. The court placed the medical problem tersely, "The types of kidney transplantations discussed in this matter were a parental homograph--transfer of tissue from one human being to another--and an isograft, that is, a one-egg twin graft from one to another. The parental homograph always presents a serious problem of rejection by the donee. Because the human body rejects any foreign organs, the donee must be placed upon a program of immunosuppressive drugs to combat such rejection. An isograft, on the other hand, is not presented with the problem of rejection. A one-egg twin carries the same genetic material, and, because of this, rejection is not a factor in the success rate of the graft." The court might also have added that the effect of immunosuppressive drugs is to lower or abolish the ability of the body to combat any infection, and as a consequence the user of such drugs is very susceptible to any routine infection which, if not readily treated by antibiotics, can be fatal. The court also pointed out that all isografts in the past few years had been successful, while the use of parental homografts had been far less so, with only about 1/3 lasting more than 7 years.

Evidence was also offered that the donor's risk was sufficiently low that life insurance actuaries do not rate up such donors above those with the usual two kidneys. Nonetheless, the physicians and the hospitals felt it necessary to obtain a court order allowing parental consent for the donation by the healthy minor twin. Again psychiatric testimony established the strong mutual identification of the twins, and made the further analysis that the graft would be of strong benefit to the donor "in that the donor would be better off in a family that was happy than in a family that was distressed and in that it would be a very great loss to the donor if the donee were to die from her illness." In addition, a clergyman was also a witness and testified that the decision by the parents to proceed with the transplant was morally and ethically sound. Further, the court appointed *guardian ad litem* for the minor testified that after consultation with all concerned he too felt the isograft should be performed. The court put the problem this way, "A further question before this court is whether it should abandon the donee to a brief medically complicated life and eventual death or permit the natural parents to take some action based on reason and medical probability in order to keep both children alive. The court will choose the latter course, being of the opinion that the kidney transplant procedure contemplated herein--an isograft--has progressed at this time to the point of being a medically proven fact of life. Testimony was offered that this type of procedure is not clinical experimentation but rather medically accepted therapy." Once having adopted this decision, the court then found authority for its decision in several other cases, including *Strunk* above. They concluded, "...To prohibit the natural parents and the *guardians ad litem* of the minor children the right to give their consent under these circumstances, where there is supervision by this court and other persons in examining their judgment, would be most unjust, inequitable and injudicious. Therefore, natural parents of a minor should have the right to give their consent

to an isograft kidney transplantion procedure when their
motivation and reasoning are favorably reviewed by a
community representation which includes a court of
equity." Thus the court granted permission for the
procedure to be carried out. But it still must be noted
that the decision to act for a minor, competent or not,
is that of a court, and not a granted right without
legal approval.

In fact, just the opposite decision was reached in
the 1973 case, *In re Richardson* decided by the Court of
Appeals of Louisiana, Fourth Circuit. Here the case was,
however, somewhat different. The case involved suit by
the husband against his wife, as natural tutrix of her
minor mental retardate (a Down syndrome individual), to
compel her to consent to the transplantation of one of the
minor's kidneys to an older sister. This was not a suit
representing bitterness between the spouses, but a suit
to establish the legal rights for the transplant.
However, in this case the medical facts were somewhat
different from the previous ones. The mother was 56
years old, the father 63, the Down child 17 (with a
mental capacity of a 3-4 year-old) and the daughter was
32, divorced and living at home because of her illness,
chronic nephritis, which resulted in her having only
about 7-10% of normal kidney function. In addition, she
suffered from Systemic Lupus Erythro-Matosus, a disease
which affected the ability of organ tissue to adhere
properly to other body tissues. This disease is
incurable, but sometimes may be helped or arrested.
However, in her case, the drugs needed could not be
applied due to the treatment for hypertension. Drugs
which would be used in the immunosuppressive treatment
for the transplant would possibly be of use in treatment

of the second disease.

Tests were made of the entire family and it was found that the Down child was the best possible donor, with a rejection probability of 4-5% during the ensuing five years as against 20-30% for all the others tested. Cadaver donation was also considered but the chance of rejection here would be far higher. Meanwhile, the 32-year-old sister was being maintained by renal dialysis, and the court found that an immediate kidney transplant was not essential for her.

The court further analysed the case on the basis of Louisiana law, and found no precedent therein. The nearest they could find was the law consisting of property gifts by a minor, and they extended this to the present case. Such gifts are specifically prohibited, especially by a minor's representative, and the court, by extension, decided that the "gift" of an organ would fall under this part of the state's laws. They also rejected the argument that by giving consent to the organ donation they would thereby allow the prolongation of the life of the 32 year old sister who, in turn, would be able to take care of the Down child after the death of the parents. The court thus gave a firm decision that there could be no permission by any concerned to allow the transplant. They also cited Strunk, but found that here none of the circumstances were similar; there was no immediacy of death, there would be no real benefit to the donor, and the court should not require, because of these reasons, that the permission for donation by the incompetent minor be transferred to his mother. A concurring opinion summed up what might have been the real issue here. "The majority, in my opinion, rightfully assumes that the court is empowered to authorize the transplantation of the kidney from the minor, provided certain standards are met, i.e., the best interests of the minor. However, I am not of the opinion that before the court might exercise its *awesome* authority in such an instance and before it considers the question of the best interests of the child, certain requirements must be met. I am of the opinion that it must be clearly established that the surgical intrusion is urgent, that there are no reasonable alternatives, and that the contingencies are minimal. These requirements or prerequisites are not met in this case. Having so determined, we are not confronted with the question of the best interests of the child." In essence, this case does not really depart from Strunk in legal terms; it merely emphasizes the need to establish both the benefit to the donor and the urgency of the need for donation. Had these two points been critical, it is possible that the court might have otherwise decided.

The case, *In re Guardianship of Pescinski incompetent*, brought before the Supreme Court of Wisconsin in 1975, shows yet another aspect of the perplexing problem of consent for an incompetent person. In this case, the proposed donor was 39 years old and in a mental hospital, classified as a schizophrenic, chronic, catatonic type. The proposed recipient, his sister, was 38 years old. She had both kidneys removed as she was suffering from chronic glomerulonephritis and was therefore totally dependent upon dialysis. In an attempt to find a suitable kidney donor, all members of her family were tested, and only the schizophrenic brother was found suitable. The *guardian ad litem* in this case would not give permission for the transplant and the lower court held that it did not have power to override his decision. The case was appealed to the Supreme Court which upheld the lower court. The basis for that decision was no benefit could possibly accrue to the perspective donor, and therefore the permission for the donation was legitimately withheld. The Supreme Court specifically declined to adopt the concept of "substituted judgment" used by the Strunk court. They pointed out that in that case it was decided that the Appellate court, not the lower court, had the power to grant permission for the transplant, and in this case the lower court in Wisconsin which also denied the petition for the transplant had no power to act. They also reasoned that here there was no benefit which could accrue to the schizophrenic and there was no basis for the court to act favorably on the petition. "We therefore must affirm the lower court's decision that it was without power to approve the operation, and we further decide that there is no such power in this court. An incompetent particularly should have his own interests protected. Certainly no advantage should be taken of him. In the absence of real consent on his part, and in a situation where no benefit to him has been established, we fail to find any authority for the county court, or this court, to approve this operation." A dissenting judge based his disagreement strongly on *Strunk*, quoting from that opinion, but adding, "I would regard this as pretty thin soup on which to base a decision as to whether the donee is to be permitted to live. In the case before us, if the incompetent brother should happily recover from his mental illness, he would undoubtedly be happy to learn that the transplant of one of his kidneys to his sister saved her life. This at least would be a normal response and hence the transplant is not without benefit to him... The *guardian ad litem* argues strongly that for us to permit this transplant is to bring back memories of the Dachau concentration camp in Nazi Germany and medical experiments on unwilling subjects, many of whom died or were horribly maimed. I fail to see the analogy--this is not an experiment conducted by mad doctors but a well-known and accepted surgical procedure necessitated in this case to save the life of the incompetent's sister. Such a transplant would be authorized, not by a group of doctors operating behind a barbed-wire stockade but only after a full hearing in an American court of law. ... With these guidelines the fear expressed that... institutions for the mentally ill will merely become storehouses for spare parts for people on the outside is completely unjustified." The dissent continues to make the point that the incompetent is considered "insane seven days a week," and there is no hope of obtaining his consent. "The majority opinion would forever condemn the incompetent to be always a receiver, a taker, but never a giver." This judge would have accepted the rule of "substituted judgment." Nonetheless the majority opinion stands and the transplant was denied. It might be added here that under other conditions, the fatally ill sister might also be judged in a sense to be "incompetent" also, but this point was not raised in the case. It appears that the court in Wisconsin, at least in 1975, was unwilling to exercise judicial action, and fell back on the fact that no legislative act enabled them to permit the graft. Other courts, in other cases and other states would almost certainly have felt no such restraint, as was the case in Kentucky.

In 1979 the Court of Civil Appeals of Texas was
faced with a case almost identical to the Strunk case.
In *Little v.Little*, there was a question of a kidney
donation from a 14-year-old child (Anne) declared of
unsound mind due to Down syndrome. Permission was
sought to use one of her kidneys for transplant to her
younger brother (Stephen) suffering from an otherwise
fatal kidney disease. The lower court had approved the
petition, and the *guardian ad litem* for the incompetent
appealed that decision. The higher court dealt with many
issues, including the Texas laws regarding such issues
and found that there was no pertinent part of the Family
Code dealing specifically with this type of case. The
court also heard psychiatric testimony concerning the
benefit to the incompetent and cited it in the opinion.
"But the testimony in this case conclusively establishes
the existence of a close relationship between Ann and
Stephen, a genuine concern for the welfare of the other
and, at the very least, an awareness by Anne of the
nature of Stephen's plight and awareness of the fact
that she is in a position to ameliorate Stephen's
burden. Assuming that Anne is incapable of understanding
the nature of death, there is ample evidence to the
effect that she understands the concept of absence and
that she is unhappy on the occasions when Stephen must
leave home for hours when he journeys to San Antonio for
dialysis. It may be conceded that the state of
development of the behavioral arts is such that the
testimony of psychiatrists and psychologists must still
be classified as speculative, but, as of today, that has
not been accepted as justifying a judicial rejection of
the value of such testimony. "The testimony is not
limited to the prevention of sadness. There is

uncontradicted testimony relating to increased happiness. Studies of persons who have donated kidneys reveal resulting positive benefits such as heightened self-esteem, enhanced status in the family, renewed meaning in life and other positive feelings including transcendental or peak experiences flowing from their gift of life to another." The court then quotes from *Strunk*, and from the dissenting opinion in *Pescinski* noted above. They preferred to go along with *Strunk* and the dissent in *Pescinski* in allowing the doctrine of "substituted judgment" to hold. For these reasons they unanimously upheld the lower court's right to permit the operation and the legality of this permission. But the court did note at the end of its opinion that legislative action would be appropriate here, and this case was not to be taken as an absolute precedent for others to come.

The decisions allowing the granting of permission for grafts from mental incompetents were all based solely upon the apparent benefit to the donor, and from this the courts then went on to apply the "substituted judgment" rule. In the cases denying the operation, of course, just the opposite reasoning was used, no benefit, and no use of the rule. It would appear that each such case continues to be on an *ad rem* basis, and some general law or judicial ruling from a higher court is badly needed to enable this type of genetically based case to be decided equitably in all courts throughout the country. Meanwhile, barring such actions, long and difficult cases will continue to appear, and it would seem that their outcome will vary from state to state and from year to year.

The cases involving sterilization of incompetents

for genetic or social reasons are also varied. The history of these cases follows somewhat closely the history of the eugenics movement in this country. Starting soon after the rediscovery of Mendel's work at the turn of the century, the original purpose of this group was a beneficial one, to improve human life by genetic means. This laudable aim, however, soon became perverted by those who sought to impose their standards upon everyone in the country. Their standards were also based on false use of genetics. The belief at the time that not only intelligence, but also social behavior, was inheritable soon led to the creation of a self-selected elite, who firmly believed that the right to reproduce should be restricted severely, namely to those of high intelligence and possessing a superior genetic background--themselves. The purported inheritance of not only high intelligence, but on the opposite end of the scale, low intelligence and criminal behavior became tenets of the movement. Not until the logical climax of such beliefs, Nazi doctrines, did the falsity of their beliefs, which never had been grounded in a true genetic science, make the entire group appear ridiculous to most thinking people. This type of eugenic movement as such has more or less disappeared , although there are still some today who cling to the belief that poverty is inherited simply because if a person were not stupid he or she would not be poor. Not only have these ideas in the past done great harm to the science of human genetics per se, but also they have bolstered the type of racial arguments which still plague much of the world. (Surely there is no single leader in the field of human genetics who would do other than ridicule such ideas.) The true facts are that even now we do not have a certain understanding of the genetics of intelligence, and there has been no agreement among geneticists just what is meant by "intelligence" in a genetic sense.

The eugenicists were undoubtedly pleased with the first major case involving genetics and intelligence in the famous (or infamous) case of *Buck v. Bell*, decided by the United States Supreme Court in 1927. The state of Virginia sought to have legal permission to sterilize Carrie Buck, an adult female in a state institution. It was claimed that she was the daughter of a feeble-minded woman in the same institution, and she had a child who also was so classified. The state, represented by Bell as supervisor of the institution, sought the sterilization on the basis that if she were so treated she would not then be required to remain in the institution and obviously would not produce more wards of the state. Virginia had a 1924 statute which was the challenged law in this case. The opinion by Justice Holmes concluded that the state did have the right to carry out the sterilization and ended with what may be the most quoted phrases in the court's history. "It would be strange if it could not call upon those who already sap the strength of the State for these lesser sacrifices, often not felt to be such by those concerned, in order to prevent our being swamped with incompetence. It is better for all the world, if instead of waiting to execute degenerate offspring for crime, or to let them starve for their imbecility, society can prevent those who are manifestly unfit from continuing their kind. The principle that sustains compulsory vaccination is broad enough to cover cutting the Fallopian tubes. Three generations of imbeciles are enough." It is perhaps sufficient to add that today there is some question as to the "imbecility" of any of the Buck family, and the confusion over mental ability, poverty and lack of education might indicate that the facts of the case would probably be viewed differently today. But the sway of the thinking of the eugenicists can be seen by reading the words just cited.

In fact, between the time of the Buck case and the one to be discussed next, many states, again under the influence of the thinking of the eugenicists, whose views supported their own racial or religious ideas, passed laws making it possible for the state to sterilize almost any recidivist felon in a state prison. Castration seemed to be the popular answer for rehabilitation in those days. However, when the case, *Skinner v. Oklahoma*, was decided in 1962, 15 years after Buck, this practice was mainly stopped by the opinion of Justice Douglas. The legal grounds for this, however, were not due to any understanding of genetics, but were based entirely upon the Due Process clause of the XIV Amendment. The problem was not only one of the right to procreate, which did form a considerable part of the opinion, but also of the unfairness of the Oklahoma statute. The legislature there had passed a law which permitted sterilization of "habitual criminals" who were defined as those who had been convicted two or more times of any crimes "involving moral turpitude," (what crimes do not?) and had been convicted a third time. But the law specifically exempted certain offenses, including embezzlement.

Presumably this type of felon was more intelligent or embezzlement is not a crime of moral turpitude. Skinner's attack on the law was on that specific basis; i.e., there was not equal protection under the amendment. It was pointed out that his first crime was

that of stealing chickens and his subsequent two crimes,
robberies, allowed the state after a jury trial to
conclude he was morally unfit to procreate and also that
the operation would have no effect upon his then good
health. As one of the old sayings goes, let a poor man
steal a loaf of bread and he goes to jail, let a rich
man steal a railroad and he goes to the U. S. Senate!

Calling the plaintiff's right one "which is basic
to the perpetuation of a race--the one to have
offspring", the Supreme Court called the involuntary
sterilization clearly unconstitutional. They pointed out
the ridiculously construed law exempting embezzlers, and
that, "Whether a particular act is larceny by fraud or
embezzlement thus turns not on the intrinsic quality of
the act but on when the felonious intent arose--a
question for jury under appropriate instructions." The
court would have no question in allowing a state to
define the classification of crimes, but that was not
the issue raised here. What the issue really was is
stated clearly in the opinion. "When the law lays an
unequal hand on those who have committed intrinsically
the same quality of offense and sterilizes one and not
the other, it has made as an invidious a discrimination
as if it had selected a particular race or nationality
for oppressive treatment. ... Sterilization of those who
have thrice committed grand larceny, with immunity for
those who are embezzlers, is a clear, pointed,
unmistakable discrimination. Oklahoma makes no attempt
to say he who commits larceny by trespass or trick or
fraud has biologically inheritable traits which he who
commits embezzlement lacks ... If such a classification
were permitted, the technical common law concept of a
'trespass' based on distinctions which are 'very largely

dependent upon history for explanation' could readily become a rule of human genetics." Two concurring opinions were entered; Chief Justice Stone in approving the decision based his opinion mainly on the basis that there had been no proper hearing in which the chicken thief-robber had been able to challenge the genetic issue. "Moreover if we must presume that the legislature knows--what science has been unable to ascertain--that the criminal tendencies of any class of habitual offenders are transmissible regardless of the varying mental characteristics of its individuals, I should suppose that we must likewise presume that the legislature, in its wisdom, knows that the criminal tendencies of some classes of offenders are more likely to be transmitted than those of others." He then points out that although Skinner was given a hearing "to determine whether sterilization would be detrimental to his health, he was given none to discover whether his criminal tendencies are of an inheritable type." The second concurring opinion, written by Justice Jackson, went further on the basis of genetics. "I also think the present plan to sterilize the individual in pursuit of a eugenic plan to eliminate from the race characteristics that are only vaguely identified and which in our present state of knowledge are uncertain as transmissibility presents other constitutional questions of gravity. This court has sustained such an experiment with respect to an imbecile, a person with definite and observable characteristics. ...There are limits to the extent which a legislatively represented majority may conduct biological experiments at the expense of the dignity and personal and natural powers of a minority-- even those who have been guilty of what the majority defines as crimes." Thus, the court struck down the law of Oklahoma on three distinct grounds. It might be noted that when the opinion was handed down, 14 states had somewhat similar laws which would now be invalid.

With these background cases in mind, we can now turn to the more modern applications. Among the first of these was the 1978 case, *Ruby v. Massey*, decided by the United States District Court, D. Connecticut. The Rubys were one of a set of three different parents each of whom sought to have a mentally defective female child sterilized. Their three girls were incapable of self-maintenance. The girls were at the time showing signs of sexual development; a 12-year-old had begun to menstruate and two others were on the verge of puberty. The 12-year-old was mentally incapable of understanding her situation and could not care for her hygienic needs. She also suffered severe pains during her periods as well as psychological distress. There is no question the fact that none of the girls would be able to understand pregnancy, nor is it at all likely that any could understand the use of contraceptives of any type. They were unable even to communicate with their physician regarding pain, and if there were to be a pregnancy they would be totally unable to self-monitor it in any way.

Dr. Massey was the executive director of the University of Connecticut Health Center where the sterilizations were requested. Both he and the Hospital had a policy of not permitting such sterilizations on the very sound policy that without informed consent by a capable person a suit could be brought, and won, against the physician and hospital for performing or allowing this type of operation to be carried out. There are two issues in this case: first the rights of the parents to give consent for the incompetent children, and secondly the legal issue of law concerning this operation itself. The Connecticut laws permitted such operations upon similar children in a State Institution following careful consideration by a committee of surgeons. The children in this case were in a private institution, and there was no provision for this procedure in such hospitals. The court held that on the issue of consent, "...The Constitution protects the freedom of even an immature teenager to decide for herself whether to bear or beget a child; no case has considered who may make the decision for the child who is mentally incapable of deciding for herself. The fact that in this case the parents seek to have the children's rights exercised in favor of sterilization, rather than against it, does not affect the character of the right. They may neither veto nor give valid consent to the sterilization of their children." Further, the court added, "This lawsuit is unmistakenly a poignant cry for help from these children uttered in their behalf by their parents. These children are what they are; they are unable to come to terms with reality sufficiently to make the decisions which are only theirs to make. That they are incapable of comprehending the consequences of their actions is clear beyond question. The fact that the demand for an 'informed' decision from each of the children is an impossible one to meet makes imperative the need for an authoritative decision on their behalf."

The court then took the necessary step to attempt to help these people. By examination of the Connecticut law upholding such decisions in public institutions, but not in private ones, the court held that statute clearly

deprived the children in such cases of their equal
protection rights. The court held that the right of
sterilization in this case was "fundamental" and thereby
rooted in the constitution. "Accordingly, it is the
order of this court that the defendants are enjoined
from refusing to provide the plaintiffs with services
identical to the [Public Institutions] to enable them to
obtain consent of a probate court to the sterilization
operations upon them." It should be added that the
physician and the hospital were both quite sympathetic
to the plight of the children, and had defended the suit
not only for their own protection but also, in a sense,
to protect the rights of the children. Other issues also
were involved; the parents wished to have the court
rewrite the Connecticut law, which of course they were
powerless to do, although the opinion does suggest the
hope that the Legislature will do so. (Thanks are
expressed here to The Center for Health Care Law, Union
College, for calling my attention to this particular
case as well as others mentioned in their newsletter.)

A very similar case, this time involving a non-
institutionalized 18-year-old Down Syndrome girl arose
in 1979 in the case, *In the Matter of Grady*, decided by
the Superior Court of New Jersey, Chancery Division. A
hospital in New Jersey, similar to the one in
Connecticut, refused to allow the sterilization
operation without the permission of the court. A
guardian ad litem was also appointed to represent the
interests of the minor. It was firmly established that
her I.Q. was about in the upper 20's to the upper 30's
range, and she was in good physical health without the
more usual secondary effects of the genetic disease. In
fact, a normal life span was predicted for her.

Incidentally, the court furnished a remarkably good history of the disease, and of the genetic causes for it as well as a history of the change in treatment for such children at home rather than in institutions. The parents had been admirable in their care for the girl and they now wished, as she was reaching adulthood, to enable her to move outward into some group-living program where special care and training would benefit her greatly. Obviously, they were concerned about the risks of pregnancy. The court turned again to psychological and medical authorities who point out that the retarded have the same drives and that the modern tendency is not to deny these simply on the basis of mental retardation. There then follows a history of the legality of sterilization from the time of Social Darwinism through the present. They also looked into the genetics and reported that a study was made showing that while there might be a somewhat reduced sex drive, those Down women who had children were found to have produced a disproportionately high incidence of children with the same affliction, or who were either retarded or stillborn. After the historical review a detailed review of the laws considering rights to privacy was made. Finally, in this careful and thoughtful opinion, the court turned to the New Jersey laws themselves. They found again the Connecticut problem of the public as against the private hospitals and further problems in the statement of the law that dealt with sterilizations. Also the rule of *parens patriae* was studied to determine whether substituted consent applied. The court concluded this lengthy discussion with the decision, unlike the Connecticut case however, that the parents had the right to make the decision for the child. "Lee Ann Grady cannot exercise her freedom of choice as guaranteed by the Constitution. If her incompetency is not to deprive her of the benefit of constitutionlly protected alternatives, her parents as the general guardians must, under the circumstances of this case, be permitted to exercise their judgment as to how that judgment should be made.

"It is not for this court to substitute its judgment for the informed consent of Lee Ann Grady, nor as has been suggested to weigh the relative advantages and risks of other methods of contraception. ...Any decision arrived at by the parents is to be protected from public scrutiny. Lee Ann has the same right to privacy as do all other persons. Therefore the decision is protected by the same privilege as all matters between physician or hospital and patient and should be arrived at privately without public disclosure.

"This decision is not to be interpreted as authorizing parents to consent to the sterilization of incompetent persons absent authorization of competent jurisdiction. Each application must be decided upon its own merits.

"Lee Ann is entitled to be treated with dignity, in a manner designed to permit her to realize her greatest potential; she is entitled to be dealt with as an individual rather than as a member of a limited class; she is entitled to participate in such activities as are enjoyable and meaningful to her. The thrust of this opinion is to enable Lee Ann to enjoy these shared common entitlements unhindered by her limitations." This opinion, it appears, shows the law and the courts at their best, attempting to mete out justice, and to interpret the law so as to reach their conclusions.

While there may be instances in this case, and in others where "stretching the law" may seem to make a poor precedent, but it would seem that the last statements of the court show judicial actions at their best. However, it must be added that when this case was appealed to the Supreme Court of New Jersey, that court, while agreeing mainly with the Superior Court, remanded the case for retrial on the basis that an insufficient standard of proof had been established to show that the procedure was in the best interests of the child.

In contrast to the lengthy opinions just discussed the opinion of the Supreme Court of New Hampshire in the 1980 case, *In re Penny N.* is brief. Again similar facts appear, a 14-year-old Down syndrome girl, and permission for her sterilization. The only question raised here was whether the probate court, where application was first made, had jurisdiction to decide upon the case. The court appointed a *guardian ad litem* who studied the facts and reported that the sought-for hysterectomy should be performed. The court ruled that the lower court indeed had such powers. "We believe, in the absence of any legislative determination to the contrary, that the 'clear and convincing' standard of proof should apply in this kind of case. 'Preponderance of the evidence' does not sufficiently protect Penny's interest in preserving her physical integrity. On the other hand, 'beyond a reasonable doubt' is overprotective. That standard is reserved for criminal cases and potential cases and situations in which the potential result is loss of liberty." The court cites *Grady*, above, and concludes that they are taking a middle course when contrasting that decision with other cases. They stress the concept that " clear and

convincing evidence" is the proper standard to be
applied.

In 1981 yet another like case, *In the matter of
C.D.M.*, arose in the Supreme Court of Alaska. Again, the
parents of a 19-year-old woman with Down were anxious
that she be sterilized. The lower court found that
indeed it would be in the interest of the 19-year old
for this to be done, but nevertheless denied the
application for lack of jurisdiction to grant the
request. The Supreme Court of Alaska held to the
contrary that it was within the Superior Court's
jurisdiction to act. In this case the afflicted person
was able to function and was receiving training as a
kitchen helper and also was working part time as a
helper in a fast food restaurant. However, it was
established that she would never be able to function without
parental or other care, and she was adjudicated as an
"incapacitated person" under Alaska laws. The petition
by the parents was made because of the increasing risk
of sexual intercourse and the likelihood of pregnancy,
an event which the court summarized as "Although she is
not currently socializing in such a manner as would make
this likely, Down Syndrome individuals are
characteristically highly susceptible to being sexually
victimized by virtue of their very innocent, trusting
and loving nature. Another factor in this case was the
fact that the girl herself was apparently sufficiently
aware of her own problems to be able to express a desire for
the operation which had been recommended by both her
family doctor and a specialist in genetics. The Supreme
Court, while remanding the case to the Superior Court
with power to act, did not act unequivocally to order
the sterilization, but required that the lower court

establish firmly whether other methods of contraception might be employed before granting the request for the sterilization. "Sterilization necessarily results in the permanent termination of the intensely personal right to procreate. Therefore, before sanctioning the sterilization of an incompetent, the court must take great care to ensure that the incompetent's rights are jealously guarded. The advocates of the proposed operation bear the heavy burden of proving by clear and convincing evidence that sterilization is in the best interests of the incompetent." The court then noted the proceedings to be carried out when the case was remanded, including whether birth control methods would not be adequate. A dissent here simply questioned whether the Superior Court had any such jurisdiction at all, and that the seriousness of the consequences of sterilization, unlike other medical procedures approved for an incompetent, are so great that only the legislature itself could properly address the issue. He also questioned the validity of the testimony of the Down child, and asked whether she had really offered this in an understanding way. He suggested that her attitudes might change in the future, and she might even marry a spouse who would be willing to face the problem of caring for any genetically defective offspring. He also cited a case not reported here, *Stump v. Sparkman*, where such a sterilization was ordered after a plea by the girl's mother; later the girl, who at the time of the operation was making normal progress through school, became married and only learned of her sterilization after she failed to become pregnant. The dissenting judge was concerned this could be the case here, although given the limited understanding of the child in

question this would seem unlikely as she had been told
plainly what the consequences of the operation were. But
the main point here is that lower courts do have the
power to determine whether a sterilization of a
genetically defective infant can be carried out,
provided the rights of the child are fully protected by
careful legal procedures.

In 1981 the same issues were faced by the Wisconsin
Supreme Court in the case "*In the Matter of the
Guardianship of Eberhardy*." The issues were almost
identical to those of other cases, and again this was
the first instance of such a case in this state. The
lower court had refused to agree to the sterilization of
a mentally deficient adult girl, and the case was
carried to the highest state court. The issues involved
the right of a judge to approve an operation of this
sort, and the intermediate court had stated, "Unless and
until the legislature confirms express power on
Wisconsin courts to authorize the sterilization of
incompetent persons under stated circumstances, the
courts are without jurisdiction to consider the same."
The Supreme Court did not agree on this issue. "We
conclude that, under the Constitution of the State of
Wisconsin, the circuit court had the jurisdiction to
approve of the proposed tubal ligation; and,
additionally, we conclude that the statutes acknowledge
the plenary jurisdiction of the Wisconsin circuit
courts." A lengthy discussion of the plenary powers of
the courts then followed, citing the previous cases,
among others, as authority upon which to reach this
conclusion. The court also goes on to give
a long and detailed history of sterilization laws, both
in Wisconsin and elsewhere. (A reader interested in this

history may be well advised to read this opinion.)

Having decided that the courts had this power, the Supreme Court then turned to the methods by which such a sterilization decision could be made. In this case there again appeared to be no legislative guide lines, and the court dealt with this on the basis of the "best interests" of the person concerned. But here they found a stumbling block. They pointed out that while most decisions made in the "best interests" could be reversed, sterilization was all but irreversible. Consequently, they did not approve the sterilization, pointing out that in the instant case other methods of contraception had not been thoroughly enough considered and the lower court record showed no evidence as to whether these would not be effective. They also felt that no record was shown that incompetents might not give birth without trauma and also might not make "good" mothers. There is a present policy of "mainstreaming" such persons, and the court here felt that there was a need for a "properly thought out public policy on sterilization or alternate contraceptive methods [which] could facilitate the entry of these persons into a more nearly normal relationship with society. But again this is a problem that ought to be addressed by the legislature on the basis of factfinding and the opinions of experts. ...This case demonstrates that a court is not the appropriate forum for making public policy in such a sensitive area."

They quote from Justice Frankfurter who once said, "Courts are not equipped for discovering wise policy. A court is confined within the bounds of a particular record, and it cannot even shape the record. Only

fragments of a social problem are seen through the
narrow window of a litigation. Had we innate or acquired
understanding in its entirety, we would not have at our
disposal adequate means for constructive solution. The
answer to so tangled a problem ...is not to be achieved
by... judicial resources." Again they quote from Justice
Cardoza who, they point out, was considered a judicial
activist and "believed it important for courts to blaze
trails where necessary to protect human rights." The
quotation is used here to point out that even Cardoza
reached limits. "The judge, even when he is free, is
still not wholly free. He is not to innovate at
pleasure. He is not a knight-errant, roaming at will in
pursuit of his own ideal of beauty or goodness. He is to
draw his inspiration from consecrated principles. He is
not to yield to spasmodic sentiment, to vague and
unregulated benevolence. He is to exercise a discretion
informed by tradition, methodized by analogy,
disciplined by system, and subordinated to the
'primordial necessity of order in the social life.' Wide
enough in all conscience is the field of discretion that
remains ...[Judges] have the power, though not the
right, to travel beyond the walls of the interstices,
the bounds set to judicial innovation by precedent and
custom."

Having said all this, in what may appear to be a
bit of piety in order to give their decision more power,
the judges reviewed constitutional rights of minors,
abortion, and the lack of laws considering rights of
mental retardates in this type of case. They finally
concluded that while the lower court did have the
appropriate legal right to make a decision, it did not
have the right to approve the requested sterilization in
this case as insufficient evidence, they felt, was given

concerning her status. They quote from Hayes (see above), "It seems to me that having clearly declared the judiciary's power to act, wisdom dictates that we should defer articulation of this complex policy to the legislature." Thus, no sterilization is authorized.

One concurring opinion would also have denied the power of the lower court to act at all in the absence of specific powers granted to that court by the legislature. Again the opinion is full of legal citations and precedents which need not concern us here.

An almost scathing dissent was also entered by one judge. "Two thousand years ago a judge, clothed with the power and authority to do justice, but sensing the political winds ('willing to content the people' as the ancient word puts it) washed his hands and said to the people: 'See ye to it.' His act resulted not in justice, but injustice. Today the majority of this court, in my opinion, withholds justice from Joan Eberhardy. It turns to the legislature, the 'representatives of the people,' and says in effect, 'you see to it.' Washing its hands and turning the demand for justice over to the legislature demeans this court, denigrates its role, and makes a mockery of its powers.

"The majority cannot be unaware that the legislature will do nothing about this matter. In today's political atmosphere, few, if any, state legislators would sponsor or support sterilization legislation. The legislature would, of necessity, have to deal with the whole gamut of when sterilization could be done. Such legislation will not be forthcoming, and for this court to tell this unfortunate woman's parents to turn to the state legislature is to leave them without relief." He reviews the circumstances of the case; the 22-year old woman with a two-year old's mental development, unable to feed or clothe herself, unable to draw a bath as she could not regulate its temperature, sexually mature, and with all concerned testifying that pregnancy would be tragic for her. In addition there was strong evidence in the testimony that her mental impairment was increasing with age. "But a majority of this court finds this pathetic, helpless, vulnerable and 'most unfortunate individual' had a *right* to become pregnant!... I think the majority opinion could have been a landmark decision on the authority of this court to act in the absence of specific statutory authorization. Unfortunately, it is weakened by an unreasonable and unjustifiable retreat with much rhetoric about 'judicial restraint' that presents no workable guidelines as to when such restraint should be exercised...This of course flies in the face of a long line of cases in which this court has acted in the great common-law tradition and fashioned remedies where facts demanded it." He continues to express his disappointment with his fellow judges, "Unlike the legislature which deals with broad issues of social policy, courts deal with individual cases. It is from the resolution of cases that the common law evolves...This case is ripe for decision. The facts are clear." He goes on to criticize almost every part of the majority opinion.He pokes fun at one citation in the opinion which went back to the "grandaddy" of long opinions, one in 1904 of some 334 pages, an "all time record for judicial verbosity in Wisconsin." He cites some of it, first saying that if the majority had "mushed on to pages 234-235 it would have found this gem." Politeness to the history of Wisconsin courts forbids that this be quoted! He ends his dissent as follows; "Maybe some day, even in Wisconsin, those with power to do justice will not ask for the washbasin." One is impressed by judicial wrath and scorn at its highest.

There was even a second dissent in this case, much more moderate in tone; however he concludes, "In sum I think it most unfortunate that the majority has chosen to defer to the legislature on this matter. While I agree the case involves a fundamental policy question, these are not strangers to this court. The people of the state have the right to apply to the courts for the protection of their inherent human rights to life, liberty and the pursuit of happiness and to have their applications for relief adjudicated. The rights of those least able to protect themselves are the rights most in need of judicial attention."

Perhaps two 1982 cases may point the way towards future decisions. The case, *In the Matter of Moe*, decided by the Supreme Judicial Court of Massachusetts, involved the petition of a mother to have her adult, institutionalized, mentally defective child sterilized. The lower court reported the matter, without decision, to the Appeals Court and through them to the present court. All of the usual arguments were made by the attorney for the child, judicial powers, knowledgeable consent for sterilization, substituted judgment, applicable standards, whether other methods of contraception could be employed, etc. The opinion again reviews all of the cases and is not a brief one. The conclusion of the court, however, in this case was favorable to the petitioner; the court decided the probate court had the power to make the decision if proper care were taken; substituted judgment was applicable, proof beyond a reasonable doubt was not needed to justify sterilization, and the judge must apply careful standards. Given these, the sterilization requested could be granted. Said the court in concluding its opinion, "We are persuaded that a conscientious judge, being mindful of adverse mental and social consequence which might follow the authorization or not of a sterilization operation, will give serious and heedful attention at all stages of the proceeding." One dissent was filed. "The court today has decided that the probate judge has the power to divine the wishes of a severely mentally retarded woman who 'currently functions at the level of a four-year-old' as to whether she should permit herself to be rendered forever incapable of conceiving and bearing a child. To say the least, this is an impossible task...The court speaks of human dignity in connection with the free choice to be sterilized. It is difficult to think of an experience more degrading to human dignity than a sterilization ordered by a judge who is empowered by the court to read the heart and mind of the incompetent ward and forever bar the ward from bringing forth a child." Despite this dissent, the state of Massachusetts adds itself to the growing number of states which will permit substituted judgment, under very carefully dictated rules and circumstances, to be used in sterilization cases where the decision cannot be competently made by a female mental retardate.

The second newer case is *Matter of Barbara C.* decided by the Supreme Court, Special Term, Kings County, N.Y. Here there was a genuine urgency for a quick decision. The woman involved was 25 years old, in her twenty-first week of pregnancy. She had a mental age of less than 2 years and was

completely nonverbal. No statement was given whether this w
a genetically defective person or her lack of mental abilit
was due to other causes; also no details of the circumstanc
involving her becoming pregnant are stated.

This was the second time that the courts were asked
to intervene. In the first attempt to allow permission
for an abortion, the case had been based on the premise
that her pregnancy was "life-threatening." The court
found then, "The evidence presented at that time was
woefully inadequate to establish any real medical
danger resulting from carrying the fetus to term," and
denied permission for an abortion on medical grounds.
Within two weeks the case was again before the court,
this time based on the legal inability of the
unfortunate woman to give her consent, and the use of
substitute judgment by her next of kin, in this case her
father. The court reviewed the pertinent state laws
dealing with abortion, rights of mentally deficient
women to abortion, and the issue of substituted
judgment. As the New York statutes deal quite
specifically with the procedure for substituted
judgment, and the father had submitted with the appeal
an affidavit consenting to the abortion, the court now
had no difficulty granting permission for the operation.
Obviously there can be no appeal from this decision, and
in fact no party would want to appeal as the whole basis
of the suit was for the benefit of the woman. The
specifics of the New York Code made this a short
decision, and an easy one. Perhaps other state
legislatures and mental health workers might want to
examine the solution to these problems found by New York
and consider them for their own state's enactment.

Several more cases showing again the disparity of

opinions have arisen. ln 1979, in *Little v. Little*, a Texas court upheld the right of parents to give consent to a kidney transplant from a Down syndrome child to her brother, fatally ill from renal disease. In 1980, in *Matter of guardianship of Hays*, the Supreme Court of Washington remanded a sterilization case on the basis that while the lower court had authority to approve the operation, no clear evidence had been shown that this would be in the best interest of the minor concerned. Similarly in 1981, in *Matter of A.W.*, the Supreme Court of Colorado acted similarly to the Washington Court and remanded a sterilization case. The court set three clear standards to be followed in the future:1) the competency of the minor to understand the nature and results of sterilization; 2) the chance of improvement in the future so a better understanding by the minor might occur, and; 3) proof that the minor is capable of reproduction. Perhaps this case may serve as the long sought for standard for the future.

New York also has permitted transplants from an incompetent, 43-year-old, to his physically sound person to his 36-year-old dying brother. In *Matter of John Doe*, 1984, the Appellate Division upheld the decision of a lower court to permit a bone-marrow transport in an attempt to arrest the brother's chronic leukemia. The court again used the *parens patriae* doctrine to justify this legal action, and found also it would be for the benefit of the retarded brother who would be supported if the leukemic sibling could be saved.

The emotional and often irrational responses to the "Baby Jane Doe" cases has been at times raucous, and often without full knowledge of the facts. In addition, the United States Government, with the express consent and encouragement of the President, originally responded by a series of acts which the courts found clearly unconstitutional, irrational, and a violation of Congressional intent. Both cases involved a child born with single or multiple genetic defects. The first case arose in Indiana where a child with Down syndrome and a deformed esophagus was born. While surgery could clearly repair the physical defect, there is no cure for the mental insufficiency. Upon careful consideration of all aspects of the case, the views of the parents, the physicians, clergy, and others, the decision was made not to perform the surgery and the child subsequently died painlessly. It was this case which triggered the intense reaction from Washington.

The second case, in New York, dealt with a child with multiple birth defects, so many that even if life-saving surgery were to be performed, there would be absolute certainty that the child would never have any true mental abilities and she would require extensive care throughout whatever short life she might lead. Again the parents and others deemed it best not to attempt radical surgery, and the United States again saw fit to try to intervene. In addition, an out-of-state lawyer took it upon himself to become appointed as *amicus curia* although he had absolutely no personal interest in the case. There were two cases involving these incidents and it is necessary to consider them separately.

The first, *American Academy of Pediatrics v. Heckler* (the latter was Secretary of the Department of Health and Human Services), was tried in the United States District Court, D.C., in 1983. The details are as follows: after the death of the child in Indiana,

Secretary Heckler promulgated an order, without any
hearings. This order required that all hospitals
receiving federal funds (this would include Medicare and
Medicaid funds) must post in large letters and in
conspicuous places such as "each delivery ward, every
maternity ward, each pediatric ward, and each nursery
including each intensive care nursery, the following
sign:

DISCRIMINATORY FAILURE TO FEED AND CARE FOR
HANDICAPPED INFANTS IN THIS FACILITY IS PROHIBITED BY
FEDERAL LAW.

*Any person having knowledge that a handicapped
infant is being discriminatorily denied food or
customary medical care should immediately contact:*
Handicapped Infant Hotline,U.S.
Department of Health and Human Services, Washington D.C.
20201 Phone 800-368-1019 (available 24 hours a day.)
In the City of Washington D.C.-863-0100 (TTY capability)
or Your State Child and Protective Agency
(address & phone number)

Had this ruling gone unchallenged, the very basis
of medical care might have been endangered. Any
anonymous or malicious call would have resulted in a
swarm of government agents descending upon a busy
hospital, demanding to see records, interfering with the
often vital needs for physicians who would have been
occupied with answering myriads of questions instead of
caring for patients, removal of medical records for
examination just at the time they might have been needed
the most, and possibly creating panic in the public, who
reading the signs and seeing the agents, might thereby
assume the hospital was guilty of nefarious practices.
As a matter of fact, several hundreds of calls were

received and not a single one revealed any valid mistreatment or failure to follow normal medical standards of care.

The case, however, was more narrowly decided against the Secretary on the basis that she had ordered the changes in Federal Regulations in absolute violation of the legal procedures spelled out by the Administrative Procedure Act, 5, U.S.C. section 706(2) (A). This section states clearly that a 30-day notice of any changes of regulations must be given and that public comment must be solicited before any change in the rules becomes effective. In the present case the Secretary's action was clearly capricious and the court was adamant in stating she had abused her powers. The court did not weigh still another provision of the proposed changes which would have required parents to remove the child from the hospital if they did not consent to surgery. The court also discussed, but did not comment in detail on, the Congressional intent when section 504 of the Rehabilitation Act of 1973, a law dealing with equal treatment of the handicapped, was passed as it found there was already sufficient reason to void the order as stated.

The second case, *United States v. University Hospital, State University of New York*, was decided by the United States Court of Appeals, Second Circuit, in 1984. This suit, in various guises, had previously been before the entire hierarchy of the New York State Courts and a United States District Court before this appeal. The facts in the case are quite similar, refusal of parental consent for surgery, although the arguments were different. The parents refused treatment of a newborn who was diagnosed as having multiple birth

defects including myelomeningocle (spina bifida),
microcephaly (an abnormally small head), and
hydrocephalus or accumulation of fluids in the cranial
vault. In addition the infant suffered from the
disabilities of being unable to close her eyes, to suck
with her tongue, spasticity, and a thumb wholly
enclosed in her fist. While some of these conditions
might be relieved surgically, there was an almost
certain prognosis that any possibility of normal brain
development or normal physical development could not
occur. Upon careful consideration and consultation
between the parents, physicians, nurses, religious
advisors and a social worker, the parents decided
against corrective surgery and chose instead a
conservative medical treatment which would entail good
nutrition, the use of antibiotics and the dressing of
the infant's open spinal sac. The courts of New York all
agreed against the surgery, and at this point the
Department of Health and Human Services began the
present action in a District Court, demanding that all
medical records be made available to them to determine
whether this was a valid medical decision or a violation
of section 504. The Hospital refused to open its records
on the basis both of the parents' refusal to allow this
as well as the Hospital's own claim that there were
"serious concerns both as to the Department's
jurisdiction and the procedures the Department has
employed in initiating an inquiry." The government
claimed that the Hospital was in violation of the laws
concerning treatment of the handicapped and demanded the
records in order to examine them with this claim in
mind. After the district court denied the government's
request, the appeal followed.

The judge in this case went to great detail to study and document not only the applicable laws, but also opinions of others regarding treatment. It was determined that no discrimination had occurred within the meaning of section 504, as the Hospital was treating this infant as it would any other infant both in accordance with sound medical practice and, of course, also with the parents' approval. Therefore, there was no basis for the government's attempt to obtain medical records which were confidential. The major point is now made that parents have the right to withhold consent for an infant incapable of doing so itself, and the government has no compelling interest in intervening in these cases.

In addition, the lower federal court imposed a fine upon the lawyer mentioned above for his intervention into a case where he had absolutely no standing to do so. The government, having twice lost in federal courts, did not choose to appeal this to the United States Supreme Court. More interesting, perhaps, is that some months later, the parents did consent to minor surgical procedures to ease what may have beeen suffering by the infant.

Despite this series of actions in this case, a contrary action was recently reported by a judge in the Bronx, New York, Family Court. According to the press (no opinion of record is found in the New York legal publications) that court ordered a life-saving surgical procedure for a Down syndrome child despite the opposition of its parents. The operation was performed to connect the esophagus of the child properly and the child is apparently, other than the Down symptoms, now in good healh. Obviously, as the operation has been

performed, there can be no appeal in this case. The fact
remains, however, that again actions brought in
different courts, even within the same state, result in
different judicial decisions. Perhaps the much lesser
defect of the child in the present case may have led to
the difference in opinions.

Another problem in granting consent to a minor
arose in Tennessee in 1984. The question was put to the
Attorney General's Office whether a 13-year-old pregnant
girl could give legal consent to testing for sickle cell
anemia because it is believed persons with the disease
may be subject to serious urinary tract infections
during pregnancy. The Attorney General ruled that the
nature of the laws in Tennessee referring to pregnancy
testing required such tests and that the minor was
capable of giving consent without having first to obtain
parental consent.

But there are still serious unresolved issues in
all these cases. At what level is the "quality of life"
of a genetically or otherwise birth-defective child
sufficiently poor as to lead to withholding of surgical
treatment? Indeed who should make these decisions,
parents, physicians, or courts? In an attempt to deal
with this type of issue, a bill is presently moving
forward in the United States Congress. The bill is not
based upon the statutes dealing with treatment of the
handicapped, but is being considered as an amendment to
the Child Abuse Prevention and Treatment Act of 1974.
While the details are still unresolved, there has been
an attempt to work out a compromise between all
interested parties. As proposed, the bill would include
exceptions to medical treatment if: "The infant is
chronically and irreversibly comatose. The provision of

such treatment would merely prolong dying, not be effective in ameliorating or correcting all of the infant's life-threatening conditions or otherwise be futile in terms of the survival of the infant, and the treatment under such conditions would be inhumane." This represents a compromise with all sides, excepting the American Medical Association, a feeling that this can be satisfactory. There may still be ambiguity as no clear definition, in medical terms, can be given to some of these conditions. In the case of the connection of the esophagus of a Down child, it is obvious that surgery would come under this amendment. But the previous case involving multiple, life-threatening conditions where the surgery would not necessarily prolong the life of the infant, may still not be clear. The many associations for the handicapped have a strong interest in this legislation, fearing that such people might receive inferior medical treatment by the very fact of their being handicapped. Many others feel, on moral or religious grounds, that prolongation of life under any conditions is desirable. Others, as mentioned, are concerned as to who determines the extent of birth defects which might lead to refusal to treat. The eventual Congressional action, if forthcoming may be as near to perfection as anything can be in this imperfect world.

Another issue, sterilization of women suffering from such birth defects as Down syndrome, has also presented a new case. In *P. S. by Hardin v. W.S.* before the Supreme Court of Indiana in late 1983, the question was whether parents may grant such permission for their minor daughter who is obviously unable to do so herself. The court decided that this permission could be granted by the parents, a decision based upon the long and miserable medical history of the child as well as upon a thorough study of Indiana law. However, one dissent was entered by a judge who conducted a similar study of Indiana laws and reached the opposite conclusion.

The issue of informed consent for minors suffering from genetic defects is therefore still a difficult one. Each case will almost certainly have to be examined as it comes forward, and the courts will more and more find themselves in a role which is both difficult and uncertain. In these cases, there probably can be no set standards and each decision will have to consider primarily the best interests of the minor in either permitting sterilization or in determining whether withholding of surgery fits that interest. These are not easy decisions, but the courts must face the genetic facts and deal with them in what they deem an appropriate way.

One may wonder at the length of so many opinions in these cases involving organ transplants, sterilization, and informed consent. There is one common thread running through the cases presented. Although in most of these the courts have cited other opinions (in cases not given here), the basic problem is that in most instances these selected cases are those of first instance in the state. The courts are faced with a real dilemma; most of the state laws are mute concerning this type of decision. In many states the legislature has enacted laws to protect the rights of the mentally incompetent, expressly forbidding any unusual, cruel, or experimental procedures being carried out upon their person, in a sense a "bill of rights" for the mentally incompetents and a necessary one to protect them against "storage" as an organ bank for others. The rising concern for the rights of such people, as individuals entitled to full benefit of the laws, is certainly commendable. Also to be considered is their rights as individuals, not as a class of persons, entitled to the protection of the law, and not subject to the whims of the state or of their parents to treat them as less than human. Their rights must be protected as strongly and as fully as those of any citizen.

Essential also to this problem is the right to privacy, and the right against invasion of their bodies by any fiat of a governmental body. Medical invasion of the body is a tort, and certainly medical invasion for sterilization is a major concern. The right to choose procreation and the alternative right to eschew procreation are constitutionally protected as part of any person's right to privacy, as attested to by many court decisions. The essential problem of informed consent is the one which troubles the courts. It has been judicially decided that sterilization by consent is a legal right of competent persons, but whether this consent can be given by any other person is the legal issue. All of the cases presented to date have been cases brought by either parents or guardians with the best of intentions for their child or ward, namely, the prevention of undue suffering which might result from a pregnancy which the incompetent could not cope with, or a consent for a transplant donation which the incompetent could only, at best, vaguely understand. The issue of who may speak for these people is simply *not* clear in many of the state laws. The question is frequently answered for those institutionalized by the state; in such cases the decision for sterilization may be made by committees or other authorities under state laws. But the decision for those genetically or otherwise incompetent who are *not* in such institutions has not been dealt with by legislative acts in most states. Therefore, the courts are forced to deal anew with each case, determining who has power to cast the decision for the incompetent, how the power is restricted, and whether such power even exists under law. Understandably, different courts, all of which try to consider the human issues and suffering involved, reach different conclusions. Some courts decide that the human issues are such that they must act, regardless of legislative fiat; others feel the danger of their decisions setting a precedent which would allow indiscriminate sterilization or permission for organ transplants of the genetically-defective child, such that they must refuse permission pending legislative action. What is sorely needed is for states to enact adequate laws which would set the conditions for the right of the court to act using either their power of "substitute judgment" or *parens patriae* in these cases. Perhaps more effectively, a United States Supreme Court opinion which could be used as a model for such cases is needed. Pending either of these two alternatives, each case will have to be dependent upon the decisions of each court. Some will, as indicated above, elect to be "activists" and try to point the way for the legislature to enact appropriate acts; others by refusing to act will also attempt to force the issue into the legislative bodies. But, as of 1985, there is a real impasse here. The rights of the incompetent must be protected. But is it a right of these persons to have decisions regarding their purported suffering, should an

incomprehensible sexual act lead to an equally totally
ununderstood pregnancy ensue, to insure by legal means
that such suffering cannot occur? We badly need guidance
from the law in these matters. With the many difficult
social issues facing society now, it is clear that a
legislator will be reluctant to come out in favor of
granting more leniency in the laws regarding
sterilization of incompetents. It might appear to an
observer that the decision here must come from
courageous judges. It would be of great benefit, one way
or the other, if some of the cases mentioned above could
come to the highest federal court. Whether the decision
would be for, or against, the powers of the courts to
deal with the two kinds of cases discussed in this
chapter is not the important issue. What is of grave
importance is the need for guidance for lower courts;
what is of crucial importance is whether humane
decisions are or are not unconstitutional. Sooner or
later, the constitutional rights of the incompetent must
be resolved as must the issue of how to protect these
rights by carefully considered actions by those best
equipped to make decisions which the incompetent cannot
make. As mentioned before, the law is mute on such
decisions; perhaps it is time that those in command of
enacting laws face up to the issue. If not, then the
courts will, willingly or unwillingly, continue to come
to diverse opinions in each case. And each decision will
continue to be long, heart-rending, and, worst of all,
useless as a precedent for other courts in other states
which must face anew one of the most difficult decisions
in human terms which the law is ever requested to make.
What might also be of help to resolve this would be a
"Uniform Consent Law," modelled after the "Uniform

Commercial Code" or the "Uniform Adoption Law" to serve
as a model solution to this emotionally laden and
distressing series of individual court decisions, each
one of which sought to find the path to truth on its
own.

Nevertheless, the problem of obtaining informed
consent from the genetically-defective person, be he
minor or adult, remains a thorny one at present. The
courts will have to struggle in each case to attempt to
determine whether the surgical procedures are in reality
of benefit to the incompetent; in cases of organ
transplantation where no direct benefit to the incompetent
can be found but failure to carry out the transplantation may
result in the death of the person needing the organ, the
courts are dealing with complex issues of law and, of course,
of life and death.

(A survey of the laws of the various states has
been made by an article in the *Albany Law Review* in
1979, demonstrating the lack of legislative guidance and
the variation from state to state of laws regarding
sterilization of the incompetent. Also included is a
suggestion of a uniform procedure, the need for which
has already been emphasized. Those anxious to study the
different state codes are urged to consult this review.)

ALPHABETICAL LIST OF CASES CITED

CASE	CITATION	YEAR
Amer.Acad.Pediatrics v. Heckler	561 F.Supp. 395	1983
Buck v. Bell	274 U.S. 200	1927
Center For Health Care, Union College	Newsletter #18	1978
Consent Issue in Sterilization	43 Alb.L.Rev.322	1979
Hart v.Brown	289 A.2d 386(Conn)	1972
In re Grady	426 A.2d 467(N.J.)	1981
In re Guardianship of Pescinski	226 N.W.2d 180(Wisc)	1975
In re Penny N.	414 A.2d 541(N.H.)	1980
In re Richardson	284 So.2d 185(La)	1973
Little v. Little	576 S.W.2d 493(Tex)	1979
Matter of A.W.	637 P.2d 366	1981
Matter of C.D.M.	627 P.2d 607(Alas)	1981
Matter of Doe	104 AD2d 200	1984
Matter of Guardianship of Hayes	608 P.2d 635(Wash)	1980
Matter of Guardianship of Eberhardy	307 N.W.2d 881(Wis)	1981
Matter of Moe	432 N.E.2d 712(Mass)	1982
Off.Atty.Gen. State of N. Carolina	Slip (NCAR)Mar 14,	1984
Ruby v. Massey	452 F.Supp. 361	1978
Skinner v. Oklahoma	316 U.S. 535	1942
Skinner v. State	115 P.2d 123(Okla)	1941
Strunk v. Strunk	445 S.W.2d 145(Ky)	1969
Stump v. Sparkman	435 U.S. 349	1978
U.S. v.Univ. Hosp, SUNY at Stony Brook	729 F.2d 144	1984

CHRONOLOGICAL LIST OF CASES CITED

YEAR	CASE	CITATION
1927	Buck v. Bell	274 U.S. 200
1941	Skinner v. State	115 P.2d 123 (Okla)
1942	Skinner v. Oklahoma	316 U.S. 535
1969	Strunk v. Strunk	445 S.W.2d 145 (Ky)
1972	Hart v.Brown	289 A.2d 386 (Conn)
1973	In re Richardson	284 So.2d 185 (La)
1975	In re Guardianship of Pescinski	226 N.W.2d 180 (Wisc)
1978	Center For Health Care, Union College	Newsletter #18
1978	Ruby v. Massey	452 F.Supp. 361
1978	Stump v. Sparkman	435 U.S. 349
1979	Consent Issue in Sterilization	43 Alb.L.Rev.322
1979	Little v. Little	576 S.W.2d 493 (Tex)
1980	In re Penny N.	414 A.2d 541 (N.H.)
1980	Matter of Guardianship of Hayes	608 P.2d 635 (Wash)
1981	In re Grady	426 A.2d 467 (N.J.)
1981	Matter of A.W.	637 P.2d 366
1981	Matter of C.D.M.	627 P.2d 607 (Alas)
1981	Matter of Guardianship of Eberhardy	307 N.W.2d 881 (Wis)
1982	Matter of Moe	432 N.E.2d 712 (Mass)
1983	Amer.Acad.Pediatrics v. Heckler	561 F.Supp. 395
1984	Matter of Doe	104 AD 2d 200
1984	Off.Atty.Gen. State of N.Carolina	Slip (NCAR) March 14, 1984
1984	U.S. v.Univ. Hosp, SUNY at Stony Brook	729 F.2d 144

While most of the issues involving artificial insemination by donor (AID) and the one dealing with surrogate motherhood may be thought of as having more moral than legal aspects, nonetheless these cases do arise in courts and should be included in this survey of "genetics in the courts." There are many reasons for both procedures, both physiological and genetic. The major cause for a decision by parents to choose these methods is often due to sterility or relative infertility of one of the parents. But, given our knowledge of genetics, good genetic reasons may also lead to such choices for achieving parenthood. For example, two carriers of a genetic disease, such as Tay-Sachs, might well elect artificial insemination by a non-carrier in order to avoid the risk of having such a child. These decisions are obviously difficult ones for parents, and at times for society, to deal with and for courts to decide. The cases where the donor is a male have been fairly unanimously decided; the issue of surrogate motherhood is still before us.

Briefly, in the case of AID, the couple desiring a child but either unable to have one due to male infertility or preferring to have a genetically sound donor due to a parent (or both parents) carrying a known genetic defect which might lead to a defective child, finds a physician who is willing to do this procedure. In most cases, signed consents from the parents-to-be are obtained, primarily one suspects, in order to avoid charges of improper conduct by the physician. A sperm donor is selected by the physician. In all cases to date

when this has been the procedure, the donor is never
revealed to the couple, and similarly they are not known
to him. For many years it has been the practice to
select males who bear some physical resemblance to the
sterile parent, have at least similar major blood types,
and who are known to have sired children of their own
who carry no identifiable genetic defect. Similarly, as
many of these procedures are carried out in association
with teaching hospitals, the donor is most frequently
either a medical student or a young intern or resident
in that hospital. The ejaculate from such a donor is
received and the sperm sample introduced into the female
genital tract at the proper time in her menstrual cycle.
Not infrequently several inseminations in successive
months may be necessary before a pregnancy is
established. The only financial transactions involved in
this procedure are the payment to the donor for the
sperm, and the medical costs to the couple seeking it.

Surrogate motherhood presents a much more thorny
issue. In this case a woman who is infertile (blocked
tubes, ovariectomized, etc.) wishes to have a child in
the family. The appropriate person, usually a lawyer,
tries to help her by finding another healthy woman,
probably one who has had children, and who is willing to
undergo a pregnancy. Sperm is obtained from the husband
and then the surrogate mother is inseminated. After the
child is born, by legal agreement, it is turned over to
the original couple and usually legally adopted in order
to assure legitimacy. The obvious difficulty here is
finding some woman who is willing to carry a child
through term and then give it up at birth to another
couple.

In order to make it possible for such women to be

found, it has been suggested that they be paid a fixed amount-$10,000 has most usually been mentioned-plus all maternity and birth costs. However, as we shall see, this procedure runs squarely against almost all adoption laws which prohibit any payment of any kind in adoptions. The obvious reason, to prohibit "black market" babies, is cited in these cases, although the reasoning is simply not the same. While this type of motherhood is not common, it is not rare either, and the courts only began to deal with it in 1981.

To date there is are only a few cases, tried at various court levels, dealing with Surrogate motherhood. But a few of the AID cases indicate the method by which the law has seen fit to treat this procedure. The problems of legitimacy, support and consent are all present in these cases.

According to Professor George J. Annas, the problem of legitimacy was nonexistent in 15 states in 1980. These states have specific laws which declare a child born by AID to be legitimate, provided both parents have given consent. Some limit the practice to physicians, and some go further in making it a crime for a donor to provide semen "if he (1) has any disease or defect known to him to be transmissible by genes; or (2) knows or has reason to know he has a venereal disease." The punishment for this crime is 30 days; the state is Oregon. Thus, in these 15 states there is no problem of legitimacy whatsoever, and cases challenging this would receive no consideration from the courts. It is only in those states without statutory laws in which cases involving legitimacy arise. Almost all are similar in detail and arise when a consenting husband wishes later to have a divorce and to avoid payment for child care.

He then claims that as the child is obviously not his, he should have no financial responsibility for child payments after the divorce. This claim has met with little success in more recent cases. However, a brief historical review of a few cases is worth looking into, and one case in particular emphasizes the court's concern for the child.

In *Gursky v. Gursky*, heard by the New York Supreme Court, Special Term, Kings County, in 1963, the facts were as pointed out above. The court pointed out that legislation legitimizing such children has twice been rejected by the legislature and that a previous 1954 case had determined that despite the husband's signed consent and the listing of him as the father on the child's birth certificate, the court had no jurisdiction to declare the child legitimate. The court did, however, find that the husband was liable for support of the child. The basis for this decision was "The husband's declarations and conduct respecting the artificial insemination of his wife by means of a third-party donor, including the husband's written 'consent' to the procedure, implied a promise on his part to furnish support for any offspring resulting from the insemination. This, in light of the wife's concurrence and submission to artificial insemination, was sufficient to constitute an implied contract." The court quoted two other New York cases, the first of which said, "A promise will be implied where the agreement is instinct with obligation and the implication is supported by the circumstances." The second basis was "An agreement may result as a legal inference from the facts and circumstances of the case, although not formally stated in words." The court thus concluded in the case at hand that the husband was therefore liable, but they mentioned that "The court does not pass upon any personal rights, including property rights, that the child may have vis-a-vis the plaintiff's husband." This of course is due to the fact that they have declared the child illegitimate, and the rights of such persons at the time in regards to inheritance and property were not clearly established.

Exactly the same circumstances arose in another New York Supreme Court in 1964. The only difference was that here there were two children born by AID with the proper consent forms. The husband in this case was shown to have a substantial income. His claim that he did not have to support the children following a divorce was rejected by quoting the passage above regarding contracts, and the court simply repeated the Gursky opinion.

The "classic" case referred to earlier was that of *People v. Sorensen* before the Supreme Court of California in 1968. Here there were a few added factors. Following the usual consent forms, Mrs. Sorensen was inseminated with a donor's sperm and pregnancy ensued. At the time of the birth of the child, she listed her husband as the father, as is always the case in these circumstances. The husband was later to raise this issue in the trial, as he claimed he had not been consulted about the use of his name here. Sometime later, the marriage failed, and in the divorce proceedings the wife did not make any claim for support as she believed she could take care of the child. However, some time later she became ill and had to depend upon public aid until she was able to resume work. At this time the district attorney of the county in which she resided demanded of the former husband that he pay support for the child, and when he failed to do so he was found guilty of violating the state penal code and was given probation for three years on the basis that he make payments of $50 per month, payable through the district attorney's office. It was from this conviction that Mr. Sorensen appealed the case.

The California court had therefore at least two issues to deal with, first the obligation to pay, and secondly the criminal charge of non-support. In addition, the court also found it necessary again to look closely at the presumption of legitimacy rules so often mentioned in the later chapter on paternity. While discussing this briefly, and noting that no legal evidence of sterility was presented, they declined to treat this as an issue and narrowed the case to the following: *"Is the husband of a woman, who with his consent was artificially inseminated with semen of a third-party donor, guilty of the crime of failing to support a child who is the product of such insemination, in violation of section 270 of the Penal Code?"* (emphasis in original opinion). The court underlines the dilemma as follows, "Under the facts of this case, the term 'father' as used in section 270 cannot be limited to the biologic or natural father as these terms are generally understood. The determinative factor is whether the legal relationship of father and child exists. A child conceived through heterologous artificial insemination does not have a 'natural father,' as that term is commonly used. The anonymous donor of the sperm cannot be the 'natural father,' as he is no more responsible for the use made of his sperm than is the donor of blood or a kidney. Moreover, he cannot dispute the presumption that the child is the legitimate issue of Mr. and Mrs. Sorensen, as that presumption 'may be disputed only by the people of the State of California * * * or by the husband or wife, or the descendant of one or both of them.' With the use of frozen semen, the donor may even be dead at the time the semen is used. Since there is no 'natural father,' we can only look for a lawful father." The court then examined the Penal Code, pointing out that at the time it was written, no consideration was given to children born in this manner. They decided that "One who consents to the production of a child cannot create a temporary relation to be assumed and disclaimed at will, but the arrangement must be of such character as to impose an obligation of support for those for whose existence he is directly responsible. As noted by the trial court, it is safe to assume that without defendant's active participation and consent the child would not have been procreated."

The possible charge that artificial insemination was a form of adultery was also laid to rest in equally strong terms.

"It has been suggested that the doctor and the wife commit adultery by the process of artificial insemination. Since the doctor may be a woman, or the husband himself administer the insemination by a syringe, this is patently absurd; to consider it an act of adultery with the donor, who at the time of insemination may be a thousand miles away or may even be dead, is equally absurd. Nor are we persuaded that the concept of legitimacy demands that the child be the actual offspring of the husband of the mother and if semen of some other male is utilized the resulting child is legitimate." Having clearly resolved this issue, the criminal offense of non-support is, of course now equally valid, and the lower court's findings are upheld.

The issue of consent arose in the 1981 case *K. S. v. G.S.* This case before the Superior Court of New Jersey, Chancery Division, involved the by now familiar successful AID and the subsequent divorce and claim by the quondam husband that he did not need to support the child. The novel part of the claim was that, although he gave his consent to the initial practice which led to a miscarriage, he had not agreed to the subsequent attempts to achieve a successful pregnancy.It was also noted that he had sired three children in a previous marriage and then had a vasectomy rendering him sterile. The man testified that after the first miscarriage he had several times protested to his wife that the costs of the AID attempts were beyond his means, and that he had told her to stop the medical process. This was obviously disbelieved by the court as data showed that he had accompanied his wife several times during her visits and her attempts to become pregnant. Two months after her pregnancy was confirmed, he moved out and four months after the child was born the woman filed the divorce action. The husband had never seen the child or contributed to its support. The court found two issues here. "Two questions are therefore presented. First does consent to AID, once given, continue until pregnancy is accomplished? Second, if consent be deemed to continue, what burdens of proof must be met to establish withdrawal of consent?" The court dealt first with the basic issue of legitimacy by citing other state laws which established the fact that a child born after AID is legitimate. They went on to deal with what they felt might be another issue, however. "From the point of view of the female partner, although the child is conceived artificially, from all other aspects it is a natural child, carried to term exactly as if conception had taken place by natural means. While there may be a lingering question in her mind during the pregnancy as to what the child's physical characteristics may be, after undergoing the painful and emotional experience of childbirth, that uncertainty will be resolved.

"For the male partner, on the other hand, it is quite possible that his perception of the pregnancy will be substantially different. He is not the natural father, as the mother is the natural parent, and must be well aware of that fact. He may experience feelings of inadequacy, resentment or other negative attitudes towards the pregnancy, as illustrated by the case at bar. Thus, from the male point of view the pregnancy and resulting issue is, in many ways, akin to an adoption of the resulting child. However, whereas society has seen fit to regulate the artificial status of parent and child resulting from the adoption process, to insure as much as possible the stability of the relationship being created, no such protections exist at this time with regard to the field of artificial insemination. Until such safeguards are supplied by legislative enactment, they must be supplied on a case-by-case basis." The court then dealt very briefly with the issue of consent withdrawal and found no reason to believe that revocation of such consent had in any way been made. They pointed out also that in the states which have statutes dealing with AID and the need for written consents, none deals in any detail with how such a consent can be revoked. They decided that in "the case at bar" no revocation had occurred and "Accordingly, defendant is declared to be the lawful father of J. S. and as such bears at least partial responsibility for the child's support." The amount was to be determined after financial data were presented to the courts, and interestingly the defendant was granted visiting rights, "should he choose to exercise it," a strange conclusion, it would seem, as he had never seen the child.

Another variation on the theme occurred in the 1981 case, *L. M. S. v. S. L. S.*. The case appeared before the Court of Appeals of Wisconsin. Again the case involved child support after a divorce. The difference in this case was that the sterile husband could not afford the cost of medical care and the time needed to travel back and forth for such treatment. "The husband suggested that his wife become pregnant by another man and agreed to acknowledge the child as his own. At the husband's insistence the wife agreed, had sexual intercourse with the surrogate father, and became pregnant." The husband for a time lived up to his agreement, being at the hospital when the child was born, listing his name on the birth certificate, and claiming the infant as a deduction on his tax returns. The surrogate father also kept his agreement, and did not establish any relationship with the wife or the child. He took the precaution further of voluntarily terminating his paternity rights in another court.

As this court put it, "Whether a husband who has agreed to his wife's impregnation by another man should be required to support a child born during the marriage, but not fathered by him, is a question of first impression in this court." Indeed, given the circumstances, it is probably of first impression in any court. The court cited the cases given previously, and decided strongly in favor of the mother's claim for support. Reviewing the agreements made in this case, they stated, "Such an arrangement imposes an obligation to support the child for whose existence he is responsible. To permit a husband's parental responsibilities under these circumstances to rest on a voluntary basis could place the entire burden of support on the child's mother and, if she is incapacitated, the burden is then on society."

There are other possible legal difficulties in AID, as given by Professor Annas in his review. Briefly there is a problem in donor selection by the physician who often does not take proper steps to assure the genetic background of the donor. He points out that the word "donor" is really improper; as the man sells the sperm he should be more properly referred to as a "vendor." But should the child born by AID carry a genetic defect, who is responsible? He also points out that for reasons which seem as a "paradigm of legalism based on fear and ignorance" many laws also require the wife of the vendor to sign a consent form as well as the other more usual parties. A further problem arises with record keeping. Most physicians surveyed keep no permanent records. What if the child, perhaps learning of the AID origin of his life should now seek information concerning the real biological father, as is the case so often in adoptions? He concludes that "The problem with AID is that there are so many unresolved problems with AID, and few of them are legal. There is no social or professional agreement on indications, selection of donors, screening of donors, mixing of donor sperm, or keeping records on

sperm donations. ....It is time to stop thinking about
legislation and start thinking about the development of
professional standards. Obsessive concern with self-
protection must give way to concern for the child."

The proposal put forth in Sweden, referred to in a
previous chapter, whereby the sperm donor's anonymity
would not be preserved after the attainment of the
child's adulthood, also presents a legal issue of
privacy and consent. Although this proposal has not been
widely adapted, it could certainly bring forth both
comments and actions on the part of the child conceived
in this manner, the parents of the child, and, of course,
the physician who performed or abetted the insemination.

Other possible legal entrapments may lie in store
for AID. With the establishment of sperm banks, the
choice of the type of sperm to be purchased is
available. Recently in California a group claiming to
have frozen semen from the most intelligent men (Nobel
Laureates, etc.) has made its wares available to the
public at large. Unfortunately while the choice of donor
to the bank may be carefully elitist, the choice of the
recipients by the bank has been somewhat less rigorous;
one of the first clients was a woman with a criminal
record, for example.

It is not clear what liability would accrue to the
sperm bank should a defective child result from their
operations. There is no assurance that men of genius may
not carry a recessive detrimental gene, and should the
recipient also carry such a gene, there is a chance of a
genetically-defective child being born. Judging from the
cases involving artificial insemination from cattle
sperm banks (see the later chapter dealing with genetics
and agriculture), a possible suit against the sperm bank

could arise.

Also, if the same donor is used frequently within a
limited community, and anonymity preserved, the chance
of having half-brother, half-sister marriages may arise,
clearly illegal in most states. And finally, the growing
choice of single women to have a child without an
established relationship in a partnering situation,
raises the issue whether this is socially desirable, and
whether this is legally proper. In this case, no father
can be listed on the birth certificate, and the child
would be obviously illegitimate under all present
laws.(This, as well as other issues of AID, is dealt
with in another law review by Jeffrey. M. Shaman, in the
1979-80 *Journal of Family Law*. The author points out
that most physicians will not perform AID on unmarried
women and that a number of states clearly forbid it.
However, he seriously doubts the constitutionality of
such laws, based on the Supreme Court's decisions that
an individual's decision to procreate is a fundamental
interest.) But, as an earlier court stated, this may
cause the law to take notice and some changes will be
necessary.

Perhaps it should also be noted that the cases
presented on this topic are few. This is simply because
most AID results must obviously be favorable and few
suits are brought except as part of the bitterness of a
divorce. The issues put forth above may come to court,
and they will have to be decided, as suggested above, on
a case-by-case basis. The increasing awareness and use
of AID, it would seem, will certainly lead to more legal
suits, and new issues to be dealt with by the judiciary
and the legislature.

In genetic terms, there should be no difference

between surrogate motherhood and surrogate fatherhood. But, as pointed out, legally there is a vast gap. While many states recognize the latter, no state laws permitting the former have been passed, and the difficulties with the adoption laws make the practice seem dubious as it is illegal in most states to pay a woman for a child to be adopted, and also no legal adoption can be made before the birth of the child concerned. A Michigan lawyer, Noel Keane, is the leading legal authority on this, and he has written both a book and a law review article as well as bringing the first reported case before the courts. The case, *Doe v. Kelley*, was first before the Wayne County Circuit Court in that state and then was appealed to the Michigan Court of Appeals. Doe is a married woman who had a tubal ligation, but she and her husband now wished to have a third child. Another woman, named in this case as Roe, was found who was willing to be impregnated by the husband's semen and to carry the child to term. She agreed to renounce all claims to the child after its birth so that it could then be adopted by the Does. Ms. Roe was to receive $5000 for her pregnancy plus all medical expenses.

Kelley was the Attorney General of the State of Michigan. The case was brought against him in an attempt to have the Michigan State Law prohibiting fees for adoption declared unconstitutional on the grounds of vagueness, and also that the application of the statute here would violate the constitutional right of privacy for the Does. The lower court dealt with the issue of vagueness of the law in detail, agreeing that vague laws are unenforceable. While the court went to some length to uphold that fact, they did not find any such vagueness in the Michigan adoption laws, citing several other cases as precedent, and finding Doe's complaint on this basis to be invalid.

The issue of the right to privacy was discussed in some detail, and the acknowledgement of such a right was made, but not for this case. Said the court, "'Baby bartering' is against the public policy of this State and the State's interest in preventing such conduct is sufficiently compelling and meets the test set forth in *Roe* [here referring to the U. S. Supreme Court abortion case, not the Roe involved in this case]. Mercenary considerations used to create a parent-child relationship and its impact upon the family unit strikes at the very foundation of human society and is patently and necessarily injurious to the community. It is a fundamental principle that children should not and cannot be bought and sold. The sale of children is illegal in all states."

Further the court cites various other authorities to point out that if a conflict arose between the mother desiring the child and the surrogate mother carrying the child, all courts would certainly allow custody to the latter. Also, as pointed out earlier, they found it illegal for the adoption release to be signed before the birth of the child. The further issue was raised as to the meaning of "voluntary" in terms of the payment to the surrogate mother, "How much money will it take for a particular mother's will to be overborne, and when does her decision turn from 'voluntary' to 'involuntary.'" The real basis for the decision however was "...that this action prohibited interjects compensation in adoption proceeding; that money *must* be paid to the biological mother before the parties will strike an

agreement." Thus the court turned down the requested
finding of the unconstitutionality of the Michigan
statutes. It is interesting to note that prior to trial
the attorney had requested an informal judicial view
from another judge who had believed that the case might
be sound and might have decided otherwise.

The appeal to the Court of Appeals of Michigan also
was of no avail. The invasion of privacy plea was
dispensed with by the additional reasoning, "The statute
in question does not directly prohibit John Doe and Mary
Roe from having the child as planned. It acts instead to
preclude plaintiffs from paying consideration in
conjunction with their use of the state's adoption
procedures. In effect, the plaintiffs' contractual
agreement discloses a desire to use the adoption code to
change the legal status of the child--*i. e.*, its right
to support, intestate succession, etc. We do not
perceive this goal as within the realm of fundamental
interests protected by the right of privacy from
reasonable governmental regulation."

Another attempt was made to circumvent the adoption
laws by bringing the issue as a paternity case. In
Syrkowski v. Appleyard, the Wayne County Circuit Court
found that the circumstances of the case were such that
it lacked jurisdiction and dismissed the attempt to have
the paternity act used for this purpose. "This court,
today, decides that the Paternity Act was not intended,
and cannot be used, as a mechanism to establish the
paternal rights of a semen donor in a 'surrogate parent
arrangement.' Neither the laws nor the public policy of
the State of Michigan permit the direct or indirect
judicial recognition and enforcement of 'surrogate
mother contracts.' The social wisdom and legal

recognition of such agreements are matters of legislative concern and not for judicial pre-emption." At the time this opinion was written such a bill had been introduced in the Michigan Legislature, but no action has subsequently been taken.

Despite these legal setbacks, the practice of surrogate motherhood continues. In his book, Attorney Keane discusses many of these in detail, and his reasons for becoming involved in this type of case. Some of the feared events, due to choice of an improper surrogate mother, have come to pass. In one instance the mother removed herself and the child to another state. Although the semen donor is listed as the child's father on the birth certificate, no attempt to obtain custody from her has been carried out. In one other state, Kentucky, where a surrogate mother clinic is operating, there was a constant threat of a state suit to shut it down. Apparently only in Hawaii is the practice not against adoption laws.

Another Michigan example has occurred. In this instance, an infant born by surrogate motherhood was mentally defective and the genetic father would neither accept the child nor pay the $10,000 cost of pregnancy and delivery as had been agreed upon. Further, the man claimed that blood tests proved that the baby was not his, and he had no obligation whatsoever under those circumstances. The biological mother does not want the child either, and it has been placed for adoption, rather an unlikely event due to the mental deficiency. The cost of maintaining the child will undoubtedly be born by the public. The case apparently has not come to court, but should it do so, the Michigan courts will have to deal again with the validity of the original "contract", obviously illegal, promising to pay costs for bearing the child. The claim that the blood tests show the child was not the result of the man's semen, contrary to the "contract," might be raised as a violation of the arrangement by the surrogate mother. However, no suit for violation of an illegal contract can be brought, so this cannot be an issue. Also, of course, if non-paternity is proved for the man, there can be no validity to any suit by the surrogate mother.

As this chapter was being finished, however, the Supreme Court of the State of Kentucky ruled that surrogate motherhood *is* legal in that state. In a Slip Opinion in the case, *Surrogate Parenting Associates Inc. v. Commonwealth of Kentucky,* on February 6, 1986, that court in an opinion headed "This Opinion is not final and shall not be cited as authority in any courts of the Commonwealth of Kentucky," ruled that as the father, the sperm donator, was the biological father of the child born to the surrogate mother, there was no need of his adopting his own child, thereby avoiding the twin issues of paid adoptions and of adoption procedures begun before the birth of the child!

It would seem that the vigorous protest against surrogate motherhood comes not from the semen donation, but from the fact that the surrogate carries the child during the pregnancy. Perhaps the whole concept is found revolting to many, and the religious aspects may play a role also, as they do in some cases of AID. But the promise of allowing a woman to obtain a child who will genetically be carrying at least one half its genes from the husband is one which probably will continue to be found attractive to some sterile wives. As Keane points out, not many women will be altruistically inclined to undergo a pregnancy for another woman, but payment might make it much more likely that suitable surrogates could be found. It certainly is likely that more attention will have to be paid to the laws involved here, and that these cases, like so many others, will simply not just go away. Meanwhile, barring legislative action, the payments either overt or covert are illegal and violate all state codes. It is also interesting to note that no appeal under the due process clause based on sexual discrimation has been made. Perhaps this is just too tenuous a distinction, but if the law says that children born by AID from a male donor are legitimate, then the possibility of similar views of a child born from an egg donor might be considered.

Particularly now that *in vitro* fertilizations (literally "in glass;" i.e. eggs are fertilized in a laboratory dish rather than within the uterus of the mother) are being frequently attempted and are becoming more successful, there is the possibility that an egg can be donated by a female, other than the sterile wife, and sperm can then be obtained from the husband. *In vitro* fertilization then can be carried out in a laboratory and the resulting conceptus can be transplanted into the uterus of the wife, and a normal pregnancy can ensue. While this type of case and its legal implications have not as yet arisen, one would expect this to be the next legal test of artificial insemination. However, as it is simply the reverse of the commonly accepted AID methods, the legality of this would seem unquestionable. The technique is, of course, fraught with more difficulties as, in this instance, fertilization *in vitro* must occur, followed by implantation of the developing embryo into the mother-to-be. But again, as this really is genetically similar to the AID cases, it should come under that rubric, and presents a compromise solution for a woman who can bear a child but cannot produce eggs either due to blockage of her tubes or other causes. However, the process of obtaining eggs is not as simple as obtaining sperm, and there may be considerable costs, bringing up again the issue of "buying" a child.

Although, except for the Kentucky case discussed above, there have been no other recent cases involving surrogate motherhood in the United States, there has been a plethora of writing on the subject. (See, for example, the June 1984 issue of *Law, Medicine & Health Care* for two such articles, and an excellent review of applicable statutes in the August 1984 Issue of the *American Bar Association Journal*.) However, recent technical advances may have made the issue moot. These techniques involved *in vitro* fertilization (recent reports indicate that more than 350 such children have been born by this method alone) followed by cryogenic (freezing) processes preserving the conceptus until such time as the recipient may be in the correct endocrinological state for the embryo to be implanted and grow. Cryogenic techniques are not new; the method has been tested thoroughly with mice and it has been shown that the rate of fertility is no lower with frozen embryos than by normal means, and that there have appeared no genetic anomalies in mice so born. This way of providing fertility by use of donor eggs, removed from a normal woman to a woman otherwise incapable of bearing children, has now been employed successfully in humans in several cases. Obviously, as this is simply the reverse of artificial insemination by donor, there may not be legal obstacles involved. However other considerations do arise.

The most publicized case is, of course, in Australia where the parents were killed in an accident, leaving their *in vitro*, and then frozen, embryos. The problem is a difficult one, as they had left several frozen embryos in an embryo bank for their possible future use, and the case is made more difficult as both were wealthy. The legal problem of the right to inheritance of the embryos, should they be implanted in, and borne by, an unrelated woman is a difficult one. The case is further complicated by the fact that the sperm used in this case were apparently from a donor, and not the husband's. (New York Times, 6/27/84). Obviously, the embryos are healthy, and their implantation into another woman is possible. The question then arises whether a child, developed in this manner, could claim inheritance from its parents. (Under the usual A.I.D. laws the husband of the recipient woman would be declared the legal father.)

While Australia had no position at that time as to the proper procedures, the American Fertility Society has issued a clear-cut "Ethical statement on in vitro fertilization" which deals with this issue. Article VI states, "Cryopreservation of concepti for the purpose of subsequent implantation into the female partner is acceptable with certain provisions. The concepti should not be retained in the cryopreserved state for longer than the reproductive life of the female donor. ..." While in this instance the reference was probably intended to mean until the menopause of the woman, it also would apply to the Australian case obviously as her life, and reproductive period, ended simultaneously.

The American Society has anticipated the issue of inheritance in such cases by stating in Article VII "After the reproductive problem has been resolved to the satisfaction of the donors, it is considered ethically acceptable to donate to another infertile couple nontransferred concepti, provided any claim to any resulting progeny is waived and provided strict anonymity between donors and recipients is assured as in any adoptive procedure." It would seem impossible in this particular case to use the frozen concepti as no waiver from the deceased exists, and anonymity might be hard to maintain in a sensational case as widely publicized as this. Also, of course, with the trend towards open adoption mentioned earlier anonymity may not even be legal in the future for any type of adoption.

The issue of the legality of what can be done with the remaining concepti not used for implantation is also not clear. Obviously simply allowing them to thaw and decay or simply discarding them is one obvious solution. But as these embryos are potentially capable of developing into children, the ethical issues are difficult and emotions are strong and there are also religious viewpoints to be considered. Both American and British groups have suggested guidelines. The Fertility Society, mentioned above, in their guidelines have dealt with "concepti in excess of those required for transfer." They make clear that no developing concepti beyond the period of 14 days following fertilization may be maintained in the laboratory for any purpose and they "may be disposed of." The British Warnock Commission came to essentially the same regulations, but here there has been considerable opposition by the scientific community; the suggestion was made that the 14 day restriction is not logical as abortion of a foetus up to 168 days after conception is legal there. In addition, the Commission went along with the United States in recommending strongly that "all surrogate contracts are illegal contracts and, therefore, unenforceable in the courts."

A different type of issue arose recently in France. Here the question of the "ownership" of sperm donated to a sperm bank arose. In this case the man, suffering from testicular cancer and fearing impotency, made a sperm donation with the expectation that should he be rendered sterile, his wife would still be able to obtain the sperm and by insemination to bear his child. Unfortunately he died soon thereafter and his widow requested that she be inseminated with his sperm so as to bear his child. The sperm bank refused, and the case went before the French Courts. They found that she was entitled to the use of the sperm, but that there would be no right of inheritance for any child so conceived. Part of this was due to the fact that it is claimed that any child born more than 300 days after the death of a father cannot be his child. Of course that legal nicety did not take into account the fact that sperm may be held in a frozen state indefinitely, thereby making the 300 day period of time for gestation inapplicable in cases of this kind.

Although none of these are United States cases, such are certain to arise in the near future. The New York Times (ibid) quotes from a prominent lawyer who once compared the law to a "killy-loo bird, a creature...insisted on flying backward because it didn't care where it was going but was mightily interested in where it had been." Or as another lawyer quoted in the same article said, "We're approaching space-age technology with Model-T statutes and cases." What both are saying is that the law has been left behind in the rush of medical advances in the fertility field, and that legislatures, and judges, must anticipate just such cases as these in order to establish guidelines when these inevitably do arise.

ALPHABETICAL LIST OF CASES CITED

CASE   REFERENCE

350 Test-tube Babies Born   Schenectady (NY) Gazette
                                Jan.21,1984
Annas,G.J. Fathers Anonymous   14 Family L.
                                Quarterly 1 1980
Anonymous v. Anonymous   246 N.Y.S.2d 835 1964
Doe v. Kelley   307 N.W.2d 438(Mich) 1981
Doe v. Kelley   Cir.Ct. Wayne Cty, Civ.
                                Act (Mich) 1980
Ethical Statement in   27 Fertility & Sterility
vitro fertilization   178 1984
Gursky v. Gursky   242 N.Y.S.2d 406 1963
K.S. v. G.S.   440 A.2d 64(N.J.) 1981
Keane, N.Legal Problems   1980 S. Ill. Univ. L.
                                Journal 147 1980
L.M.S. v. S.L.S.   312 N.W.2d 853(Wis) 1981
Legal Rights of Embryos   New York Times, 6/27 1984
People v. Sorensen   437 P.2d 495(Ca) 1968
Shaman,J.Legal Aspects of AI   18 J. Family Law 331 1979
Surrogate Parenting Assoc,Inc. S.Ct. KY Slip
v. Commonwealth of Kentucky   85-SC-421-DG 1986
Syrkowski v. Appleyard   Cir.Ct. Wayne County
                                Civ.Act.(Mich)1981
Test-tube Embryos in the Dock   220 Science 606 1984
The Stork Market   70 Amer.Bar Assn

CHRONOLOGICAL LIST OF CASES CITED

YEAR	CASE	CITATION
1963	Gursky v. Gusky	242 N.Y.S.2d 406
1964	Anonymous v. Anonymous	246 N.Y.S.2d 835
1968	People v. Sorensen	437 P.2d 495(Ca)
1979	Shaman,J.Legal Aspects of AI	18 J. Family Law 331
1980	Annas,G.J. Fathers Anonymous	14 Family Law Quarterly 1
1980	Doe v. Kelley	Cir.Ct. Wayne Cty, Civ. Act (Mich)
1980	Keane,N.Legal Problems	1980 S. Ill. Univ. L. Journal 147
1981	Doe v. Kelley	307 N.W.2d 438(Mich)
1981	K.S. v. G.S.	440 A.2d 64(N.J.)
1981	L.M.S. v. S.L.S.	312 N.W.2d 853(Wis)
1981	Syrkowski v. Appleyard	Cir.Ct. Wayne County Civ.Act.(Mich)
1984	350 Test-tube Babies Born	Schenectady(NY) Gazette Jan.21
1984	Ethical Statement in vitro fertilization	27 Fertility & Sterility 178
1984	Legal Rights of Embryos	New York Times, June 27
1984	Test-tube Embryos in the Dock	220 Science 606
1984	The Stork Market	70 Amer.Bar Assn J.50
1986	Surrogate Parenting Assoc. v. Commonwealth of KY	S.Ct. Ky Slip 85-SC-421-DG

The geneticist recognizes two types of radiation damage to humans. First is somatic damage, that harm done directly to the person receiving dosages of radiation. Burns, induction of cancer, radiation sickness are some of the effects found here. The second type of damage is genetic, i.e., no effect may be found by the recipient of the irradiation, but his or her offspring may be defective due to mutations induced in the germ cells of the parents. The Nobel Prize in Physiology and Medicine was awarded to Herman J. Muller for his discovery of the genetic mutations induced by Xrays in fruit flies. The effects of Xray and most other forms of radiation (ultra-violet, gamma, alpha or beta rays) has been found to occur universally in all living things, bacteria, plants, animals and undoubtedly humans. As all of these cannot be detected by any of the human senses, but only by instruments such as Geiger counters, the person exposed has no awareness of the exposure at the time it occurs. A further complication is that the genetic effects may not be noted until years after the radiation has been received. For example, an exposed young child may not bear children showing the mutagenic effect until many years after the initial radiation has been received. A further complication may arise in that once a person is born with the mutation, there is a likelihood that all further generations may carry this mutation so that the effect of the initial irradiation may be expressed many generations later when the original source of the mutation may have been forgotten.

It must also be recalled that almost all mutations are most likely to be deleterious. This is because of

two factors. First, humans are believed to carry an
already high mutational load, and any further mutations
may tip the balance to the point of non-survival.
Second, to use an analogy used by Dr. Muller, a
volunteer for an experiment may be sought. The
experiment is simple: ask for a volunteer to give you
his watch for the following purpose: you will open the
watch and randomly give it a jab with a screwdriver,
close the case and return it to its owner. The question
is to determine whether the watch now runs better than
before. It is quite obvious that the number of
volunteers of watches for this type of test will not be
high. But, compared to a watch, with its limited number
of parts, an organism is vastly more complex and
complicated. As mutations are almost always harmful, the
mutation in the organism is comparable to the jab with
the screwdriver which no one would want to try on his or
her watch. Humans are among the most complex of living
things, and the analogy is clear.

A vast amount of literature concerning the effects
of very low doses of irradiation is available. There are
two schools of thought here. First, the geneticist knows
that in certain organisms the effect of irradiation is
cumulative; the effect of small amounts given over a
long period of time are equal to the effect of the same
total amount of irradiation given either as a pulse or
in short periods of time. This was part and parcel of
Muller's work. The second school, supported by advocates
of such things as atomic power, holds that there is a
threshold below which there are no harmful effects of
radiation. The difficulty here is that the study of this
problem with higher organisms is a statistical one, and
would require vast numbers of organisms exposed to the

same total dosage of irradiation, but administered in varying amounts during a time period. Obviously such experiments cannot be carried out on humans, and even to do this with rodents would require many years of study and enormous numbers of animals. While the geneticists mainly believe in the cumulative theory, as they do not see man as belonging to a different biological category from other organisms where the data support the cumulative theory, proof of this remains difficult, if not all but impossible. Most feel that if there is to be error, it should be on the side of caution; the theory that all unnecessary radiation is to be avoided seems to be the safe one. Add to this problem the fact that the effects of the irradiation may not show for one or two decades after it has been received, and one understands the caution even more.

It is also known that most carcinogenic agents are mutagenic, and vice versa. Indeed, one of the standard tests for carcinogenicity of a new compound is the *Ames Salmonella* test whereby the suspected agent is applied to cultures of bacteria and its potential mutagenic activity is screened. Those compounds showing a positive *Ames* test are then further tested for carcinogenicity. Other tests using fruit flies or other organisms with short reproductive periods also depend upon the relationship between the induction of cancer and of genetic mutations. While this chapter will deal mainly with those cases in which the genetic effects may occur, many of the actual cases involved presumed induction of cancer by irradiation as well as presumed induction of birth defects due to mutagenicity caused by exposure to excess amounts of radiation.

MEDICAL USE OF IRRADIATION

The misuse of irradiation by negligent physicians or others falls quite easily into standard tort and malpractice law, and while alarming, would form the subject of another type of book. At times some of these cases, however, will have to be referred to as background for the strictly genetic cases; it is these latter cases, relatively few in number which concern us here. (Also, there is another form of mutagenic agent, chemical mutagens, whose effects are equally damaging and due, ultimately, to the same cause as the radiation ones, damage to the DNA of an organism.)

Among the earliest cases in which the possibility of radiation caused defects is mentioned was brought for an entirely different reason. The case, *Chiropractic Association of New York v. Hilleboe* was dealt with by the Court of Appeals of New York (the highest court in that state) in 1962. The basis of the case was the constitutionality of the enactment of a law which forbade the use of x-rays by chiropractors. Their association brought suit against the Commissioner of Health, Hilleboe, to prevent the enforcement of the law. The court upheld the validity of the law, despite claims made that anyone who was adequately trained in the use of irradiation, physicians or their agents, could use x-rays while chiropractors, however well trained in this field, could not. The chiropractors claimed this was unconstitutional discrimination. It is obvious from the opinion and the facts of the case that the law was expressly passed to interfere with the practice of chiropractors. Nonetheless, the court held that the law was valid. The part of the opinion of concern here is that which states, "Both courts below having found, upon adequate evidence, the existence of deleterious consequences, as a matter of fact, from the use of Xrays upon human reproductive cells and hence the necessity for controlling their use, that issue is not open in this court." What is of even more immediacy in our context is the paragraph which begins, "The science of genetics dates from Gregor Mendel, the Austrian monk whose experiments were published in 1865 and came to the general attention of biologists in 1900. Not until the advent of nuclear warfare in 1945 did scientists or the public become aware of the hereditary effects of the exposure of the genital organs of the human body to nuclear fallout and other forms of ionizing radiation. The catastrophic effects of Hiroshima and Nagasaki brought this subject into the foreground of popular discussion and, on the scientific side, the studies and experiments in the structure of the atom pointed toward physiological chemistry (and physics) leading prominent scientists to the conclusion that ionizing radiation causes changes in the male and female genital organs capable of producing deleterious effects upon future generations. The details of these interesting and important concepts and experiments are for scientists rather than judges and lawyers." The court then goes on to list the effects and pointed out, "...that taking Xray pictures of the full spine, such as are commonly used by chiropractors, exposes the reproductive organs of the male or female patient to the direct, primary Xray beam in quantity sufficient to damage the hereditary material of the individual from which the generations as yet unborn are to come." The Commissioner replied in his defense, "...his department concerned itself 'with every conceivable source of

ionizing radiation. We are not interested in just
chiropractors, doctors or hospitals or factories. We
want to cut out every single bit of unnecessary
radiation.'" A lengthy discussion of the amounts of
radiation used by the chiropractor and the almost
routine use of Xrays was part of the opinion. "From
data of this nature it was concluded that the benefits
derived from this kind of exposure was out of proportion
to the hereditary damage to be anticipated from the
wholesale use of X ray in this field." It was pointed
out that the use of fluoroscopes for shoe fittings had
also been banned by law as well as the amount of radium
used in watch dials, and that the Commission had the
appropriate powers to regulate such matters. The court
agreed that the statutes were not especially designed
against chiropractors as a class and upheld the
Commissioner. Today, all of this would be moot in New
York, as a few years ago the legislature passed a law
licensing chiropractors and thereby gave medical
standing to the practice, automatically allowing a
trained member of this group to use diagnostic Xrays as
part of medical examinations.

ATOMIC TESTING AND THE MILITARY

The cases to be dealt with next all deal in one way
or another with the exposure of either servicemen or
civilians to purported damages caused by either exposure
to atomic radiation during the Nevada or the Bikini
Islands series of atomic tests. They show a variety of
different causes for suit, and several will be examined
in considerable detail as they shed light on both
genetics and legal reasoning. The cases are numerous

and an attempt to be selective has been made, but
frequently the facts or the claims are sufficiently
different as to warrant closer examination of each
case.

The case of *Punnett v. Carter* came before the
United States Court of Appeals, Third Circuit, in 1980.
The suit was brought against not only the then President
of the country, but also against a host of other
officials, cabinet members, members of the Nuclear
Regulatory Commission, etc. Among the plaintiffs was
Howard Hinkie who also had been at the Nevada atomic
test site. The basis for the suit was the alleged claim
by the plaintiffs for injunctive relief for themselves
and all others who supposedly had received mutagenic
damages from fallout during the series of atomic tests
in Nevada some twenty years previously. The plaintiffs
claimed that the government had withheld information
concerning the effects that such tests might have and
that they now wished to enjoin the government from
withholding public dissemination of all studies and
documents concerning the effects of these tests on the
health of all military personnel participating in these
tests. They also sought to prohibit the participation of
any person in further tests unless the risks were made
clear to them first. As an expert witness for the
plaintiff put it, a warning should be given to all
concerned in these tests: "You may have an increased
risk of having children with birth defects because of
your participation at atomic bomb tests at the Nevada
test site while you were in the service. If you decide
to have children you may wish to seek genetic counseling
or other medical advice before your children are
conceived and while they are developing inside their

mothers and even after they are born. The purpose of
this extra precaution for your family would be to
minimize the risk of your child being born with
unexpected birth defects and to determine the risk of
such defects under all the medical and family
circumstances of your particular experience and
situations." The Hinkies were one of a series of
families of servicemen who had been at the Nevada test
site and all of whom claimed they had personally
suffered from symptoms of irradiation sickness. In the
case of the Hinkies themselves, they had two children
with birth defects which they attributed to irradiation
effects upon the father while he was in military service
and stationed at the site of the tests.

The court was not in agreement with the plaintiffs.
It held that there had been no scientific proof that
injury had occurred and that the Hinkie childrens'
defects were not directly traceable to the exposure
claimed by their father. It was held they did not offer
proof that the risks to exposed service personnel were
greater than the general risk of the population as a
whole. A U.S. District Court had also ruled, "...the
effects of a warning from the United States
Government that the recipient had been exposed to
radiation which may cause mutagenic defects in his
children could be thoroughly devastating....Public
interest does not favor the dissemination of this
information, at least not until there has been much
greater exploration of its potential effects...in light
of the unconvincing nature of plaintiffs' scientific
evidence, their failure to show that they will be
irreparably harmed, without the probability that other
interested parties may be injured...the plaintiffs have

not shown any likelihood that they will prevail on the merits of this suit."

The Court of Appeals decided this case on other bases than the question of irradiation damage. The first legal point was the question of the enjoinment mentioned above. The court held that a preliminary injunction could not be granted unless there was a reasonable probability of ultimate success on the merits of the litigation and that the movant would be irreparably harmed if relief were not granted. The court examined in detail the evidence presented by the plaintiffs' expert witness concerning the dosage they may have received and the effects of such dosage. The evidence was highly technical, but nonetheless the court determined that testimony was subject to recalculations in terms of the true amount of radiation received and therefore was not valid proof of the plaintiffs being damaged by being at the test sites. "The plaintiffs' contentions are without merit." The court examined all of the mass of data concerning the exposures claimed, including the possibility of radioactivity being carried by aerosols formed at the time of the test as well as direct exposure. They found that the exposure dosage calculated was based on the maximum possible dosage, and that calculation was certainly invalid. Thus on the first issue, that of the probability of a permanent injunction, the plaintiffs had no chance of winning their case. The court also dealt with the irreparable harm issue. They considered the plaintiffs' claim that there is no level of irradiation which will not increase the risk of mutations (the linear theory discussed previously). The court's conclusion was that such a risk had not been established as even the dosage was in

question and the linear theory itself was not proven.
They then went on to consider the public interest in the
asked for warning to all participants also mentioned
previously. The conclusion here stated, "Plaintiffs have
not shown what impact such anxiety would itself have on
an existing pregnancy. However, a risk of unnecessary
abortion is present... [Test] participants may be
unreasonably rejected as marital partners. Families that
have long existed may be broken up. Recrimination and
doubt may follow. Couples may make unnecessary family-
planning decisions which they would not otherwise have
made." The court thereby ruled that there was no
possible cause for the injunctive relief sought and
indeed that such relief would be harmful rather than
beneficial to those who might receive such a warning as
the one proposed by the plaintiffs. In essence, what the
court said was that there was no possibility of a legal
decision favoring the plaintiffs, based on the
impossibility of proving the genetic claims of damage.
Therefore, barring this possibility, no injunction would
be granted. But it is to be noted that this appears to
be the first case in which a court attempted to look
into the genetic controversy concerning linear v.
threshold dosage effects and that the court could not
make that type of decision. Barring proof of the linear
theory no action could be cognizable, and even if such
proof could be made, no accurate evaluation of the
amounts of radiation received by the plaintiffs could be
made at the present time.

A different legal basis for a somewhat similar
claim was made by the United States District Court,
District of Columbia, in 1981. This case, *Lombard v.
United States*, was also based on a serviceman's alleged

genetic damage due to exposure to radioactivity. The claims involved the United States Federal Torts Claim Act, as well as violation of the First, Fifth, Ninth, and Tenth Amendments to the Constitution.

As in the previous case various U.S. agencies were named in the suit. Mr. Lombard claimed that he was exposed to radioactive agents while in the service without his consent or knowledge and as a result he suffers from various physical injuries and genetic damage which his children inherited. The exposure apparently occurred when he as handled radiooactive materials while assigned to the atomic bomb project.

The judges here fell back upon a well-established precedent, the so-called *Feres* doctrine. This doctrine stated in 1950 that "The government is not liable under the Federal Torts Claim Act for injuries to servicemen where the injuries arise out of or are in the course of activity incidental to service." Here it is necessary to digress for a moment. The case *Feres v. The United States* was decided in 1950 by the United States Supreme Court. It dealt with the consolidation of several suits against the government for claimed negligent acts committed against military personnel incident to their military service. Several reasons were given by the Court to establish this doctrine. First, the court felt it necessary to review the purpose of the Federal Tort Claims Act which was to allow suits for damages against the United States Government. Three reasons for the application of the act and exemptions from it were given. First, the diversity of state tort laws is great and Federal Courts must take into account the nature of that state's law in which the claims first arise; obviously a Federal Law incorporating the states' laws would make all claims from any state come under a single Federal act. Second, as the court put it, "The relationship between the Government and members of its armed forces is 'distinctively federal in nature.'" However, there is a different and special relationship between the military personnel and those giving them orders. Were any member of the armed forces able constantly to question the legality of an order from a superior then military discipline would completely break down. Any order, even if negligent, therefore cannot be the basis for a suit. No matter how negligent the order might appear, such personal determination by the subordinate as to whether an order was legitimate would make it impossible for any military system to function. In addition, there exists a special pathway for some claims, namely the Soldier's and Sailors' Civil Relief Act of 1940, and some claims can be processed under its provisions.

The court applied the *Feres* doctrine to the *Lombard* case. The attempt by the Lombards to circumvent this doctrine was ingenious; their claim was that the government did not inform him of the risk *after* he was discharged and as a civilian he was not informed of the risk and also that as a civilian he no longer was under the *Feres* doctrine. The court did not accept that claim either, as the purported injury came while he was in service and not afterward, and the issue of a warning was moot. A further attempt to avoid *Feres* was made by having his wife and children also enter a claim; while they had not been in the military and they could claim to have been damaged by the alleged radiation of Mr. Lombard. Citing earlier cases, to be discussed next, the court ruled that "...even though the plaintiff had never

been a member of the military, her injuries had their
'genesis' in her father's exposure to radiation
sustained 'incident to the performance of military
service.' Consequently, the suit was barred by *Feres*....
to hold otherwise might open the door for governmental
liability to countless generations of claimants having
ever diminished genetic relationship to the person
actually injured." The court goes on to say "When these
factors are applied to the Lombards' allegations, it is
clear that *Feres* bars this suit. The claims are
'distinctly federal in character,' for they are no
different from claims by the descendants of soldiers who
might be exposed to radiation on the battlefield during
some future limited nuclear war, claims surely barred by
Feres." But it was pointed out that an alternative
method of relief might be available, namely to appeal
for relief under the Soldiers' and Sailors' Civil Relief
Act, an act specifically designed for this type of case.

The constitutional claims were also ruled to be
invalid. Lombard had claimed a "conspiracy" to violate
his constitutional rights; the court could find neither
any such action, nor any violation whatsoever of the
Lombards' rights based on the amendments cited above,
and as a result dismissed the case "for lack of subject
matter jurisdiction." While the decision may seem highly
legalistic and harsh, the apparent need of the *Feres*
doctrine has been pointed out.

In 1981, the Hinkie family again brought suit
against the United States, this time on their own behalf
(for the wife, son, and deceased son), a suit based upon
different claims. In *Hinkie v. United States*, brought to
the United States District Court, E.D. Pennsylvania, the
claims were on the basis of specific chromosome damage

to Mr. Hinkie which caused the children to have birth
defects, and also for emotional anguish and miscarriages
by Mrs. Hinkie. This time the court went into more
details than the previous opinion involving the Hinkies.
Mr. Hinkie was involuntarily involved in seventeen or
eighteen nuclear tests as a linesman laying and
retrieving communication wires and also as a switchboard
operator prior to the tests. After them he walked to
"Ground 0" after the explosions. He was never informed
of the dangers of such tests, although furnished a film
badge, but not given protective clothing. It was further
established in the record that the government either
failed to read or to preserve his film badges. Further
it was noted that the government had assured his
"safety" with a film badge program that it knew was
inadequate. In addition the facts disclose that at no
time was Hinkie given any information concerning his
risk. "Plaintiffs allege that these negligent acts
caused perturbations in the molecular substance of
Hinkie, Sr., that is breakages in the chromosomes as well
as other forms of chromosomal alterations such as
inversions, partial displacements [translocations], and
deletions not amounting to a total breakage of the
chromosomes. These perturbations in molecular substance
of Hinkie, Sr. eventually formed parts of the bodies of
Hinkie's deceased son (Timothy) and his son (Paul) that
manifested themselves as birth defects at the time of
their births...caused the minor plaintiff (Paul) to
suffer birth defects including Rubenstein-Taybies
syndrome,lack of joints in his thumbs, constant
uncontrollable twitching of his eyes, severe mental
retardation and photophobia." The other boy was born
without an esophagus and other defects which led to his

early demise. This time, the court found reason for the
suit. Hinkie did not sue in his own behalf; instead the
suit was brought in behalf of his wife and of the birth-
defective children. The court now found precedent from
other cases so that the *Feres* doctrine did not apply,
pointing out that it did not pertain to suits brought by
others. They added that the alternative use of the
Veterans' Benefits Act, as suggested in the first case,
also was not of significance here as the plaintiffs made
clear that the act did not cover this kind of damage to
Hinkie, Sr. They easily found that the usual conditions
applying to cases under the *Feres* doctrine did not apply
for the most part. The court did have some difficulty
with the question of adverse effects upon military
discipline, but in this particular case found that "The
plaintiffs' injuries, concededly derived from those to
Hinkie, Sr. when in military service, manifested
themselves decades after his discharge. The program of
which the Teapot Dome Tests [the series of nuclear tests
in which Hinkie was involved] were a part has long since
ended. The undermining of discipline or refusal to
follow orders present less of a problem because of the
time lapse here...Moreover, adverse effects on military
discipline must not flow inevitably from the judicial
scrutiny of a military order, for such scrutiny is no
less involved in other suits under the FTCA which
Congress has not barred. An example would be an action
by civilians injured by nuclear testing near their
residences." Thus the court allowed that a suit could be
brought. They recognized the difficulty the plaintiffs
would have in establishing that the presence of the
father at the tests was the cause of the genetic
defects, but they also recognized the right of the

Hinkies to attempt to establish this in a trial court,
and that if such a claim could be verified, the Hinkies
would have justification for collecting damages from the
government. In brief, this seems on the surface to be
the opposite of the first case in which the Hinkies were
involved; on careful consideration, the charges are
different, the establishment of facts is more clear, and
Feres seems not to be applicable. What will be of utmost
interest in this case when it now comes to retrial will
be the establishment of the genetic evidence. For
example, chromosome tests should be made of both the
father's and the mother's body cells and also the living
child. It is possible that the parents might be carrying
the chromosome defect without showing overt signs of it.
While of course tests cannot be made of germ cells,
tests of somatic cells can show that either parent
already has the chromosome anomaly and that the
radiation received was not the cause. As the presence of
such a chromosomal anomaly in the somatic cells would
conclusively demonstrate that the alleged change was not
radiation-induced, but was present prior to that effect,
the case would be lost on genetic evidence, should the
court be willing to accept it. As the decision discussed
here only occurred a short time ago, the disposative
evidence will not be available until sometime from now
when the case is again brought forward.

    Two decisions, handed down by the same court
on the same day in 1981, illustrate the difficult
decisions regarding the *Feres* doctrine. The first,
Broudy v. United States, decided by the United States
Court of Appeals, involved a suit brought by the widow of
a former Marine major who was involved in the Nevada
bomb tests. He died from cancer 20 years after his

exposure to radiation while involved in the Nevada
tests. His wife and children brought the suit for
"wrongful death," with part of the alleged claim being
that the service had known that such exposure was likely
to cause cancer, and that part of the experimental
procedures of the test was to see how well combat troops
could withstand atomic blasts and the resultant fallout.
However, the legal issue in this case dealt not with
that claim, but rather with the applicability of *Feres*.
The appellants made a strong issue that *Feres* itself was
wrongly decided. That claim was not validated. But the
court did find that negligence existed *after* the
discharge of the officer, and to that extent there might
be a case which could come under the FTCA. For this
reason the court remanded the case to determine whether
this could be established. "At this point the
Appellant's allegations concerning the Government's
knowledge of the danger to Major Broudy are somewhat
confused, and do not indicate any post-service negligent
act. However, we realize that this confusion may result
from an inability to gain necessary information from the
Government. We vacate the district court's order of
dismissal with prejudice and remand to the district
court for such proceedings as may be necessary in light
of this opinion." In other words, if the Broudy family
can establish that the negligent act was failure to
inform and treat following discharge, then *Feres* does
not apply.

The difficulty with this doctrine was underscored
again in the opinion handed down by the same court on
the same day in the case, *Monaco v. United States*, the
case which gives the subheading to this chapter. Here
the case did not deal with the Nevada tests, but with

events long before they were conducted. During World War
II, for almost 3 years, Daniel Monaco was stationed in
Chicago where he daily took part in calisthenic
exercises at the University's football field. What was
not made clear to him, or to any other participant in
those exercises, was that the field was located directly
above the Fermi nuclear pile, the world's first atomic
reactor. Subsequent to discharge, Daniel himself
developed radiation-induced cancer of the colon in 1971,
25 years after the exposure. He also learned at the same
time that this was probably the cause of the birth
defects of their daughter, Denise, "to be born with a
birth defect known as arterio-venous anomaly in the
brain. This defect has induced three brain hemorrhages,
aphasia and other permanent injuries." The suit was
brought both by the father and separately in behalf of
his daughter alleging that the government "knew or
should have known of the dangers posed by exposing
military personnel to atomic radiation, and that the
Army was negligent in permitting exposure to radiation."
The two actions were consolidated upon appeal after the
District Court for the Northern District of California
had dismissed the claims. Again at issue was the *Feres*
doctrine, and again, as in the previous case, the court
reviewed this decision and other applicable cases. They
cited an earlier opinion to the effect that, "[T]he
protection of military discipline...serves largely if
not exclusively as the predicate for the *Feres* doctrine.
Although the [Supreme] Court has woven a tangled web in
its discussion of the 'distinctly federal' notion and of
the alternative compensation system, it has not wavered
on the importance of maintaining discipline within the
armed forces. The Court has found it unseemly to have

military personnel, injured incident to their service, asserting claims that question the propriety of decisions or conduct by fellow members of the military. Only this factor can truly explain the *Feres* doctrine and the crucial line it draws." The court then decided that the injury to Daniel Monaco, and the subsequent mutational damage to his daughter, had occurred while he was on active duty and not subsequent to his discharge and "...the proper focus in applying the *Feres* doctrine is not the time of injury, but the time of the negligent act. Whether Denise's injury 'occurred' when she was born with a birth defect or when her father suffered chromosomal change, the allegedly negligent act drawn into question was performed while Daniel was in the service. Thus, just as Daniel could not recover on the ground that his injury was post-service, we cannot grant Denise recovery on these grounds. Denise's second argument is that *Feres* does not block her claim because she was never a member of the armed forces. She argues that her claim can thus have no effect on discipline between soldiers and their superiors, and points out that she can recover no benefits under the Veterans' Benefit Act for her injuries." It should be recalled that this was the alternative suggested for possible recovery in an earlier case. The court rejected both arguments. They also pointed to the basic problem with *Feres*—"Denise's claim differs...in that she seeks relief for an injury to herself rather than indemnity for losses due to injury to her father, but this does not change the substantive analysis: the court still must examine the Government's activity in relation to military personnel on active duty. It is precisely this type of examination that the *Feres* doctrine seeks to

avoid. --The *Feres* doctrine today stands on shaky ground
with its precise justification somewhat confused. But
the doctrine has been applied consistently to avoid
examining acts of military personnel which were
allegedly negligent with respect to other members of the
armed services. While the injuries claimed by the
Monacos may well have become manifest long after Daniel
left the service, a court could not rule on either claim
without examining acts occurring while Daniel was in the
service. The results in this case disturb us,
particularly with respect to Denise. In her case, the
price of avoiding examination of events long past, and
involving her behavior in no respect, appears to be
complete denial of recovery. If developed doctrine did
not bind us we might be inclined to make an exception in
cases such as this. Unfortunately, we are bound, and the
decision of the district court must accordingly be
affirmed." Clearly the court was empathetic to the
claim by Denise but, as in so many other instances, felt
compelled to comply with the decision of the highest
court and could do nothing to acknowledge what they must
have felt was most probably a justifiable cause of
action for a helpless infant.

What impresses one, again, is not only these court
decisions, but also the almost deliberate lack of
disclosure on the part of the military of the dangers
involved. At the time most of the claimed injuries
occurred, the dangers of radiation damage, both somatic
and genetic, were well established. There seems to be
little doubt that the conductors of the tests at Nevada
knew of these risks, but either chose to ignore them, or
deliberately not to give full information or protection
to military personnel involved in monitoring or setting

up the tests. It must be admitted that at the time of
the setting up of the Fermi pile, while the effects of
irradiation were clear, it is most probable that no
thought was given to the use of the field above the
reactor. Even if such considerations might have been
made, the absolute top secret classification might not
have permitted the banning of its use, lest suspicion of
the activity under the field might occur.

The number of cases involving irradiation of
servicemen and the use of the *Feres* Doctrine continues
to rise. But more remarkably, there has come to light
full evidence that the military personnel exposed
deliberately to dosages of irradiation which were
harmful. These are fully documented in a series of
Congressional Hearings before the Committee on Veterans'
Affairs of the House of Representatives in May, 1983.
Before discussing the individual claims and cases, it is
necessary to examine the information which was presented
there, indicating that standards of protection of
military personnel were consistently violated, and that
knowledge of the danger of exposure was available at the
time the men were exposed. These are based on the
discovery of the personal papers of Colonel Stafford L.
Warren, then Chief of the Radiological Safety Section at
Operations Crossroads, held at Bikini Island in 1946.
These papers, originally classified as "top secret" or
"secret," were subsequently declassified. For some time
their existence was not known, but following the death
of Colonel Warren his estate apparently gave them to the
University of California Library where they were
discovered some time later. The documents make
distressing reading. Not only were the dangers of the
effects of radioactive fallout fully known, but there is

consistent evidence that they were ignored by the naval officers at the time. Men were sent into the danger area immediately after the underwater blast, which turned out to be much more dangerous than anticipated. These men were allowed to stay for hours or overnight in areas where some received the maximum permissible dose in less than five minutes. Further the equipment used for determining accurate dosages was entirely inadequate, either failing completely or else in not being sufficiently calibrated to register dosages off scale on the instruments. Further no previous experience in attempts to decontaminate ships was available, and about two-hundred ships were exposed at the time. Some were so seriously contaminated with radioactivity that they had to be sunk. There was simply no way to decontaminate these vessels, many of which were to be towed thousands of miles away for further work. No one had taken into account the extent of alpha emission from plutonium, and no meters for detection of alpha particles were even available. No adequate monitoring of beta emissions was made, and no film badges were available for determining individual exposure to gamma rays. To quote a letter from one of the safety monitors to Dr. Warren, "I am not an alarmist, Dr. Warren. I do believe that many of us probably received much more penetrating ionizing radiation than the instruments of very low beta sensitivity were able to record." Dr. Warren himself is quoted as stating, "I never want to go through the experience of the last three weeks of August [1946] again." Despite this evidence documented as well as possible with the limited dosimetry available, the Defense Nuclear Agency (ironically using the acronym DNA) denied all of the data and claimed no hazardous conditins existed. Yet Colonel Warren pointed out that reduction of 90% or more of contamination on surfaces still left "large and dangerous quantities of fission and alpha emitters scattered about...

"Contamination of personnel, clothing, hands, and even food can be demonstrated readily in every ship... in increasing amounts day by day. Some radioactive ships were even cleared *before* decontamination was begun simply because Operation CROSSROADS had contaminated so many ships that the Pacific Fleet was experiencing shortages. In addition testimony shows that at least one naval officer had at best a cavalier attitude "...like Captain Maxwell who insists upon a 'hairy-chested' approach to the matter with a disdain for the unseen hazard, an attitude which is contagious to the younger officers and detrimental to the radiological safety program."

These and other damning facts were presented to the Committee by Dr. Karl Morgan, based on a report by Drs. Mahijani and Albright, on May 24, 1983. If their testimony is to be credited, there can be no doubt that the dangers of radioactive exposure were ever-present, and that they were well-known to the Navy at the time the men were exposed. According to these hearings there appears to be no doubt that there was deliberate exposure of naval personnel to high dosages of radiation and naval radiation specialists were well aware of this.

Readings of the evidence presented to the Committee are both shocking and frightening. If the papers preserved by Colonel Warren had not come to light, perhaps none of these facts would have been spread before the Committee. It is also notable that the DNA still maintains that no dangerous exposures of the type so fully documented by Colonel Warren existed.

Despite this recent information, claims for injury at the Bikini tests, are still being denied under the *Feres* Doctrine. A liturgy of cases will be examined briefly, but for the most part the results in the federal courts have been similar in that suits under the FTCA have consistently been denied on the basis of *Feres*. In *Laswell v. Brown*, the family of a serviceman at the tests mentioned above sued to recover from claimed irradiation damage leading to his development of Hodgkin's Disease (cancer of the lymph nodes) and to his ultimate death from a heart attack which may have been induced by the disease. In addition his children sued for damages due to their own "exposure to 'an unusually high risk of latent cellular or other genetic defects either in themselves or in their offspring.'" Several federal issues involved violation of the Fifth Amendment by making Laswell "an involuntary guinea pig in nuclear radiation tests;" further Fifth Amendment violations were claimed because of failure to warn the deceased and his descendants of the dangers and to offer protective treatment and negligence on the part of the government to carry out these procedures. The Appellate court, in this case the United States Court of Appeals, Eighth District, in 1982, dismissed the claims for damages on the basis that the subsequent, post-service illness and death was a "continuing tort," i.e. the claimed cause of injury occurred while Laswell was in the service and *Feres* banned such claims. Similarly, the Fifth Amendment Claims are equally barred. The suit by the children was also dismissed on the basis that "the complaint is conspicuously void of any allegation that the children have sustained any damage other than the exposure to a higher risk of disease and cellular damage." In other words, the fact that they were apparently healthy and no illness was claimed makes them ineligible to bring suit. Thus the entire claim by the Laswells was dismissed.

An almost similar set of circumstances was involved in *Seveney v. United States Government, Department of the Navy*, also in 1982, but this time before the U.S. District Court, D. Rhode Island. Seveney also participated in the Bikini tests and also subsequently died from what were claimed as illnesses caused by his irradiation. In this case, however, his daughter was born with birth defects which the Seveneys attributed to the radiation damage, and her son claims to suffer mental injuries brought about by the observation of the "harrowing physical effects of his grandfather's exposure to irradiation." A legal point was raised by the government in that it was claimed the suit should have been brought not under the FTCA but under the Public Vessels Act. The point is of little importance as the court decided that *Feres* would apply in either case. The court also reluctantly dismissed the childrens' suit on the same basis. However, the court did find that a further claim for later failure to take medical precautions which might have alleviated the ensuing "adverse effects" of Seveney's exposure might be allowed. The Court here stated, "Given this factual predicate, it would take the most tenuous skein of logic to justify the further extension of *Feres* to civil suits brought by civilians and founded upon supposed negligent acts or omissions occurring independent of, and at a time subsequent to, that ancestor's active duty. Seveney's relationship with his superior officers is

not, on these facts, in issue; military discipline is
not jeopardized; no inhibition exists as to the giving
of duty orders; and no suggestion is made that the
Veterans' Benefit Act can be extended to compensate for
harm proximately caused by post-discharge torts." The
court ordered that an amended complaint, dealing only
with the post-separation claims, may be filed within
twenty days of the decision; if no such complaint is
entered the court will award full judgment to the
defendants. This decision is of interest for two
reasons. First, if it is sustained upon appeal, or if
the amended complaint is filed, the "failure to warn"
might apply to almost all ex-servicemen involved in the
Bikini tests and a vast number of such cases may be
brought forward. Second, it deals with specific genetic
claims by the female offspring, and indeed by claims of
even the third-generation son who witnessed all of the
suffering. Truly, as mentioned in the main text, the
propagation of irradiation damages may continue
indefinitely through many generations.

The third case arising from the Bikini tests is
that of *Hampton v. United States of America*, before the
United States District Court, W.D. Arkansas, in 1983. In
this case, only the serviceman claimed injuries and also
the failure to warn after discharge. In this case the
judge again felt compelled by *Feres* to dismiss all
complaints, including the failure to warn, thus, it
would seem, putting himself in direct opposition to the
Rhode Island case previously discussed. The opinion
deals with the history of sovereign immunity and points
out that while the FTCA does grant some rights to sue
the government, those rights are preempted by Seres. His
opinion is in complete accord with the 1981 case, *Kelly*

v. United States, decided on the same facts and issues by the United States District Court, E.D. Pennsylvania, including the determination that the later failure to warn ex-servicemen of possible dangers to their health from the atomic exposure were not valid cause for suit as this was again found to be a continuing tort traceable to the original claimed service-related injury.

A different facet of the Bikini Island tests arose in the case *Juda v. United States*, before the U.S. Court of Claims in 1984. This time suit was brought for damages by the original natives of that atoll. The facts are not pleasant. First, the U.S. Government evacuated all of the Bikini natives to another island prior to the test. The chosen refuge turned out to be a disaster for the natives; not only was there a water shortage, but there were no suitable harbors for their fishing boats, and they depended heavily upon that occupation for survival. Further, the species of fish found there, similar but not identical to the one they were used to harvesting, turned out to be relatively poisonous. The Government eventually moved these people back to Bikini, claiming that it was now safe for occupancy. It was not. For example, "In April 1978, a medical team examination of islanders on Bikini showed an 'incredible' 1-year seventy-five percent increase in body burdens of radioactive cesium-137, causing U.S. scientists to conclude that the people likely had ingested the largest amount of radiation of any known population." It should be noted that cesium is one of the most dangerous of all isotopes as it replaces potassium in the body and therefore enters into all cells and most biochemical reactions. As a result of these tests, the government had once again to move the natives to yet another island.

The suit was complicated by the Government's defense that first this was the wrong court in which to bring the islanders' case, and also because of a difficulty in determining whether a treaty signed with other Marshall Islanders applied here. The Government further claimed that there could be no Fifth amendment plea by the plaintiffs for just compensation. However, the court carefully examined the entire history of the treaties with the Marshallese, the relationship between the Government and the plaintiffs, and the court's own right to decide the case. Having decided the latter in favor of the plaintiffs, the court then examined the entire history of the treaties with the Marshall Islanders and concluded that the Bikini residents had every right to bring suit for damages under the Fifth Amendment. The case was therefore decided in favor of the Bikini natives. There are many other legal facets to this case, all dealing with technicalities of the law, but the result was that the Bikini people could bring suit for damages.

Following this, an out of court settlement of a large sum was made to the natives, and they received recompense for seizure of their properties and for damages to their health.

In addition to the Bikini atoll test cases, many more irradiation cases have arisen from the Nevada tests mentioned earlier. Some of this type of cases have been mentioned earlier, and as almost all of them gave the same results, it is not necessary to go into detail in each and every one. A typical case, *Fountain v. United States*, gives the type of facts which most of these cases also offer. The opinion of the United States

District Court, W. D. Arkansas, Fayette Division, states them as follows: "In 1954, while Mr. Fountain was a citizen and resident of Arkansas, he enlisted in the United States Marine Corps. In 1955, while he was on active duty, he was ordered to Nevada. On March 22, 1955, while he was in customary military attire, he was ordered to crouch in a field as the government exploded an atom bomb approximately 3500 yards from him. As soon as the explosion commenced he was ordered to march toward the center of the explosion. He was able to approach within 500 yards of the center of the explosion before the heat and wind stopped him. When the drill was over he was swept with a broom to remove some of the radioactive dust. The sweeping was the only measure taken by the government to protect Mr. Fountain from the effects of radiation.

"When Mr. Fountain was made to participate in the atom bomb testing the government knew atomic radiation had lethal effects. Mr. Fountain's participation in the test was to increase and refine the government's knowledge of the harmful effects of radiation ...

"In 1979 Mr. Fountain was diagnosed as having chronic myelocytic leukemia." Mr. Fountain had married in 1956, and he was joined in this suit by his wife both in the claim that the government had taken no steps to warn him of the effects of his exposure, and also for her loss of consortium as well as her medical expenses in caring for her husband.

In *Jaffe v. U.S*, the case dealt with the effects of a soldier ordered to stand in a field in 1953, with no protection, while an atomic bomb was exploded a short distance away. Jaffe developed inoperable cancer in 1977 and brought suit under various amendments of the

Constitution for the harm done him. His case appeared
before the United States Court of Appeals, Third
Circuit, En Banc, in 1979 with the opinion rendered in
1981. Again *Feres* was applied; in this instance other
issues led to some division among the concurring judges
as a state claim had also been made by Jaffe. A rather
blistering dissent was entered, comparing the action of
the United States in these cases to violations of
International Codes such as the Universal Declaration of
Human Rights, the Geneva Convention, The Declaration on
the Protection of all Persons from Being Subjected to
Torture and Other Cruel, Inhuman or Degrading Treatment
of Punishment, and the Nuremberg Code. The dissenting
judge also had much to say about what he deemed the
"macho" attitude of the majority of the court. It is of
interest that his dissent was noted and specifically
denied in the majority opinion.

The facts are again similar in *Sheehan v. United
States*, heard by the United States District Court, S.D.
Mississipi, in 1982, and the injury here was claimed to
be "Raynoud's Disease and Degenerated Disc."

Targett v. United States decided by the United
States District Court, N.D. California in 1982, again
with much the same factual information was a bit more
detailed in that following exposure he suffered loss of
body hair (a common radiation-induced symptom) and
subsequently developed a pituitary tumor which has lead
to two brain operations and extensive therapy. Again he
claimed post-discharge failure to warn him of the
possibility of harm from the atomic blast. In this
instance the court was sympathetic to this claim,
stating, "We agree that if appellant can allege and
prove an independent, post-service negligent act on the

part of the Government, her [Targett's wife] claim would
be cognizable under the FTCA," an opinion based on the
Broudy decision mentioned earlier. The Court also
dismissed governmental claims of the statute of
limitations, ruling that it did not begin to toll until
the discovery of the claimed injury.

Two other servicemen, Gaspard and Sheenan, also
took part in these tests. The suit, under the name
Gaspard v. United States, was decided in 1983 by the
United States Court of Appeals, Fifth Circuit. Both men
became seriously ill years after discharge with
"violently painful afflictions sustained as a direct
result of their in-service radiation exposures. These
included the breakdown of their immune systems, leukemia
and other cancers, and many other afflictions. Gaspard
died of these diseases in July, 1982." Again
constitutional rights were claimed under the so-called
Bivens ruling. That case did not involve servicemen, but
the unauthorized behavior of agents of the Federal
Bureau of Narcotics, *Bivens v. Six Unknown Named Agents
of Federal Bureau of Narcotics* (403 U.S.388, 1971). It
was held that the present case was sufficiently different
in fact so that the *Bivens* rule did not apply here.

The case *Mondelli v. United States*, decided by the
United States Court of Appeals, Third Circuit, in 1983,
is still another example of the *Feres* doctrine. In this
instance suit was brought by a woman whose father was
exposed in the test. She claimed that as a result she
was born with retinal blastoma, "a genetically caused
cancer of the retina." As a result of the disease she
lost her left eye. It should be noted that this disease
is frequently fatal as well as causing blindness. Again
no recovery was allowed.

The facts in *Cole v. United States* in 1985 were somewhat different. Here the claim of radiation injury was made by the daughter and widow of a deceased veteran who had served on a submarine in 1946. The ship had been involved in the Bikini tests earlier, and the claim was made that residual radiation led to the development of cancer and death of the serviceman. The court, as usual, fell back upon *Feres*, but decided that a post-discharge failure to warn the serviceman of possible injuries could be brought forward.

In 1985 another similar suit, this time brought by the widow of a member of the United States Coast Guard, was based on that Service's failure to warn and to take administrative action when requested by her. This case, *Shipek v. United States*, before the United States Court of Appeals, Ninth District, was decided in her favor by the court's holding that she had properly exhausted her administrative appeals and could bring suit in the courts for post-discharge injury.

The appeal of *Hinkie v. United States*, was decided in 1983 by the same court as before. Here, as may be recalled, the complaint was that of a wife, her surviving son and a diseased son both of whom were claimed to have received genetic damage due to their father's exposure to the atomic tests. The court stood by its former decision although admitting, "We are forced once again to decide a case where 'we sense the injustice ... of [the] result, but where nevertheless we have no legal authority, as an intermediate appellate court, to decide the case differently."

A different approach was taken when *Punnett v. United States* was again brought forward. This time, the suit before the U.S. District Court, Eastern District of

Pennsylvania, was brought in order to compel the Government to issue a warning to all atomic veterans of an alleged increase of genetic risks to any of their offspring. The Court dismissed the suit on the basis that the Army had not flatly denied their requests, and also the DNA was at that time in the process of investigating and after the *Jaffe* decision had distributed such a warning notice to over "15,000 newspapers, radio stations and television stations. Almost 1700 of these media outlets confirmed publication or broadcast of the notice." As the DNA had begun to take the steps the Punnetts requested, the court found no need for further action.

There are also several other cases dealing with radiation effects upon military personnel, but they all involve essentially the same general facts already presented, and they all have been decided essentially in the same manner as the cases just presented.

All of these cases may leave one with a strong feeling of dissatisfaction. It appears apparent that by the time of most of these tests and the directed exposure of the personnel concerned, it was known that radiation was dangerous, and it should have been apparent at that time that the inductions of cancers might take scores of years to occur following the tests. Yet the *Feres* doctrine, a court and not a legislatively enacted "law," has been elevated to the status of an act of legislation. None of the cases here cited has gone forward to the United States Supreme Court where a reexamination of the original *Feres* decision might be considered, and perhaps it is now time for either Congress or the court to change the law to allow for compensation for what can only be considered deliberate

negligence in these types of cases. Indeed the
Congressional hearings mentioned above might some day
lead to action by that body. Meanwhile there appears to
be no restraint upon the exposure of military personnel
to known dangerous amounts of atomic radiation, or for
that matter of any other type of negligence by the
military.

However, any changes will have to come from Congress
as the United States Supreme Court in June, 1985,
reaffirmed *Feres*. This case did not involve radiation,
but was brought against the government by the
mother of a murdered serviceman who claimed the death of
her son was due to negligent release of another
serviceman known to the government to have "dangerous
tendencies." The lower federal court had denied the
claim, and the intermediate court had allowed it.
However, in *United States v. Shearer*, the U.S. Supreme
Court overruled the Court of Appeals and restated the
Feres doctrine and denied any relief to the mother.

A further blow to the hopes of the atomic veterans
has just been received by an opinion written in August
of 1985, and reported by the press (the text is not
available at the time of this writing). In this instance
a U.S. District Court Judge ruled that all but one of 43
consolidated cases were invalid. He based his opinion on
a little known rider to the Defense Appropriations Bill
of 1985 in which Congress specifically forbade such
suits by the veterans against either the United States
or its independent contractors. The suit before the
federal judge was to have the Congressional Act held
unconstitutional. The judge ruled otherwise and held
that the acts leading to the suits fell under the
"discretionary function" of the government and as such

were not subject to legal challenge. The atomic
contractors working on the testing programs were also
exempted from suit because they are "acting as
'instruments' of national policy." The veterans claimed
the Congressional Act was unconstitutional on the
grounds that it violated the "separation of powers"
between Congress and the Courts by dictating the outcome
of pending law suits. They intend to appeal this
decision and also to work to repeal the Congressional
Act.

ATOMIC TESTING AND CIVILIANS

Ironically, claimed radiation induction of
cancer in non-military persons are of a somewhat
different nature and do not, of course, come under the
test of *Feres*. The first major decision in *Allen v.
United States* was decided by the United States District
Court, D. Utah, C.D. in 1981. This was a class-action
suit on behalf of more than 1000 persons for injuries
allegedly sustained during the aerial tests of the
atomic bomb in the Nevada series. There were two
original issues: did the action of the government come
under the "discretionary function exemption" of the
FTCA, and did the statute of limitations apply as the
cases were not brought until long after that two-year
limit. The judge in this case wrote in his long opinion
a review of the discretionary function and found ample
cases to cite and concluded that that exemption did not
apply in this case. In addition, he ruled that the time
of discovery of the injury, (developments of cancers can
frequently take a quarter of a century or more after the
causative event) rather than the actual date of the
discovery of the alleged injury, would be determinative
and therefore the statute of limitations did not apply

to these cases. The opinion then allowed the claims of damages to be brought forward and to be the basis of another trial.

On May 10, 1984, the same judge who wrote the previous decision ruled that the exposure to fallout from the tests had contributed to causing ten cancers, nine of which were fatal. The class-action suit was not carried forward, but specific claims of radiation-induced cancer by a number of these persons were the basis for the decision. The judge ruled that only leukemia, and not solid cancers, was covered in this suit. The judge relied heavily upon testimony from experts and also upon the fact that one of the radiation monitors had "gone off scale" after a test, and no local warning was issued by the government. The government was specifically held to blame for not issuing this warning which might have led to preventive measures. The testifying experts also seriously questioned the data concerning internal exposure to radiation from the fallout given by governmental experts. The opinion is of particular interest, not just for its length of over 200 pages, but for the fact that the judge essentially wrote an introductory text to atomic physics, atomic chemistry and radiation biology. Contained within the opinion are such data as the table of the atoms, listing of all radioactive elements and their decay products with the radiation emission spectra of each and their half-lifes, and a series of diagrams showing chromosome abnormalities induced by radiation.

The case seems certain to go forward to the United States Supreme Court, and it now may be seriously damaged for the plaintiffs; there have been two recent studies indicating that the increased number of cases of leukemia reported in the exposure area may be based upon faulty data. It is claimed that the increase in numbers of such cases is spurious in that the earlier baseline for the number of cases expected in the population exposed is in itself wrong as it reports an unusually *low* number of cases for that area and the purported increase is therefore not real. With these reports, which appeared in *Science* this past year, the plaintiffs' case may be in jeopardy. The amount granted, and the government's interest in estoppel of further cases, make it most likely that the case will be appealed and whether *certiorari* and an opinion from the Supreme Court will be given is at present not decided.

An entirely different series of cases all dealing with irradiation concerns the effects of radioactive tailings or dumpings on personal property. In the first, *McKay v. United States*, a group of landowners brought suit against the government for alleged damages to their property bordering a nuclear weapons manufacturing site. They claim that the soil surrounding the plant and part of their landholdings has become sufficiently contaminated so that the property is worthless. In addition, further damages were claimed to have resulted from two fires which occurred in the manufacturing plants. There was much discussion of the proper jurisdiction of the federal courts concerning the case as it had originally been brought against the operators of the plant, and refused on the basis that they were acting as agents of the government. The lower court had denied the suit on the basis that the decision to manufacture atomic weapons was a "political" act and fell under the discretionary powers of the government. The present court, The United States Court of Appeals felt in 1983 that under applicable state laws there might indeed be a recognizable suit for negligence and it should also be determined whether the manufacturing companies were in reality independent contractors or employees of the United States, and the case was remanded to a lower court for trial on these issues.

Prior to the case coming to trial, the plaintiffs and the defendants agreed to an out of court settlement. The amount was several millions of dollars, borne in part by the United States, and in part by counties and town wherein the facility was located. In addition, the manufacturer agreed to take steps to bring the radioactivity on some 2,000 acres down to a level within state guidelines. The settlement reflects clearly that improper practices had been followed and the plaintiffs had a valid tort claim, as suggested by the opinion cited.

Somewhat similarly, in 1984, the United States District Court, D. Colorado, had to deal with an alleged loss of property due to uranimum mill tailings contaminating the plantiffs' property. Further, the extent of radiation was such that the plaintiffs in *Brafford v. Susquehanna Corporation* had to vacate their home and sued not only for this but also for "present physical injury, for their drastically increased risk of cancer and other diseases and consequences of radiation, including present and future medical costs, and for mental grief and anxiety stemming from their radiation exposure." The major claim was their "forcible ejection" from their home and their exclusion from the use of their property. The defendant corporation claimed that under the appropriate state law "forceful eviction" was only applicable to cases involving ejection by physical force. The plaintiffs also sought treble damages as a punitive measure against the defendant. The court here examined appropriate state laws and found no reason to deny the suit, agreeing that the high level of radiation produced a safety hazard which was equivalent to "physical force." Of more interest to the genetic issue was their statement that, "...experts of national renown who express their opinion that the extent of subcellular damage resulting to plaintiffs because of their exposure to the radiation constitutes a present physical injury. ...'the damage has been done' and the 'trigger' has been cocked.....given the strength and consensus of the experts' deposition, and given the levels of radiation to which plaintiffs were allegedly exposed, plaintiffs have raised a question of fact with respect to whether a present injury in the form of chromosome damage was suffered by the plaintiffs as a result of their exposure to the radiation emitted by the mill tailings." The case was thereby remanded for retrial on the basis of the alleged claims by the plaintiffs.

Still another issue involving radiation and its effect on persons as well as the environment was raised in several cases, all of which raised the issue of damage to the environment by proposed or actual atomic tests. The cases, *Growther v. Seraborg*, *The Aleut League v. the Atomic Energy Commission*, and *The Committee for Nuclear Responsibility v. Seaborg*, date from 1970 and 1971 and were heard by various federal courts. All are in accord that the plaintiffs' cases are not valid, and that the decision to carry out the tests fall within the government's discretionary powers and as such are not subject to litigation.

Another involved case, *Johnston v. United States*,

was decided in 1984 by the U.S. District Court, D.
Kansas. This case involved the claimed exposure of
several persons to radiation from radium dials from
aircraft instruments being reprocessed by a company in
Kansas. The court went into a lengthy study of the
effects of radium emissions, the time from exposure to
effect, and the total amount of exposure each of the
several plaintiffs claimed was responsible for either
death or injury from induced cancers. The court
essentially again wrote a text book of irradiation
effects of this type, deciding that the level of
radiation, from 10.2 to .05 microcuries, was
insufficient to cause any claimed damages. In addition,
the damages claimed by the plaintiffs occurred in a much
shorter time than could be expected if their cancers had
been induced by the radiation. Thus the failure to warn
of possible radiation effects by the government which
used the plant for reprocessing, could not in any way
have induced the injuries claimed, and the court
rejected all of the plaintiffs' claims for relief. It
might be pointed out that the court was well aware of
the earlier radium effects upon watch dial painters
during World War I, and found no similarities here.

EFFECTS OF FALLOUT ON SHEEP

A case involving the effects upon animals during
and following the open air nuclear tests in Nevada in
the 1950's came before the courts. The sheep case,
Bulloch v. United States, was originally decided against
the sheep owners who claimed that fallout from an atomic
test had caused the death of most of their herd.

Some two decades later, a federal judge allowed the

case to be reopened ruling that there had been "fraud against the government," in that evidence had either been suppressed or distorted by the government.

In *Bulloch II* (95 F.R.D.123), the court found four basic reasons to believe fraud and misrepresentation had occurred, including an incorrect comparison of earlier experiments dealing with the effects of radiation on sheep, misleading information on radiation dosages, governmental pressure upon witnesses to testify in a certain way, and misleading answers on the part of the government to interrogatories.

The appeal by the government, *Bulloch v. United States of America*, was decided by the U.S. Court of Appeals, Tenth Circuit, in 1983. The court here found no evidence for any of these four claims, deciding that all of the necessary information was presented in a fair way in the original trial, including controversy over the dosage of radiation received by the sheep, veterinarians' reports, etc. "We find nothing to demonstrate that misleading answers were made by the Government in *Bulloch I*. The plaintiffs were familiar with all the background data...with this familiarity they chose not to seek additional answers or clarifications." In addition, some of the original witnesses have died, and there is no longer any physical evidence available. It would appear from the court's decision that the 25-year lapse between *Bulloch II* and *Bulloch I* is somewhat inexplicable. The only factor which the court could find for reopening the case was that a Congressional Hearing concerning the tests was held in the interim, in 1979. At this hearing only representatives of the plaintiffs testified. The present decision, correctly, concluded that such a hearing was not part of the record in the original trial, and therefore cannot be a basis for the present case. As a result, they dismissed the plaintiffs' claims, stating, "The recent increase in the heat, the light and awareness of atomic energy, is all worthwhile, but we are unable to find in it any reason, as the plaintiffs apparently would have us do, to overturn the considered judgment of the court reached 25 years ago."

AGENT ORANGE

What certainly are among the most complex legal cases are those arising from the Viet Nam war. These cases, all involving the usage of the herbicide, Agent Orange, involve questions of federal jurisdiction, various constitutional amendments, *Feres*, and many other legal points. The basic problem is simple in terms of the facts upon which the cases are based, the use of that herbicide as a defoliant in Viet Nam and the exposure of what may be millions of veterans to it. The defoliant contained a compound known as dioxin, one of the most deadly compounds in the modern arsenal of commercially manufactured chemicals. Various suits have been brought against the five major chemical companies which produced the compound. The suits were for amounts which, if allowed, would have exceeded the total assets of these five huge manufacturers. Dioxin can produce a wide variety of damage, both somatic and probably genetic. The veterans involved in this potentially mammoth class action suit, *In re Agent Orange*, originally brought suit in federal courts which they claimed had jurisdiction in such suits. A Federal District judge in the United States District Court for the Eastern District of New York found reason in the law for federal

jurisdiction over the suit. In a lengthy analysis of the
federal jurisdiction standards, however, the United
States Court of Appeals, Second Circuit, reversed the
lower court judge and held there was no federal
jurisdiction in the case, citing comparable cases at
length, and concluding that under those cases the suit
should not have been brought in federal courts. It
should be pointed out that suits similar to these,
claiming damages, including possible genetic damages,
had been brought forth in a number of states and this
was an attempt on behalf of the veterans concerned to
consolidate the issue into one federal issue. The
opinion is replete with discussion of federal cases
involving jurisdiction, and might serve as a review
article concerning that issue. It also was probably no
comfort to the veterans to know that one Appellate judge
filed a dissenting opinion, pointing out that the
numbers involved in the suit could reach 2,400,000
persons; already there existed suits in more than
twenty-five judicial districts. As pointed out by the
dissenting judge, "To the non-legal mind, it would be an
odd proposition that this litigation, so patently of
national scope and concern, should not be tried in
federal court." To an observer of the law, this
certainly is true. The narrow interpretation of the
legalistic majority opinion seems to be just that,
judicial refusal to become involved in the enormity of
the case. The dissenter pointed out parallels to laws
involving federal prisoners who could sue through
federal courts if injured by county jailors who were
under federal contract. He pointed out that this type of
duty to criminals should extend to veterans! "Perhaps no
relation between the Government and its citizens is more

distinctly federal in character than that between it and
members of the armed forces...This obviously federal
relationship does not depend primarily upon any
particular statute, but rather inheres in the federal
government's exclusive capacity to wage wars....The
United States has a clear interest in the protection of
its soldiers from harm caused by defective war
materials. What other interests does the United States
arguably have that might conflict with this clear
interest?... In short, in the case before us the
paramount interests of the United States are in the
welfare of its veterans and in their fair and uniform
treatment." Nonetheless, this was only a lone dissenting
opinion, and the majority held that the three standards
necessary for federal jurisdiction, "(1) the existence
of a substantial federal interest in the outcome of a
litigation; (2) the effect on this federal interest
should state law be applied; and (3) the effect on state
interests should state law be displaced by federal
common law," did not apply.

The issue was far from over, however, as the
veterans' group found other ways to attack the case. They
were rebuffed once under *Feres* in another case of the
same name, *In re Agent Orange* in 1980, and made still a
third attempt in *Ryan et al. v. Cleland* in 1982. It
might be pointed out that Cleland was the Administrator
of the U.S. Veterans' Administration Bureau. All of the
cases involving agent orange came before the same U.S.
District judge; one can imagine that by now he was
somewhat conversant with the facts. But this third case
showed considerable imagination, to say the least. Suit
was brought against the V.A. on the basis of practically
all of the Amendments in the Bill of Rights. The claim

was that the Veterans' Bureau, by not properly informing and caring for victims of the herbicide violated the veterans' rights under the Fifth, Eighths, Ninth, and Fourteenth Amendments as well as a miscellany of statutes and an article in the Constitution itself. The judge patiently dealt with each of these, eliminating them as claims against the governmental agency and again denying the suit. One ingenious plea was that the veterans were "prisoners" of the V.A. system, and that they had received cruel and unusual punishment by the use of various hospitals of what were claimed to be psychotic drugs in an attempt to cure them. The court dismissed the cruel and unusual punishment plea by pointing out that the purpose of the amendment was to protect the rights of those convicted of crimes, and obviously was not applicable to this case. The other constitutional claims were equally analysed and found wanting.

It was clear that this issue would not simply disappear. One could predict with confidence that more legal attempts and further appeals would continue to come forth in this particular case. The fact that dioxin is so deadly will probably lead to a host of individual claims by persons who believe that they have medical proof of their direct injury; it was the consolidation of these cases in the first instance which attempted to avoid myriad a of individual suits.

That dioxin *is* dangerous perhaps can be made more clear by pointing out that a certain reagent containing dioxin used extensively by electron microscopists has now been taken off the market, forcing these scientists to look for a substitute which did not pose the dangers of dioxin. As yet no individual suits by the investigators have been brought, but one wonders when the first of such cases will appear.

The abuse of military and civilian personnel by various governmental agencies which were supposedly protecting them against both physical and genetic harm seems hardly to be in doubt. However, unless specific legislation should be forthcoming, which seems quite unlikely, it looks as if the combination of cases studied will for the most part bar further suits of those harmed by the willing or unwilling actions of the military. Certainly a closer look at the *Feres* doctrine will have to be taken in the future if the rights of the military personnel are to be preserved.

The cases involving Agent Orange have now risen to such a number that an entire chapter might have been given to them. A list of cases from the last two years alone shows that at least twenty-nine various court opinions dealing with many different aspects of this difficult problem have been handed down, many by the same judge who now must qualify as the world's expert in this legal field. Indeed, it is hard to believe that the U.S. District Court, Eastern District of New York, would have found time to decide any other matters. There are more than 2600 separate documents in the case file running into hundreds of pages for some. According to the Schenectady Gazette (NY) the index of the file is over 140 pages in itself.

However, surprisingly, on the night before the class-action suit which would have determined liability or lack thereof of the manufacturers of the defoliant was to begin, it was suddenly settled for $180 million dollars. This sum, to be put aside by the companies which manufactured the agent, will be used to provide individual relief for those proving claims of harm from its use. (While this is being determined, the money is accruing interest at the rate of $60,000 daily.) What may have brought this settlement about was a decision by the United States District Court, E.D. New York, on February 16, 1984. In this case, named like so many before it *In re "Agent Orange" Product Liability Litigation*, the judge had ruled that the *Feres* doctrine did not bar dependent claims by veterans' wives and children although the doctrine did bar claims by the veterans themselves, and derivative claims by their families. The legal point is summed up as follows: "Again the failure of the government to warn former servicemen and their wives of the hazards of conception when a serviceman has been exposed to Agent Orange presents a special complication. Whatever discretionary function was involved in battlefield decisions, subsequent treatment and non-warning decisions seem to fall within the class of ordinary malpractice long encompassed by the F.T.C.A.

"For the reasons stated, in the forthcoming trial involving a number of claims of miscarriages and fetal deformations, the government will be a third party defendant....It should be emphasized that this memorandum is tentative to assist the parties in preparing for trial. The government may renew its motion to dismiss at any time before or during trial as further evidence and legal developments suggest.

The actual case would have involved a selected group of some nine persons who claimed birth defects caused by Agent Orange. It would have been a long trial, possibly involving an advisory jury who would have to listen to volumes of testimony by experts on both sides. The chance for a jury awarding even larger sums may have

been one of the factors leading to the manufacturers'
settlement of the case prior to trial.

In reality the actions are only beginning. Each
person will have to prove that there was actual exposure
to the agent and that there have been real personal
injuries caused by it before sharing in the settlement.

The settlement has not pleased everyone; both sides
are somewhat bitter; the veterans feel that the sum set
aside is far from sufficient, while the manufacturers
are unhappy both because of the way the apportionment of
the costs are to be distributed and also because they
continue to feel that no proven damage, other than skin
lesions, has been found to be caused by the agent.
Indeed, some veterans have refused to be parties to the
settlement and may well continue their cases. It would
appear that the book is not closed on Agent Orange, and
may not be for years to come as more birth-defective
children are born and more suits are brought on their
behalf.

TRANSPORTATION OF RADIOACTIVE MATERIALS

A case in which the use of irradiation was not the
major issue, arose in 1976 and was decided by the Court
of Appeals in Georgia. This case, *Value Engineering
Company v. Gisell*, involved a suit brought by an airline
passenger against the airline for alleged possible
genetic damages arising from the shifting of a drum
containing radioactive materials during flight. The
issue of genetics was, however, secondary; the main
issue was whether Georgia had jurisdiction over the
case. The plane, owned by Delta Airlines, had made an
intermediate stop in Atlanta on its way to Baton Rouge,
Louisiana. The radioactive material was shipped by Value
Engineering; their claim was that they had no knowledge

of the routing of the shipment and that the courts of Georgia had no jurisdiction over the case. The court held otherwise: "...Appellant here took affirmative action in introducing an allegedly defective and dangerous article into the stream of commerce and now it wishes to close its eyes (and have this court do likewise) to the injuries claimed to have been sustained as a direct consequence of its own election to ship the material by interstate carrier. That it 'intended' the material never to enter Georgia cannot belie the fact that it did so enter because of the appellant's action of shipping it on a Delta flight which landed in Atlanta. Appellant started the material on its journey and cannot 'wash its hands' of responsibility when damages are claimed as the result of those materials being in interstate commerce because it 'intended' the ultimate destination to be Louisiana and not Georgia. "Thus the court held it had jurisdiction in this case. No determination of damages or other results of alleged radiation injury or genetic harm were dealt with in this case.

Possibly as a result of this case, most airlines now forbid the carrying of radioactive materials on passenger flights. Such regulations may have at times hindered research as some of the more common radioisotopes used in biological research and medical treatment have a short useful life, and only air shipment is possible if they are to arrive in time for their appropriate usage.

A case which does not involve alleged direct personal injury as part of the court's decision arose in 1982 when the United States District Court, Southern District of New York, dealt with *The City of New York v.*

United States Department of Transportation. This
enormously long opinion dealt with a suit brought by the
City, the State and a town against the U.S. DOT
regarding regulations for transportation of radioactive
materials through the areas mentioned in the suit. Of
concern was the possibility of an accident with massive
exposure to radiation. The length of the opinion is
again due to the welter of regulations concerning such
transportation; was a study of environmental impact
made; were alternate, less hazardous routes considered,
etc.? Reading this opinion in its entirety is almost a
herculean task, and the intricacies of the involvement
of many federal regulations and many federal agencies
might make a fine study for students in law schools. The
basic problems, however, are clear and involve genetic
risk as well as nuclear waste disposal. Brookhaven
National Laboratories wished to dispose of spent reactor
fuels by shipping them via a trucking agency which
planned to route them through the territories controlled
by the plaintiff governments. The governments sought to
prevent such shipments and claimed all of the necessary
studies, the environmental impact, the search for less
hazardous routes (such as shipment by water and not
through city streets or on state roads) and other
necessary regulations had not been properly conducted by
the defendant agency. Claims of Tenth Amendment
violation of states' rights (New York has laws governing
such situations) were also made. The lower court decided
that the claims by the states and localities were valid.
"DOT may well have reached its conclusion...because of
its genuine conviction that highway transportation is
safe. It has implemented its judgment, moreover, with a
courageous willingness to accept responsibility for the

decision. The national interest no doubt sometimes
requires courageous regulation in the face of public
opposition on issues commanding extraordinary media
attention. But the laws being administered by DOT do not
demand regulatory valour, so much as they demand care in
carrying out its duties...It may be good policy in the
agency's view to override state and local efforts that
prevent any reasonably safe use of a transportation
mode, and to force an unwilling public to bear risks
that are reasonable compared to others commonly
accepted. But these are not Congress's policies. Congress
has mandated that important and reasonably avoidable
risks not be borne, and specifically recognizes the
propriety of local efforts to increase safety. DOT has
acted without sufficient study and is beyond its
authority." The decision therefore is that such shipment
cannot be made without compliance to all environmental
problems and was made after almost a 100 page opinion
attempting to decipher the regulations and the role of
various agencies in applying these regulations. There
certainly is no doubt that the issues are most complex,
and that, simplification of both the regulations and the
fixing of responsibility in some single agency will have
to occur so that cases such as this involving several
governmental agencies and their separate regulations can
avoid having the courts spend countless hours and
innumerable pages of opinions in trying to decipher the
maze of regulations and to decide the agency responsible
for the type of problem arising here. However this
may be, the case again, *City of New York v. U.S. Dept.
Transportation*, was appealed in 1983 to the U.S.Court
Appeals, Second Circuit, and the higher court in a 2:1
opinion reversed the lower court and the long opinion

and work of the lower court was in vain.Ironically,
following the decision, the City and Brookhaven National
Laboratories independently reached a compromise after
the appellate court decision. Shipments of the spent
fuel will be delayed for a year while the laboratory
investigates the use of larger shipping containers,
thereby reducing the number of trips through the city,
and also it was agreed to explore the use of barges for
shipment in order to bypass New York City. Sure to be
followed by law suits, some of which will involve
international law, are the shocking events involving a
major and serious nuclear spill beginning in Juarez,
Mexico and crossing into the United States. Two firms in
Mexico and one in the United States received scrap metal
to be reprocessed. Unfortunately, the scrap, unbeknownst
to the recipients, contained hundreds of pellets of
highly radioactive cobalt-60, presumably from the
discard by a hospital no longer using this type of
treatment. Some claim the machine containing the cobalt
was stolen from the hospital. In any event the material
was taken from the Mexican steel plant by a truck from
which they apparently spilled. The material entered the
United States when a "hot" pick up truck dusted by the
spill was found. Apparently the material was spread by
adhering to tires of other vehicles. The pick up truck
was returned to Mexico but remains so radioactive that
Geiger counts 300 yards away are able easily to detect
the radiation. Two Mexicans are reported to have
received up to 10,000 rem and have developed radiation
injuries as a result. The legal implications of this
seem vast; the fixing of responsibility for the spill
will involve suits against the hospital for illegal
disposal, against the steel plants for allowing the

spill, as well as personal claims by those injured. Obviously this will involve international law as the origin of the material occurred in this country, and the spill itself involved both Mexico, and ultimately the United States when the truck entered this country. The predicted suits will obviously keep lawyers and the courts busy for years to come.

PSYCHOLOGICAL HEALTH

In May, 1982, the United States Court of Appeals, District of Columbia Circuit, dealt with the issue of whether psychological health was a necessary factor in determining under an Environmental Impact Statement the effects of rebuilding and restarting the Three Mile Island reactor which been "seriously damaged in the worst nuclear accident Americans have yet experienced." In this case, *People Against Nuclear Energy v. United States Nuclear Regulatory Commission*, the plaintiffs sought to require that any consideration concerning the reopening of the facility must include "potential harms to psychological health and community well-being," factors which were not considered in the original 2:2 decision by the Commission to permit eventual restarting of the repaired facility.

The court's opinion is complex, with partial concurrence and partial dissent by several of the judges. It is further complicated by the fact that the court took into cognizance that there would be a considerable delay before the necessary repairs could be completed and the fact that the real issue here appears not to be the damaged reactor, but the reopening of a similarly designed unit (Unit TMI-1) which had been in a shutdown condition at the time that Unit TMI-2 suffered the near catastrophe. The Court far from unanimously

ruled that "... PANE contends that under the National
Environmental Policy Act (NEPA) and the Atomic Energy
Act the Commission must take into account potential
harms to psychological health and community well-being.
We hold that these environmental impacts are cognizable
under NEPA. Therefore the Commission must make a
threshold determination, based on adequate study,
whether the potential psychological health effects of
renewed operation of TM1-1 are sufficiently significant
that the NEPA requires preparation of a supplemental
environmental impact statement.

However, the Court also found that "Today this
court also holds that the Atomic Energy Act does not
require the Commission to consider potential harms to
psychological health." It is on the basis of the latter
statement that the main part of the dissenting opinion
was written. The majority opinion later states,
"Although we are not aware of any cases that have
considered the cognizability of post-traumatic
psychological health effects under NEPA, it is not
surprising that this is an issue of first impression.
Americans have never before experienced the
psychological aftermath of a major accident at a nuclear
power plant, one that arouses fears of a nuclear core
meltdown and led to mass evacuation from the surrounding
community."

The Court, after its opinion requiring that these
factors should be studied, then issued an Amended
Judgment. "It is ORDERED and ADJUDGED by this court
that, for the reasons stated in the opinion, for the
court issued this day, the record in this case is
remanded to Commission for a determination whether,
since the preparation of the original environmental

impact statement for the nuclear facility at Three Mile
Island Unit 1 (TMI-1), significant new circumstances or
information have arisen with respect to the potential
psychological health effects of operating the TMI-1
facility. The Commission may choose the procedures by
which it makes this determination. If the Commission
finds that such significant circumstances or information
exist, it shall prepare a supplemental environmental
impact statement which considers not only effects on
psychological health but also the effects on the well-
being of the communities surrounding Three Mile Island."
The court also vacated its injunction against the
restart of TMI-1 as serious difficulties arose with this
reactor and no schedule for its operation seem to have
been made. The case was appealed, and the United States
Supreme Court in a rare unanimous decision reversed the
Appeals Court and found that no EIP dealing with
psychological impact is necessary for the reopening of
TMI-1. The opinion of that court stated in part, "We
think that the context of the statute shows that
Congress was talking about the physical environment--the
world around us, so to speak. If a harm does not have a
sufficiently close connection to the physical
environment the law does not apply." Thus, as no
property damage was claimed here, the harm only being
possibly to people, the court rules that no cause of
suit exists. As mentioned, the case is complex and
illustrates the difficulty which both the courts and the
public have in assessing the possible risks and benefits
of atomic power.

The struggle between the two sides over the
restarting of the reactor continues with unabated vigor
by both sides. The power plant was to have been allowed

to start up, but delays by its opponents have so far managed to keep it inactive. The eventual outcome of this legal "battle" cannot at present be predicted.

PROPERTY DAMAGE

A further legal issue in the Three Mile Island accident arose in *Commonwealth of Pennsylvania v. General Public Utilities Corporation*, decided in 1983 by the United States Court of Appeals, Third Circuit, did deal with supposed real property damages. Here the state of Pennsylvania sought recovery for the costs of the accident in terms of the expense of evacuation, claiming that "The activities of defendants as alleged hereinabove render the Three Mile Island facility a public nuisance and have caused plaintiffs and class members irreparable harm." The lower court found in favor of the defendants on the basis of failure to state a claim upon which relief can be granted. None of the four alleged damages, "overtime and compensatory time and other personnel costs...operational expenses and emergency purchases incurred in responding to the nuclear incident...lost work time as a result of the nuclear incident...other expenses incurred as a result of and/or in response to the nuclear incident." The courts rule that "public expenditure made in the performance of governmental functions are not recoverable in tort." Thus the operating companies were not held accountable for the considerable expenses which resulted from the State or community costs as a result of the accident. A further factor in the decision was that no claims for damage to any property were made and the losses claimed were "all purely economic losses."

The claim was made that the release of radioactivity made their public buildings unsafe and therefore physical harm had resulted, but as there was no proof offered of this the court declined to express an opinion on this matter. The Appeals Court actually did not refuse all claims but decided that " ...the district court properly granted summary judgment as to any claims for damages for reduction of real estate tax revenues and that the district court properly dismissed, for failure to state a claim, any claims based upon state common law or federal common law nuisance theories. To the extent that the district court dismissed plaintiffs' other damage claims the decision of the district court will be vacated and remanded." Thus the case will be retried and real evidence of actual harm or losses due to temporary loss of use of property will now have to be shown.

GENERAL CONCLUSIONS

The cases involving nuclear incidents illustrate well the problems of the courts in attempting to deal with new scientific discoveries. Obviously no laws or decisions concerning the use of atomic devices, either for military or civilian usage, could be made before the first test of an atomic weapon some decades ago. Following the use of the bomb and the then seriously believed dangers of another war, testing of new and improved devices was felt to be mandatory, giving rise to the host of cases discussed. Unfortunately, there seems to have been inadequate consideration of personal injuries implicit in the test plans, even though from the evidence given such potential should have been recognized and suitable precautions taken. Indeed, had this been done in a rational manner, neither courts nor

Congress would have to wrestle with the legal aspects of recovery for damages to veterans fatally injured by neglect of simple safety precautions.

Similarly, civilian usage of atomic power also could not have been anticipated until reactors were built. As may be realized from the Three-Mile Island near disaster, adequate safety precautions were apparently not taken. Indeed, since that accident, evidence of many more near accidents and improper operation of these plants is coming forward. The legal problems dealing with these are foreshadowed by the later cases described, and may just be beginning.

While most of the suits have been for deaths allegedly due to irradiation-induced cancers, several have also claimed mutagenic effects on their defective children. As pointed out earlier, carcinogens and mutagens are closely related, and it is necessary to examine the cancer cases for an understanding of the cases dealing with alleged mutagenic effects caused by radiation. Thus, the length of this chapter and its relationship to *Genetics in the Courts* should be clear.

ALPHABETICAL LIST OF CASES CITED

CASE	CITATION	YEAR
Aleut League v. A.E.C.	357 F.Supp. 534	1971
Allen v. United States	527 F.Supp. 476	1981
Allen v. United States	588 F.Supp.247	1984
Braff v. Susquehanna Corporation	586 F.Supp. 14	1984
Broudy v. United States	661 F.2d 125	1981
Bulloch v. United States	721 F.2d 713	1983
Childhood Leukemia & Fallout Chiropractors' Association	223 Science 139	1984

of NY. v. Hilleboe   12 N.Y.2d 109   1962
City N.Y. v. U.S.Dept. Trans.   F. Carr.Cases 58792
                                                             1982
City N. Y. v. U.S. Dept. Transport.715 F.2d 762   1983
Cole v. United States   755 F.2d 873   1985
Com. Pa. v.Gen.Public Util. Corp.   710 D2d 117   1983
Comm. for Nucl. Resp. v. Seaborg   463 F.2d 788   1971
Crother v. Seaborg   312 F.Supp.1205 1970
Disputed Study on Cancer Increases NYT, Jan. 13   1984
Feres v. United States   340 U.S. 135   1950
Fountain v. United States   533 F.Supp. 698 1981
Gaspard v. United States   713 F.2d 1097   1983
Hampton v. United States   575 F.Supp.1180 1983
Heilman v. United States   731 F.2d 1104   1984
Hinkie v. United States   524 F.Supp. 277 1981
Hinkie v. United States   715 F.2d 96   1983
Honicker v. United States   465 F.Supp. 414 1979
In Re Agent Orange   580 F.Supp.690   1984
In re Agent Orange   560 F.Supp. 762 1980
In re Agent Orange   635 F.2d 987   1980
Irradiated Personnel during   House Veterans
        Operation Crossroad   Affairs Comm. 1983
Jaffe v. United States   663 F.2d 1226   1981
Johnston v. United States   597 F.Supp. 374 1984
Juarez, an unprecedent radiation
                      accident   223 Science 1152 1984
Juda v. United States   6 Cl.Ct. 441   1984
Kelly v. United States   512 F.Supp. 356 1981
Lab. & City agree on Plan to delay NYT June 6,   1984
Laswell v. United States   683 F.2d 261   1982
Lombard v. United States   530 F.Supp. 918 1981
McKay v. United States   703 F. 2d 464   1983
Monaco v. United States   661 F.2d 129   1981

Mondelli v. United States	711 F.2d 567	1983
People Against Nuc.Energy v.		
U.S.Nuc.Reg. Commission	678 F.2d 222	1982
Punnett v. Carter	621 F.2d 578	1980
Punnett v. United States	U.S.D.Court E.PA.	
	79-29 1984	
Radiation & Preconception Injuries	28 S.W. L. Journal	
	414 1974	
Report could undercut Atomic		
Veterans' Cases	NYT August 5	1983
Ryan v. Cleland	531 F.Supp. 724	1982
Sheehan v. United States	542 F.Supp. 18	1982
Shipek v. United States	752 F.2d 1352	1985
Targett v. United States	551 F.Supp.1231	1982
Tentative Agent Orange Settlement	224 Science 849	1984
United States v. Shearer	U.S. Supreme Court,	
	Slip 1985	
Value Engineering v. Gisell	140 Ga. App. 44	1976

CHRONOLOGICAL LIST OF CASES CITED

YEAR	CASE	CITATION
1950	Feres v. United States	340 U.S. 135
1962	Chiropractors' Association	
	of NY.v.Hilleboe	12 N.Y.2d 109
1970	Crother v. Seaborg	312 F.Supp. 1205
1971	Aleut League v. A E.C.	357 F.Supp. 534
1971	Comm. for Nucl, Resp. v. Seaborg	463 F.2d 788
1974	Radiation & Preconception	28 S.W.Law
	Injuries	Journal 414

1976 Value Engineering v. Gisell	140 Ga. App. 44
1979 Honicker v. United States	465 F.Supp. 414
1980 In re Agent Orange	560 F.Supp. 762
1980 In re Agent Orange	635 F.2d 987
1980 Punnett v. Carter	621 F.2d 578
1981 Allen v. United States	527 F.Supp. 476
1981 Broudy v. United States	661 F.2d 125
1981 Fountain v. United States	533 F.Supp. 698
1981 Hinkie v. United States	524 F.Supp. 277
1981 Jaffe v. United States	663 F.2d 1226
1981 Kelly v. United States	512 F.Supp. 356
1981 Lombard v. United States	530 F.Supp. 918
1981 Monaco v. United States	661 F.2d 129
1982 City N.Y. v. U.S.Dept. Transportation	F. Carr Ca.58792
1982 Laswell v. United States	683 F.2d 261
1982 People Against Nuclear Energy v. U.S.Nuc.Reg. Comm.	678 F.2d 222
1982 Ryan v. Cleland	531 F.Supp. 724
1982 Sheehan v. United States	542 F.Supp. 18
1982 Targett v. United States	551 F.Supp. 1231
1983 Bulloch v. United States	721 F.2d 713
1983 City New York v. U.S. Dept. Transport.	715 F.2d 762
1983 Com. Pa. v.Gen.Public Util. Corp.	710 F 2d 117
1983 Gaspard v. United States	713 F.2d 1097
1983 Hampton v. United States	575 F.Supp. 1180
1983 Hinkie v. United States	715 F.2d 96
1983 Irrad. Personnel during Operations Crossroad	House Veterans Affairsd Comm.
1983 McKay v. United States	703 F. 2d 464
1983 Mondelli v. United States	711 F.2d 567
1984 Allen v. United States	588 F.Supp.247

1984 Braff v. Susquehanna Corporation   586 F.Supp. 14
1984 Childhood Leukemia & Fallout   223 Science 139
1984 Disputed Study on Cancer
                     Increases   NYT, Jan. 13
1984 Heilman v. United States   731 F.2d 1104
1984 In Re Agent Orange   580 F.Supp. 1242
1984 In Re Agent Orange   580 F.Supp. 690
1984 Johnston v. United States   597 F.Supp. 374
1984 Juarez, an unprecedent
        radiation accident   223 Science 1152
1984 Juda v. United States   6 Cl.Ct. 441
1984 Lab. & City agree on Plan to delay NYT June 6,
1984 Punnett v. United States   U.S.D.Court
                            E.PA. 79-29
1984 Report could undercut Atomic
              Veterans Cases   NYT Aug. 5,
1984 Tentative Agent Orange Settlement 224 Science 849
1985 Cole v. United States   755 F.2d 873
1985 Shipek v. United States   752 F.2d 1352
1985 United States v. Shearer   U.S. Supreme
                            Court, Slip

Chapter 6
SEX AND SOCIETY
THE PATERNITY CASES

"It took the testimony of a sage, an oracle, a drunken party goer, a messenger, a sheepherder and his own wife before Oedipus could figure out who his father was." This quotation cited in *Matter of Beaudoin v. Tilley* by a family court judge in New York in 1981, perhaps sums up the complexity of cases involving paternity suits.

"Paternity proceedings were unknown at common law and paternity statutes must hark to the Elizabethan Poor Law of 1576 (18 Eliz I, ch 3) which provided the government with a means for recouping funds advanced for a child's support (cited from *Matter of Lydia*).

"Subdivision (b) of section 517 of the Family Court Act [New York] reflects the determination that the conservation of the public purse outweighs the potential for 'gross unfairness' and 'emotional harshness that may adhere to respondents in particular situations." (ibid).

"Testimony from the sitting judiciary hearing paternity cases revealed to the Commission...that the evidence in most cases consists of an accusation by the woman and a denial by the defendant. Under such circumstances the judges feel constrained to enter a finding of paternity. Not even the slightest corroborating evidence is required. * * *"

"In these circumstances, it is not surpising that a study based on blood tests indicated that in a group of 1,000 cases of disputed paternity, 39.6 per cent of the accused men were not actually the fathers of the children in question. The problem even reaches beyond *disputed* paternity matters: The putative father's admission that he is the father of a child is not necessarily credible. Again on the basis of blood tests Sussman and Schatkin estimated that fully 18 per cent of a group of men who voluntarily admitted paternity were not in fact the fathers of the children in question. Combining these figures and applying them to 4200 paternity cases that arose in New York City in 1959, Dr. Sussman estimated that as many as 645 men (more than one in seven) were erroneously held liable as the 'fathers' of someone else's children.

"In sum, current paternity prosecution practice in many metropolitan areas is abhorrent. Blackmail and perjury flourish, accusation is often tantamount to conviction, decades of support obligation are decided in minutes of court time and indigent defendants usually go without counsel or a clear understanding of what is involved.

"The paternity defendant, or course, has a substantial interest in the accuracy of the adjudication. He has a direct financial interest, for as an adjudicated father he will be ordered to contribute to the support of the child throughout its minority. Similarly, in light of recent case law, the adjudicated father's estate can be burdened by the child's claim to inheritance, workers' compensation benefits and insurance proceeds. In addition to his financial interests, the defendant, if found to be the father, is also indirectly threatened with loss of liberty, since incarceration may be imposed for criminal nonsupport. Finally the social stigma resulting from an adjudication of paternity cannot be ignored.....

"Because the child's rights to inheritance, workers' compensation benefits and insurance proceeds flow from an adjudication of paternity and because of the

increasing recognition of an illegitimate child to equal
protection under the law, it is must now be accepted
that the child's interest in an accurate determination
of paternity at least equals that of the putative
father. Moreover, since the adjudicated father may also
assert rights to custody superior to the rights of any
person except the mother and must give consent before
the child can be adopted, the child's interest may
arguably exceed those of the putative father.

"More than three hundred and ninety-eight thousand
illegitimate children were added in 1970, 360,000 in
1969, 339,000 in 1968, 318,100 in 1967, 302,000 in 1966,
for a total exceeding 1,700,000 in just those five
years. Moreover, not only has there been an increase
in the absolute number of illegitimate births, but the
rate has been accelerating and now exceeds ten per cent
of all births. In many urban areas illegitimacy stands
at forty per cent and in some cases it exceeds fifty
percent." The citation, with references omitted, is from
Hepfel v. Bashaw.

It should also be pointed out that many times the
accused male may be threatened with the criminal charge
of rape by the prosecution, and is told that if he
admits paternity this charge will not be brought.
Needless to say, given the difficulty of disproving such
a charge, the male, even if completely innocent, is
likely to accept the lesser burder of paternity with its
costs, rather than face trial on such charges. This
threatened charge is, of course, one way for the legal
system to avoid trials, and in a sense is a type of
"plea bargaining," i.e., no charges of rape in exchange
for admitting paternity.

Part of the reasoning for this somewhat unappealing

procedure is to avoid court trials for paternity, and to simplify the work of the prosecutor and of the trial judges. In fact, as will be apparent, much of the willingness of the courts to admit genetic evidence was on the basis that pretrial scientific investigations of paternity probabilities would lessen the burden of the courts and remove many of these cases from the trial docket. Unfortunately, as will also be evident, this has not been the case; despite the genetic evidence tending to prove or disprove paternity, the number of cases actually going to trial has in no way decreased, and the dispute over the admissibility of evidence of the type to be presented has replaced the simple issue of paternity so that the well-meaning attempts to lower the work of the courts has not been in any way successful.

As a result, by far the majority of cases dealing with *Genetics in the Courts* consists of paternity cases. These occur in the hundreds or more, in all states, and in all levels of courts. They arise in the lower courts such as family courts, surrogate courts, or other courts. But a large number of them are appealed for various reasons and the decisions of the higher courts, like all others so far, are varied and inconsistent. One cannot begin to cite all of the cases in a single chapter, and again those have been chosen which seem best to illustrate the underlying role of genetics in legal systems.

As pointed out by the introductory citations, paternity cases also involve many areas of the law, and are brought for a wide variety of purposes. Among the legal issues which arise are admissibility of evidence, child support, inheritance rights, constitutional questions, and at times criminal law is also involved.

Not only do putative parents have their rights, but the state also has an interest. No state wishes to burden its taxpayers with undue costs of child support; if a paternity suit can be resolved so that the putative father is shown to be the actual father, then child support shifts to him, and failure by him to furnish such support can lead to criminal charges and possible imprisonment if such support is withheld. The male wishes to establish non-paternity in order to avoid such costs; in some cases this becomes part of a divorce procedure. Conversely, there are times when the male wishes to establish that he is the true father in order to obtain custody of the child. In other cases the guardian of the infant brings suit to establish paternity for the child's sake, possibly a share in an inheritance. The variety of reasons for bringing a paternity suit is great, and many cases involve complex legal issues in addition to the establishment of paternity or non-paternity by genetic or other means.

Among the most traditional rules of English Laws is *Lord Mansfield's Rule*. This deals with the almost absolute determination that any child born to a married couple is legitimate, provided that the husband is "within the 4 seas" (in those days the Western, including the Scotch and Irish Seas, the Northern, or North Sea, the Eastern or German Sea, and the Southern or British Channel). It must also be shown that the husband was not impotent and that he had "access" to the wife at the time of conception. Given these conditions, regardless of flagrant infidelity by a wife, any child born to the married couple was assumed legitimate. Another biological determination will thus have to be made by the courts as to the time of conception;

obviously this will involve prematurely born children and those resulting from longer than normal pregnancies, and a good deal of medical testimony regarding the presumed time of conception enters some paternity cases. This is based on proving that the husband indeed met the conditions above; obviously if the time of conception can be determined as an interval when the husband did not have "access" then the rule would not apply. Similarly, in many of the more recent cases, modern methods of disproving paternity will be used in attempts to overthrow the Rule.

Another major point of law involved in paternity cases is the determination of the admissibility of evidence. Here, paternity law and criminal law diverge. In the latter instance, all evidence tending towards proof or innocence is admissible and it is up to the jury to determine the weight of the evidence in finding the accused guilty "without a doubt." In paternity cases, the statutes were formerly quite different. Until recently, only evidence from the results of blood tests which showed without doubt the male could not be the father was admissible; evidence showing he could be the father could not be used, and evidence which demonstrated without a reasonable doubt that he was the father was not allowed. With the recent advances in genetics, blood tests which can now show in some cases better than a 99% probability that the accused man is the father present a problem in terms of admissibility under these standards. Many states have recently amended their codes to permit introduction of such evidence (New York in May of 1981, for example) while as of this writing some have refused to make this change in paternity law. The problem here is one of statistical inferences. While it is almost impossible to prove absolute paternity, when the data show a probability of that order, and the accused male had access at the time of conception, few reasonable jurors would fail to consider the weight of that evidence in reaching a decision that the accused is the true father of an infant. This raises, therefore, one of the major themes of this study, at what time do the laws and courts allow new scientific discoveries to be admissible. There is also a danger of a circular argument here, to be shown in later cases. It might be interpreted that if the statistical probability of paternity is high, then "access" must have occurred. But there is no legitimate reason to examine probability of paternity unless the problem of deciding whether "access" as well as taking advantage of that feature occurred.

The basic legal precedent for admissibility of scientific evidence was laid down in 1923 under what has come to be termed the *Frye* rule derived from *Frye v.. United States*. (That case did not involve blood tests, but the use of lie detectors.) There the opinion stated clearly, "Just when a scientific principle or discovery crosses the line between the experimental and demonstrable stages is difficult to define. Somewhere in this twilight zone the evidential force of the principle must be recognized, and while courts will go a long way in admitting expert testimony deduced from a well-recognized scientific principle or discovery, the thing from which the deduction is made must be sufficiently established to have gained general acceptance in the particular field in which it belongs." The burden of admissibility of evidence now shifts to a basis of whether the evidence is accepted by the scientific community, and this will lead many judges to have to determine what the vague term, "general acceptance," should mean.

Given these general rules concerning paternity tests and the problems of the use of various genetic techniques for establishing the fact of paternity, we can now turn to the cases, again as noted previously, picking only those of the most general interest in studying the interaction of genetics and judicial decisions.

In tracing the methods by which new genetic evidence becomes acceptable, there are three main issues to be discussed. First the early use of identification evidence, that is the bringing in of the child to demonstrate its purported similarity, or lack of similarity, of obvious traits to those of the purported father. Second, the discovery of the various red cell typing tests, first the A,B,O system by Landsteiner and then a battery of others, at least led to some tests which were exclusionary. However, in no instance could these tests be introduced as evidence for paternity as they could, at best, only exclude a relatively small proportion of the accused men. Finally, with the introduction of the so-called HLA tests and other new tests to be discussed, not only can exclusion be shown in a much higher number of cases, but due to the rarity of some of the HLA types, definite evidence of paternity can be brought forward, leading to a high degree of certainty of paternity.

A third issue involved with the use of the HLA and other systems is the admissibility of statistical probabilities as valid evidence in paternity suits. Many cases, claiming the use of statistics is not legally valid, will be cited to show the difficulty of introducing this type of evidence into the legal system.

PHYSICAL TRAITS AND LENGTH OF PREGNANCY

The attempt to display an infant in court to point out its purported resemblance to an accused man in paternity cases has arisen often in such cases. Most states do not allow such a procedure, based on the possible bias which might be induced in the minds of a jury. As the judge in one case, denying the attempt to use testimony of a 9-year-old child, stated, "In my view, the display of an infant to demonstrate a supposed resemblance to the putative father does not rise to the stature of evidence. Rather, it is an injection of speculation and conjecture regarding real or fancied resemblances existing in the eye of the beholder with a dash of old wives' tales in lieu of any basis in fact, scientific or otherwise. If such conjecture emanated from the lips of a witness it would clearly be objectionable....

"Even if we then further assume that the trier of the facts is fully informed as to the workings of Mendel's law of genetics and the functioning and interplay of the genes in the parents in relation to such feature and trait (knowledge of which would apparently place the jury head and shoulders above authorities in the field) we are still left, under the hypothetical assumptions stated, with a mere possibility that the putative father is the father. I do not regard such a possibility as being evidence. (*Tennessee Department of Human Services v. Braswell*, Court of Appeals of Tennessee, W. Section, Slip Opinion 1984).

Interestingly enough, however, as early as 1892, in *Bullock et al v. Knox*, the Supreme Court of Alabama essentially overruled Lord Mansfield's rule in a case involving a genetic issue, that of race, and the genetic fact of the skin color of an infant. The case was brought on behalf of a mulatto son who wished to share in an inheritance of the estate of two white married parents using the presumption of legitimacy clause. The court summarized this briefly by saying that "where a mulatto child is born of a white woman, whose husband is white, it may be shown that it is contrary to the laws of nature for both of the parents of a mulatto to be white." The son was therefore adjudged to be illegitimate, and his claim for a share in the estate of the deceased whites was denied. He was adjudged to be "...treated as a bastard, and not an heir of his mother's husband, deceased." This case, then, is of interest both for the genetic reasoning, and for the fact that at that time illegitimate children had no rights of inheritance.

Just the opposite was found in 1978 when *County of San Diego v. Brown* was heard by the Court of Appeal, Fourth District, Division 1, of California. Here the issue was the legitimacy of a child born to an "open marriage." Paternity was assigned to Brown by the lower court and he appealed on the basis of his race. Brown was black; his then wife was white and the child in question had blue eyes, fair skin and red hair, as does the mother. Brown had relied heavily on a textbook of human genetics which the court studied and found not supportive of his claim of the impossibility of his being the true father. The court then cited another anthropological work which the court claimed supported its opinion that Brown could not bring the physical features of the child as a defense; as a result Brown was declared the father. Despite the two texts cited, it still seems illogical that a child of that physical appearance could be attributed to Brown's paternity, and today he would probably been acquitted of the charge, especially since there was evidence that the mother did not restrict her favors to Brown alone during the possible time of conception of the child. A factor in this decision may also have been a reliance upon Lord Mansfield, although this is not specifically stated in the opinion.

The timing of pregnancy, as mentioned above, also applied to another early case, *In re Walker's Estate*, decided in 1919 by the Supreme Court of California. Here the issue was whether twin children born about 5 months after their presumed father's death and nine months after their parents divorced could be considered legitimate for purposes of inheritance. Another older daughter brought suit to deny the inheritance on the basis that the twins were illegitimate. The suit was brought twice; in the first instance a jury found the twins to be illegitimate but a higher court remanded the case. The second jury found them to be legitimate and the appeal to the Supreme Court eventually followed. It turned out that despite the divorce, the father had had intercourse with the mother, or as the court wryly stated, "It turns out that Walker [the deceased] was not entirely absent from his wife during the period of possible conception." The issue of impotency also arose, and the court dealt at length with the history of English Law on this issue, particularly with an English case where the presumed father was 80 years old and presumably not capable of fathering a child. However that may be, this was not the main issue; rather, access at the time of conception, some 10 months before the birth of the twins formed the basis of the court's decision that they were legitimate. There was also a lengthy discussion in the opinion as to whether the possibility of access in itself proved that intercourse had occurred, and whether the conditions of meeting of the parties in itself were such as to afford such an opportunity. To cloud the issue, further it was also shown that the mother had an illegitimate child before her marriage, and the attempt was made to show that she had promiscuous tendencies, to say the least. In any event, the court ignored these peripheral issues, although previously discussing them, in its final decision that the minor twins were entitled to their share of the inheritance on the basis that they were presumed legitimate.

The opposite result occurred in 1984 in *Estate of Cornelious*, in California before the Supreme Court there. In this instance, a 27-year-old woman sought to be declared the executrix of an estate left by the late Willis Cornelious. At the time of her birth, her mother was married to another man, and continued to be so until the man's death. Despite that the mother admitted to the daughter that Willis was her true father, although she never gave this information to her husband who continued to believe he *was* the father until his death. Willis and the 27-year-old established a relationship in which it was shown that he acknowledged his paternity. The court referred to the presumption of legitimacy and ruled that the plaintiff had no legal share of Willis's estate. What is of concern to a geneticist is not only the adamant stand in favor of Lord Mansfield, but the additional *genetic* evidence proving that the woman could not have been the daughter of the late husband. She was a carrier for sickle cell trait, and neither her mother nor the husband were carriers. Barring mutation, an exceedingly rare and most unlikely assumption here, there is no way the plaintiff could have been the biological child of the deceased husband.

The issue of the length of a pregnancy arose in the 1942 case, *Dazey v. Dazey*. Here the District Court of Appeal, Second District, of California, dealt with the determination of which of two husbands fathered a child. The first husband was divorced and the mother remarried less than a month afterwards. Subsequently she produced a child, born 225 days after the second marriage. A year later the mother died and the second husband then remarried, and he and his wife then sought to adopt the child on the basis that it was not his and must have been fathered by husband number one. The court relied upon the presumption of legitimacy as noted above, and added in a different vein, "In the mysteries of birth, and of life, and of death, Mother Nature follows no exact pattern. In dealing with man's length of life, the Bible speaks of three score years and ten as but an approximate average; our life expectancy tables are but averages. Living men and women and children are, within limitations, each different from the other. Indeed no two of us are, or ever have been, exactly alike. When we deal with mankind, there is no fixed standard of height or weight or conduct, or time. And so, in dealing with the mystery of the beginning of mortal life, we find no fixed number of days from conception to birth." They go on to cite cases where 304 days were held to be too long

a period for conception and 105 days too short. They
then do considerable arithmetic in this case to
determine that all things considered conception must
have occurred within 28 days of the mother's last
menstrual period, and conclude that conception most
probably had occurred the night of the remarriage of the
woman to her second husband. As the child was held to be
legitimate, the second husband is the legal father and
there is no need for adoption . What is amazing in this
case is that at no time was the issue of blood tests
raised which certainly could have been performed on the
second husband and the child.raised. Although of course,
no tests could have been obtained from the mother, there
is a chance that had certain patterns been seen in the
child and father exclusion might have been found.
Obviously in the earlier case involving length of
conception blood tests were not known, but here there is
the possibility, although somewhat remote, that
performance of such tests could add useful information
concerning the possible paternity of the second husband;
indeed tests of the first husband, had he been willing,
might also have been useful. One would have expected,
after the court's statements concerning the
individuality of each person, that such tests might have
been suggested. Also, the court overlooked   the fact
that not all individuals are genetically different;
identical twins must have identical genetic traits.

INTRODUCTION OF BLOOD TESTS
Before beginning to consider the paternity cases
involving the usage of various blood tests, it is
important to jump ahead in time for a moment to examine
the constitutional issues involved in blood testing. The
case which will be cited in all later cases involving

constitutional rights was one which did not involve
paternity at all. This case, *Schmerber v.California*
decided by the United States Supreme Court on appeal
from California in 1966, did not involve a genetic
issue, but dealt with the legality of a blood-alcohol
testing of an alleged drunken driver. The driver claimed
that taking of blood for this test was a violation of
his right of due process, his right to counsel, his
right against self-incrimination, and his right not to
be subjected to undue searches and seizures, in other
words most of the amendments to the Bill of Rights. It
should be pointed out that the tests showed the driver
to be intoxicated and this evidence was sufficient to
convict him. It is equally obvious that had the evidence
been the other way Schmerber probably would not have
appealed the case! The opinion, delivered by the court
in considerable detail, carefully examined each issue
and determined that no violation of Schmerber's rights
occurred.

The due process claim was rejected on the basis of
a previous opinion in which the blood sample had been
taken from an unconscious driver with due care that it
was done properly and that it "did not offend that
'sense of justice' of which we spoke [in a previous
case]" The self-incrimination claim was similarly
rejected. "We hold that the privilege protects an
accused only from being compelled to testify against
himself, or otherwise provide the State with testimonial
or communicative nature, and the withdrawal of blood and
use of the analysis in question in this case did not
involve compulsion to these ends."

The right to counsel claim was not as easy;
Schmerber had obtained a lawyer prior to the blood

tests, and the lawyer had advised him not to submit.
Nonetheless, the tests were carried out against counsel's
advice. "Since petitioner was not entitled to assert the
privilege, he has no greater right because counsel
erroneously advised him that he could assert it." The
search and seizure claim was dealt with as follows:
"History and precedent have required that we today
reject the claim that the Self-Incrimination Clause of
the Fifth Amendment requires the human body in all
circumstances to be held inviolate against state
expeditions seeking evidence of crime. But if compulsory
administration of a blood test does not implicate the
Fifth Amendment, it plainly involves the broadly
conceived reach of a search and seizure under the Fourth
Amendment.

"Because we are dealing with intrusions into the
human body rather than with state interferences with
property relationships or private papers--'houses,
papers, and effects'--we write on a clean slate...We
begin with the assumption that once the privilege
against self-incrimination has been found not to bar
compelled intrusions into the body for blood to be
analyzed for alcohol content, the Fourth Amendment's
proper function is to constrain not against all
intrusions as such, but against intrusions which are not
justified in the circumstances, or which are made in an
improper manner. In other words the questions which we
must decide in this case are whether the police were
justified in requiring petitioner to submit to the blood
test, and whether the means and procedures employed in
taking his blood respected the Fourth Amendment. ...
Similarly we are satisfied that the test chosen to
measure petitioner's blood-alcohol level was a

reasonable one. Extraction of blood samples for testing
is a highly effective means of determining the degree to
which a person is under the influence of alcohol. Such
tests are a commonplace in these days of periodic
physical examinations and experience with them teaches
that the quantity of blood extracted is minimal, and
that for most people the procedure involves virtually no
risk, trauma, or pain. Petitioner is not one of the few
who on grounds of fear, concern for health, or religious
scruple might prefer some other means of testing, such
as the 'breathanalyser' test petitioner refused." Thus
Justice Brennan dealt with, and disposed of the civil
rights issues. Justice Harlan concurred, and stated,
"While agreeing with the Court that the taking of this
blood test involved no testimonial compulsion, I would
go further and hold that apart from this consideration
the case in no way implicates the Fifth Amendment."
Chief Justice Warren entered a brief dissent, while
Justice Black with the concurrence of Justice Douglas
entered a lengthy one, based mainly on the Fifth
Amendment, stating in part, "To compel a person to
submit to testing [by lie detectors, for example] in
which an effort will be made to determine his guilt or
innocence on the basis of physiological response,
whether willed or not, is to evoke the spirit and
history of the Fifth Amendment. Such situations call to
mind the principle that the protection of the privilege
'is as broad as the mischief against which it seeks to
guard.'" Justice Douglas adds a brief paragraph, ending,
"No clearer invasion of this right of privacy can be
imagined than forcible blood-letting of the kind
involved here.' Likewise, Justice Fortas dissents
briefly, saying in part, "As prosecutor, the State has

no right to commit any kind of violence upon the person, or to utilize the results of such a tort, and the extraction of blood, over protest, is an act of violence." The lengthy quotations from this series of opinions is necessary as the problem of the constitutionality of enforced blood tests for paternity will be a serious and continued objection to their use in many of the paternity cases which arise after this 1966 opinion. It will appear that, despite the dissenting opinions, this case will serve to permit such tests and constitutional grounds for objecting to the use of blood tests will not stand the scrutiny of the courts.

We can now turn to the actual paternity cases, both before and after *Schmerber*. In a mixture of paternity and criminal law, the case of *State v. Damm* arose twice in the Supreme Court of South Dakota, first in 1933 and again in 1936. This case, one of the first to invoke the legal use of blood tests for paternity, also involved a charge of second degree rape by the defendant upon one of his adopted daughters. Damm was convicted and sentenced to sixteen years' imprisonment. In this early case involving blood tests, the appellant asked for such tests to be carried out to demonstrate his non-paternity. The court discussed blood tests in some detail, and even included a table of possible exclusions based on genetic findings. However, expert witnesses were found who testified that the tests were not at that time generally accepted as infallible. The court cited British cases again, where the refusal to order such tests was made. There were other legal issues involved, which will appear in the next case, but the court's refusal to order blood tests was the main interest in this discussion. It seems somewhat peculiar that in this opinion the judges turned to foreign law, and in no way referred to the *Frye* doctrine.

A complete reversal by that court occurred in the second time the case came before them. After referring to their earlier decision, and in a sense more or less apologizing for it (there had been considerable criticism in this country concerning the court's inability to accept scientific evidence), the court then said, "We therefore say, without further elaboration or discussion, that it is our considered opinion that the reliability of the blood test is definitely, and indeed unanimously, established as a matter of expert scientific opinion entertained by authorities in the field, and we think the time has undoubtedly arrived when the results of such tests, made by competent persons and properly offered in evidence, should be deemed admissible in a court of justice whenever paternity is in issue." The Constitutional issue arose whether taking blood for such tests violated personal rights but the court determined that there would be no violation of the rights of the parties. "We perceive no valid reasons why courts of record may not require of any person within their jurisdiction the furnishing of a few drops of blood for test purposes when, in the opinion of the court, so to do will or may materially assist in administering justice in a pending matter... As has been often said, a trial judge should not be a mere umpire presiding over a game of skill conducted by counsel. It is difficult to see any reason why a trial judge might not of his own motion in a given case, if he thought it likely to be helpful, order the making of a blood test by a competent person and a report thereon

under oath subject to cross-examination by both
parties." Unfortunately for the defendant, the court
then decided there was no evidence that, "if the
requested order was granted defendant could and would
have the tests made by a competent, capable and
experienced person. The making of these tests
(particularly for those for agglutinogens M and N...) is
an appreciably more difficult and complicated matter
than the mere matching of blood for the purpose of
transfusion....The record before us embraces no
satisfactory showing that the medical practitioners whom
defendant proposed should make the tests were
experienced in so doing or were competent technicians
for the purpose." For those reasons, and those alone,
the court for a second time refused to order the blood
tests necessary and, guilty or no, Mr. Damm remained in
prison. The case is of great interest, as it is one of
the earliest to recognize the power of blood tests, and
it does suggest, in the second decision, the eating of
much judicial crow. But the final decision seems to be
one of *stare decisis* to the nth degree; by denying the
tests on the basis of the incompetence of the testers,
they could uphold the earlier guilty decision. That the
court could not have suggested the ordering of the tests
by competent, if necessary court-appointed or
recommended, physicians seems a rather peculiar
interpretation of court duties. Certainly, such a
decision based on incompetence, alleged or otherwise, by
the person selected to conduct the tests would seem
ridiculous to most today, and in a sense does not even
follow the part of the opinion quoted above.

One of the cases decided in the year between the
two Damm cases is *Beuschel v. Manowitz*. The case arose

in the Supreme Court of New York, Kings County, in 1934. The defendant in this case was accused of carnal assault which resulted in the birth of a child by the plaintiff. The defendant naturally denied all charges and moved for an order requiring the plaintiff and the child to undergo blood tests. One of the interesting sidelights to the case was also the argument by the plaintiff's counsel that the child be exhibited to the jury so that the presumed likeness of the father and child be observed, something clearly prohibited by the rules of evidence in New York. (See above cases dealing with personal appearances.) The court, having denied this illegal request, then turned its attention solely to the matter of the request for blood typing, and whether this was a clearly established medico-legal procedure. In dealing with this, the court established its philosophy by saying, "Law and jurisprudence, which are something more than the dry tomes of the past, can be understood by considering fundamental principles not only of government and economics but also at times by giving consideration in particular cases to sociology, medicine, or other sciences, philosophy and history. New concepts must beat down the crystallized resistance of the legally trained mind that always seeks precedent before the new is accepted into the law. Frequently we must look ahead and not backwards.

"In the instant application the court is not proceeding to anticipate scientific facts, but rather to act judicially upon the basis of scientific facts already ascertained. Speaking before the New York Academy of Medicine, BENJAMIN N. CARDOZA, then Chief Justice of the Court of Appeals and now Associate Justice of the United States Supreme Court, said: 'We

turn at times to physiology or embryology or chemistry
or medicine--to a Jenner, or a Pasteur, or a Virchow, or
a Lister, as freely or submissively as to a Blackstone
or a Coke.'" The judge then continues in a later part of
the opinion, "Naturally the courts will not permit the
application of scientific tests which have not attained
definite and dependable results accepted generally by
those qualified to judge." Obviously the reference is to
Frye. Further the court in ordering the blood tests
stated, "...It is not claimed that either plaintiff or
child would, in the slightest degree, be injured by
being submitted to a blood test...The plaintiff's
affadavit frankly states that she 'would for the benefit
of science, as well as for the purpose of aiding the
progress of law, be willing to submit herself and the
child to a test, provided certain astringent conditions
are laid down.'" The court turned to an opinion by the
Supreme Italian Court of Cassation which upheld the
reliability of such tests in excluding paternity,
although admitting the proof of paternity could not be
made. As a result the court ordered the blood tests to
be made.

However, on appeal, the Appellate Division reversed
this thoughtful opinion. Later in 1934, they stated, in
what may seem to be an epitome of judicial ignorance,
"Plaintiff may submit or not to the taking of her own
blood, but it plainly determines nothing. She asserts,
and no one would gainsay it, that she is the mother of
this child. A blood test of the defendant and the child
may possibly determine his nonpaternity, but it is not
claimed, as we understand the record, that such a blood
test would determine the defendant's paternity. The
child is not a party to this action; and while a court

of chancery has an inherent jurisdiction over the welfare of an infant, a ward of the court, nothing in this case indicates in the slightest that the welfare of this infant is in any wise involved or that the blood test could possibly be beneficial to the infant." Thus the order for the test was overruled. While one might agree with the decision that the child might not benefit, it would seem that the "welfare" of the defendant, when accused of carnal assault,who might be acquitted on the basis of blood tests of being the father of a resultant child, has been completely overlooked by the Appellate Division. The decision seems arbitrary and inherently unfair on this basis.

Another famous case, *Berry v. Chaplin*, also in the District Court of Appeals, Second District, arose in The lower court case involved, as one would expect, a good deal of showmanship, including having the most generously endowed Ms. Berry holding the child, next to Chaplin, in front of the jury who were most probably convinced that she was certainly well capable of nourishing the child. Chaplin had generously offered to pay all of the medical care and costs of the pregnancy provided Berry agree to submit herself and the forthcoming child to blood tests after its birth, a stipulation agreed to and carried out. In this case, blood tests were ordered by the court and proved conclusively that the famous comedian was not the father. Nevertheless, perhaps due in part to the above circumstances, the jury found Chaplin guilty and assessed him $75 per week for support of the child. The fact that Ms. Berry may have had relations with other men during the time when conception was likely also did not deter the jury from reaching its verdict and ignoring the blood test data. The court here cited an earlier California case in which it had been stated, "But the blood tests were not conclusive evidence. It was so declared in the only reported case in California in which blood tests were used for the purpose of attempting to determine the parentage of a child. In that case, the court held that the evidence concerning the blood test 'is expert opinion because the conclusions reached by the examiner are based upon medical research, and involve questions of chemistry and biology with which a layman is entirely unfamiliar,' but that such tests and the evidence thereof are not conclusive because not so declared by the code; and further, that expert testimony is to be given the weight to which it appears to be justly entitled." They also added that, "We are not impressed with the thought advanced by counsel for defendant that the sympathy of the jurors could have been aroused by the juxtaposition of the three parties in interest. The apprehension expressed by counsel that plaintiff in the arms of her mother caused compassionate visualization of the ancient masterpieces of 'Madonna and Child' is dispelled by the character of the evidence in the case which kept the minds of the jurors fixed on the unspiritual and terrestrial affairs of the mother and defendant." A concurring opinion put the matter into proper prospective. He concurred only on the basis of *stare decisis*, and then went on to say that were it not for that he would have decided otherwise. Indeed, the concurrence is in a sense more a dissent than a concurrence. "But modern science brought new aids. The microscope, electricity, Xray, psychology, psychiatry, chemistry and many other scientific means and

instrumentalities have revised the judicial guessing
game of the past into an institution approaching
accuracy in portraying the truth...If the courts do not
utilize these unimpeachable methods for acquiring
accurate knowledge of pertinent facts they neglect the
employment of available, potent agencies which serve to
avoid miscarriages of justice.

"In the case at bar a widely accepted scientific
method of determining parentage was applied. Its results
were definite. To reject the new and certain for the old
and uncertain does not tend to promote improvement in
the administration of justice." The judge here had
previously gone into detail upon the genetics of the
situation; obviously he would have decided oppositely
had he not felt bound by the earlier decision. It would
appear that this dissent, like so many others, was
written for the record, and to attempt to set a
precedent upon which later courts, dealing with similar
paternity suits, could then rely. The higher court thus
let the jury verdict of guilty stand and allowed the
jury to ignore completely the genetic fact that Chaplin
was *not* the father.

Another oddity in the early use of blood tests
arose in the case *Jordan v. Mace,* decided by the Supreme
Judicial Court of Maine in 1949. In this bastardy case,
blood tests were carried out upon non-identical twins,
their mother and the purported father. The tests showed
conclusively that the man could not be the father of one
of the twins, but might be the father of the other. The
court pointed out the tests had been carried out eleven
times with these same result. Said the court, "Second,
the father of twins must be one and the same man." They
went on to add, "We are not disposed to close our minds

to conclusions which science tells us are established. Nor do we propose to lay down as a rule of law that triers of the fact may reject what science says is true; for to do so would be to invite at some future time a conflict between scientific truth and *stare decisis* and in that contest the result could never be in doubt." The court thus decided that the accused was not, in fact, the father of the twins. The blood tests used here were both the standard A B O tests as well as the M and N tests. The fact that the woman had almost certainly had intercourse with the accused and stated firmly that she had not had sexual relationships with anyone else was also not pertinent in light of the blood tests. This case is cited here as a similar case will arise in the future where the introduction of more modern scientific methods of paternity may well indicate that this case was, in light of present day knowledge, wrongly decided. The blood tests here may very well have allowed a guilty party to escape the consequences of his actions.

The entire issue whether a court may require blood tests in paternity suits also arose early in the history of such cases, particularly in 1950 in the case *Cortese v. Cortese*, decided by the Superior Court of New Jersey, Appellate Division. In this case the wife brought suit to compel the husband to support an infant child which he claimed was not his. He in turn requested that the court compel the woman to undergo blood tests of herself and of the child. She refused to do so, and the lower court held that it had no power to compel her to take such action. The Appellate court reversed, holding that "in the absence of any reason for refusing to take a blood test, denial of husband's motion for an order to compel wife and child to take a blood grouping test was

an abuse of judicial discretion." There was nothing in
the record to indicate why the woman had refused the
tests. The opinion went on to state, "The citizen holds
his citizenship subject to the duty to furnish to the
courts, from time to time, and within reasonable limits,
such assistance as the courts may demand of him in their
effort to ascertain truth in controversies before them."
Interestingly, this citation is from State v. Damm,
mentioned previously. The court goes on to distinguish
this case from a criminal case and the privilege there
for refusing on the basis of self-incrimination. The use
of blood tests is shown. "As far as the accuracy,
reliability, dependability--even infallibility--of the
test are concerned, there is no longer
any controversy. The result of the test is universally
accepted by distinguished scientific and medical
authority. There is in fact no living authority of
repute, medical or legal, who may be cited adversely.
Furthermore, that the weight of enlightened legal
authority is in favor of according decisive evidentiary
effect to reliably reported blood test exclusions, is
shown by the favorable comment on instances of judicial
acceptance of exclusions, and the critical, deprecatory
view of judicial disregard of conclusions which have
appeared in law reviews throughout the country.

"We are not called on to resolve this question at
this time. The issue can arise only when and if a blood
test is made and the result excludes defendant's
paternity. It is sufficient we hold the interests of
justice require the trial courts to reconsider
defendant's motion in light of or opinion to the end
that valuable evidence relevant to the case may not be
overlooked if it may be obtained in a manner consistent

with the rights of the parties and of the child through
blood test properly performed." Thus the appellate court
in essence ordered that the blood tests be administered,
and the thrust of the opinion was clear; blood tests are
of major importance and are legally necessary in suits
of this kind.

A case in Ohio in 1957 supplies what may be an all-
time favorite citations from a court. This case, *State
v. Gray*, decided by the Juvenile Court of Ohio, Cuyahoga
County, was based on the right of a putative father to
be a witness against himself in a paternity suit. All of
the facts, except for the blood tests, indicated that
the man could have been the father. The case is of
scientific interest as the usual A-B-O tests, the M-N
tests, the Rh tests, and the Hr tests were inconclusive,
but the data on exclusion by the Rh factor C test showed
clearly that the accused man could not be the father.
The judge explained all of this with the following
statement, "The operation of blood tests to prove nonpaternity may be reduced to this simple homely
illustration: If an apple is found under an apple tree
the natural inference is that it came from that or some
other apple tree.

"However, if a pear is found under an apple tree,
it is clear under the laws of nature that some other
tree with pear-producing factors produced that pear, and
the apple tree under which the pear was found must
definitely be excluded as the parent-tree.

"At present, practical application of blood
grouping in bastardy cases is restricted to ruling out
paternity rather than attempting to establish it.
Applying this principle, it is apparent that the apple
could have come from any apple tree. Therefore it is

difficult and unsafe to establish which is the parent apple tree. It could have been one of several. But it is comparatively simple to exclude the apple tree as the parent of the pear found under it. Accordingly we must look for a tree with pear-producing factors as the parent tree.

"Great advances have been made in this science since 1939 when the Ohio law was enacted. In 1939 only two blood group systems were known. Since then at least seven blood-grouping systems have been discovered to supplement the earlier knowledge, and their inherent patterns have reached the point of reliability that there can hardly be any question about them. These advances have resulted in more and more courts in states with statutes similar to Ohio's recognizing the virtual infallibility of properly conducted blood grouping tests which exclude paternity. Increasingly the courts are giving judicial recognition to the reliability and accuracy of such tests."

"In accordance with enlightened judicial acceptance of the high value of blood-grouping tests properly conducted, I hold that in the absence of any competent proof that blood-grouping tests establishing non-paternity were not properly made, the results of such tests, scientifically conducted and objectively made by doctors expert in such field, should be given such great weight by the court that the exclusion of the defendant as the father of the child follows irresistibly." Thus the court determined the non-paternity of the defendant and despite the testimony of the mother that the accused must be the father because she had no sexual relations with any other male, Gray was acquitted on the basis of the scientific evidence that the battery of blood tests proved definitively that he could not be the genetic father of the child in question.

A 1960 case, *Kusior v. Silver*, decided by the Supreme Court of California, was probably the first in which blood tests were employed to overcome the presumption of legitimacy of a child born to a married couple. In this case, the child was born nine days after a final degree of divorce was entered. The couple had separated over eighteen months previously; although the husband visited the plaintiff and their daughter frequently after that, he testified that intercourse between the plaintiff and himself did not occur after the initial separation. There was also considerable evidence that several other men had spent time with the plaintiff. The plaintiff as usual testified that she had intercourse with the defendant and no others.

Blood tests showed conclusively that the child in question here could not have been sired by the former husband and did not rule out the defendant as a possible father. The main case for the defendant was based on the presumption of legitimacy rule and despite the blood tests the child should be declared to be that of the former husband. The court examined this claim thoroughly, and decided that the blood-typing tests were of sufficient scientific merit to overrule the presumption of legitimacy. There was a legal difficulty here, however. At this time the Uniform Paternity Code had been formulated, but when adopted by the California Legislature one section had not been included in the statutes. The section omitted read, "The presumption of legitimacy of a child born during wedlock is overcome if the court finds that the conclusions of all the experts, as disclosed by the evidence based upon the tests, show that the husband is not the father of the child." The court found another route to reach its decision. "We conclude that the result of blood tests taken under the Uniform Act may not be used to controvert the conclusive presumption of paternity created by subdivision 5 of section 1962 of the Code of Civil Procedure, but under the language of section 1980.6, if 'the conclusions of all the experts * * * are that the alleged father is not the father of the child, the question of paternity shall be resolved accordingly; where the tests so taken establish that the mother's husband could not be the father of the child the rebuttable presumption of paternity are conclusively rebutted." Thus the court found statutory basis for making a correct decision, and there is now a scientific method to overcome the formerly almost unshakeable presumption of legitimacy. All of the various testimony concerning access and behavior of the mother would not have been necessary, if this issue had been decided early in the opinion instead of at the end.

Nonetheless, in a case in 1961, only a year after the previous case, the District Court of Appeal, 2d District, in California, essentially reversed the previous decision. This case, *Wareham v. Wareham* involved a divorce also, and the birth of a child after the divorce. The main difference the court found here was that the court used a different interpretation of the meaning of the word "cohabiting" in its approach to the problem of the possible access of the husband to the wife. As in the previous case, blood tests showed conclusively the husband could not be the father of the child. The interesting point is that this court took its stand on the basis of the first part of the code cited previously, and not the latter part which formed the basis of the decision in *Kusior*. The facts of the two cases are so similar-divorce, admission that intercourse with another man, not the husband, had occurred during the presumed period of conception, the blood tests conclusively proving the husband could not be the father-that it is hard to believe that exactly opposite conclusions were reached. A concurring opinion reviewed all of the previous cases cited, plus scores more, and then agrees reluctantly with the majority opinion by saying, "In this case we have followed the ruling which denies to the husband the most potent, accurate and competent evidence known to man to insure the ascertainment of the truth; because there is no evidence known to the judicial process which is more lucid and scientifically certain than the blood grouping test when used to negate paternity.

"This whole situation is a perfect example of what occurs when truth and justice are not equated. Any law or decision which bypasses, ignores or disregards a manifest truth should be changed forthwith.

"We are placing our stamp of approval upon an outdated fiction of the law in the face of and against inexorable scientific facts. I do so reluctantly and because, as a member of an intermediate court, I am constrained to follow the law as it has been stated by the Supreme Court of this state." Here, there seems to be some confusion. It would seem that the *Kusior* court had not refused to find a way of avoiding the misuse of the presumption of legitimacy and the lower court in the present case did not seize upon that lead. As *Kusior* was so similar, it seems strange that a decision in that case which ruled that the husband was not the father and the decision in this case that the husband was the father is a fine example of legal legerdemain, and the two cases continue to be baffling. Furthermore, as later cases depend to a large degree on earlier opinions, future courts have no clear guidelines and are free, in a sense, to use either opinion as precedent.

Lest it be thought that early paternity cases only arose in the west, a 1954 decision by the Supreme Court of New Hampshire should be mentioned. In this case, *Groulx v. Groulx*, the court ruled that a petition by the mother annulling her marriage to her child's alleged father and for an order of custody and support of the child should not be granted. The genetic interest in this case came from the evidence showing nonpaternity by the use of the "S" factor, probably the first such case. The standard A-B-O, M-N, and Rh tests did not eliminate the putative father, but the tests showed that the blood cells of both mother and accused father did not contain the S factor, while the cells of the child did, a genetic impossibility. Much was made concerning the validity of the tests and the fact that the original testimony concerning the conclusiveness of the test was modified somewhat to be less strong, after consultation with Dr. Norman Weiner of New York, the discoverer of many blood types. Nonetheless the physician testifying was unshakeable in his belief that the tests were accurate and conclusive and the court accepted the fact that such tests were admissible as evidence and that the decision of the lower court based in part on them, as well as on the usual issue of access, was upheld. Another point was made that the accused father's name appeared on the birth certificate, and the court ruled that while this under law makes the town clerk's record of birth "*prima facie* evidence of that fact," it is essentially reputudiated by the blood tests.

Again in California, in 1966, the problem of admissibility of evidence from blood tests arose. In this paternity suit, *Huntingdon v. Crowley*, the issue was whether recently discovered blood tests should be excluded on the basis of *Frye*. The usual blood tests did not exclude paternity, but a newly found typing method, the Kell-Cellano tests, did. At that time only a few hundred such tests had been performed and the serum for such tests was rare, and in some instances might give misleading results. In fact at the time of the case the American Medical Association had recommended that these tests not be used in routine medicolegal practice but be reserved for special cases only. The court did not allow the results of the tests to be offered as evidence. However the higher court did remand the case on another basis, the fact that there was ample evidence, not fully utilized, to show that the mother had no lack of opportunity to have had intercourse with other men. A statement in the opinion is worth considering: "As sexual intercourse is seldom conducted in the presence of witnesses, however, that evidence is usually circumstantial. One such circumstance, of course, is the simple physical opportunity of the parties to have intercourse. That condition is a necessary one, but it is not sufficient in and of itself. The law also recognizes that animal instincts do not prevail in modern society, and that acts of sexual intercourse do not occur spontaneously when a man and a woman are merely brought into each other's presence; there must ordinarily be an element of disposition or desire on the part of each to have sexual relations with the other. The latter element, however, was omitted from the relevant instruction given to the jury in the case at bar." On this basis, the case was remanded, and the blood type evidence was ignored.

Another example of evidence admissible in these cases is derived from the 1966 case, *Matter of Crouse v. Crouse*. The Family Court of Nassau County, New York, had a support case before it and the married father refused support of one of the three children supposedly born of that union. A blood-grouping test was requested and performed. The test showed unequivocally that Mr. Crouse was not the father of that child. Objections to the introduction of the evidence, and the claims of the mother under the usual rules that children born to married parents are legitimate, formed the basis of the case. The judge not only ruled that the blood tests were admissible, but called Dr. Norman Weiner, mentioned previously as the foremost expert on certain blood tests, to testify in court concerning the tests. Dr. Weiner, who must have delighted in the opportunity, not only testified, but performed the actual tests in the court then and there, with full explanation, and with the expected results excluding paternity. The court accepted this evidence and overruled the presumption of paternity. Mr. Crouse willingly paid support for the two other children, but rightfully did not have to pay for the child in question. One minor gem from the opinion might be quoted here: "In such cases the courts should accept the decisiveness of a nonpaternity finding properly arrived at as it would accept the demonstrable fact that a mixture of blue and yellow colors will produce varying shades of green, but never a red color."

The judge obviously was not color blind.

Another aspect of the admissibility of blood tests came in the case *Anonymous v. Anonymous*, decided by the Court of Appeals of Arizona in 1969. Here the issue was again the paternity of a child presumably conceived before a divorce. However, the real issue hinged upon the fact that the tests were carried out in a laboratory, and unlike the previous case the physician responsible was not present when the tests were made. This was a case of first impression in Arizona, and the court took pains to examine all of the facts carefully. The opinion contains an explanation of Mendelian Genetics dealing with the situation: "A person's biological makeup (color of hair, physical characteristics, blood type, etc.) becomes fixed at the time of impregnation of the mother's ovum by the father's sperm. The inheritance of these biological substances is governed by genes, which occur in rod-like groups or lumps called chromosomes. ..." The judge continues to explain genetics at some length. The exclusionary blood evidence was also explained by the testifying physician:"The putative father was grouped as a little 'c'. The putative mother was grouped as a big 'C' and a little 'c' and the child was grouped as a big 'C' * * *.

"The child has two big 'Cs'; consequently there was no place for this child to have gotten the two big 'Cs'. He got--we know where he got one, from the mother because it was demonstrated there but it must have gotten the big--other big 'C' elsewhere." The mother had previously agreed by stipulation to accept the results of the blood test, with no objection to how it was done, but the court felt the need for caution in this first case and also found the need to be sure of the manner in which the tests were conducted. As the physician was not present at the time of the tests, the court felt that the case should be remanded for this testimony and also decided that the stipulation by the mother was not a legally valid one. The court had no objection to the use of blood tests to overrule the presumption of legitimacy in this case, but wished to be certain that proper procedures in carrying out the blood tests had been performed. After all, mistakes can be made in laboratories, and in later cases have been made (see for example *Curlander* cited in the Wrongful Life Chapter). Particularly in a case of first impression the court wished to be absolutely certain of its legal grounds, and the remanding of the case was only for the opportunity to present adequate testimonial information concerning the carrying out of the tests.

Still another issue which arises in paternity cases is the time at which they may be brought, in other words the statute of limitations. This was the principle issue, along with admissibility of blood test evidence in the case, *Matter of Mores v. Feel*, decided by the Family Court of New York County, New York, in 1973. Here the statutes of New York concerning these cases does not permit the initiation of such a paternity suit after the infant is two years old if the initiator is one of the alleged parents. However, the statute runs for 10 years if the petitioner is a public welfare official or if the alleged father has previously admitted paternity. The issue here is a case brought for paternity of three children; the plaintiff claimed the defendant had admitted paternity and even though more than two years had elapsed, the statute of limitations was tolled thereby. Needless to say, the alleged father vigorously denied such an admission. The suit was brought by the Commissioner of Social Services and the mother sought permission to join as copetitioner. Considerable legal maneuvering took place in the case, including a request for blood testing by the alleged father. The court cast a cool eye on this request. "Respondent's main source of exculpatory evidence in such a case will be the exclusionary results of blood-grouping tests. But even with the recent discovery of new blood types, these tests are still capable of exonerating 6 out of every 10 incorrectly accused men. This leaves 4 of every 10 innocent respondents threatened with a false finding of paternity unless they can marshall some evidence to rebut petitioner's claim." Obviously this judge was not aware of the genetics of the situation, and while his statement is correct, it could be looked at in a different light, i.e., if the putative father is indeed innocent there was a 60% chance, at that time, of exclusion. Nonetheless, the main issue was the question whether the mother could join the suit as copetitioner, and no decision was made concerning paternity at this time. The decision was to allow her to be a copetitioner.

There appears to be no further information as to the outcome of this case, possibly as the names of the participants were assumed names in order to protect the interests of the child.

Another New York case dealt with the doctrine of "laches," or as Black's Law Dictionary puts it, "...the maxim that equity aids the vigilant and not those who slumber on their rights." If an action is not brought in proper time, then the "estoppel by laches" applies,;

i.e., the case can not be brought because of failure to bring it at a proper time. It is somewhat similar to the statute of limitations, but the latter is fixed by statute. The case here, *Hansom v. Hansom*, was dealt with by the Family Court of Richmond County, N.Y., in 1973. It again dealt with the paternity of children born after the separation (but not the divorce) of the parents, and with blood tests establishing the nonpaternity of the male. There were three children at the time of separation, and the woman continued on to deliver three more children in the subsequent nine years. The man never acknowledged paternity of the three latter children and never supported any of the children except when under court orders to do so. The wife kept adding each child born after separation to the demands for child support. Unfortunately the husband, without benefit of counsel, and apparently without understanding the procedures at all, orally admitted paternity of the last three children. Finally, in 1973 he obtained counsel and obtained-blood grouping tests for the additional three children. He was shown not to be the father of the first two. The woman claimed that despite the separation, her husband visited her regularly and that sexual relations between the two continued during the times when she conceived. She claimed she had no such relations with any other man. He claimed that he had no such sexual relations with her at any time following the separation. The court had demanded support from him for all six of the children, although no filiation procedures had been instituted. The woman was on welfare in addition to obtaining a minimum support from the man. The claim was now made that he was prevented from making his nonpaternity case by the estoppel of laches. The issue was thus whether he could now institute proceedings to prove nonpaternity. The contested claims of the times of sexual relations led the court at one point to state, "To pinpoint with complete and unerring accuracy the number and frequency of sexual relations almost 20 years ago would strain the credulity of even a benighted backwoodsman to the breaking point." The court ruled that the basic legal issue was that this was not a nonpaternity claim on his part, but rather a suit brought under the laws dealing with child support, and this would not be subject to laches as the Family Court has *continued jurisdiction* and may order a blood-grouping test at any time before such proceedings are concluded with *finality*. The court ruled that "Uncontradicted evidence of a blood-grouping test is conclusive as to nonpaternity. In the instant case, the doctor who conducted the test testified. His qualifications as an expert were conceded and, despite a long cross-examination by the Corporation Counsel, there was no effective rebuttal of his testimony and report." The court then decided that despite the presumption of legitimacy the man could not be the father of two of the later children, and he could not be excluded from being the father of one of them. The man had agreed that if the tests did not exclude him, he would be bound to support the children. The court added that the 2-year limit of the statute of limitations, previously referred to, did not hold here as this was not a paternity but rather a support case. The doctrine of estoppel by laches did not hold either. The final decision was therefore that he would support the three children born before separation, and one of the three children born subsequent to that time. Of course, as noted previously, the parents were never divorced, and the decision here again used blood tests to overcome the presumption of legitimacy.

Another issue arose in *Matter of Brian v. Johns*, tried before the Family Court, New York County, N.Y., in 1974. Here the issue was whether a court had the power to order a second blood test when the requester of the original test was dissatisfied with its results. The woman in this case wished to declare the man the father of her illegitimate child. He requested, and received, the right for blood grouping and it was determined that he could not be the father. The woman then sought a court order for a second test. The court at first was willing, but upon second thought denied the request on the grounds there was no authority by statute or court decision to permit such an order. The court agreed that the reliability and accuracy of the first test was a fit subject for determination, citing the case in which Dr. Weiner had appeared, and stated, "The trial is to proceed and the respondent is to produce the doctor who performed the test as a witness to explain the basis for his conclusion so that the petitioner may question the test and the analysis of the result. The court recognizes the fact that there may be a margin of error in the making of these tests. Since the legislature has not made a blood test exclusion the only evidence on the issue of paternity, petitioner will therefore be given the opportunity to rebut this evidence." In other words, the woman may have her day in court, but she will have to produce overwhelming evidence that the blood test is inaccurate and that the accused man was the father of the child.

Two cases in North Carolina involved the same

defendant and the same facts. The first, *State v. Camp*, decided by the Supreme Court of North Carolina in 1974, was a straight-forward case of non-support of a child alleged by the mother to be that of Camp's. The blood-grouping tests absolved him of paternity; however, at that time the state laws did not give full weight to such tests, but stated that they were only part of the evidence which could be used. The claim was that when the lower court judge charged the jury he stated this fact. The only witness in the case for the State was the mother who testified that she had had intercourse only with the defendant and no others. The sole defense witness was a physician who testified as to the validity of the blood tests. Apparently the jury, aided by the charge of the judge, felt, that the blood tests notwithstanding, Camp was the father. The Appellate court ordered a new trial on the basis of the charge and the appeal by Camp came to the Supreme Court. The Court of Appeals had declared the instructions to the jury as being erroneous, and said, "...under the laws of genetics and heredity a man and a woman of blood group "O" cannot possibly have a child of blood group "A", and that if they believed the testimony of the doctor and believed the tests were properly administered, it would be their duty to return a verdict of not guilty." The Supreme Court reviewed many cases from other jurisdictions, but fell back upon the North Carolina statutes which, as noted previously, only stated that the results of such exclusionary tests were only part of the evidence and were not by themselves sufficient. The issue of public policy again was raised by the court; if the results of exclusionary evidence from blood tests were to be conclusive, it was up to the legislature to change the statute, and not the duty of the court to do so. A dissent was entered: "The majority opinion depicts the present state of the law and is unquestionably correct unless we are prepared to take judicial notice of the laws of heredity. I think we should judicially recognize these hereditary laws and my dissent is based solely on that ground. The medical profession apparently admits that, theoretically, due to possible mutation of the genes, two parents with type O blood might produce a child with type A blood in one out of 50,000 to 100,000 cases. Notwithstanding this 'theoretical exception,' when the quantum of proof required to convict is 'beyond a reasonable doubt,' the possibility of error is so infinitesimal that the tests should be accepted as infallible when they exclude the defendant as father of the child. The administration of justice is not aided by a rule of evidence, such as ours, which permits a jury in its unbridled discretion to ignore scientific facts and base its verdict on testimony which, according to Mendel's Law, is false 49,999 out of every 50,000 times! It is my view that Mendel's Law of Hereditary Characteristics is so notoriously true as to exclude reasonable dispute and its accuracy and reliability has been demonstrated by readily accessible scientific sources of indisputable accuracy." The dissent, notwithstanding, still left the defendant as the legal father of an infant which was not his.

In 1980 the case arose again in the same court. The problem here was that when all was said and done, Camp had been found guilty and placed on probation for the crime of bastardy in lieu of a six-month sentence. In addition he was required to pay $15 per week for the support of the bastard. Apparently he made only two such payments and refused to make more on the ground that he was not the father. After being cited many times for nonpayment, he was haled into court, his probation was revoked and he was sentenced to serve his term.

The court again reviewed the facts presented in the first case, but the major reason for the decision, overturning revocation of probation, was that for five years he was available, and made no attempt to conceal himself, and that the five-year probationary period (the maximum allowed by law) had expired. Interestingly, the same judge who filed the dissent in the first case wrote the unanimous opinion in this case, holding that the trial judge had no right to revoke the probation after it had expired and when no effort had been made to do so earlier. Thus, in the long run, justice triumphed; Camp went free and did not have to pay for the upkeep of a child not his. One must remark that it is a rather roundabout method of obtaining justice. Another sidelight is the fact that for some reason in the first case the defendant was referred to as Samuel Lee Camp, and five years later as Sammie Lee Camp. Perhaps, based on the second opinion, familiarity removes contempt.

The question of how many times a blood test should be performed, along with other issues, arose in the case, *Torino v. Cruz* (again fictitious names), decided by the Family Court, Bronx County, N.Y., in 1975. The case had originally begun in 1971 but, as can be seen, was delayed repeatedly. This was a straight-forward paternity case, in which Torino accused Cruz of being the father of her child. A blood test was ordered in 1971, and paternity was excluded by the M-N series. There was apparently some question as to these results and the court in 1972 ordered a second test, this time performed by our star performer, Dr. Norman Weiner. This test did not exclude Cruz from paternity. At this point, again in 1972, Cruz requested a third test which confirmed Weiner's non-exclusionary results. The court, still not certain, then ordered a fourth test again in 1972. This test again did not exclude Cruz. It is of interest to note that three of the four tests, including the first which had shown exclusion, were performed by the same physician. The problem with the first test was admittedly the use of weak serum. Still another test was made in 1974 by this physician and again he received results which excluded Cruz. By this time, all must have been confused, to say the least. Testimony from both Dr. Weiner and the other physician was heard, and there was no apparent resolution and the court held that the net result of the series of tests was such that paternity could not be excluded. However, the court heard other evidence upon which it ultimately based its conclusions; first in order for the paternity claim against Cruz to be valid the child would have to have been born after a 300-day pregnancy. The physician called on by Torino stated that this was possible, but "that a 300-day pregnancy did not occur 'every day in the week' and that the probability became less with each day after the average gestation period of 240 to 280 days." This lengthy pregnancy would be necessary for Cruz to be the father as it was established that he was in Puerto Rico while Torino was in New York at the supposed time of conception, given a normal term pregnancy. Besides this there was evidence that Torino had not been exactly sexually inactive while Cruz was away. In addition, the plaintiff had not offered in evidence hospital records concerning her pregnancy even though they had been subpoenaed by her and were in court. All in all, the court decided that Cruz was not the father based on the weight of evidence, and the confusion regarding the blood tests was thereby not the main source of the decision. But the case does point out the necessity for accurate blood tests, and the need for physicians to be certain that their sera used for these tests are active.

Another legal point arose in the case *Matter of Esther T.*, decided by the Surrogate Court, Nassau County, N.Y., in 1976. This case involved paternity determination so that a child of a deceased woman could inherit. In addition, there was a problem in proper identification of the woman in question. It was claimed that the child was neither the natural nor the adopted child of this family; however, the testimony from a brother and a sister of the deceased indicated that this was her son. The basic problem of identification arose as apparently the hospital records referred to an Esther Du, while she was known to others as Esther Dru. There was also a conflict in the hospital records concerning blood typing, with one record reported as type O and another as type AB. In addition, there was testimony from those who knew Dru that she had never been observed to be pregnant. The court, confronted with this series of confusions, decided that the dead woman was not the same person for whom the hospital had records, and based on all the other evidence the child for whom the inheritance was sought could not have been hers and was therefore not entitled to a share of the inheritance. This case has been used to show how, again, uncertainty in the records can lead to confusion in the courts.

A different point of law arose in *Franklin v.*

District Court, tried before the Supreme Court of Colorado in 1974. Here, the issue was not the results of blood typing tests, but rather who should pay for them. The defendant in a paternity case filed a motion that the court should order such tests and that the cost should not be borne by him as he was an indigent. The lower court agreed that the tests should be performed, but held he must pay for them. The appeal to the Supreme Court was based on the Fourteenth Amendment, under which he claimed his rights were violated as he could not pay and therefore the tests would not be carried out. The Supreme Court agreed with the plaintiff and held that the court should have appointed a physician to carry out the tests, and if the plaintiff were truly indigent the costs should have been made a county charge. "The right to have blood tests performed cannot be denied an indigent defendant without violating the equal protection clause of the Fourteenth Amendment of the United States Constitution.

"Accordingly, the rule is made absolute, and the district court is directed to determine whether the petitioner is, in fact, indigent and unable to pay for blood-grouping tests and, if so, to order the blood-grouping tests be made at county expense." The court in this brief opinion also made a strong statement concerning the use of blood-grouping tests and pointed out, "If the results of the blood-grouping tests are undisputed and in the defendant's favor, the defendant cannot be found to be the father. It has been estimated that with the aid of blood grouping tests, based upon each of the three blood type classification, A-B-O, M-N, and Rh-hR, a man falsely accused has a 55% chance of proving his nonpaternity." The right for blood tests in

paternity suits would now seem to be established as a
constitutional privilege not to be denied those who
request it.

The case of *Serafin v. Serafin*, decided by the
Supreme Court of Michigan in 1977, should have caused
the demise of Lord Mansfield's rule. This paternity
case, while blood types were not involved, is important
in dealing with the basic problems of access and the
presumption of legitimacy. The case involved a divorce
and the claim by the former wife for support of a child
presumedly born in wedlock. The basis of the defense
was nonaccess by the husband. But more importantly the
entire basis of the rule was challenged. The court made
many remarks, which, while adding length to this
chapter, are necessarily quoted for future
determinations by this and other courts. Among the early
phrases is the reference to the rule, "The law of
England is clear; that the declarations of a father or
mother cannot be admitted to bastardize the issue born
after marriage. ***As to the time of the birth, the
father and mother are the most proper witnesses to prove
it. But it is a rule, founded in decency, morality,and
policy, that they shall not be permitted to say after
marriage that they have no connection, and therefore
that the offspring is spurious. ...The Mansfield
rule...works for the peace of the community and society
generally. ...we think the prime reason for the rule is
as stated--that it is against public policy to give
testimony bastardizing their issue. We are satisfied
that further adherence to Lord Mansfield's rule cannot
validly be premised on the assertion that it operates to
prevent increased enrollment on public welfare lists.
But even assuming that it has such an effect, and apart

from the due process objections that might be raised
against such a policy, we say with the Supreme Court of
Maine: 'We are not persuaded that the public treasury
should be protected by foisting upon a husband the
support of a child obviously not his.'...familial
tranquility might be more readily destroyed by forcing a
husband to support a child that in fact is not his,
while protecting his wife and her paramour who engaged
in extramarital activity in gross violation of the
marital relationship." In other words, evidence should
be allowed to speak for itself, rather than having the
court bound by the rule. The court then states, "Further
when a court voluntarily blindfolds itself to what every
citizen can see, the public must justifiably question
the administration of law to just that extent." The
equal protection clause of the Fourteenth Amendment is
cited to "invalidate the arbitrary distinction drawn
between illegitimate and legitimate children as regards
substantive rights." The court is here referring to
rights of inheritance..."Michigan statutes now provide
that an illegitimate may inherit from his or her
mother." A concurring opinion was entered. " I write
separately to emphasize the interests of the children
who will be affected by the rule's abolition and to
emphasize the continuing strength and validity of the
presumption of legitimacy. ...Despite these enlightened
advances, there still are, unfortunately, social
distinctions made between the legitimate and illegitimate
child which continue to stigmatize the illegitimate
child and scar his or her psychological development. We
need no learned treatise to know that many children
brandished as illegitimate suffer painful and sometimes
crippling emotional damage at the hands of cruel or

thoughtless peers and adults. The word 'bastard' has not yet lost its sting to the children against whom it is too often applied. ...It is no accident that many of these children strike back by committing antisocial or criminal acts. ...The presumption speaks for the child whose future is at stake,even though he or she is not a party to the legal proceedings. It is distinct from Lord Mansfield's Rule and is not affected by today's decision. ...In short, even after today's decision abolishing Lord Mansfield's Rule, a very strong presumption of legitimacy will continue to protect the otherwise defenseless child. The husband who seeks to establish that he is not the father of a child born during the marriage still faces a formidable task.

"Two other minor aspects of this case deserve some attention. First, the trial judge refused to order a blood test for the purpose of determining whether Mr. Serafin could be the father of the child. Such an order should be granted on remand if Mrs. Serafin continues to allege that Mr. Serafin is the father and if such an order is requested. Second Mr. Serafin's military duty records and evidence indicating the weather conditions during the alleged time of access were not admitted at the first trial. These items, if otherwise admissible, should be admitted on remand." Although this information is not cited in the case, it would seem that the claim of lack of access due to being on military duty, and prohibited by meteoric conditions from being present at the time of conception were made in the lower court. What is important here is that the Michigan Court specifically noted that its decision would and did overthrow many earlier cases which had been therefore wrongfully decided by the rule. In addition, the court, it would seem, implies that blood tests must be required and that the rule cannot stand inviolate in those cases when all other evidence shows that the putative father cannot be rightfully accused of paternity. The proof of illegitimacy must be "very convincing," but if it can be made then the age-old rule must fall to modern evidence.

Another aspect of paternity law and the red blood cell tests deals with cases involved with immigration. The exemplar case here is *Oi Lan Lee v. Director of Immigration and Naturalization Service*, decided by the United States Court of Appeals, Ninth Circuit, in 1978. Here the issue was whether an oriental couple, claiming a child still abroad, be granted preference status for immigration. The Immigration Bureau was suspicious of the claim of paternity by an oriental couple in the United States.

The Immigration officials agreed to the preferred status, provided all parties submitted to blood-grouping tests. The two "parents" both were type O; the child in question was type AB; obviously it was impossible for the child to be theirs, and the Immigration Bureau then denied preferred status for the child. The couple brought suit on the basis that the blood-grouping test order was not proper, and that the results of the tests were not conclusive. The court denied both claims, first on the basis that the blood grouping tests were all but conclusive, and secondly that the Bureau had not acted beyond its powers in ordering such tests, and the claims by the alien residents, the Lees, were denied. It is obvious that such procedures, while somewhat different from the ordinary investigations in such cases, were necessary here; had the blood grouping tests shown no exclusion the Bureau would undoubtedly have permitted

the entrance of the child, but given the evidence which
now faced them, they had no choice, under their mandate,
but to deny permission. Blood-grouping tests are
henceforth valid for determination of immigration status
when claims for paternity are made.

The case *Matter of Carol B. v. Felder*, decided by
the Family Court, New York County, N.Y., in 1978, raised
yet another legal point in paternity suits. This suit
began in 1973 and a blood test was administered in 1974
which appeared to preclude paternity on the part of the
accused male. A second test was requested and denied on
the basis that, while a court had the right to order one
blood test, there was nothing in the law permitting the
ordering of a second test. The case was continuously
adjourned for three years due the illness of the mother's
attorney. When it became apparent in 1977 that the case
would have to be further adjourned, the original
petition was dismissed without prejudice. In 1978, the
petitioner requested that the order of dismissal be
vacated and requested a second blood test. The court
agreed to the first request and also granted the request
for a second blood test. The basis for the decision was
that a change in the New York State Law had been made
and the request for the second test was no longer barred
by any interpretation of the new statute. The court
stated, "Such exclusionary testing results, if reliably
and accurately achieved, are not subject to varying
expert interpretation nor are they colored by the
personal interest and emotion which pervades testimonial
evidence; rather such results are firmly premised upon
immutable laws of genetics and constitute scientific
fact.

"It is this nearly conclusive nature of

exclusionary blood test results which requires this
court to scrutinize carefully the testing methodology
employed to assure that the results are reliable and
trustworthy. Given the long passage of time since the
blood test herein was conducted, the simplest and most
reliable method of confirming its accuracy is through a
second blood test performed by another doctor. This
court is cognizant the multiple testing has, on
occasion, disclosed erroneous initial findings of
exclusion resulting from subpar testing procedures." The
court cited *Torino* as its authority for the last
statement. It also rejected summarily the claim by the
accused father of any violation of his Fifth Amendment
right against self-incrimination. The major point here,
of course, would be the difficulty of establishing the
validity of the tests made many years ago, and the long
delay in this case, through no fault of the petitioner,
was held to be of sufficient importance as to almost
require a second test. Of course, the change in the
statute in the interim also mitigated for the petitioner
as there was no longer any legal bar to permitting the
second test to be made. It might seem somewhat .
extraordinary that a change in the law made after a case
had been originally decided and then delayed should have
been accepted by the court, as the usual procedure might
have been first to determine if the legal change was
meant by the legislature to be retroactive; no
consideration of the intent of the legislature in this
regard was mentioned in the opinion.

By 1979 it was estimated that over 10% of all
births were illegitimate. Consequently, due to the huge
amounts of money needed for child support, the Federal
Government ordered that all states develop plans for the

ascertainment of paternity if the states wish to
participate in Federal Programs. The types of tests
needed are well cited in *State v. Ortloff*, decided by
the Supreme Court of Minnesota in that year. The case
itself is relatively simple; no blood tests were
requested by the putative father, and evidence of his
failure to request such tests was accidentally (?)
entered during cross-examination of the father. The
lower court sustained an objection by the father's
lawyer, but the fact that the question was even asked was
claimed to be prejudicial to his trial. The Supreme
Court decided it was not. But the main reason for citing
this case is the long citation concerning blood tests
given by the deciding judge. "The Committee recommends
the use of up to seven different blood tests, using more
general tests first, with the testing on the samples
proceeding in stages, stopping only after the putative
father has been excluded as the father or all seven
tests have been used. If the blood is subjected to all
seven of the screening tests recommended by the
committee, there is a 91-to 93-percent probability that
if the putative father is in fact not the father, one of
the seven tests will have eliminated him as the father.
In those cases in which the use of all seven tests does
not exclude the putative father and he still maintains
that he is not the father the committee recommends that
certain scientific formulas be used to calculate the
likelihood that he *is* the father. Basically what the
expert does is estimate the likelihood that the putative
father is the actual father by taking the frequency of
the blood constellation of the putative father among
real fathers for that mother/child combination and
compares it with the frequency of the father's genetic

constellation in the general population." The judge goes
on to point out that where there were previously few
laboratories in the country where these tests could be
done, there are now many federally established regional
laboratories throughout the country, including one in
the state here involved. This case is of importance as
it may foreshadow the admission of evidence pointing to
the proof of paternity and the establishment of the use
of the HLA antigen test for determining paternity as
this is the first in which such a possibility is
mentioned. It will be remembered that even then evidence
pointing toward paternity was not admissible, and this
is the first suggestion that it may become so in the
future.

Another case in which the putative father failed to
obtain blood tests arose in The Court of Appeals in
North Carolina in 1979. In *State v. White*, a husband was
living with his wife up to 265 days before she gave
birth to a child. Three days after he left her, she
began an adulterous relationship with another man, with
whom she thus had been living for 262 days prior to the
child's birth. There was also evidence that the husband
was not impotent, and the only question was which of the
two men was the true father. The court pointed out that
no blood tests, which might have resolved this quite
readily, were requested, and that the presumption of
legitimacy was therefore conclusive in this case. Again,
the importance of using all available evidence,
particularly blood-grouping tests, is shown.

Another attempt to introduce actual probability of
paternity was made in *Dodd v. Henkel*, decided by the
California Court of Appeal, First District, in 1978.
Here the appellant made an in *camera* offer to produce an

expert who would testify that the man was "among the 14 1/2 to 15 percent of the population that could be the father." The judge decided that this evidence was of no probative value and might unduly influence a jury and ruled the evidence could not be introduced. The appellant sought to have it admitted, pointing to the valid use of such evidence in criminal cases. The Appeal Court noted the difference between criminal cases and paternity cases and upheld the lower court's decision not to admit the evidence. The law of the state did say that such evidence might be admitted at the discretion of the trial judge, but did not make it mandatory and the higher court simply agreed with the discretionary power of the lower court.

Another case which illustrates the need to establish paternity by some means arose in The United States Supreme Court in 1978. In *Lalli v. Lalli* the issue was whether a grown son could share in the estate of his deceased father. Unfortunately no steps had been taken to establish paternity under the law during the lifetime of the father. Despite the fact that the deceased had referred to the child as his son in some non-legal document, the Supreme Court ruled against the son's claim for a share of the estate on the basis that this claim, however legitimate it might be in this case, would allow any individual to come forth after the death of a person and claim to be that person's child. All sorts of suits might arise and the courts could become clogged with fraudulent claims of this kind. Again, the need for some type of filiation proceedings during the lifetime of the parent is apparent. As a matter of fact, the father could simply have gone to court, declared his paternity and the whole matter would have been resolved.

There would be no need for blood grouping tests or other evidence if the deceased Lalli had taken this step.

HUMAN LEUCOCYTE ANTIGEN TESTS

The HLA (Human Leucocyte Antigen) test is the most sensitive of all blood-typing tests. There are a very large number of such antigens, at least forty and the reliability of such tests is far higher than any other test known. The tests had originally been developed for use in organ transplants in an attempt to be certain that rejection of the organ by the recipient would be less likely to occur; at the time of the development of this sensitive test no thought of its use for paternity cases was made. However, as will be seen, the HLA tests, and later blood enzyme tests, will prove to be the most potent methods for both establishment of, and disproof of, paternity. The great variety and relative rarity of various combinations of these types of antigens make the tests highly probative. The rarer the frequency of the various different possible antigens, the greater is the confidence in these tests. The presence of an antigen in a child and the finding of this type in a parent are almost certain indications that the suspected father is indeed the father; conversely the lack of the antigen type in the child makes nonpaternity all but certain. It is possible to ascertain with a high degree of accuracy (over 90% or more) the probability of a man's paternity. As the frequency of these plus other antigens varies from population to population, it is also necessary to use separate probability data for each ethnic group involved, and tables of frequencies of these antigens for most groups have been prepared from large population surveys.

Perhaps a further word on the issue of probability
should be given. If the probability of one event is
independent of another event, the total probability for
the two events is the product of the two probabilities.
To give an example all are familiar with, the
probability of a coin landing with "heads" up is 1/2.
The probability of the next flip of the coin is also 1/2
for heads. Thus the probability of having two successive
tosses result in heads is simply 1/2 x 1/2, or 1/4. If
the probability of each of two antigen types is rare,
say 1/1000, then finding the two antigens in an
individual, if they are inherited independently from one
another, would yield a probability of 1/1000 x 1/1000, a
very small number. Given the known probability of each
antigen type in a general population of putative
parents, then the probability of finding that exact
combination of antigens in an offspring can readily be
calculated. Further, if that number is known, and the
putative father had access to the mother, the chances
that he is not the true father are indeed slim. Again,
it should be emphasized that these probabilities for
various antigen types vary from group to group and
this must also be taken into account in the
determination by the courts. In some racial groups
various frequencies of different antigens are found, and
to show the true probability of the father being justly
or unjustly accused the racial group should, and is,
also taken into account.

In *State v. Meacham*, decided by the Supreme Court
of Washington in 1980, the constitutional issue of blood
withdrawal for paternity testing again arose. Actually
this is a consolidated case arising from two separate

cases, in which similar issues were involved. Two cases
of paternity arose and in each the court ordered blood
tests to be taken. Both men protested that such tests
violated their constitutional right to personal privacy,
freedom from unreasonable search and seizures and
freedom of religion. It is also of interest that this
case was among the first which recognized the so-called
HLA test for paternity. The test at the time of trial
was relatively new to the courts. However, the issue
here was not the validity of the tests, which will be
dealt with in later cases, but the constitutionality of
such tests and whether the ordering of them violates the
rights claimed by the putative fathers in this case. The
court here falls back, as was suggested earlier, upon the
decision in *Schmerber v. California*, referred to
previously. The court, citing another case also, stated,
"First Amendment religious freedom embraces two
concepts: freedom to believe and freedom to act. The
concept of ordered liberty precludes allowing every
person to adhere to his or her private standards of
conduct in which society has important interests." But
equally important is the right of the child. "We
recognize the lurking problems with respect to proof of
paternity. Those problems are not to be lightly brushed
aside, but neither can they be made into an impenetrable
barrier that works to shield otherwise invidious
discrimination. ...Neither of these appellants has
denied having sexual intercourse with the particular
mother concerned at about the time conception is alleged
to have occurred. Had such a denial been made, it would
have been incumbent upon the court to hold a hearing to
determine that issue prior to ordering submission to a
blood test. The trial court should be satisfied, at

least *prima facie*, of the fact of sexual intercourse during the appropriate time period as a condition to requiring submission to a blood test. That is, however, not an issue in these cases." Although this other issue is raised, it is clear that the basic issue was that of the religious objections the two purported fathers raised. The court, obviously following *Schmerber*, refused to recognize this as an issue in the case, and rather turned its attention to the question of access, which was not contested at all in this case. Thus the issue of the constitutionality of ordering of a blood-grouping test was not a plea recognized by the court. Of most interest will be the role of HLA antigen cases arising in future cases, and the influence these will have on court decisions both as to establishing nonpaternity, and, more importantly, upon establishing the basis of paternity in cases to come.

A landmark case concerning the use of blood tests arose in *Malvasi v. Malvasi*, decided in 1979 by the Superior Court of New Jersey, Chancery Division (Matrimonial). A putative father sought court order to have the mother submit to an HLA antigen test, and to use such medical evidence in the determination of paternity. Again, it is necessary to quote at length from this short opinion as the court defined the circumstances so well. "Human Leucocyte Antigen (HLA) typing, a method of tissue-typing, detects antigens on white blood cells (leucocytes). An antigen is any substance which can stimulate antibody production when introduced into another individual. Antigens are produced under genetic control by genes. This phenomenon permits the affirmative use of blood test evidence to show parentage. With blood-grouping tests the

probability of a nonexcluded male being the actual
father is usually over 90%. This high degree of
discrimination in either excluding or including, with a
high probability, a given male is a result of the
extreme diversity of HLA types in the population."
Citing from Teraski (q.v.) the court adds, "Most people
are 'rare' types because only about one out of a
thousand people will have a similar HLA type.
Consequently this relatively rare type can be looked for
in the child of any given mating. If the child has the
same rare type as the putative father, the man is likely
to be the actual biological father. On the other hand,
if the putative father is wrongly accused, he can
usually be excluded because the child would have
inherited a different rare form from the actual father."

"The courts are willing and desirous to utilize
scientific techniques to aid them in the discovery of
truth and the administration of justice. However, before
such new scientific techniques can be utilized, it is
essential that they achieve general recognition in
science as to their accuracy. "The acceptance of HLA
testing in the scientific community as reliable and
accurate is evinced in the joint *AMA-ABA* guidelines for
serological testing in paternity cases where the HLA
test was recommended as the most powerful single
paternity test for exclusion....In view of the
scientific community's recognition of the reliability
and accuracy of HLA testing and its important probative
value where paternity is in issue, the court will
consider such medical evidence, together with other
proofs, in deciding parentage."

Almost the entire opinion in this case is cited as
it will be the basis for changes in the laws of most

states in the years to come; it is the first case in which evidence pointing *toward* determination of paternity rather than only exclusion of paternity by blood grouping tests is deemed to be legally acceptable. Further like the somewhat veiled reference in other cases, the HLA test is now definitely admissible in court cases. Interestingly enough, although the decision concerning admissibility closely follows the *Frye* rule, that decision was not cited. But nonetheless, evidence establishing paternity with a high degree of accuracy is now, at least in New Jersey, admissible.

The same issue arose, also in 1979, in *Cramer v. Morrison*, in the Court of Appeal, Fourth District, in California. Here the issue was again the use of HLA antigen tests in a paternity case. The lower court granted the defendant's motion to exclude the results of this test as performed by Dr. Paul Teraski, mentioned above. There was no claim of lack of expertise here, but the claim was made that the laws of California did not permit use of such evidence, and the appropriate code dealing with use of blood tests to exclude, but not prove, paternity was cited. After the exclusion of this evidence, the putative father was acquitted. The testimony of Dr. Teraski determined that "...there was a 98.3% percent probability that defendant was the father, i.e., only 1.7% of the population could be the father of this child and defendant was within that group." Dr. Teraski noted also that this test, which he had performed in some 260 cases, was widely used in Europe in paternity cases. He also noted that the test had originally been developed for use in typing for organ transport operations and was in general medical use. The entire issue centered upon the admissibility of the

evidence presented by the HLA tests. The court found a
novel way around the statutory prohibition of probative
evidence. "In our opinion, the drafters of the Uniform
Act did not have in mind tests of the nature of the HLA.
Initially, the terminology which the statute employs in
referring to blood tests--'blood types'-is that commonly
applied to the Landsteiner series of red cell grouping
in tests. ...As Dr. Teraski's testimony shows, the HLA
test is not one of the Landsteiner series, nor is it
even a red cell blood-grouping test. It involves tissue
typing of the white blood cells and results in far
higher probabilities of paternity than those yielded by
any of the blood grouping tests." The court deals at
length with the fact that the California legislature, as
mentioned early, had omitted the section dealing with
the use of blood tests to establish paternity, but
concluded, "Even if we were to accept defendant's
contention that omission of the blood test section from
the Uniform Parentage Act indicates that the Legislature
meant to exclude blood test evidence to prove paternity,
we would interpret the omission to refer not to tests
such as the HLA, but to the standard Landsteiner blood
grouping tests. In 1975 [when the legislative action was
taken] the Landsteiner tests were still the standard
tests for paternity. Expert testimony in this case shows
that the HLA test was not in use for paternity testing
in this state in 1975. ...We conclude that California
law does not compel exclusion of the results of the HLA
test to prove paternity." The second major issue in this
case was the use of statistical probability. The
accused, naturally, in his claim of nonadmissibility of
the evidence, objected to the use of this type of
evidence on the basis that it was not probative, but only

showed a statistical probability of his parentage. The higher court felt evidence of this kind should be admitted. "Paternity is now determined by extraordinary flimsy evidence for which there is no quantitative measure of value." Citing an article in the *Journal of Legal Medicine*, the court adds, "Rather than relying on such evidence, the law should not ignore readily obtainable genetic evidence that can provide a precise and objective basis for deciding such an important question as the paternity of a child.

"We are not aware of any public policy which would require exclusion of highly probative scientific evidence on the issue of paternity in the absence of a statutory mandate. On the contrary, all the policy considerations advanced by California commentators and case law argue for broad inclusion of such evidence in paternity proceedings. These considerations include protecting children from the stigma of illegitimacy, preservation of the family, and insuring that individuals rather than the government bear responsibility for child support. Further, we note that the recent and growing recognition of the right of illegitimates to equal protection with legitimates before the law has served to heighten the significance of paternity litigation throughout the United States."

The final issue dealt with in this case is the acceptability of the HLA by the scientific community. The issue was dismissed, after citing *Frye*, as it had not arisen in the original suit, and the trial judge's decision not to admit the evidence was not based on its validity, but on the problem whether evidence tending to prove paternity was admissible under law. Nonetheless, the Appeals Court felt it necessary to state, "Here the question whether the reliability of the HLA test to prove paternity has attained general acceptance in the scientific community presents a mixed question of fact and law to be determined from the testimony of qualified experts in the field, by reference to legal and scientific publications and journals on the subject, and from relevant judicial decisions." The court then goes on to cite several such journals and points out that even if the issue of validity of the tests had been raised, there would have been no dearth of evidence to show that HLA antigen tests are indeed valid.

The case for use of HLA tests now has been made in two states in 1979. The courts have been willing to accept by their ruling that these tests have met the *Frye* standards. Also the courts have admitted evidence dealing with probability, another major step towards providing justice to those involved in paternity suits.

Despite these two cases, other states were averse to accept this kind of decision. In 1979, the Supreme Court of Colorado in the case *Figueroa v. Juvenile Court* was not as yet ready to depart from the rule to admit only that evidence which would disprove paternity. Although the case at hand dealt mainly with other issues, namely the costs of the tests and who should pay for them, judicial recognition of the exclusionary provision was given.

Similarly, other cases involving payment for tests continued to arise in 1979. *Matter of Russo v. Hafner*, decided by the Family Court of Monroe County, N.Y. dealt with the issue of whether a minor, accused of paternity should pay the costs or whether his parents were responsible. The minor was an unemployed high school student and could not pay. The court found that the

costs of such tests were legally no different from the
usual costs of feeding, clothing, housing, etc., of a
minor, and withheld judgment on the case until a
determination was made to see whether the parents could
afford the payments.

Another issue involving paternity arose in
California again in 1979. In this case criminal law was
also involved as the accused in *People v. Thompson* was
found guilty of non-support of a child supposedly
fathered by him. He denied the paternity, but offered no
real proof of this. The only evidence he claimed was
that if his wife were called upon to testify she would
admit to having doubts that he was the father. No blood
tests were either requested or ordered. The court had to
consider whether the presumption of legitimacy extended
from paternity cases to criminal cases, and the fact
that in the latter cases guilt must be proved "beyond a
reasonable doubt." The court so decided, and Thompson
was thereby not only ruled to be the father, but was
guilty of the criminal charge of non-support.

The question whether an indigent was entitled to
have the costs of HLA testing paid by Department of
Social Services in New York arose in *Lascaris v. Laredo*,
heard by the Family Court of Onandaga County in 1979.
The court studied the facts that the standard blood
tests did not exculpate the accused, and that the
additional costs of the HLA tests were relatively high,
at the time $100. The court reviewed the use of the
additional test and found that the probative value of
this test in excluding the accused was sufficiently high
that the costs must be assumed by the public. It should
be noted, however, that the tests were ordered for
exclusionary evidence only.

The case of *Thompson v. Thompson*, brought before
the Court of Appeal in Maryland in 1979, dealt mainly
with the constitutionality of a two-year statute of
limitations dealing with the bringing of paternity
suits. There was a further involvement with blood
grouping; however, the court upheld the statute of
limitations as not violating the Fourteenth Amendment
and dealt only briefly with the blood-grouping issue.
The mother had failed to bring the case in due time and
claimed that if she were allowed to do so she would have
tried to introduce blood tests, implying, but not
specifically naming the HLA tests. The court, having
decided that she could not bring her case at all did,
however, comment briefly on the blood tests, and pointed
out the inadmissibility of anything except evidence
pointing to exclusion. They were thus really not faced
squarely with the issue of blood tests of this type, and
it is not at all clear what the court would have done
had this been an issue.

The issue of the right of cross-examination of
experts arose also in 1979 in the case *People v. Yarn*,
decided by the Appellate Court of Illinois, First
District. Here the initial test had excluded the
putative father but a second test was requested and the
putative father refused. The initial test was made by a
hospital which, as a matter of policy, refused to
provide their personnel as witnesses in such cases. Both
parties were aware of this policy at the time the tests
were made. The court here pointed out that regardless of
the policy of the hospital the lower court had the power
to summon these hospital persons, and that should have
been done. As a result, the higher court remanded the

case and said that the lower court must allow the
presence of expert witnesses according to Illinois
statutes, and the second blood test should be carried
out so as to allow this. Thus the plaintiff is accorded
those rights.

The Court of Appeals of Wisconsin in 1979 decided
that the HLA tests to establish paternity were not
permissible evidence under the then laws of the state.
In *J.B. v. A.F.*, the lower court permitted an expert to
testify that the use of the HLA test established
paternity of the accused with a 99% certainty. The
appellant, in this case the accused father, argued that
HLA tests are blood tests because blood is drawn; the
counter argument was that they are not blood tests as
they could have been taken without drawing blood. (This
is because any tissue is now known to demonstrate the
presence of the various antigens). There was also some
discussion that "It is not blood itself that is tested.
It is rather the constituents of blood which are
determined and from which conclusions are drawn based on
Mendelian laws of inheritance. ...HLA testing involves
antigens found in most tissues of the body, including
the liver and kidneys, and not just the blood as in the
ABO and other systems. The fact that the HLA test as to
plaintiff and Joshua [the child in this case] is based
upon their blood samples is not decisive as to whether
an HLA test is a 'blood test.'" Having neatly avoided
deciding just what type of test the HLA is, the court
then turned to the legality of admissibility of evidence
and found that regardless of the type of test, evidence
tending to prove paternity was prohibited. But the court
did take legal notice of the dilemma. It pointed out at
length the value of the HLA system and then stated,

"Births out of wedlock in Wisconsin rose from 1,174 in 1935 ...or 2.6% of total births, to 8,446 in 1978 or 12.4% of total births.

"It may be that the restrictive approach in Sec.885.23, Stats, to medical tests for paternity should be reviewed because of medical advances and changed social conditions between 1935 and the present. We cannot, however, construe the negative use permitted by the state to include a positive purpose which the statute forbids. Such a construction would work an amendment which may not be made by the court of appeals." The court, although given the opportunity, did not see fit to adopt the approach taken by the California court which essentially did amend the statutes by ruling that blood tests as construed were restricted to the Landsteiner typing.

The court then reviewed the rest of the evidence, including the fact that two men had ample opportunity to father the child and finally that ruled the bulk of the evidence, HLA not being admissible, showed that A.F. could be the father.

The admission of HLA evidence continues to play an increasing role in paternity cases as can be seen in *Matter of Goodrich v. Norman* which occurred in Family Court of New York County in 1979 also. There were three issues actually involved. "(1) Does a petitioner have the right to institute a paternity proceeding when he is not certain he is the father; (2) does the petitioner have a right to a court order for the Human Leucocyte Antigen (HLA) test in addition to the standard ABO blood grouping test; and (3) when are the results of the HLA test admissible as evidence when the test does not exclude paternity." Part of the problem was that the mother in this case was deemed emotionally unable to care for the child and if the putative father was not the real father then the state would be forced to care for it. The putative father was not trying to shirk his responsibility; indeed he agreed to assume full support should the child be his. The court decided affirmatively on the first point, his right to a determination; the mother had not requested it, and despite a Federal requirement dealing aid for child support, the Commissioner of Social Services had not instituted paternity proceedings either. As the court put it, the man was "left in limbo." The second issue, that of HLA testing, was also favorably treated by the court which recognized that there was a better than 90% chance of his being excluded should he not be the father and the tests were carried out. The court pointed out, "One of the elementary scientific rules of heredity is that a person may not possess a genetic factor which is lacking in both of his or her real parents. The doctor found no hint of maternal exclusion and therefore assumed that Alice Normal was the true mother. However, the doctor found that the child had at least three antigens in the HLA system and one in the RH system, none of which were present in either adult. When these factors were considered with the results of routine blood tests performed by the doctor, he concluded that paternity was definitely excluded."

The court then did not have to deal with the admission of evidence showing paternity. Nonetheless, foreseeing the likelihood that such cases would arise, and considering then existing statutes of the New York Laws prohibiting the use of such evidence, the court urged the legislature to reexamine the laws on the basis that as written they might well be held a violation of due process of law.

A case involving contempt of court and HLA tests arose in *Rachal v. Rachal* before the District of Columbia Court of Appeals in 1980. Here the matter dealt with was the refusal of the mother to appear for court-ordered HLA tests. The man sought divorce on the grounds of adultery and a declaratory judgment that the child was not his son. The lower court held the woman in contempt for failing to appear for the ordered test, and in addition did the same when she failed to appear again at an appointed date for the tests on her and the child. In addition, she was also denied the right to visit her older child now in the father's custody. The Appeals Court struck this order on the basis that the lower court did not have the power to hold the mother in contempt and also that denial of visitation rights would be to the detriment of the older child. The basis for this was the statute that a court could not hold a person in contempt for failure to appear for a physical examination, and the interpretation that HLA tests were indeed a "physical examination." They also held that there was not any attention given to the best interests of the older child in prohibiting the visitation, and indeed in awarding custody to the man involved. There seems here a most peculiar set of laws involved, as in other cases the right of an accused to request blood tests has never been refused on the grounds found in this case. Nevertheless, the higher court refused to uphold the contempt order. It also seems somewhat peculiar that no attention was paid to the decision in *Schmerber* by the U. S. Supreme Court ruling that blood tests were legal. The decision here that "physical" examinations were exempt would seem not pertinent in that the high court had ruled that blood tests were not a violation of any rights of an accused. One can only wonder whether this decision was known to the deciding court in this case.

Further use of HLA evidence occurred in *Matter of Moore* before the Family Court of Westchester County, N.Y., in 1980. The plea was for a second blood test to attempt to prove nonpaternity made by an accused male. The problem here arose from the fact that he wished the test performed at no cost to him. At the time of the initiation of the trial, neither the defendant nor his counsel objected to the use of the usual blood-grouping tests, nor did they request the HLA test. By now, the costs of such testing had risen from the $100 mentioned previously to $500, and the court pointed out that, "It is clear that the burden to be borne by public assistance would become insurmountable if the cost of a second blood-grouping test was paid by the county at the request of each respondent in a paternity proceeding who was not excluded by the standard test. Judicial notice is taken of this court's official records which indicate that 1,416 suits were filed in Westchester County Family Court in 1978 and 1,190 were filed in 1979. While many of the proceedings did not involve a blood-grouping test, it is readily apparent that if the county were held responsible for the additional expense of the HLA test in each case where a respondent claimed he was unable to pay this substantial expense, the potential liability to the public purse would be enormous... While this court agrees with the general proposition that one major purpose of paternity proceedings is to shift the burden of support of the child from the State

to the putative father, it cannot agree that additional
blood tests are valuable in protecting the public from
'collusion between a respondent with empty pockets and a
person in need of public assistance.'" Nonetheless,
after this fiat, the court then decided that this case
fell within the rule of "special circumstances" and,
while denying the motion for payment by the county on
its face, allowed the second test to be carried out with
a further hearing to determine whether the respondent
was able to pay these costs, and presumably if he was
not to allow the costs to be borne by the county. The
reasoning was interesting, as the court took into
account the fact that at the time of the initial
paternity trial the accused was 14 years old and that
the case had been delayed by unforeseen circumstances
for a number of years, the "special circumstances"
referred to above. Had the accused not been a minor, and
had the request for the second test, at the county's
expense been made by an adult, such circumstances would
not have been applicable; it might be suggested that the
court here interpreted the law on an *ad hoc* basis.

In *De Weese v. Unick*, decided by the Court of
Appeal, Second District in California in 1980, the issue
arose whether a case could be reopened for an HLA test
four years after it had been decided. The accused father
had been adjudged to be the father, and had been ordered
to pay support for the child. The accused had not been
excluded by standard blood tests and had then stipulated
that he was indeed the father and agreed to pay child
support. The mother, some four years later, requested
modification of the amount of support and at this time
the father requested a new blood test using the recently
developed HLA system. In the meanwhile, the father had

agreed to pay the increased amount. The man then claimed
that despite this agreement on his part, he was entitled
to demand a new blood test. For support, he cited *Cramer
v. Morrison* and the California court's acknowledgment of
the superiority of this test. His request was denied on
the basis of *res judicata*, i.e., the matter had already
been settled and therefore he had no legal claim to
reopen the case now. He would have had the usual legal
right of appeal within 60 days after the previous
decision, but his failure to do so excluded a reopening
of the case almost four and one half years later.

The acceptance of HLA tests for proof of paternity
was definitively accepted in New Jersey in *Camden v.
Kellner* by the New Jersey Court of Juvenile and Domestic
Relations in 1980. While the court here cited *Malvasi v.
Malvasi*, mentioned above, it pointed out that that case
was not clearly defined in terms of the use of HLA tests
and went into some detail as to the by now general
acceptance of the results of such tests by the
scientific community. It more or less used the same
logic in comparing the Landsteiner grouping tests for
red blood cells and the HLA test for white cells as had
the Cramer court in California, and decided that the
probability of positive evidence of paternity would be
legally acceptable. The probability of paternity in this
case was determined to be 99.88%!

Recognition of the HLA tests in the state of
Florida came also in 1980 in *Simons v. Jorg*, tried
before the District Court of Appeals, Second District.
The lower court had ordered such tests in a paternity
suit and granted the motion of Simons to compel the
test. Jorg refused to do so and this appeal came on the
basis that such evidence would be inadmissible. The

Court of Appeals upheld the requirement for the HLA
test, but avoided the direct issue of whether the use of
the test would result in evidence to be admitted by the
court, although they noted that the California and New
Jersey courts had allowed its use.

They noted, however, "With HLA testing, if a male
is not excluded as the father, the probability of his
being the father is usually over 90%. In 16% of the
cases, the probability exceeds 99%. ...Therefore we
agree with the trial court that good cause was shown to
compel petitioner to submit to the blood test. At this
stage of the case, we may not and do not reach the
question of the admissibility of evidence of the HLA
test results at trial, nor can we anticipate what weight
the trier of fact would give such medical evidence, if
admissible, together with other proofs which may be
offered in deciding the issue of paternity." In essence
what the court did here was to say that the evidence
must be obtained and then there could be a later hearing
on whether it was to be used.

Twice, on the same day, the Supreme Court of
Minnesota rendered opinions concerning the use of HLA
tests. The second, *State v. Denny*S, dealt with a
paternity case in which no blood-typing tests of any
kind had been introduced as evidence, and the only issue
was the impeachability of a defense witness who claimed,
against the statement of the plaintiff, that he had
intercourse with her at about the time of her
conception, and the putative father had stopped
cohabiting with her prior to that time. The court was
somewhat angry, stating, "We have repeatedly stated that
recently developed blood types are the most reliable
means for making the determination of paternity more

accurate. ...Use of these blood tests may make trial unnecessary in many cases by facilitating settlement before trial. Thus, we strongly reiterate our belief that recently developed blood tests should be ordered whenever possible to obtain them in adjudications of paternity."

In the above opinion, the court referred to its first opinion in the case, *Ramsey v. S.M.F.* Here, while the issue was somewhat beclouded by problems of the irrelevancy of testimony seeking to determine the sex life of the mother, the court, ruling that the questions asked of her should not have been allowed, went on to deal with the HLA issue. By now, 1980, the Minnesota laws had been changed with certain alterations of the Uniform Paternity Act so that the new act read in part, "...Genetic and blood-test results, weighed in accordance with evidence, if available, of the statistical probability of the alleged father's paternity." The court noted that "This legislation makes the results of the more sophisticated blood tests admissible as evidence. It also establishes a procedure by which the trial courts can direct that such tests be taken. Under the statute, the court may, on its own motion, require such tests, and must order such tests if any party requests them; or if a party refuses to accept the court's pretrial recommendation.

"We can imagine no situation in which it would not be in the interest of a paternity plaintiff, whether it be the county, the mother or the child, to have blood tests taken. When such reliable evidence is available, it is no longer sensible to rely solely on customary, less reliable evidentiary techniques. We therefore believe that in every paternity case, the party bringing

the action should request the court to order blood tests
as early as possible in the litigation. ... Settlement
of these paternity cases prior to litigation or at the
earliest possible stage of litigation would be to the
benefit of the county, the child, the father, the
mother, the putative father, the courts and the public."
The court is thus trying to unclutter its docket, and
that of lower courts from suits which, if the HLA tests
are, used would most often be useless. There is therefore
an important new use of the HLA tests from a legal
viewpoint, that of reaching speedy results without
having to hold a lengthy trial which more often than not
might be appealed on some procedural grounds and thereby
lengthen the time it took to reach what should have been
a quick and proper decision. Unfortunately, this has not
turned out to be the procedure in most cases.

Nevertheless, some weeks after the Minnesota cases,
the Supreme Court of Utah remained unconvinced. In
Phillips and State of Utah v. Jackson, the Supreme Court
in that state held that the admission of HLA evidence
which showed a 97% probability that the man was the
father of the child in question was improper. There was
no question concerning the proper administration of the
tests nor of their relative reliability. The entire
question was one of admissibility of this evidence.
Despite citing the *Cramer*, *Malvasi*, and *Laredo* decisions
noted previously, the court here also found no statutory
objections to the admissibility of the evidence, but
applied the *Frye* standard once again and after a lengthy
opinion concluded that the plaintiff failed to establish
the general acceptance of such tests and rejected their
use as evidence. In light of what the opinion had to say
concerning the tests, one would have been at first

inclined to believe that the court was about to uphold their use, but, in a rather difficult to understand decision, declared them nonadmissible and remanded the paternity suit for retrial without this vital evidence. One judge did dissent and said essentially the same. "It is upon the basis of what is said in the main opinion and what has been said herein that in my judgment that there was no prejudicial error, because the receipt of such evidence was well within the latitude of discretion which should be allowed the trial court, and that, consequently, the judgment should be affirmed."

Late in 1980 the District Court of Florida, First District, in *Carlyon v. Weeks*, decided that HLA tests could be used in paternity cases. There was a specific reference to *Simon* (1979), and it was pointed out that this case differed sufficiently so as to make a different decision. In the former case the order for the HLA tests had been objected to; here no such objection was made, and the tests had been carried out. Carlyon not only was not excluded from paternity, but the court further stated, "The odds that a random white man is the father...are 1194 to 1. Stated another way, the plausibility of paternity is 99.9%, making paternity 'practically proved' in the terminology of Hummel." The court decided that the admissibility of the evidence was within the trial court's discretion and the HLA tests are now valid for proof of paternity in Florida.

The Florida Courts again recognized HLA testing in *McQueen v. Stratton* in another District Court, this time the Second District Court of Appeals. The circumstances were somewhat different in that a man adjudged to be the father and assessed the costs of future and past child support appealed that the tests should not have been admitted as evidence. The court now contrasted the two previously cited Florida opinions (*Simon* and *Carlyon*) and adopted the opinion of the latter, upholding thereby the use of HLA tests for a second time.

Colorado was the site of another HLA case in 1980. The Supreme Court in *R. McG. and C.W. v. J.W. and W.W.* had a much more complex case to deal with. Here the "natural" father claimed paternity as against the claims of a husband and wife that the child was theirs. In addition, the problem of the timing of the suit was an issue, as well as the legality of a male seeking determination of paternity instead of the usual claim against a man. The Uniform Paternity Act allowed a woman up to four years to bring such a suit, but did not specify such rights for a man. The court dealt with this by declaring that this was a violation of the equal protection clause and the man was therefore entitled to his claim within the 4-year period. The HLA test showed that the probability of the claimant being the father was 98.89%. However, insufficient cells were obtained from the husband to make the HLA tests. In addition, the claimant stated that he had regular intercourse with the mother and that both she and he had plans to divorce their respective spouses and then to get married. The married parents who by now obviously had been reconciled claimed that the suit would violate their constitutional rights of privacy. The court did not see this case as such a violation. The final decision was to remand the case to the juvenile court with order to proceed "in a manner consistent with the views expressed herein." Both a concurring opinion, based on the constitutionality issue, and a dissenting opinion, based on the presumption of legitimacy, were filed. In addition, the dissent disagreed on the interpretation of the constitutional issue. "It may be objected that to draw a line between any factual circumstances in which a third-party father cannot be denied a hearing to challenge the presumption of the husband's paternity, and the presumption of the husband's paternity and factual circumstances in such a hearing would be a difficult task. This is undoubtedly true. Litmus test certainty in application has never been the criterion for adoption of rules of constitutional adjudication. Development of limit of constitutional protections on the basis of case by case determinations is a traditional role of the courts. We should not shirk the task with respect to the issue involved here." Within the long debate on constitutional rights is hidden the fact that the court took judicial notice of the HLA tests, and, although not clearly stated, it would appear that such evidence is now to be considered by the courts. The troubling issue here was the granting of court permission for a third party to challenge the presumption of legitimacy and thereby the stability of a marriage and of a family. If any person could make a claim such as the one here, the courts might find themselves filled with such cases. However, if the court of original jurisdiction were to insist on HLA tests prior to trial, as suggested previously, many would either not be brought, or could be disposed of without trial.

In a case of disputed paternity, *Hrouda v. Winne*, before the Appellate Division in New York in 1980, a man, not the husband, applied for custody of the child supposedly born to a married couple who had been divorced. The woman died suddenly subsequent to the

divorce and the former husband took custody of the
child. Hrouda brought suit, originally in 1972, claiming
paternity. For some reason the court-ordered blood tests
at that time were never taken, and the presumption of
legitimacy prevailed. Hrouda, continuing to maintain
that he was the true father, learned of the HLA tests in
1979 and reopened the case, requesting that the test be
administered to the surviving husband and the child. The
lower court directed the tests to be carried out.
Meanwhile, during the 8 years between the cases, the
N.Y. law had been changed concerning who was entitled to
request blood tests. The law formerly read, "the
respondent," but now stated that "any party" could
request such tests. However, the law did not specify
that it was to be retroactive. The court, in refusing
Hrouda's appeal, stated quite simply, "...We perceive no
reason in law or logic that would warrant the reopening
of a final order upon the mere promise of increased
exclusionary accuracy that might be derived from new
scientific developments in the field. To allow such
procedures would place a child, the custodians, and
putative parents in continual limbo." A tangential issue
that the suit had been brought the second time in the
wrong court was also noted. But the interesting thing
here is the court citing the HLA tests only for
exclusionary purposes, and no mention at all is made of
the possibility or probability of paternity. In any
event, Hrouda's right to reopen the case was firmly
denied.

New Jersey again in late 1980 reaffirmed the use of
the tests. In *J. H. v. M. H.*, tried before the Superior
Court, Chancery Division, the issue of the admissibility
of the tests against the principle of presumption of

legitimacy again arose. In this case, which also was a divorce suit, the wife moved to have the court order the husband to submit to such a test. The lower court, treating this as a civil action, and not a paternity case, granted the request and ordered the test. The case also involved custody as the wife claimed the child was not sired by the husband, but by a third person, not named, and not a party to the case. The woman submitted to HLA tests on herself, the child, and the man she claimed was the true father. The evidence showed that the probability of this being true was 98.7%. The husband protested that the presumption of legitimacy, plus the New Jersey law prohibiting the use of blood tests for positive results, forbade introduction of the HLA results. The court looked at the decisions in *Malvasi* and *Cramer* and decided that the use of such evidence was substantially based. They went further in discussing the sociological issues, quoting an earlier judge who had said, "One would expect that a child has a natural yearning to know his true parentage. Every child has the need to feel rooted, to find himself, and to know his true origins. When such knowledge is denied, the child may resort to fantasy to fill the void. As the links to his past disappear with time, the search for his identity will become more difficult. The anxiety to learn what was in his past may be pathological, making it more difficult for the child to lead a useful life and to form meaningful relationships." The quote, as one may gather, was from a decision in an adoption case. The court went on to deal with the issue of legislative intent concerning the use of blood-test results. "Recognizing that...[New Jersey] made existing scientific blood testing available as an aid to

evidential proof in paternity actions, it would be
anomalous to conclude that the legislature thereby
intended to limit litigants and the courts of this state
from the benefits of future scientific tests developed
after the passage of that act in 1939. 'It is frequently
difficult for a draftsman of legislation to anticipate
all situations and to measure his words against them.
Hence cases inevitably arise in which a literal
application of the language used would lead to a result
incompatible with the legislative design.'" It would
appear again that the court was taking an activist
position in reinterpreting the legislative act, but
nonetheless the order for the HLA tests was upheld.

Legal constraints seem to be the main obstacle in
the New York Family Court, Kings County, finding that it
could not order further HLA tests in the case *Matter of
Edward K. v. Marcy R.*, again late in 1980. Here the
petitioner sought to have the court make a declaratory
judgment that a child born to a married couple was in
reality his. The standard Landsteiner tests did not
yield any useful evidence of exclusion of either the
petitioner or the husband, and the former requested
further blood tests, and also the declaratory judgment
of his paternity. He claimed the test would definitively
prove him to be the true father and would exclude the
husband, thus overcoming the presumption of legitimacy.
The court avoided the entire issue of the tests by
deciding that it did not have the power to make this
legal judgment and referred petitioner to the Supreme
Court. But the court did recognize the utility of the
HLA tests by saying, "Should the petitioner choose to
proceed in the Supreme Court, then of course the final
arbiter of the evidentiary use to be made of the HLA

test would be that court. While the law should proceed with deliberation it need not mark time while science marches on. Unfettered by the anachronistic statute handcuffing this court, it is hoped that the Supreme Court will recognize the valuable scientific use of the HLA test and exercise an expanded modernistic approach to its application to paternity proceedings." What this apparently amounted to was a plea to other courts to act where this court felt itself powerless.

Michigan still remained adamant against the introduction of evidence tending to prove paternity. In *Cardenas v. Chavez*, the Michigan Court of Appeals ruled that as the legislature had not changed the act pertaining to the use of evidence from blood tests only to prove nonpaternity, the court was bound by the laws and did not have authority either to amend or repeal an act of the legislature. The court agreed that the argument to use the HLA tests was very convincing; they simply felt bound to follow the law as written. Apparently they did not feel, as was the case in *Cramer*, that they should take the lead in finding their way around the legislative barriers. "Until such time as the Legislature deems it appropriate to amend the provisions of the paternity act and permit the results of the HLA tests to be admitted at trial, we must enforce the act as written."

Pennsylvania also in late 1980 was faced with the problem of how to deal with these tests. *Commonwealth of Pennsylvania v. Singleton*, decided by the Superior Court of that state, was also a case in criminal law as the putative father was adjudged guilty of nonsupport of a child. There was also much debate over whether the fact that he had visited the child and had at one time taken

the mother and child to lunch at Woolworths' constituted
intention of support! There was also an issue of the
statute of limitations as the child was now more than
two years old and support had been withheld since birth.
The plea that a single lunch was admission of the need
to support was considered not to be very good evidence
of willingness to support and the court held that the
man was not guilty on the grounds of the statute of
limitations. But the use of HLA tests did form the basis
of a dissenting opinion, based on *Commonwealth v. Young.*
The dissenter found numerous other cases in which the
statute of limitations had not been applied and stated:
"Here, again, the introduction of the HLA blood test
serves to minimize the effect of the passage of time on
paternity determinations by shifting the weight of
evidence to one side. However, our treatment of the
statutory issues makes the constitutional question
academic." The constitutional issue was, of course, the
denial of due process.

The last of the 1980 cases, *McGowan v. Poche*, was
decided by the Court of Appeal of Louisiana, First
Circuit, and dealt only with the use of the standard red
blood cell tests. The evidence showed that there was a
double exclusion of the accused on the basis of two
tests. However, the issue here was not the results of
the test, but again was a narrow legal issue dealing
with the rights of a lower court, to which the evidence
was submitted, to use the procedural device of summary
judgment or whether there must be a full trial with the
introduction of expert witnesses and the chance for
cross-examination. The Court of Appeal held that the
lower court did not have such power and remanded the
case for full trial. Thus the entire purpose of blood

tests, to expedite paternity cases and to avoid the costs and delays of such trials was denied. There seems little doubt that given the evidence, the results of a full trial could not but be the same as the summary judgment, but the higher court felt the laws of their state demanded such a trial.

The years 1981 and 1982 continued to produce novel legal issues involving blood tests as well as repetition of some of the earlier cases. Again, selectivity will be exercised in presenting those cases of major interest or those which illustrate different legal aspects of paternity suits.

In *Shults v. Superior Court, etc.*, the novel issue was the attempt to introduce blood tests in a welfare fraud case. The mother of two children was charged with receiving welfare fraudulently as there was a man living in the house. The lower court ordered blood tests on all to determine whether he was the parent of the children, and she filed for a writ to prohibit such tests under the Fourth Amendment. Additionally, by the way, the mother was also charged with no less than thirty-five counts of perjury in the case. The Court of Appeal, Third District, in California, dealt with this case in which there was also an *amicus curiae* which argued that the blood tests had nothing to do with the determination of fraud and were therefore of no relevance in the case. The majority felt differently and ordered that the writ to prohibit the tests be denied. The case was based on the assumption that blood tests were only a minor intrusion and concluded that the blood tests here, including HLA tests, were necessary. While they might not be relevant as to the issue of fraud, they would possibly be of value in the perjury issue. The court cited *Cramer*, above, as precedent for using blood tests, and thus presumably, although not mentioning it specifically, the HLA test would be included. There was a dissent, claiming that blood tests of the child involved would violate her Fourth Amendment rights as she was not a party to the case. Thus attempts to take blood from her in order to determine perjury of her mother were, in his mind, clearly violations of this Amendment. He cited many cases, but basically felt that the paternity issue should be separated and, if desired, tried as a regular civil suit rather than be part of what was here essentially a criminal case.

The Supreme Court of Minnesota in 1981 admitted the HLA tests as probative evidence. In *Hennepin County Board v. Ayers*, the lower court had entered a verdict of acquittal for the putative father, and the mother and the County Board appealed on the basis of the use of HLA tests as admissible evidence. The tests had shown that the probability of the defendant's being the father was 99.9%. The tests were made with the advice of the defendant's counsel who felt that there was nothing to lose as they might prove exclusion and could not be used if they showed paternity. However, his advice was poor, as the Supreme Court, citing changes in the Minnesota Laws, ruled the tests were admissible, and remanded the case for rehearing. They pointed out that this was a case of first impression in this state, and took care to cite the pertinent law, "The court may, *and upon request of a party shall, require the child, mother, or alleged father to submit to blood tests or genetic tests, or both*. The tests shall be performed by a qualified expert, appointed by the court" (emphasis in original opinion). Thus, Minnesota is added to the list of states

in which HLA tests may be used for determination of
paternity.

An attempt to introduce the HLA test to reverse
decisions made earlier arose in *Johnson v. Johnson*, in
the District Court of Appeal of Florida, Second
District, in 1981. Here two cases, with the same facts,
were consolidated. The issue dealt with the request by
the putative fathers who had agreed to child support in
1976 to reopen the case on the basis of the newly
discovered use of these tests. The court acknowledged
again the legality of such tests for determination of
the probability of paternity, but refused to order or
allow the tests on the basis of *res judicata*, holding
that the husbands should have come forth at the time of
the original decree of divorce and not some 3 years
later. "To allow former husbands to come into court long
after entry of final judgment and challenge the
legitimacy of children born during their marriages would
be chaotic at best. In addition, to require former wives
and their children to submit to blood tests would, in
many instances, be a humiliating experience for them."
The real paternity of these children may never be
determined, of course, but in this and similar cases it
is essential that charges of nonpaternity be brought at
a reasonable time, and new scientific discoveries may
not be used long after adjudication and final decisions
have been made.

A series of cases in New York in 1981 illustrate
further novel attempts to use the HLA tests. In *Matter
of Virginia E.E. v. Alberto S.P.*, tried before the
Family Court of Queens County, the request for blood
tests came as part of a custody procedure, not a
paternity case *per se*. The mother wished to reacquire

her child from the father who had the child living with him. There was some dispute about the length of time the child had been in his custody and initially some difficulty in locating him at all. Nevertheless, the court found that the sole issue was "whether a court can order a human leucocyte antigen blood tissue test...in a custody proceeding on the motion of one party." The court took cognizance of the fact that New York had changed its laws regarding the use of such tests one month earlier, but felt this was only for direct paternity proceedings and denied the request for ordered blood tests. The law which had been changed was only in regard to that type of proceeding and did not apply here. The Family court noted an earlier decision by the Appellate Division which stated, "Common sense, public policy, reason and the overriding consideration for the welfare of the child will bar a wife from bastardizing her child where, as here, she lived with her husband as his wife during the period of conception and birth of the child ***all the while concealing from him the adultery to which she now confesses for the sole purpose of securing custody of the child." Earlier the court had stated clearly, "In the instant case, the court finds that illegitimating a seven-year old child who would lose all rights vis-a-vis the only father she has ever known would certainly fall into the category of objectionable consequences. The court refuses to be a party to such action." One cannot be but impressed that the court was moved by ethical and humanitarian reasons, as well as by legal precedents, in reaching its decision.

In the second New York case, *Matter of Jane L. v. Rodney B.*, before the Family Court of New York County in

1981, the issue was whether the recent change in the
law, two months previously, could be applied to a case
already in progress. The case was apparently bitterly
contested, and the last minute attempt to introduce HLA
evidence equally so. The judge ruled first on the
admissibility and the constitutionality of the new
tests. (It may be of some surprise to find a Family
Court judge dealing with issues of constitutionality.)
The admissibility of the HLA evidence was upheld by
referring to the statement, "The principle is well
established that 'the procedure in an action is governed
by the law regulating it at the time any question of
procedure arises *** the legislature may change the
practice of the court and *** the change will effect
pending actions in the absence of words of exclusion.'"
The Family Court judge then went on to cite *Schmerber* as
providing a constitutional basis for the ordering of the
HLA tests, and concluded its reasoning, "Thus a
justification for the imposition of the HLA test is the
State's deep, pervasive, and abiding interest in the
welfare of its children. Any evidence aiding the
accurate determination of parentage, with a consequent
fixing of paternal responsibility, undoubtedly
contributes to the welfare of children.

"Besides the general public interest in
establishing a child's paternity, in many cases, as in
the case at bar, the public fisc is also a concern
because the child without a known father receives public
assistance in lieu of paternal support...Finally, even
if only the private interest of mothers is considered,
the ends of justice are served by opening the courts to
reliable evidence, and by equality between parties with
regard to relevant examinations of each other." The

court also noted another area where these tests would serve to resolve doubt. Although here the accused admitted to sexual relations with the woman, he claimed that three other males also had such intercourse with the plaintiff, the usual defense in many paternity cases. The court commented, "The ancient truism that a charge of having sexual intercourse 'is so easy to make and so hard to defend'---which is the rationale for the mother's heavy burden of proof in a paternity case--also is logically applicable to testimony as to her intercourse with others than respondent. The HLA test results, if favorable to petitioner, would thus be in the nature of rebuttal evidence that would be particularly appropriate because of its independence from the conflicting testimony." It is assumed, although not specifically stated, that the blood tests will be carried out and the decision of the court withheld until that evidence is available.

In the third New York case in 1981, *Matter of Carmen L. v. Robert K.*, the issues were essentially the same, the change in the New York State law after the initiation of the case, and the constitutionality of the new law. The Family Court of Kings County, reached the identical decision on both grounds as in the immediately preceding case and ordered that the tests be done. It might be noted that here the claim was brought under the Fifth and Fourteenth Amendments rather than the Fourth Amendment, but this is not an important issue as the same had been tried in *Schmerber*.

The rights of a man who sought blood tests to determine whether he was the father formed the basis of the case *In re Mengel*, heard by the Superior Court of Pennsylvania in 1981. The man had sought to have blood

tests carried out, but the woman objected to his
petition and the lower court sustained her objection.
She also claimed, despite earlier statements, that the
real father was a married man whom she refused to name.
The putative father appealed to the Supreme Court, and
this court dealt with several issues. The major basis of
the case was whether he had the standing to obtain the
relief he sought through a petition for declaratory
judgment. The issue was really one of what the rights of
unwed fathers should be. "What appellant seeks to
establish is his status as parent, his legal relation to
the child and the protection of those rights that accrue
from this relationship." This was also a case of first
impression in Pennsylvania, and the court turned to
other states and cases for reference, particularly
Stanley v. Illinois in which the Federal Courts held
that unwed fathers had rights. (That case dealt with the
right of an unwed father to custody of children after
the death of the mother and his right to stop adoption
procedures.) After determining that Mengel had this
right, the court then dealt with the question of blood
tests. The question whether all "possible" other men
should be tested also arose during the appeal. The court
denied this claim, noting that the case dealt only with
the purported paternity of Mengel and not that of any
others who might have had the opportunity to father the
child. The court then ordered the case remanded for the
necessary blood tests to determine the issue of Mengel's
paternity. A concurring opinion raised another
interesting issue regarding the possibility of other
fathers: what would be the legal decision if, after
blood tests established the paternity of Mengel, another
man stepped forward and claimed that he, not Mengel, was

the father? He also added, in regard to the rights
issue, "Since a man has rights, which will be legally
protected, incident to his status as father of a child,
and not just obligations, such as support, which will be
enforced only at the initiative of the mother or the
state, it must follow that an unmarried woman does not
have the unqualified right to refuse to cooperate in
legal proceedings to determine the paternity of her
child." These points are of interest, but what is more
important is that the dissent clearly acknowledged the
importance of using HLA tests, and that they would be
valid evidence in establishing paternity in this case.
The concurrence also notes that the decision to allow
this evidence as admissible had not been clearly made in
previous cases.

A peculiar quirk in legal proceedings was involved
in the case of *Ross v. District Court of Eighth Judicial
District*, considered by the Supreme Court of Montana in
1981. The man in this case had been held in contempt of
court for failing to submit to blood tests, and the
issue was one of whether the proper legal procedures
needed to require him to take these tests had been
followed. The tests would have included the HLA series.
There was also the by now usual claim of rights under
the Fourth Amendment. This was readily dismissed by
citing other cases. However, the court did decide that
the order for blood tests was improper as there had been
no pretrial examination of the case, as should have been
done under the Uniform Parentage Act. As no such hearing
was held, the order for the blood tests was not legal
and the man could not be held in contempt for failing to
comply with an illegal court order. Without the proper
procedure there was a violation of the Fourth Amendment

and the contempt citation was ruled invalid. The case was thus thereby returned to the lower court for a pretrial hearing. Presumably, if there was still reason, after the hearing to suspect paternity, the order for blood tests could then be given, and failure to obey on the part of the accused would this time be a genuine contempt of court.

A legal nicety prevented the use of the HLA tests in an Arkansas case in 1981. In *Story v. Hodges*, the Supreme Court of that state dealt with an appeal from a Circuit Court which found that "allowance of blood test in evidence only to exclude paternity is not denial of equal protection." The appellants had sought to declare the Arkansas law which allowed only proof of nonpaternity (the accused had been found not guilty on the basis of the standard ABO tests) to be a denial of the mother's and child's rights. The trial court had found no denial of equal protection and so ruled. The Supreme Court disposed of the issue by simply declaring that the "evidentiary ruling is not an appealable order. The general rule is that appeals lie only when and if there is a judgment." Thus the mother and child could not bring a case in this instance as they had hoped.

In *Hayward v. Hansen* several other legal issues beclouded the use of blood tests when the case was heard by the Court of Appeals of Washington, Third Division, in 1981. Here a default judgment had been entered against the defendant and he challenged the legality of this on the basis that the court had no jurisdiction to find paternity as the child involved had not been served and brought before the court as a party to the case. Further, Hansen claimed that he should have been arrested before a default judgment could be entered. He also challenged the fact that the judgment was entered much too late after the initiation of the case. In addition, he claimed, 3 years later, that he was sterile, despite the fact that he showed a sperm count of 15,000,000 per cc. The court noted that a count of 10 million would be medically accepted as being fertile, and also that the sterility claim was also made after three years. The Court of Appeals had no sympathy for the variety of claims made and refused to set aside the judgment which declared him to be the father. However, one judge did not agree and in his dissent stated he would have given Hansen the opportunity for blood tests in order to settle the matter. "He has a direct financial interest since he could be contributing to the support of Alisha [the child] throughout her minority. In addition to financial interests, Mr. Hansen, if found to be the father, is also indirectly threatened with loss of liberty since criminal prosecution for nonsupport has been initiated, which may result in incarceration. ...I would set aside the default judgment and remand the case to allow Mr. Hansen the opportunity to provide proof of nonpaternity. ...The interests of the child, parents, society and our system of justice are not furthered by denial of an opportunity for an accurate determination of nonparentage. As stated in a well-known English case where a blood test proved adultery: 'There is nothing more shocking than that injustice should be done on the basis of a legal presumption when justice can be done on the basis of fact.'" The dissent notwithstanding, Hansen remained the legal father of Alisha.

The Supreme Court of Iowa also dealt with the issue of a default judgment of paternity in *State v. St. John*

in 1981. St. John, through counsel, filed a request
prior to the default judgment, seeking blood tests. The
state agreed but the court ordered St. John to pay for
the tests. When he learned of their cost, he decided to
forego them as he could not then afford to pay. The
court ordered the tests and the concerned parties had
them carried out. St. John refused to pay, and the court
then entered its default judgment. St. John then
requested public assistance for the $210 cost of the
test. This was denied and the court declared him the
father. However, fortunately for him, the amount of
support was not fixed at that time. Because of this his
appeal to the Supreme Court could be considered as the
order for payment was not a final judgment in the case.
They also noted that "We have therefore a finding of
paternity for failure to pay $210. We conclude that the
court abused its discretion by imposing the harsh
sanction of a default judgment of paternity for failure
to pay $210 for blood tests." The Court here also was
faced with several constitutional issues, but felt they
had not been sufficiently raised in the prior trial so
as to make it necessary to deal with them here. What
seems somewhat peculiar, however, is the fact that
nowhere in the case are the results of the blood tests
given, although, since they were administered, these
facts must have been available, and presumably when the
case is brought again they will be introduced as
evidence. It would seem that if an appeal had been based
on these in the first place, the legalities of the
situation might have been secondary to the facts if the
blood tests had shown either exclusion or decisive proof
of paternity.

The issue of paternity arose in what must be one of

the most bizarre cases of this type in *Happel v.
Mecklenburger*, in 1981, before the Appellate Court of
Illinois, First District. The facts are as follows: the
"paramour" of Mecklenburger brought suit against the
mother and her former husband for declaring that the
husband was the father of her child. The case was
complicated by the fact that the former husband was
deemed as having a low sperm count, and after
examination and consultation, the wife was artificially
inseminated with a mixture of sperm from her husband and
an unknown donor. She underwent the insemination process
on three successive days; on the night of the first
treatment she had intercourse with her husband. After he
had gone to sleep she then proceeded to have intercourse
with Happel. (Happel had been the baby sitter for the
couple's 3-year-old adopted child). As a result of these
various activities on the part of the woman, she was
diagnosed soon afterwards as being pregnant and
delivered a child at the normal time after that
memorable evening. Later, the Mecklenburgers separated
and the husband was adjudicated to be the father of the
child. The wife then moved in with Happel for a time
until she ordered him out after a quarrel. The plaintiff
then entered the action seeking the court to declare
that he, not the former husband, was the true father.
The court ordered blood tests, and the normal
Landsteiner series did not exclude him. He then
requested that HLA tests be administered, but no such
tests were ordered. To complicate matters still more,
after the initial decision, finding no paternity for the
plaintiff, the state of Illinois changed its laws so
that HLA tests might have been admissible as evidence
for paternity.

The Illinois court went through all of the evidence and the legalities with great care. The problem of presumption of legitimacy was, of course, the ruling factor as was also the plaintiff's claim under the Fourteenth amendment. After weighing all factors, the court found that there was no legal basis to Happel's claim of paternity. There was also some question of the fact that he claimed that he had been supporting the child, but that was found not to be the case either. However, one dissent would have allowed him to use the HLA tests. He cited evidence not given in the majority opinion to the point that Happel had sought these tests and they had been carried out, but the trial court refused to allow the testimony of an expert witness on the basis that the law did not permit the use of these tests in a case such as this. The dissenter, citing *Cramer*, would have allowed this important evidence to be admitted. The majority opinion, however, held sway, and to this day it is doubtful if anyone can tell who the true father of the child may have been.

The North Carolina Court of Appeals in 1981 dealt with the issue of the use of HLA tests to overcome the presumption of legitimacy in *Wake County v. Green*. The case was brought as the county was providing support for the child involved and sought to demonstrate the paternity of Green in order to order support from him. The mother was married and separated from her husband; she had moved to North Carolina 6 years before the child was born. She testified that she had intercourse with no other man than Green at the time of conception of the child, born by now some 7 years after her separation. Further testimony indicated that she had no contact with her former husband since separation, and indeed court attempts to locate him either in North Carolina or in New Jersey where they had previously lived were in vain. The only defense in the case was the presumption of legitimacy as there had been no final divorce in this instance. The HLA test showed a presumption of paternity on the part of Green of 97.6%. The Court of Appeals rendered an historical treatise on Lord Mansfield's rule and declared not only that it might no longer be valid, but, more importantly, that as the facts were clear that she had no access to her former husband for 5 years, he could hardly be adjudged the father of her child. The court, however, did not discuss the HLA tests, but simply remanded the case for rehearing for determination of paternity by Green. Presumably the issue of the use of that evidence will occur in the retrial. If the evidence is admitted, as it probably will be in the present legal climate, then Green will be properly found to be the father.

Two Michigan cases decided in 1981 dealt with the status of HLA tests. In the first, *Varney v. Young*, a paternity case had been decided *sua sponte* by a lower court judge without hearing proofs which the prosecutor held would have shown paternity, namely the HLA tests. The decision of the Court of Appeals in this case was simply that the judge had abused his powers, and that no opportunity was granted for a ruling on the use of the HLA evidence. They did not disagree with the lower court's insistence that a proper foundation be laid before the use of the evidence would be deemed proper, but they felt that no opportunity was offered to do so.

However, this point would seem moot as the same Court of Appeals later in 1981 in the case *Klein v. Franks* decided that the Michigan statute permitting only

admission of evidence to show nonpaternity was not
unconstitutional, and that statutes take precedence over
rules of evidence invoked by courts. There is, however,
recognition that the statute here might be obsolete:
"The Paternity Act does not distinguish between various
blood-testing procedures. ...was adopted well before the
HLA testing procedure had progressed to its current
state of the art. Perhaps it is time for the Legislature
to reevaluate the integrity of the newer testing
procedures and reassess their usefulness in paternity
cases. However, it is not the function of this Court to
upset statutes which have been held procedurally and
constitutionally sound by our Supreme Court." Thus for
the second time the lower courts have felt bound by the
Supreme Court of Michigan, and for the second time they
have appealed to the Legislature of that state to bring
the state laws up to date.

Three additional 1981 California cases dealing with
paternity show the relationship of HLA tests with other
facets of the law. The first, *Vilasenor v. Vilasenor*,
tried before the Court of Appeal, First District, dealt
with the fact that the evidence demonstrating a
probability of 98.95% paternity by the HLA tests was not
properly used as it had been presented by a medical
technologist who admittedly did not have the training or
ability to make such a probability decision. No question
as to the use of the evidence was raised, but simply
whether the assignment of probability was proper and
sufficient. Had the evidence been properly presented, it
would appear that it would have been legally acceptable.

The second California case, *In re Marriage of
Stephen and Sharyne Y.*, dealt with the relationship of
the HLA tests to the statute of limitations. The case

was not brought to the trial court until more than two years after the birth of the child involved. The Court of Appeal, Second District, heard the case after the lower court had denied the accused husband the right to use HLA tests to prove his nonpaternity more than two years after the child was born. The reason for the claim arising at the time it did was that a divorce proceeding was underway and the husband wished to avoid child support costs as part of the settlement. The higher court examined the statute of limitations specifically stated for such cases and found that introduction of this type of proceeding was specifically barred. The basis for such a law was presumably to protect the integrity of the family and to prevent suits such as this from disrupting the bond between father and child. One might wonder how strong such a bond could be if the father denies paternity, but the law does not apparently consider this.

The court also found no constitutional issue under the equal protection right. "If appellant was given the right to introduce the blood tests, other husbands now or later involved in marriage dissolution actions could seek the same relief, even though the minor involved might have reached her seventeenth birthday. In such a situation the public policy reasons for the statute, which we have held are constitutionally valid, would be defeated."

The third HLA involvement with other laws arose in *Ferguson v. Ferguson*, tried before the Court of Appeal, Second District, in California. Here the question was: who has a right to request the HLA tests? The suit was brought on behalf of a minor seeking to establish a man as his natural father. The presumption of legitimacy

also was part of this case. The court ruled that the
child could not bring such a motion for the tests as the
law read that only the husband or the mother could move
to have these tests ordered. The court applied the
reasoning for such a legal point, "As noted, section 621
permits only the husband to make the motion for blood
tests. As noted, in the instant case he is not the
movant. Nor can it be successfully argued that there is
no public policy served by such limitation. Not to
attempt to consider all possible third party situations
in an action to establish paternity, but rather
restricting ourselves to the facts of the instant case,
the results, if Derek [the child] is permitted to attack
his family's integrity, effectively charges his mother
with infidelity and adultery and changes his status from
legitimate to illegitimate as well as deprecating his
legal father. We see no constitutional right for the
child to choose his parents, regardless of the financial
or other benefits which, to the minor, might be
beneficial."

In contrast to the statute of limitations case in
California, just the opposite was held in *State of W.
Virginia ex rel. S. M. B. v. D. A. P.* This 1981 case,
decided by that state's Supreme Court of Appeals, dealt
with a 3-year statute of limitations on the bringing
of paternity cases. The court here noted the power of
the HLA tests in such issues and held that part of the
state law limiting the bringing of suits which would
involve such tests was unconstitutional. The argument
was brought before them that there was a legal
requirement for a parent to support children during the
child's minority and that to deny paternity suits after
3 years would negate that need. Further it was

pointed out in the appeal that if support was not
allowed for illegitimate children, but was for
legitimate children, a constitutional violation would
occur.

The court waxed literary in their opinion: "In this
regard we note that the incidence of illegitimacy rises
in this country every year; furthermore, a few women,
perhaps following the example of T.S. Garp's mother in J.
Irving's *The World According to Garp* (1978). They
deliberately choose to have their children out of
wedlock because they consciously decide that they want a
child but not a husband. ...While we hardly find this
either an intelligent or an appropriate approach to the
sound upbringing of children, nonetheless we must
recognize the existence of new patterns of life. The
difficulty, of course, with eccentric lifestyles is that
when they fail to yield the results which were intended
the ultimate burden of compensating for individual's
lack of foresight ultimately falls upon the inadequate
resources of the West Virginia Department of Welfare.

"Frequently in this age men and women will enjoy a
protracted family-like relationship over a course of
many years without any solemnization through marriage.
When people are living together and a father voluntarily
supports his illegitimate child or children there is
little thought of commencing an adversarial procedure to
establish paternity. In fact, a woman who either aspires
to an eventual legal marriage or merely to a continued
stable relationship would be quite foolish to bring ill
will upon herself by initiating a paternity suit....In
effect if women were aware of the pitfalls which this
statute [of limitations] creates, that awareness would
actually achieve an entirely irrational result. If,

cognizant of these pitfalls, a woman initiated legal actions against her mate, that adversarial proceeding in and of itself would obstruct collateral attempts to regularize an initially illicit relationship through marriage--a result which ultimately must inure to the detriment of the child." This opinion has been quoted from at some length as it shows how the modern courts are trying to adjust to changes in social mores, and to make justice adapt to them. What the court is here doing is not to approve these changes, but to adapt to them, always with the interest of the child foremost. By considering the social aspects of justice, the courts are attempting to temper it to the times, always, of course, within the constitutional provisions of Federal and state laws. There may be many who disagree with this decision and the interpretation of morality which is implicit in it, but the law must deal with the world as it finds it, and not with some imaginary legal ivory tower.

Yet another facet of the power of the HLA tests and another point of law arose in *In the Matter of F. J.F. and F.M.F.*, tried before the Supreme Court of South Dakota, still in 1981. The problem here was the neglect and nonsupport of twins and the accusation of paternity. The lower court had ordered termination of the parental rights of both the mother and the alleged father. Sometime later, a Mr. Henkel, not the alleged father, appeared and sought to reopen the dependency and neglect proceedings for his claimed paternity, and custody rights of the twins. He claimed that he had not been properly informed earlier or he would have come forward at that time. He requested, and was granted permission to have, HLA tests performed upon himself, the mother and the twins. To everyone's surprise, the tests revealed a 97.27% likelihood that he was the father of the male twin, and a 0% probability that he was the father of the female twin! The lower court determined thus that he was the father of one of the twins, but not the other; however custody remained with the Department of Social Services until the appeal by the State could be decided. The Supreme Court dealt with the issue whether Henkel could reopen an already decided case; the decision was based on his XIV Amendment right as protected by the Due Process Clause. Also, the court pointed out that as the originally accused man had denied paternity, "When Mr. Henkel appeared on the scene, no one, including Mr. Henkel, knew for certain who was the father of the twins." The Supreme Court also held that since the tests provided "new and substantial" evidence the case should be reopened and the decision of the lower court concerning reopening the case and the establishment of paternity was upheld. The higher court did remand to the lower court the decision as to who should pay for the costs of the tests as the state had done so, and it appeared to them that Henkel should bear this expense. But the interesting point here is the comparison of this case this to the much earlier *Jordan v. Mace* case in Maine where the court, without having the benefit of HLA, tests had declared, on the basis of the ABO tests only, that if the father could be shown not to be the genetic father of one twin, he could not be the father of the other. The refinement of paternity decisions is nowhere else better shown than it is by contrasting these two cases.

The state of Louisiana has in its laws regarding

paternity made a legal defense the proof that the woman involved is a "woman of dissolute manners." Two cases involving this defense arose in 1981, both decided against the father. In one, *Callahan v. Landry*, the use of HLA tests, not specifically named as such, showed a 99.06% proof of paternity by the accused. The Court of Appeal, Third Circuit, examined this evidence and also the woman's history. The defense brought in a number of men who testified to having sexual intercourse with the woman, but the court found that none of their testimony was to be believed and that the only history of sexual behavior by the woman occurred many years prior to her taking up with Landry. "This one prior relationship is not sufficient to establish the plaintiff is a woman of dissolute manners. To do so the defendant must show the plaintiff is 'loose from restraint; especially loose in morals and conduct; debauched.' The defendant has not shown that the plaintiff possesses these traits."

The second similar case, *State of Louisiana v. Watson*, before the Second Circuit Court of Appeals, dealt with welfare support and the determination of paternity. Apparently no blood tests were ordered, but the court adjudged paternity and after examining the woman's past also found no evidence to accuse her of wanton behavior. The State had sued for both past and future support of the child, but the Appellate Court held that no past support could be claimed as the man had not as yet been adjudged to be the father. He would, however, have to assume future support of the child.

The problem of the statute of limitations again arose in *State of Montana v. Wilson*, consolidated with *State of Montana v. Fatz*. The lower court had ordered blood tests in these paternity cases and the defendants

refused on the basis that the 3-year statute of
limitations was long past when the cases were brought,
both over 3 years, and one 8 years after the
birth of the child. The Supreme Court of Montana decided
that the statute of limitations, as applied to these
cases, was unconstitutional in that it differentiated
between the support rights of legitimate as against
illegitimate children. However, the court also found
that the statute as applied to State Agencies was
constitutional. "The interest of the State in these
matters is economic, and the power of the State to
continually threaten its citizens in paternity suits
must always be thoroughly examined and not taken
lightly.

"The State is not a child. In reality, it cares not
so much about the relationship of father and child but
more about economic reimbursement for welfare and other
dependent aid. The rights given to the State are not
equal to the rights and interests of the child or the
reasons of necessity for finding the child's father. The
statute of limitations, therefore, provides a protection
against the inadvertence and delay of the State in
actions for paternity. The paternal parent protection
offered by the statute as against the State has a
substantial relation to the intended purpose and,
therefore, is constitutional." In other words, the
tolling of the statute is constitutional for the
bringing of a paternity suit, but not if the State is
the mover in such suits.

As might have been expected from the cases given
above, a group of new cases arose in New York after the
initial ones cited previously. The first, *Matter of
Linda K.L. v. Robert S.*, was a filiation procedure heard

by the Family Court, Queens County, in 1981. The court
had ordered an HLA test for all parties with the costs
to be borne jointly by the accused and by the Department
of Social Services. The man objected on the ground that
this test had not been shown to be scientifically
accurate and also that taking the test would violate his
Fifth Amendment rights. The court simply deferred to the
enactment of the paternity law by the legislature, and,
as these tests were now part of the law, felt no need to
comment upon their scientific validity. The second issue
was equally easily disposed of by citation to *Schmerber*.
As no issue over the costs had been raised, the court
felt no need to deal with this issue either. However,
they did raise the point that prior to the change in the
New York law, when blood tests were only useful for
exclusion, there could be great benefit to the man to
pay for such tests. Now that positive evidence of
paternity is admissible, having the man pay for tests
which could prove his guilt seemed to the court somewhat
peculiar. It was also pointed out that the legislature
had not addressed this point when the law was changed
and it might be well for them to do so. Nevertheless,
the court ruled that the man must submit to the tests
and the case was thus suspended until he followed the
court order.

Somewhat more complicated was the case *Lee v. Lee*,
tried in the Family Court of Suffolk County. Here the
issue was whether a California resident could challenge
a ruling in that state of paternity and support in the
New York Court system. The original marriage of the
couple had been in this state and the couple had moved
to California and then divorced there after the birth of
a child. The California courts had issued a final decree

of divorce and support; however the respondent had never resided in California "or had any connection with that State other than the presence of *res* of the marriage which the petitioner carried there." The basic issue in New York was whether blood tests, including HLA tests, could be ordered by a New York Court after the final decree in California. The court here found that there had been no determination of paternity by the California court, and therefore it could, and did, act to order the tests.

The case *Matter of Beaudoin v. Tilley*, tried in the Family Court, Rensselaer County, is the source of the quotation regarding Oedipus used at the beginning of this chapter. The case was a simple one of a husband demanding the use of HLA tests in response to a demand for child support from the county social services agency. The only issue of note was again who should pay for the tests. The court examined the financial position of the respondent and found him able to pay and so ordered. However, this court also appealed to the legislature to reexamine the statute in regards to costs of the tests so that the law would deal with these cases and each one would not have to be challenged on that basis.

The issue of who should pay for tests also arose in *Matter of Richardson v. Gibson*, in the Family Court, Monroe County, still in 1981. Here the matter raised was whether the parents of a 17-year-old boy accused of paternity should pay for the tests. It was pointed out that parents are responsible for all costs during the minority of their children, and the court determined that the costs here were part of that support. However, due to the legal technicality that the parents were never served with a summons and therefore had not been a party to the case, and no *guardian ad litem* was appointed for the child, the Bureau of Social Services (of which Richardson was Director) was barred from requiring the parents to pay these test costs. The parents were thereby held not liable for that expense. The court pointed out that the Bureau could bring a civil suit against the parents under the law and thus were not totally without possible remedy in the matter.

Stephen K. v. Roni L. presents still another aspect of paternity cases. Here paternity *per se* was not an issue, but the means of achieving it were. This 1980 California case, before the Court of Appeal, Second District, involved charges by the father of fraud, negligent misrepresentation, and negligence. The facts were simple. The mother of the child had informed the father that she was using birth control prior to his having had intercourse with her which resulted in the birth of a baby girl unwanted by the father. The lower court dismissed this suit, and the Appeal Court did likewise, essentially upon the grounds that what went on in someone's bedroom was none of their business. "The claim of Stephen is phrased in the language of the tort of misrepresentation. Despite its legalism, it is nothing more than asking the court to supervise the promises made between two consenting adults as to the circumstances of their private sexual conduct. To do so would encourage unwarranted governmental intrusion into matters affecting the individual's right to privacy." The court went on to suggest that if Stephen had not desired a child he too could have practiced birth control!

An second case of this type, arose in 1981, in the

Family Court of New York County. In *Matter of Pamela P.
v. Frank S.* the facts are quite unusual. Here there was
evidence that the woman involved had desired to have a
child and an earlier male friend had left her rather
than sire such an infant. She then took up with Frank S.
and at the appropriate time told him she was using the
pill. She was not, and their intercourse led to the
birth of a child. Paternity tests, using the HLA system
showed high probability of his paternity. The whole
matter of the case was whether her deceit was sufficient
grounds to deny child support on the part of the father.
There seems to be no legislative act dealing with this
kind of case. There is also an issue of fraud and deceit
on the part of the woman's deliberately informing the
man that she was using contraceptive methods when, in
truth, she deliberately did not do so in an attempt,
successful here, to have a child. The court was somewhat
puzzled in its rendering of the opinion in this case. At
one point it is stated, "Petitioner's planned and
intentional deceit here bars her, in this court's
opinion, from financial benefit at respondent's expense.
... Petitioner's wrong towards respondent precludes her
transfer to him of her financial burden for the child
she alone chose to have." But the court then continued
on to examine the financial status of the two parties.
Both the woman, an airline stewardess, and the father
had considerable means. The man claimed that he should
not pay, and if the woman had insufficient income the
state could then take over the additional burden.
The court did not agree to this, and stated,
"Because need above a subsistence level is a flexible
concept, standards of living must be considered. The
child is entitled in this court's opinion to no less a

standard of living than his father's, because it would deprive him of the likely level the child would enjoy if he had been born into the still-prevalent circumstances of an intact family or a father willingly sharing his custody and care. However, the father's standard would under this reasoning set a maximum as well as a minimum limit on his duty of child support, even if his standard is more modest, as in this case it may be, than his means require." In other words, he pays up to his ability. The court referred to the earlier California case, mentioned above, *Stephen K. v. Roni L.* as the only precedent for such a case as this.

The court also recognized an important constitutional issue in the suit. It examined the recent cases involving women's right to determine whether or not to have a child, and decided that as the respondent here had specifically asked whether the woman was using contraceptive methods and she had falsely said she was, his right to have or not have a child would also be violated under the various decisions. They also pointed out that no one had ever brought a support case against a sperm donor in artificial insemination cases, and in a sense likened his position to that of such a donor. In any event, this case of first issue in New York allowed the court to examine these issues and to contrast the excuse of fraud against the welfare of the child, reaching the conclusion that full support was not required, but supplemental support to bring the child to his standard of living would be necessary.

The Appellate Division of the New York system reviewed in 1982 the lower court's decision in the case of *Pamela P. v. Frank S.* For the first time the name of the defendant was used. Frank S. turned out to be Frank

Serpico, the former New York City policeman who had been
instrumental in exposing corruption in that city's
Police Department. The Appellate Division unanimously
overturned his novel defense that his need to pay
support was balanced in part by the fraud and deceit of
the mother. Although that decision had found that he
must pay $790 a month as his share of the child support,
the Appellate Division, acting in behalf of the child
increased the amount to $945 a month. The case was
eventually appealed further to the Court of Appeals,
which upheld the intermediate court, and the novel
defense was rejected. Both courts also ruled that there
was no constitutional issue involved as had been
suggested by the lower court.

Another New York case in 1981 concerned the
question whether an accused man who refused the HLA
test can have this refusal used as evidence against him
as admission of paternity. In *Moon v. Crawson*, before
the Family Court of Delaware County, the accused had not
been eliminated as a possible father by the standard
tests, but for some reason was unwilling to undergo the
more sophisticated HLA tests, and this was to be held
against him. His claims of non-intercourse with the
woman in question were found to be inadequate, and the
court found that paternity had been established by
"clear and convincing proof." During the hearing the
claim was made by Moon, acting as Commissioner on behalf
of the child, that failure to submit to the tests was
evidence that the man was assuming the tests would have
been unfavorable to him. The court here denied the
petitioner's request to use this denial as proof of
paternity simply on the grounds that no such provision
was enacted by the legislature in the paternity laws and

also that the power to order such tests by the courts existed in the law and the court had not seen fit to do so. Thus the refusal was not admissible as evidence of paternity.

A case in Texas involved the dual issue of impotency and blood tests. In 1981 the Fourth Court of Civil Appeals, San Antonio, dealt with this in *M. D. L. R. v. L. V. D. L. R.*, decided in 1981. The court held here that although the man claimed to have had a vasectomy, the defendant had not made proper legal request for blood tests, and that the presumption of legitimacy would hold. No genetic evidence was asked for, apparently, and there also appears to be no mention of sperm counts or other evidence to indicate that the man could not have been the father. For these reasons the judgment of paternity was upheld, although it would appear that the failure of the man to establish his innocence, if such were the case, was based more on ignorance on his part than any attempt to avoid proper legal procedures.

Also in 1981, an Ohio Court dealt with the issue of who should pay for the costs of blood tests. In *Anderson v. Jacobs*, the Appellate Court held that, based upon a United States Supreme Court decision (*Little v. Streater*, q.v.), an indigent had the right to have such tests administered and therefore denial of such tests was a violation of the Fourteenth Amendment.

The just mentioned United States Supreme Court opinion dealt with a case originated in Connecticut. The law in that state permitted blood tests but at the expense of the putative father only. As the accused man was indigent, no tests were carried out because the state refused to pay their costs. The Connecticut

Supreme Court upheld this part of the law and refused
the appeal, which was then accepted by the U.S. Supreme
Court. For a change, that court's opinion, delivered by
Chief Justice Burger, appears to have been unanimous.
The court took firm judicial notice of the efficacy of
blood tests and held that denial of the privilege of
obtaining these tests solely on the grounds of inability
to pay for them was a clear violation of the Due Process
clause of the Fourteenth Amendment. They pointed out
again the three factors necessary for this clause:
"First the private interest that will be affected by the
official action; second the risk of an erroneous
deprivation of such interest through the procedures
used, and the probable value, if any, of additional or
substitute procedural safeguards; and finally, the
Government's interest, including the function involved
and the fiscal and administrative burdens that the
additional or substitute procedural requirement would
entail." They then went on to cite Mendel's laws of
heredity, and, although not mentioning the HLA tests,
pointed out that the several other tests could exclude
positively some 91% of erroneously accused Negro males.
It is clear that the second of the three grounds
mentioned here would apply readily to this case. In any
event, the State of Connecticut was, as are now all
other forty-nine states, prohibited from exclusion of
blood tests on the grounds that the accused is not able
to pay for them. As a side note, it is quite clear the
accused in this case was unable to pay due to the fact
that he was in jail at the time the case arose, and the
support amount of several thousands of dollars would
also obviously be beyond his means.

Citing this case, the United States District Court,

D. Utah, had no difficulty in reaching a similar decision. In *Nordgren v. Mitchell*, decided later in 1981, the court consolidated no less than 5 cases, all brought also by indigents in the Utah State Prison and all defendants in paternity suits. Here the issue was not whether to pay for blood tests, but whether the various defendants had the right to appointed counsel in such cases. Based on *Little*, the U.S. District Court had no difficulty in deciding that this was a right given to them by the Fourteenth Amendment. The decision went somewhat further in citing another U.S. Supreme Court opinion, *Lassiter v. Department of Social Services*, in which the right to counsel existed only when there was a risk of deprivation of physical liberty. The court there had stated that they could not establish a general rule in every parental termination procedure. (That case had dealt with the rights of an incarcerated woman who was protesting the decision to remove her parental rights and have her child placed for adoption.) As a result of these cases, the court here then examined the right for blood tests, although this issue had not been raised originally. Determining that under proper conditions the battery of tests, including HLA, could reach a probability of over 90% in excluding falsely accused men, the court noted that such tests should also be used. Presumably an appointed counsel for the incarcerated inmates would have properly demanded such tests, and under Little they would have to be given at the expense of the state if the men were shown to be indigent.

These, then, are some of the major decisions in 1981. Doubtless there are countless other cases, not cited here, in which many of the same facts and decisions were reached. But the variety of laws involved, and the statements by the courts here noted, show that the issue of paternity is not easily resolved in many cases.

The fact that Supreme Court opinions do not always quickly change lower courts' views can be seen in the case of *White v. State*, tried before the Court of Appeals of Georgia during 1981 but decided in the first month in 1982. Here the accused was found guilty of willful abandonment of a minor illegitimate child. The lower court had acquitted him of this charge but found him to be the biological father of the child. He objected to the second charge and requested blood tests. He was unable to pay for these as he was at the time unemployed and the tests were not carried out. The Georgia statute specifically stated that in such a case the costs would have to be initially borne by the putative father. The issue here was that as he had been found innocent of abandonment, he had nothing to lose by being declared the father. The court avoided the issue of costs by deciding that finding him the father in no way changed his status, and until such time as he was again brought to court for other charges, such as support, there was no harm done to him by the jury's decision regarding his paternity. Apparently the decision in *Little*, above, did not appear to be an issue to the court in this case.

Louisiana continues to stand out in its laws regarding the wanton charges against a mother. In *State of Louisiana v. Wiggins*, the issue was that the blood tests here did not exclude him as a possible father. It is apparent that the HLA tests were not used in this case. The mother testified that at the time she was 16 years old and living at home with her parents. She admitted to having other male friends during the time, but that she had intercourse only with the accused. Particularly she was frequently in the company of a Mr. Nutes and the defendant testified that they broke up over that issue. In addition, the paternity suit was not brought until seven years after the birth of the child involved in the case. During that time the mother had since lived with two other men and produced another illegitimate child. Further, the defendant had married subsequently and had two legitimate children of this marriage. The court weighed all the facts and, finding some evidence that the mother had relations with other men at about the time of conception, ruled that the decision against the defendant was improperly reached, thereby relieving him of the charges of paternity.

Effective in 1981, the State of Iowa amended its laws to permit introduction of proof of paternity by blood tests. Accordingly in 1982, in the case *State of Iowa v. Vinsand*, before the Supreme Court of Iowa, it was ruled that a lower court had acted improperly in refusing to admit evidence of this type in a paternity hearing. The mother had brought the paternity suit and all types of blood tests were carried out, with the result that the probability of the accused being the father was estimated to be "greater than 98.059%" (This degree of statistical accuracy to the third decimal point may seem a bit unnecessary, but this is what was stated in the opinion.)

The relevant statute, which is cited as one model for such laws, reads, "...If a blood test is required, the court shall direct that inherited characteristics, including but not limited to blood types, be determined by appropriate testing procedures, and shall appoint an expert qualified as an examiner of genetic markers to analyze and interpret the results and to report to the courts. Blood-G test results which show a statistical probability of paternity are admissible and shall be weighed along with other evidence of the alleged father's paternity. If the results of blood tests or the expert's analysis of inherited characteristics is disputed, the court, upon reasonable request of a party, shall order that an additional test be made by the same laboratory or by an independent laboratory at the expense of the party requesting additional testing." The law goes on to stipulate the proper procedures for these tests. In the present case, the court ruled that this law held, and that the determination of paternity of the accused without allowing the evidence should not have occurred. The lower court had found the accused not to be the father, and in the light of the evidence which was not allowed, such a decision seemed quite improper. There was some claim concerning the applicability of the law to this case in terms of the fact that the statute had been changed during the time of the initial case and the appeal, but the Supreme Court found that the law was "remedial and procedural" rather than "substantive." Such laws thereby do not come under the attack that the law was either "prospective or retrospective," and therefore the statute quoted applied to the case at hand. The case was thereby remanded for retrial, and, although no decision is as yet available, it is obvious that paternity will be established here if the expert witnesses testify to the validity of the procedures used for the HLA tests.

Prior to 1973, Texas recognized no right of support

for an illegitimate child from his biological father.
In that year, the United States Supreme Court in *Gomez
v. Perez* held that any state which recognized the need
for support by a father of a legitimate child must also
recognize the rights of illegitimates also for this
support. Despite this decision, based on Due Process,
the legislature simply enacted a statute which would
allow fathers of illegitimates voluntarily to legitimize
their children and thereby support them. The Texas
Supreme Court found this inadequate and the legislature
had another go at modernizing their laws. This time that
body acknowledged that fathers of illegitimates must
support them, but only if paternity proceedings were
brought within one year after the birth of the child. The
United States Supreme Court, in *Mills v. Habluetzel*, was
again forced to examine this statute, and in 1982
concluded that Texas law was still in violation of the
federal rights under the same amendment as before.
Consequently they ruled that this law discriminated
against illegitimates and such children must be given
the same rights as those born to married couples. This
opinion, written by Justice Rehnquist, also pointed out
that it would be to the State's advantage to allow
paternity suits at a time later than one year after
birth as this would then permit paternity decisions
which indicated the actual father to remove the children
from welfare rules. The claim by the state that this
time period was enacted to avoid frivolous suits was
considered, but the court felt that the equal protection
clause made the act unconstitutional. A concurring
opinion was entered by Justice O'Conner. She pointed out
that the customary practice in Texas was now to allow
four years for such suits to be brought, and wished to

make it clear that the case at bar was not to be considered as approval of this practice either.

The issue whether blood tests may be ordered by courts again arose in 1982 in *State of Minnesota v. Graham*. The Supreme Court of that state quickly dealt with the claims of unconstitutionality made by the defendant by citing *Schmerber* and *Little* discussed above. They examined the state laws which required such tests and also required that a refusal of the man to undergo them was admissible evidence and found no constitutional violation of rights. It is interesting that such a case should even arise at this time and one cannot help but wonder why a lawyer, presumably knowing the precedence set by the cases mentioned, would believe that a valid case could be made here.

Also in 1982, Massachusetts, again citing many of the cases already listed, decided in *Symonds v. Symonds* that the presumption of legitimacy under Lord Mansfield's rule could be overcome by blood tests, thus joining the other states which had done so previously. The case, decided by the Supreme Judicial Court in that state, involved a divorce and a request for annullment. The husband claimed that he had been tricked into marriage by the claim of the woman that she was carrying his child prior to the marriage. At the time of the divorce proceeding, he now claimed it was not his child at all, and he would not have married the defendant were it not for this false claim. The trial judge had refused to order blood tests to make such a determination. The higher court ruled that as the divorce had been uncontested it should stand, but that proper procedure should be to carry out the tests to determine paternity, presumably to serve as a basis for the need or lack of need on his part to support the child. If the tests proved to exclude the father, then the evidence could overrule the presumption of legitimacy. Interestingly, the court made no mention of what would happen if the man were not excluded, and it is not clear whether the tests would also be used as proof of paternity. From this case one is not certain whether that state really recognizes the use of HLA tests for such a purpose.

The question of the right of an accused father to a jury trial arose in the 1982 case, *State ex rel. Stevenson v. Murray*, in the Supreme Court of Ohio. At the time of origin all parties had agreed to blood tests. Subsequently Stevenson, the alleged father, disagreed with the lower court's interpretation of the test results and demanded a jury trial. That court had entered judgment against him on the basis of the stipulation made at the time of agreement to the tests, and the man appealed on the basis that he was entitled to the trial by jury requested. The Supreme Court found that there was no set procedural basis or policy regarding paternity cases of this sort and as the demand for such a trial was made late, there was no need for such a procedure. Thus there is in Ohio, at least, no constitutional provision for the right to a jury trial in all cases.

Two cases involving constitutional rights were decided simultaneously by the Supreme Court of Appeals of West Virginia in 1980. The first, *Graves v. Daugherty*, involved a state statute that stated "... the alleged father is then arrested and required to post bond for not less than $500 nor more than $1000...If an accused is financially unable to post the bond, he is committed to jail." There was no statutory authority for appointment of a counsel for a man who could not afford one. The case here involved a suit against a putative father who had filed an affidavit of indigency but nevertheless had been jailed in default of a $500 bond. The Supreme Court of Appeals held that the failure to appoint a counsel for him was in violation of the due process clause of the Fourteenth Amendment and ordered that such counsel be provided by the court. It is interesting that no mention was made here of the Sixth Amendment which specifically states "...and to have the Assistance of Counsel for his defense."

The second case was *Gue v. Dunbar*. (It perhaps should be noted that in both cases the defendants were Circuit Court Judges whose actions were challenged, and the suits were brought in behalf of the plaintiffs by the state itself.) In this instance, the trial court had agreed to the motion for blood tests, but refused to allow the state to pay for them. The suit here was on the basis of the Fifth and Fourteenth Amendments. The state law expressly stated the accused in a paternity case should bear such costs. The court found reason to allow the costs of such tests in another part of the state code, namely, "In each case in which an attorney is assigned under the provisions of this article to perform legal services for a needy person, he shall be compensated for actual and necessary services rendered. ...Expenses of the attorney in rendering such services, including but not limited to *necessary expenses for transcripts, investigative services and expert* witnesses shall be reimbursed ... The expense of blood-grouping tests is a necessary expense to be borne by the State." Again, the court as it had in *Graves*, agreed that the defendant had the right to an adequate defense.

Another case in which a mother refused to permit a
blood grouping test for herself and her child arose in
1980 before the Supreme Court, Rensselaer County, N.Y.
In *Ball v. Beaudoin*, the issue was whether a mother and
her child were to be removed from the public assistance
roll. The court had ordered a second blood test, and
when she refused on the basis that such a test would be
harmful to the child, the Department of Social Services
removed her from eligibility for support. Her claim was
that she had not been properly informed that such an
action would be taken if she refused, and that she would
have made a proper claim with medical verification if
she had been duly informed. The court ruled that the
proper notice indeed had not been given, and therefore,
despite her earlier statement that she would have
dropped the entire case rather than have the tests
performed, she was entitled to due notice and should not
have been arbitrarily removed from the assistance rolls.
Although the legal basis for this decision may be sound,
it would appear to anyone reading this case that the
statements made by her indicated clearly that she
understood the consequences of her refusal, and that the
court used a legal technicality to decide in her favor.

The problem of what the law should do in the case
that a putative father had already admitted paternity,
and agreed to a settlement cost and then later demanded
blood-tests arose in *Polk v. Harris*, decided by the
Court of Appeals in Maryland, again in 1980. The case
was unusual in that the requested blood tests were
carried out and did not exclude the putative parent. The
suit was brought by the mother to force the father to
support the child. The issue here was whether the
agreement of paternity and the amount of cost of support

was a valid one. The Court decided that the issue of
paternity was not involved, but remanded the case on the
basis that the chancellor should have first held a
hearing to determine the needs of the child and the
ability of the parents to support the infant. The court
felt the case should have been tried as a civil
paternity proceeding under the state laws, and ordered
that this be done. As there was no real contest
concerning the paternity of the accused, the only issue
was how much he could and should pay for the support.

The presumption of legitimacy rule also arose
again in 1980 in the case *Santiago v. Silva*. The
Illinois Appellate Court in this instance used the
determination of nonpaternity on the part of the woman's
husband and the assignment of this paternity to another
man, in this case the woman's obstetrician. Tests showed
conclusively that the husband could not be the father,
and although no evidence of blood tests on the physician
was offered, he had had access and therefore was adjudged
to be the true father. The presumption of legitimacy was
thus again overcome by the use of blood tests.

A case involving not only paternity but also both
bigamy and child support arose in *Rowbatham v.
Rowbatham*, before the Court of Appeal of Louisiana,
Fourth Circuit, again in 1980. The mother claimed to
have been married to the defendant and to have borne him
two children. She subsequently filed for legal
separation and for support for the two children.
Subsequent investigation showed the defendant to have
been married to another woman during the time he was
supposedly the spouse of the plaintiff. Further, blood
tests were administered, and it was shown that the
defendant could not be the father of the younger of the

two children, but that he was not excluded from paternity of the older child. The court ignored the bigamy issue, and noted that the defendant had claimed both children on his income tax, he had told other people that they were both his children, supported both including their education in a Catholic School, and entered them on his health insurance policy as his dependents. The woman also swore he was the father of both. The case was further complicated as the woman had relinquished custody of both children to the defendant, and he had then taken this document to his legal wife and placed the children with her for care. The court waded through all of this legal entanglement, and then decided Mr. Rowbatham was the father of the older child and awarded the sum of $1000 per month for its care. The higher court agreed to this amount after determining that the defendant had an income of $77,000 annually. Thus his plea that he was not the father of the older child was turned back. It is interesting to note that no mention of what was to become of the younger child, definitely not his, was made in this decision and what was to happen to it is not clear in the opinion.

A case where the timing of a suit became more important than the blood test evidence arose in *Redmon v. Redmon*, again before a Court of Appeal in Louisiana, this time in the Second Circuit. Here a father had appealed for declaration of nonpaternity and the blood tests supported his claim. However, the child in question was now 6 years old when the claim of nonpaternity was made. The Louisiana Laws are specific in that, unless there are circumstances which prevent a case from being brought within six months of birth, no action can further ensue. As a result, despite the blood test evidence, Mr. Redmon was declared to be the legal father and he would have to assume his paternal duty to support the child.

New York, as pointed out previously, changed its law in 1981 to allow admission of HLA as probative evidence of paternity. In the years following this change, the courts have been flooded with cases involving the constitutionality of this change; reading through them is similar to listening to old changes being rung on the same bells. All of the issues decided elsewhere have arisen again, only to be handled in the same manner. A few such are cited here as evidence of the ingenious attempts to thwart the HLA laws. After reading these, one cannot help but admire the ingenuity of members of the legal profession!

In *Matter of Pamela Kirk v. Leslie Kirk*, heard before the Appellate Division in 1983, the issue was not the admission of blood tests *per se* but the failure of anyone to order such tests. Both parties had appeared without counsel and the lower court had determined against Leslie. The Appellate Division remanded the case for proper hearing, and the HLA tests.

Matter of Commissioner of Social Services v. Martinez, also before the Appellate Division, upheld a finding of nonpaternity despite inconclusive blood tests. The basis for this was the failure of the mother to establish clearly the possible paternity of the putative father, despite her testimony that she had intercourse only with him. There was a great discrepancy in the testimony between the two participants, disagreeing upon where they met, the length and depth of their consortium, and a failure of the plaintiff to call witnesses whom she said could verify her case. This led the court to decide that, inconclusive blood tests notwithstanding, she had failed to make "clear and convincing" proof of her allegations.

Again before the Appellate Division in 1983, the case *Jeanne M. v. Richard G.* dealt with the failure to request HLA tests at the time of a divorce action, and a subsequent request by the male for such tests and an appeal against having to pay increased support amounts asked for by the woman. The court denied this late request by the man for such tests, on the basis of *res judicata* and upon further examination ordered increased monetary support. The failure of the man to request HLA tests at the time of the divorce now prohibited him for raising the paternity issue.

A Fourth Amendment plea was made in *Matter of Social Services v. O'Neil*, again before the Appellate Division in 1983. Here the Court cited *Schmerber*, and also went further in pointing out that before blood tests should be ordered, proof of sexual intercourse, not provided in this case, should precede the tests. As no such proof was adequately provided, the court denied the request by the mother for HLA tests of the accused man.

The case of *Social Services v. Kenneth N.* was somewhat more complex when it arose before the Family Court of Suffolk County in 1983. In this instance, the HLA tests showed a probability of 97.6% for paternity. However, the woman admitted also to having sexual intercourse with another man at the presumed time of conception, and the accused claimed this other man could therefore be the father. However, the whereabouts of the other man could not be established, so no HLA test could be given him. The court, going further than previously, used not only the 97.6% figure, but also for the first time investigated the "combined paternity index," another way of determining probability of paternity. "The respondent's index of 38.6 is basically an odds calculation which means if we divide the index figure 38.6 into a random sampling of one 1,000 males (sic), 26 males including the respondent would possess the necessary genetic markers needed to produce the genetic results found in the child herein." The court goes on to the now familiar analogy of drawing a particular colored jelly bean from a sample of 1000 beans with 974 red and 26 green beans, and points out the odds of drawing a green bean are indeed slight. The respondent argued that as long as there is a 2.6% chance of the missing male being the father, he should not be found to be the father. But the court pointed out "He has confused being the father with the potential of being the father. He contends there is a 2.6%...chance that H [the other, missing man] fathered Mark [the child]; but 2.6% is only the potential chance of H. being the father. To determine actual mathematical likelihood of this being the case, the 2.6% figure must be halved." Of course, this is correct as, if both men were found to have identical HLA genes, and both men had intercourse with the woman, there would be an equal chance of either being the real father. The Court then properly found that the respondent was indeed the father and an order of filiation was issued. Once more, this case not only indicates the importance of HLA testing, but the realization by judges of mathematical probabilities properly calculated.

Res judicata also formed the basis for an appeal to the Appellate Division in the 1983 case Merrill v.

Ralston. Here the change in the New York law concerning the use of HLA tests came between the time of the original suit and the time of this appeal. The Court held in this instance that the case should be reopened at the Family Court level and the respondent was ordered to have an HLA test taken within 30 days of the handing down of the decision. It decided that the amended law should be applied retrospectively as the change occurred between the time of the finding by the Family Court and the appeal.

A case involving the presumption of legitimacy arose before the Appellate Division, again in 1983. In *Dawn B. v. Kevin D*, the issue was whether a child born to the mother was fathered by her husband or by another man with whom she was also having sexual intercourse at the time conception occurred and to whom she was subsequently married. HLA tests taken, but for some reason not introduced at the original trial before a Family Court, showed that the respondent had a probability of 99.8% of being the father. The Court held that despite the failure to introduce this evidence at that time, the case should be remanded in the "best interests of the child."

A case involving for the first time in New York the issue of First Amendment rights and HLA tests arose before the Family Court of Kings County ln 1983. In *Martine S. v. Anthony D.* it was found that the refusal of the accused man to take an HLA test, based on his claim of violation of his rights, was completely unfounded, as he simply asserted that his refusal to the taking of blood was a personal belief, and he could cite no recognized religious convictions against this. The Court was particular to point out, "It should be noted

that the respondent is a member of the New York City
Transit Authority Police. It is hard to believe that he
would refuse to give blood if any of his fellow officers
were in need of it as a result of being shot, stabbed or
otherwise injured. It is similarly hard to believe he
would refuse to have his blood drawn in connection with
the physical exams that police officers must regularly
undergo." The Court did deal with the First Amendment
rights more seriously, and did not believe the defendent
had established any genuine claim for such protection.
The respondent's refusal was found only to be an attempt
to avoid guilt, and the court decided that the refusal
was due to fear that the tests would show him to be the
father rather than any true religious belief and the
First Amendment claims were simply an ingenious attempt
to circumvent being found to be the father. (A similar
decision was reached in a New Jersey Court where refusal
to allow HLA tests by a respondent based on his
religious beliefs, in this case an established religion,
Christian Science, was also denied (*County Divison of*
Welfare v. Gilbert Harris).

The interweaving of the HLA tests and the statute
of limitations formed the basis for *Lascaris v. Hinman*
in the Family Court, Onondaga County, in 1983. The
defendant claimed, 9 years after the birth of a child
and after his having been declared the father, that he
was not the father and sought an HLA test to prove his
claim. During this interval the respondent had never
appealed the case, and only did so now because a
judgment of $3,540 in arrearage for child support was
made against him. The court quickly disposed of the case
by refusing to reopen the issue of parenthood and
ordered the sum to be paid.

A case of the use of statistics based on racial origin arose in *Matter of Smith v. Jones* in the Family Court of New York County in 1983. Here the respondent challenged the HLA test finding, a 99.9% probability of paternity on the basis that the data for Hispanics were not clear as to the frequency of various HLA genes in that population. Admittedly this group, probably of heterogeneous origin, may be genetically more diverse than others, but even taking this into account, the court found no valid basis for his claim. The opinion is long and deals at length with the problems of probability, but concludes that even if the figures for Hispanics are difficult to obtain, the HLA tests are valid. The defendant's claims were also not helped by the fact that he admitted a long-term relationship with the mother both before and after the birth of the child!

The case of *Social Services v. Bart D.* show the ultimate in use of statistics for HLA testing. As part of the opinion, the following mathematical formula based on Bayes Theorem is given.

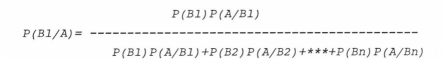

$$P(B1/A) = \frac{P(B1)P(A/B1)}{P(B1)P(A/B1)+P(B2)P(A/B2)+****+P(Bn)P(A/Bn)}$$

To his great credit, the judge explains this quite well! Indeed, much of the opinion is a discussion of statistics and their use in the HLA tests. Along with this was the fact that the judge found the woman's testimony to be highly creditable, and the reliance by the accused on various records which he claimed showed he could not have been present at the time of conception (he was a police officer and claimed the log of the precinct showed him to be on duty at such time) to be much less reliable. The judge then entered an order of filiation.

A more difficult matter arose in *Michaella M.M. v. Abdel Monem El G.* In this Appellate division decision in 1984, the question was whether to order an HLA test following a divorce. The plaintiff sought to deny the paternity of the defendant, as well as to change the surname of her child and to retain custody of the child. The court realized that granting this request would make the child illegitimate, and this was a case of first instance where a mother sought a blood test to disprove paternity of her then husband. The Court wrestled with these issues, suggesting that should paternity be disproved the "child may never learn who his real father is." Nevertheless, the Court felt compelled to remand the case for trial with the appointment of a *guardian ad litem* to represent the child and to determine whether an HLA test would be in the best interests of the infant.

Another New York Case is that of *Social Services v. Thomas J. S.* This case involved most of the HLA issues, rights under the Fourteenth and Fifth Amendments, presumption of legitimacy, statute of limitations, etc. The opinion reads as if it were indeed a Law Review concerning these issues. Of peculiar interest is the citing of an expert that HLA cases can establish paternity up to "99.999999999%" (sic). The Court took all of the various issues into account and particularly pointed out that the case was brought by the Social Services Agency and the New York State law specifically allowed ten years from birth for the Agency to bring such cases. Therefore, this was not a violation of the equal protection clause of the Fourteenth Amendment. It also dismissed the claim under the Fifth Amendment and reaffirmed the validity of the New York State laws dealing with these cases. Perhaps with this case, the various issues dealt with will now no longer be used to attempt to defend against paternity charges. However, this is doubtful; as one of my friends in the law has pointed out "there is no limit to the attempts of some lawyers to obfuscate a case!"

Other states have fallen in line with those using, or allowing, HLA evidence to help establish paternity. Michigan changed its law in 1982, and a typical case in that State is *Pizana v. Jones* heard before the Court of Appeals in 1983. Here the question arose as to whether a child born to Jones was fathered by Pizana or whether another man (Moore) was involved. The plaintiff testified that she had sexual relations with both men and Moore believed that he was the true father. The use of the various blood tests plus the HLA test established that the probability of Pizana being the father was 95.56%. In addition there was a study made of the menstrual cycle of the woman at the various times of intercourse, although this issue was not clearly determined. But the court, in admitting the HLA evidence, took cognizance of Michigan's amended laws "...the mother, child and alleged father shall submit to blood- or tissue-typing tests which may include, but are not limited to, tests of red cell antigens, red cell isoenzymes, human leucocyte antigens, and serum proteins to determine whether the alleged father is likely to be, or is not, the father of the child." It should be noted here that the legislation is all-encompassing in that it recognizes all present means of establishing paternity and also, by setting no limits on the tests to be used, will allow any new methods, such as possible chromosome identification, to be used when developed.

A possible flaw in the use of HLA tests as evidence was raised in *Everett v. Everett*, in the California Court of Appeal, Second District, in 1984. This case, which involved a famous television personality, dealt not only with HLA, but also with whether the use of this test was in itself circumstantial evidence of sexual intercourse between the mother and putative father. There was considerable question over the judge's instructions to the jury over this point. The basic issue was whether before considering the HLA evidence the jury should have decided first whether intercourse took place, and only then, if their decision was that it did, could the HLA evidence be used. The danger here, of course, is that there is a circular argument; HLA shows probability of paternity, therefore probably intercourse took place. However the Court of Appeal found that the instructions given the jury were clear enough to avoid this possible pitfall and the original jury's finding of nonpaternity was upheld.

Other states continue to fall into line with the use of HLA. In Idaho in 1983 the case *Crain v. Crain* was decided by the Supreme Court of that state. Here the decision of the lower court which found nonpaternity was remanded as there was prejudicial error in not allowing the HLA evidence (98.98% probability of paternity) to be introduced. The Court reviewed many cases in other states and decided that the Idaho Legislature had not meant to exclude tests such as this which as yet were undiscovered at the time of the original legislative acts dealing with establishment of paternity.

Two 1983 Missouri cases show some degree of

hesitancy about the HLA tests, but tend to allow them. In *State of Missouri ex Rel D.K.B. v. W.G.I.* before the Missouri Court of Appeals, Eastern Division, it was held that the lower court did not commit error by admitting results of HLA tests which established a 91.5% probability of paternity. However, a similar case before the Western Division of the same Court found reversible error in allowing evidence which showed an 89% probability of paternity. Thus east does not meet west over a 1% difference! However, the establishment that probabilities of less than 90% are not admissible evidence seems to be an important point. It should be noted that this figure is based solidly on the *Joint AMA-ABA Guidelines* for use of this type of evidence.

A similar use of the 90% figure occurred in 1983 in the neighboring state of Kansas. Here the series of blood tests including HLA seem to show only a 70% probability of paternity in *State ex rel Hausner v. Blackman*. The Supreme Court in that state also found that values less than 90% were inadmissible evidence to prove paternity.

Another legal difficulty in the use of HLA tests arose in *State ex rel Hodges v. Fitzpatrick*, heard by the Court of Appeals of Iowa in 1983. The HLA test showed a 99.636% probability of paternity; however the results of the test were improperly submitted to the assistant district attorney, rather than directly to the trial court itself. The use of this evidence presented in that way was found to be inadmissible, but there were sufficient other proofs of paternity so that reversible error had not occurred and the trial court's original finding of paternity was affirmed.

Another paternity suit, *In re E.G.M*, occurred in

1983 before the Court of Appeals of Texas, San Antonio
Division. Here nonpaternity was decided by a single
judge of a lower court without a jury trial. In this
case the HLA tests had shown a 98.9% probability of
paternity; all of the other evidence also clearly
pointed to the likelihood of paternity. Despite this,
the lower court judge found nonpaternity. The Court of
Appeals disagreed, finding that decision clearly to be
"against the weight and preponderance of the evidence"
and reversed and remanded the case, stating, "We do
think, however, that evidence of a high probability of
paternity can amount to strong corroboration of a
witness's story on the material issues and when taken
together with proper undisputed facts can preponderate
in favor of a finding of paternity. This is such a
case."

Similarly, the Court of Appeals of South Carolina
in 1984 found the same reasoning in *South Carolina Dept.
of Social Services v. Bacot*. Here again, there was no
reasonable doubt of access (the couple had opened a
joint bank account and Bacot took the mother to the
hospital when she was about to deliver.) The HLA tests
showed a probability of 95.9% for paternity. The claim
that this evidence should not be admitted was overthrown
by the Court of Appeals, and the finding that Bacot was
the father was sustained.

Also before the same court in 1984 the case *Corley
v. Rowe* arose. Here there were three issues: one, the
question of access, two,the use of the HLA tests and
the weight given them, and three, the amount of support.
The main defense claim was that although the woman in
question slept in a bed only 6 to 10 feet away from the
bedroom shared by the man and his wife, he could not

have had intercourse with her as his wife was a light sleeper and would have been awakened by the sounds of such activity, especially since there was apparently no door between the two rooms. The court stated, "We do not regard as 'incredible' evidence that a husband could have sexual intercourse with another woman while his wife slept ten feet away. An act of that nature would be risky, perhaps, but it could happen." The possibility of access, along with the HLA test indicating failure to exclude the defendant at the 99.2 percentile, allowed the court to decide against the defendant. However, the lower court's finding that he should contribute $50 per week for support was deemed to be too high, considering the financial situations of all parties, and the sum was reduced to $35 per week.

A case before the Court of Special Appeals in Maryland dealtng with the admissibility of HLA evidence in this state. In late 1983, *Haines v. Shanholtz* showed a critical error by the defendant. "Appellee denied that he had ever had sexual relations with Appellant and requested, to his eventual dismay, a blood test. The results of that test indicated a 98 percent probability that Appellee was the father. Appellee promptly reversed his field and argued that the test results were inadmissible..." The court, needless to say, supported the mother; as the lower court had found the tests inadmissible, the Court of Special Appeals reversed and remanded the case for retrial.

By mid-1985 many more states have admitted the HLA evidence and the statistical evidence as valid for proof of paternity. Louisiana, Ohio, Minnesota, Michigan, Kansas, Pennsylvania, Illinois, Indiana, Arkansas, Oregon, Maryland, South Carolina, and Oklahoma have decided that this evidence is admissible, and doubtless other states have done the same in joining California's original decision and the change in New York's paternity laws in 1981. Indeed, it is quite probable that all fifty states will come in time to recognize the utility of these tests. But, unfortunately, the use of the tests has not, as originally hoped, lessened the burdens of the courts as each time a state law is changed it is challenged, often in multiple and nearly identical cases, and lower courts as well as the appellate courts are still flooded with paternity trials and their appeals.

An example of one state apparently still holding out is Alabama, where in the case *Calloway v. Calloway*, before the Supreme Court of that state in 1984, it was decided that the accused father, an indigent, was not entitled to the HLA tests at state expense. "Because we find no requirement...for the proposition that any particular blood test must be administered to an indigent putative father, and because Calloway has already received one blood test, although perhaps not the most accurate available, we find that he is not constitutionally entitled to a second such procedure." Calloway's appeal for the test was thereby denied and he was judged to be the father of the child in question.

Despite all of the cases given in this chapter, new attempts to use the HLA system or to avoid it, keep arising. A few of these will form the last part of this chapter. The cases are varied and cover some of the same issues already discussed, but from a sufficiently different approach as to merit their being included in this chapter.

In 1982, the Family Court, Kings County, New York,

took the unusual step in *Matter of Catherine H. v. James S.* of finding that the use of HLA tests *without* first using the standard red blood cell tests was improper. The HLA tests had shown a "conclusion of paternity of .923 plausibility of paternity." Despite this the judge felt that the use of other tests might either provide a higher probability, or even possibly exclude the accused and ordered the simpler tests to be carried out. As it is usual to do the red cell tests first, and then turn to the HLA tests only if no exclusion is shown, the judge's logic seems correct.

In *Salico v. Salico*, before the Supreme Court, Special Term, Queens County, N. Y., in 1984, the issue was whether voluntary HLA tests, not ordered by any court, were admissible. The mother and father had both undergone the HLA tests and had them taken by the child as well. The tests definitively excluded Mr. Salico as the father. (The mother had admitted that she had intercourse with another man at about the time of conception, and the husband claimed that he was sterile.) The mother sought to have the test results excluded on the basis that they were voluntary. The court dealt with this novel claim as follows: "Moreover it appears to this court that such voluntary tests are unquestionably admissible, highly probative, and perhaps in that they were undergone *voluntarily* should be given greater weight in the over-all determinations than those tests performed pursuant to court order." The evidence was thereby used and Mr. Salico was determined not to be the father of the child concerned.

The earlier issue of proof of intercourse before using the HLA tests again arose in 1983 in *Matter of Social Services v. O'Neil*. The mother had offered no sworn affidavit that she had intercourse with the defendant, and the court therefore refused to order the tests until such time as the mother came forth with the claim that she had intercourse with the defendant.

New York continues to present new issues in blood testing for paternity. The 1983 case, *Matter of Alicia C. v. Evaristo G.*, the issue was which of two men was the father. The maternal grandmother brought this case as the actual mother had died prior to the bringing of the suit. The deceased and Evaristo had separated and another man had befriended the woman and after the birth of the child had allowed his name to be entered upon the birth certificate. Sufficient evidence was presented, however, showing that the woman and Evaristo had consistently cohabited during the presumed time of conception, and HLA tests showed a 98.5% probability that Evaristo was the genetic father. Part of the difficulty was that a racial identification of Evaristo as Caucasian had been made, and while not stated, it is probable that the statistics derived for the HLA probability was from that ethnic group. The court decided that this was not sufficiently in error to invalidate the HLA results and Evaristo was determined to be the father.

A prize example of hair-splitting formed the basis for a defense in the Louisiana case, *State through Department of Health v. Smith*, before the Court of Appeals. The defendant had been adjudged to be the father and he appealed on the basis that the state law said that the testimony of *experts* was to be deemed admissible, but in this case only one *expert* had appeared. The court was not bemused by this novel claim, and despite the explicit wording of the statutes found that while only a single expert testified, his report was signed by two other *experts* one of whom would have been available to testify had he been called upon to do so.

Similarly, in an attempt to place the burden of parenthood upon the defendant in *Bradley v. Houston*, the plaintiff claimed that the expert testimony was received from a chemist who testified concerning the findings of the blood tests which indicated that the accused was probably guilty of the charge. There was also some dispute over the offering of another expert, a college registrar, and the use of his evidence as to the proportion of blacks in the county, a figure which the chemist then used to form the conclusion that the probability of the defendant's being the father was 95.83%. Despite this, the lower court held that he was not the father. The Court of Appeals of Arkansas nonetheless upheld the lower court's finding of non-paternity on the basis that there had been no reversible error by the lower court. It might appear here that an injustice was done by the court's upholding the decision of the lower court, but the case was based on legal issues only, and the probability figure was not deemed of sufficient importance to cause a reversal.

In March, 1985, a lower court in Minnesota found a man guilty of paternity charges by default. In the case *State v. Ketter*, the man had twice failed to appear at the designated time for taking blood samples, and the lower court must have interpreted this as a sign of guilt. The Court of Appeals, however, decided that guilt by default was too heavy a penalty to pay and remanded the case so that it would go to a full trial. Evidence of the defendant's refusal to take the tests could then be introduced at the trial.

Another attempt to avoid guilt arose in *Baker v. Wagers*, before the Court of Appeals, Second District, of Indiana in 1985. Here the issue was whether the results of blood tests were admissible as a *business record* and, as the physician testifying was not present while the tests were being carried out, whether his testimony was therefore hearsay. There is an exemption from hearsay laws if the matter is simply one of showing records kept in the ordinary conduct of business. The court therefore found no difficulty in stating that the results of the blood test fell under the business exemption clause and the evidence of the tests was admissible.

In *Aroonsakul v. Flanagan*, heard by an Appellate Court of Illinois, in 1984, the blood test evidence showed conclusively that the defendant was not the father of the child concerned. The appeal by the plaintiff was made on the grounds that the defendant at the time of the tests had been taking medication and had been receiving blood transfusions which interfered with the results of the tests, giving false results which were used to exonerate him. However valid this claim might be, testimony showed that the defendant again, six months after medication had ceased, took the same tests with the same exclusionary results. Interestingly not only the defendant himself, but also four of his five brothers and his parents all took the tests. Neither the defendant, nor his family, showed the presence of a particular antigen found in the child, and there was no difficulty in sustaining the verdict of not guilty found by the lower court.

In still another peculiarity of the paternity suits, a case, *Dix v. Laux*, arose in Ohio in 1984 before the Court of Appeals, Montgomery County. The case was first brought to trial *before* the birth of the child concerned. The case was properly deferred until after the child was born and appropriate blood tests could be made. Surprisingly even then the defendant did not request blood tests and none were carried out. The defendant was found guilty of the paternity charges and appealed that his rights had been denied by the lower court's refusing his request for the court to order the tests. The defendant further claimed that he had not known of the tests and the court should have made them mandatory. The higher court upheld the verdict of guilty based on the fact that the request for blood tests came too late and the case had already been decided.

In *Miller v. Kriner*, decided in 1985 by the Superior Court of Pennsylvania, another novel issue arose. HLA tests were taken both of the defendant and of his brother. The defendant was found to have a slightly lower probability of paternity than his brother. The expert witness testified that this was not unusual as brothers inherit from the same parents and are very likely to have quite similar blood factors. (Indeed this is the basis for the use of HLA testing in organ transport, i.e., the similarity of antigen types in siblings or relatives and therefore the lower chance of organ rejection by the recipient.) The basic problem was then one of access, and there was considerable dispute concerning this fact. Nonetheless, the court agreed that all of the evidence pointed to the accused being the father and they upheld that decision.

Usually, of course, it is well known who the mother of a child is. But in a case which might also be included in the chapter on criminal law, this was an

important issue. The facts in the 1985 case, *Davis &
Davis v. State of Indiana*, before the Court of Appeals
of Indiana, Second District, are different from the
usual paternity cases. Here, an abandoned baby a few
hours old was found beside a gravel pit. An anonymous
phone call suggested that the Davises were the parents.
Blood tests were taken of the presumptive parents and
showed that "the chance that they are the parents of
this child versus a random set of parents in the
community ... comes out to be ninety nine point nine
eight eight six seven percent (99.98867%) or just based
on tissue typing alone odds of eight thousand eight
hundred and twenty eight (8,828) to one that Mr. and
Mrs. Davis are the parents." The Davis family also
raised the usual constitutional objections so often
referred to previously, but the court had no difficulty,
given the results of the HLA tests, in finding the the
Davises to be the true parents and to be guilty of the
criminal charge of child neglect.

As predicted, the use of chromosome analysis of
parents and children has recently been admitted as
evidence in the courts. The first case appears to be
Robinson v. Jones, in the Circuit Court of Richmond,
Second Division. (15 American Journal of Human
Genetics, 47, 1983.) Furthermore, the article refers to
an anonymous case in Oregon presently being settled out
of court on the basis of chromosome analyis. Blood and
enzyme tests in this case had shown the probability of
paternity to be low, but there was no definite
exclusion. As a result chromosome analyses of the
putative father, the mother, and the infant concerned
were carried out. By use of staining techniques, it
could be shown that one chromosome each of 5 of the

twenty-three chromosome pairs of the infant came from
the mother, but in all 5 of the pairs, the homologue
could not have come from the accused male. The article
illustrates the results and discusses the methods and
uses of such tests so that they can be clearly
understood by a non-geneticist. The final statement may
be worth quoting: "As the legal profession becomes more
aware of the value of chromosome analysis in
establishing paternity, and the results are accepted by
the courts, there will be a demand for qualified
cytogenetic laboratories to provide this service."

Many facets of such interplay have been observed.
First, the introduction of new types of evidence into
the legal system and their import on changes in
legislative acts. The first, of course, was the use of
the standard Landsteiner tests for exclusion of
paternity, and their impact on the long-standing Lord
Mansfield's Rule concerning presumption of legitimacy.
The issue of the admissibility of this evidence versus
the weight of the evidence was seen in many early cases.
Following these tests, and the introduction of many
blood tests besides the standard ABO series, such as the
Rh, Dufy, Kell, etc. tests and their use, the courts
then had to face the problem of how to deal with the HLA
series of antigens and *Frye* was often applied concerning
their scientific validity; today few courts would
challenge their accuracy or their use. But along with
their introduction came the difficulty of admitting
them; it should be remembered that most state laws in
the past, and some even today, would not permit evidence
determining positive paternity to be allowed in
paternity cases. The ingenious solution, first in *Cramer*
where the court simply declared that HLA tests were not

blood tests at all, has been replaced now by statutes
which not only permit this evidence to be used, but many
of which also make their use all but mandatory. Indeed,
their use has been approved by most states simply to
lighten the load of the courts as almost all paternity
cases could be resolved without the need for lengthy,
and costly, trials. Thus this new scientific discovery
has caused both legislatures and courts to alter laws
and procedures.

Also, as seen, at least some courts are now
inclined to use chromosome analyses in addition to the
blood tests. If this becomes widespread, the courts will
have still another genetic tool with which to settle
disputed paternity cases accurately and quickly.

Along with the introduction of these tests as
evidence there arose many constitutional issues,
particularly those dealing with the First, Fourth,
Fifth, and Fourteenth Amendments. These numerous claims,
still arising in 1985, were effectively dealt with in
Schmerber when the United States Supreme Court found no
violation of religious freedom, no violation of the
search and seizure clause and similarly no violations of
the privilege of self-incrimination or of due process.
It is also of interest to note that it was another
scientific discovery, not dealing with paternity but
with the blood-alcohol content of a drunken driver,
which led to this decision. Once it was determined that
blood could be taken for evidence, the application to
paternity tests was obvious.

The problem of due process was also complicated not
only by the blood tests themselves, but also by the
decisions as to the rights of illegitimate children as
compared to legitimate children. Earlier when

illegitimates had no rights of support, inheritance or other dependency upon their fathers, paternity cases were of little import as even definite proof of paternity would have no effect in these matters. As the laws and society changed and the biological father was held responsible for his act, the paternity cases took on a much more important role on behalf of the children concerned. The interest of the state in removing them from the welfare rolls became a new reason to bring such suits and to assure that the biological father supported his offspring. With this new regard for the rights of illegitimates, the paternity suits became much more embittered and contested. Fortunately for the courts, this occurred at about the time the HLA evidence became usable and many cases could be readily decided.

Still another issue, the statutes of limitations in various states, also plays a role in these legal decisions. The tolling of this statute for minors has allowed paternity suits to be brought many years after the birth of the child, and the claim for equal support rights for legitimate and illegitimate children through their minority has been established, again under the Due Process clause of the Fourteenth Amendment.

Further, with the changes in mores in the present society, more and more illegitimate children are being born. Some claim that close to 50% of all children born in the District of Columbia are in this category. Were all of these infants to receive public aid, the costs would be staggering. The determination of paternity by scientific means now seems essential to avoid this burden to the taxpayer.

These, then, are some of the issues with which paternity cases now deal. Doubtless other new types of cases and new issues will arise in the future. Fortunately, there is now a solid block of past cases upon which courts can base future decisions. It will, however, be interesting to see whether the use of the evidence proving paternity with a high degree of certainty will eventually bring about the desired decrease in such suits and free the courts for other actions involving more complex legal principles and actions.

One last case has been held to end this litany of the use of genetic evidence in paternity cases. Here the issue was one of inheritance in this case, *Estate of Cornelious* before the Supreme Court of California in 1984. An adult woman asked to be made the executrix of a deceased man, not the husband of her mother. The man had died intestate and other relatives claimed that this presumed daughter had no rights to any of the estate. During the time of the marriage of her mother to her presumed father, also deceased, she had lived with them as their daughter, only to learn at the age of fifteen that the husband was not her biological father and that Cornelious was the real father. The husband died without ever learning of this development. The daughter visited Cornelious until the time of his death, and Cornelious told many of his friends that she was indeed his daughter.

Further genetic evidence showed conclusively that the husband could not have been the true father as the daughter carried the genetic trait for sickle cell anemia and neither of the married couple did so. It is not clear how this was determined for the deceased husband, nor whether it was ever determined for Cornelious either.

In this case, despite the genetic evidence, the

court held that the daughter had no right to the estate and that the presumption of legitimacy should hold. The court here went into a long historical comment upon the presumption of legitimacy, noting the fact that Shakespeare "was familiar with the rule, for he made reference to it in King John, act I, scene 1: 'King John.--Sirrah, your brother is legitimate; Your father's wife did after wedlock bear him; And, if she did play false, the fault was hers; Which fault lies on the hazards of all husbands That marry wives.'"

With this quotation, it is perhaps best to leave it to the bard to end this long discourse upon the use of genetic evidence in paternity cases.

ALPHABETICAL LIST OF CASES CITED

CASE	CITATION	YEAR
Anderson v. Jacobs	68 Ohio St. 2d 67	1981
Anonymous v. Anonymous	460 P.2d 32(Ariz)	1969
Aroonsakul v. Flanagan	464 N.E.2d 1091(Ill)	1984
Baker v. Wagers	472 N.E.2d 218 (Ind)	1984
Berry v. Chaplin	169 P.2d 442 (Ca)	1946
Beuschel v. Manowitz	151 Misc. 899	1934
Bradley v. Houston	676 S.W.2d 746 (Ark)	1984
Bullock v. Knox	11 So.339(Ala)	1892
Callahan v. Landry	402 So.2d 782(La)	1981
Camden Cnty.Bd of Soc. Sev. v. Kellner	24 FLR 2413 (N.J.)	1980
Cardenas v. Chavez	103 Mich App 646	1980
Carlyon v. Weeks	387 So.2d 465(Fla)	1980
Chumley v. Hall	601 S.W.2d 803 (Tex)	1980
Cnty Hennepin v. Brinkman	364 N.W.2d 458(Minn)	1985
Comm. of Pa .ex Rel. Atkins v. Singleton	422 A.2d 1347 (Pa)	1980
Commonwealth v. Blazo	406 N.E.2d 1323(Mass)	1980
Commonwealth v. Young	419 A.2d 57 (Pa)	1980
Corley v. Rowe	312 S.E.2d 720	1984
Cortese v. Cortese	76 A.2d 717 (N.J.)	1950
County of Ramsey v. S.M.F.	298 N.W.2d 40 (Minn)	1980
County of San Diego v. Brown	145 Cal.Rptr. 484	1978
Crain v. Crain	662 P.2d. 538 (Ida)	1983
Cramer v. Morrison	153 Cal.Rptr. 865	1979
D.W.L. v. M.J.B.C.	601 S.W.2d (Tex)	1980
Davis and Davis v. State of Indiana	No.2-1083 A356 (Slip) Ind	1985
Dawn B. v. Kevin D.	96 AD2d 538	1983
Dazey v. Dazey	122 P.2d 308(Ca)	1942

Matter of Goodrich v. Norman	100 Misc 2d 33	1979
Matter of Jane L. v. Rodney B.	108 Misc 2d 709	1981
Matter of Jeanne M.		
v. Richard G.	61 NY Reports 2d 638	1983
Matter of Lydia L. v Vidal L.	95 Misc 2d 507	1978
Matter of Martine S.		
v. Anthony D	120 Misc 2d 567	1983
Matter of Moore v. Astor	102 Misc 2d 472	1980
Matter of Mores v. Feel	73 Misc 2d 942	1973
Matter of Pamela Kirk	95 AD2d 888	1983
Matter of Pamela P.v. Frank S.	110 Misc 2d 978	1981
Matter of Richardson v. Gibson	110 Misc 2d 20	1981
Matter of Russo v. Hafner	100 Misc 2d 841	1979
Matter of Smith v. Jones	120 Misc 2d 834	1983
Matter of Virginia E.E.		
v. Alberto S.P.	108 Misc 2d 565	1981
McGowan v.Poche	393 So.2d 278(La)	1980
McQueen v. Stratton	389 So.2d 1190(Fla)	1980
Merrill v. Ralston	95 AD2d 177	1983
Michaella M.M. v. Abdel		
Monem El G.	98 AD2d 464	1984
Miller v. Kriner	Slip No.00349 (Ohio)	1985
Miller v. Smith	6 FLR 2661(Ill)	1980
Mills v. Habluetzel	50 LW 4372(Tex)	1982
Moon v. Crawson	109 Misc 2d 902	1981
Nordgren v. Mitchell	524 F.Supp. 242	1981
People v. Thompson	152 Cal.Rptr.478	1979
People v. Yarn	392 N.E.2d 606(Ill)	1979
Phillips by & Through Utah		
v. Jackson	615 P.2d 1228(Utah)	1980
Pizana v. Jones	339 N.W.2d 1 (Mich.	1983
Polk v. Harris	420 A.2d 1004(Md)	1980
R. McG. v.J.W.	615 P.2d 666(Colo)	1980

Rachal v. Rachal	412 A.2d 1202(D.C.)	1980
Ramsey Cty Pub.Defend.Off.		
v Fleming	294 N.W. 275(Minn)	1980
Redmon v. Redmon	391 So.2d 916 (La	1980
Robinson v. Jones	15 Am.J.Hum Genetics	1983
Ross v. Dist.Court,		
8th Judicial Dist.	628 P.2d 662 (Mont)	1981
Rowbatham v. Rowbatham	389 So.2d 877(La)	1980
S.Car. Dept.Soc. Sv.		
v. Johnson	266 S.E.2d 878(S.C.)	1980
S.Car.Dept.Soc.Sv. v. Bacot	313 S.E.2d 45 (S.C.)	1984
Salicco v. Salicco	125 Misc 2d	1984
Santiago v. Silva	90 Ill. App.3d 554	1980
Schmerber v. California	384 U.S. 757	1966
Serafin v. Serafin	258 N.W.2d 461(Mich)	1977
Shults v. Superior Court, Etc.	170 Cal.Rptr. 297	1981
Simons v. Jorg	384 So.2d 1362 (Fla)	1980
Social Services v. Bart D.	121 Misc 2d 425	1983
Social Services v. Thomas J.S.	100 AD2d 119	1984
Stanley v. Illinois	405 U.S.645	1972
State ex rel. Graves		
v. Daugherty	266 S.E.2d 142(W.V.)	1980
State ex rel.Gue v. Dunbar	266 S.E.2d 142(W.V.)	1980
State ex rel D.K.B. v. W.G.I.	654 S.W.2d 218 (Mo.)	1983
State ex rel Hausner		
v. Blackman	662 P. 2d 1183 (Kan)	1983
State ex rel Hodges		
v. Fitzpatrick	342 N.W.2d 870 (Iowa)	1983
State ex rel.Stevenson		
v. Murray	431 N.E.2d 325(Ohio)	1982
State of Iowa v. St. John	308 N.W.2d 8 (Iowa)	1981
State of Montana v. Fatz	634 P.2d 172 (Mont)	1981
State of Montana v. Wilson	634 P.2d 172 (Mont)	1981

State of W.Va.Ex Rel S.M.B. v.D.A.P.	284 S.E.2d 912(W.V.)	1981
State on Behalf of Hastings v. Denny	296 N.W.2d 378(Minn)	1980
State through Dept. Health v. Smith	459 So.2d 146 (La)	1984
State v. Camp	209 S.E. 754 (N.C.)	1974
State v. Camp	263 S. E.2d 592	1980
State v. Damm	252 N.W. 7(S.D.)	1933
State v. Damm	266 N.W. 667(S.D.)	1936
State v. Graham	(Slip Opinion) S.Ct. Minn. No.81-846	1982
State v. Gray	145 N.E.2d 162(Ohio)	1957
State v. Meacham	612 P.2d 795(Wash)	1980
State v. Vinsand	(Slip Opinion) S.Ct. Iowa 109/66704	1982
State v. Watson	403 So.2d 1249(La)	1981
State v. White	256 S.E.2d 505(N.C.)	1979
State v. Wiggins	409 So.2d 1264(La)	1982
State,on behalf of Ortloff v.Hanson	277 N.W.2d 205(Minn)	1979
Stephen K. v. Roni L.	164 Cal.Rptr. 618	1980
Story v. Hodges	614 S.W.2d 506(Ark)	1981
Symonds v. Symonds	432 N.E.2d 700(Mass)	1982
Thompson v. Thompson	404 A.2d 269(Md)	1979
Varney v. Young	308 N.W.2d 276(Mich)	1981
Villasenor v. Villasenor	177 Cal.Rptr. 839	1981
Wake County ex rel Manning v. Green	279 S.E.2d 901(N.C.)	1981
Wake County v. Townes	53 N.C. App. 619	1981
Wareham v. Wareham	15 Cal.Rptr 465	1961
White v. The State	288 S.E.2d 574(Ga)	1982

CHRONOLOGICAL LIST OF CASES CITED

1978	Lalli v. Lalli	439 U.S. 259
1978	Lee v. Dist.Director	
	of Imm.& Nat.Serv.	573 F.2d 592
1978	Matter of Carol B. v. Felder	94 Misc.2d 1015
1978	Matter of Lydia L.	
	v Vidal L.	95 Misc 2d 507
1979	Cramer v. Morrison	153 Cal.Rptr. 865
1979	Figueroa v.Juvenile Court	595 P.2d 223(Col)
1979	Hepfel v. Bashaw	279 N.W.2d 342 (Minn)
1979	J.B. v. A.F.	285 N.W.2d 880(Wis)
1979	Lascaris v. Laredo	100 Misc 2d 220
1979	Malvasi v. Malvasi	401 A.2d 279(N.J.)
1979	Matter of Goodrich v. Norman	100 Misc 2d 33
1979	Matter of Russo v. Hafner	100 Misc 2d 841
1979	People v. Thompson	152 Cal.Rptr.478
1979	People v. Yarn	392 N.E.2d 606(Ill)
1979	State v. White	256 S.E.2d 505(N.C.)
1979	State on behalf of Ortloff	
	v.Hanson	277 N.W.2d 205(Minn)
1979	Thompson v. Thompson	404 A.2d 269(Md)
1980	Camden Cnty.Board of Social	
	Services v. Kellner	24 FLR 2413(N.J.)
1980	Cardenas v. Chavez	103 Mich App 646
1980	Carlyon v. Weeks	387 So.2d 465(Fla)
1980	Chumley v. Hall	601 S.W.2d 803(Tex)
1980	Com.ex Rel. Atkins	
	v. Singlet	422 A.2d 1347(Pa)
1980	Commonwealth v. Blazo	406 N.E.2d 1323(Mass)
1980	Commonwealth v. Young	419 A.2d 57(Pa)
1980	County of Ramsey v. S.M.F.	298 N.W.2d 40(Minn)
1980	D. W. L. v. M. J. B. C.	601 S.W.2d 475(Tex)
1980	De Weese v. Unick	162 Cal.Rptr. 259
1980	Hrouda v. Winne	77 AD2d 62

1980	J.H. v. M. H.	426 A.2d 1073(N.J.)
1980	Matter of Ball v. Beaudoin	103 Misc 2d 1000
1980	Matter of Edward K.	
	v. Marcy R.	106 Misc 2d 506
1980	Matter of Moore v. Astor	102 Misc 2d 472
1980	McGowan v.Poche	393 So.2d 278(La)
1980	McQueen v. Stratton	389 So.2d 1190(Fla)
1980	Miller v. Smith	6 FLR 2661(Ill)
1980	Phillips by & Through Utah	
	v. Jackson	615 P.2d 1228(Utah)
1980	Polk v. Harris	420 A.2d 1004(Md)
1980	McG. v.J.W.	615 P.2d 666(Colo)
1980	Rachal v. Rachal	412 A.2d 1202(D.C.)
1980	Ramsey Cty Pub.Defend.Off.	
	v Fleming	294 N.W. 275(Minn)
1980	Redmon v. Redmon	391 So.2d 916(La)
1980	Rowbatham v. Rowbatham	389 So.2d 877(La)
1980	Car. Dept.Social Services	
	v. Johnson	266 S.E.2d 878(S.C.)
1980	Santiago v. Silva	90 Ill. App.3d 554
1980	Simons v. Jorg	384 So.2d 1362(Fla)
1980	State ex Rel. Graves	
	v. Daugherty	266 S.E.2d 142(W.V.)
1980	State ex Rel.Gue v. Dunbar	266 S.E.2d 142(W.V.)
1980	State on Behalf of Hastings	
	v. Denny	296 N.W.2d 378(Minn)
1980	State v. Camp	263 S. E.2d 592
1980	State v. Meacham	612 P.2d 795(Wash)
1980	Stephen K. v. Roni L.	164 Cal.Rptr. 618
1981	Anderson v. Jacobs	68 Ohio St. 2d 67
1981	B. S. H. v. J. J. H.	613 S.Wl2d 453(Mo)
1981	Callahan v. Landry	402 So.2d 782(La)
1981	Ferguson v. Ferguson	179 Cal.Rptr. 103

1981	Happel v. Mecklenburger	427 N.E.2d 974(Ill)
1981	Hayward v. Hansen	628 P.2d 1326(Wash)
1981	Hennepin Cty.Welfare Bd. v. Ayers	304 N.W.2d 879(Minn)
1981	In re Mengel	429 A.2d 1162(Pa)
1981	Johnson v. Johnson	395 So.2d 640(Fla)
1981	Klein v. Franks	314 N.W.2d 602(Mich)
1981	Lassiter v. Dept.Soc. Serv.	101 S.Ct. 2153
1981	Lee v. Lee	110 Misc 2d 623
1981	Little v. Streater	68 L Ed 2d 627(Ohio)
1981	M.D.L.R. v. L.V.D.L.R.	(Slip Opinion) 4 Ct. Civ. Appeals 16470
1981	Marriage of Stephen and Sharyne B.	177 Cal Rptr. 429
1981	Matter Carmen L.v.Robert K.	109 Misc 2d 259
1981	Matter of Linda K .L. v. Robert S.	109 Misc 2d 628
1981	Matter of Beaudoin v. Tilley	110 Misc 2d 696
1981	Matter of F. J. F.	312 N.W.2d 718(S.D.)
1981	Matter of Jane L. v. Rodney B.	108 Misc 2d 709
1981	Matter of Pamela P. v. Frank S.	110 Misc 2d 978
1981	Matter of Richardson v. Gibson	110 Misc 2d 20
1981	Moon v. Crawson	109 Misc 2d 902
1981	Nordgren v. Mitchell	524 F.Supp. 242
1981	Rose v.Dist.Court 8th Judicial Dist.	628 P.2d 662(Mont)
1981	Shults v. Superior Court, Etc.	170 Cal.Rptr. 297
1981	State of Iowa v. St. John	308 N.W.2d 8(Iowa)
1981	State of Montana v. Fatz	634 P.2d 172(Mont)

v. Martinez   96 AD2d 496

1983	Matter of Comm.Soc.Sv.	
	v. O'Neil	94 AD2d 480
1983	Matter of Jeanne M.	
	v. Richard G.	61 NY Reports 2d 638
1983	Matter of Martine S.	
	v. Anthony D	120 Misc 2d 567
1983	Matter of Pamela Kirk	95 AD2d 888
1983	Matter of Smith v. Jones	120 Misc 2d 834
1983	Merrill v. Ralston	95 AD2d 177
1983	Pizana v. Jones	339 N.W.2d 1 (Mich.)
1983	Robinson v. Jones	15 Am.J.Hum.Gen. 1983
1983	Social Services v. Bart D.	121 Misc 2d 425
1983	State ex re Hodges	
	v. Fitzpatrick	342 N.W.2d 870 (Iowa)
1983	State ex rel D.K.B.	
	v. W.G.I.	654 S.W.2d 218 (Mo.)
1983	State ex rel Hausner	
	v. Blackman	662 P. 2d 1183(Kans)
1984	Aroonsakul v. Flanagan	464 N.E.2d 1091 (Ill)
1984	Baker v. Wagers	472 N.E.2d 218 (Ind)
1984	Bradley v. Houston	676 S.W.2d 746 (Ark)
1984	Corley v. Rowe	312 S.E.2d 720 (r)
1984	Dix v. Laux	Case CA8140 (Slip) Ohio
1984	Estate of Cornelious	674 P.2d 245 (Ca)
1984	Everett by Warner v. Everett	198 Cal. Rptr. 391
1984	Ex Parte Calloway	456 So.2d 308 (Ala)
1984	Haines v. Schanholtz	468 A.2d 1365 (Md.)
1984	Michaella M.M.	
	v. Abdel Monem El G.	98 AD2d 464
1984	S.Car.Dept.Soc.Sv. v. Bacot	313 S.E.2d 45 (S.C.)
1984	Salicco v. Salicco	125 Misc 2d

1984	Social Servs v.Thomas J.S.	100 AD2d 119
1984	Spring v. Bias	No.Ca-3073 (Slip) (Ohio)
1984	State through Dept. Health v. Smith	459 So.2d 146 (La)
1984	Tenn.Dept. Human Sv. v. Braswell	Ct.App. Tenn, (Slip)
1985	Cnty Hennepin v. Brinkman	364 N.W.2d 458 (Minn.)
1985	Davis and Davis v. State of Indiana	No.2-1083 A356 (Slip)(Ind)
1985	Miller v. Kriner	No.00349 (Slip) (Ohio)

Chapter 7
"YA GOTTA BELIEVE"

(Note, in this chapter I have deliberately departed
from the form of writing used in the rest of the text.
As I am a geneticist, and feel qualifed to comment as a
scientist on the issues which follow, I have used the
first person where explicitly referring to my knowledge
of the field.)

The single unifying thread in all of the fields of
contemporary biology is the theory of evolution. Whether
the field be molecular genetics, taxonomy, anatomy, or
any other subdivision of this science, the assumption of
the relationships of organisms is essential.
Particularly in the field of genetics, one must
understand this concept. Thus the geneticist chooses an
experimental animal, plant, bacterium or virus with the
knowledge that mechanisms found by its use will be
directly applicable to all other forms. One does not do
research on bacteria or protozoa because one wants only
to know about these organisms; such knowledge, while of
interest in itself, is sought because of its
applicability to all other organisms, including man. One
certainly cannot do genetic or biochemical experiments
on people; perforce other organisms are chosen either
because of the knowledge, through the facts of evolution,
that discoveries here will be useful in the study of
man, or because the organism of choice lends itself to
the type of manipulation needed in the experimental
protocol. If evolution were not a fact, then general
discoveries in biology could never be made. Mendel's
laws, for example, would apply only to peas; the fact
that DNA is the hereditary material in a bacterium would
apply only to that form; the discovery of biochemical

pathways, for example amino acid synthesis, would have to be rediscovered in every one of the millions of species which exist today. While, of course, in the vast passage of time, some organisms show minor variations from the general pattern, there remains no other plausible hypothesis other than the evolutionary one to explain the fact that biological laws are general and can be applied to any and all living organisms.

In all science no credence is given to an untestable theory or hypothesis. If the proposed ideas are not subject to test, and must be accepted only by faith, they are by that fact not a part of science. The very heart of any science, and its progress is the willingness to modify an hypothesis, or discard it, should experimental results demonstrate it to be either incomplete or incorrect. There are many scientists who are deeply religious in their faiths, but not in their laboratories; the two simply are not dealing with similar sets of circumstances. For a biologist not to accept the facts of evolution would be as great a sin in the laboratory as any cardinal sin is to his religion.

Another important factor which seems to be playing a role in the present revival of anti-Darwinism is the increasing public suspicions of science and scientists. Perhaps this is due to their role in the creation of the atomic bomb or the sometimes apparent withholding of information concerning radiation already referred to. But it must be underscored that these political or military decisions are not made by the scientists; the use or misuse of scientific discoveries most frequently is made by governmental or industrial non-scientists whose decisions are obviously not made on scientific grounds. This is not to imply that scientists themselves

are always above reproach, but to point out that
the public use of scientific discoveries is an entirely
different type of decision and procedure than was the
process of making the discovery. While scientists are
beginning to assume more public roles, they are doing so
based on their special knowledge as scientists; when a
scientist speaks out on issues which do not pertain to
his own area of expertise, he speaks as any citizen, not
as some sort of Godhead.

Another part of the increased anti-scientific
feelings is the growth of belief in the occult. It
was once stated by a newspaper editor that if the paper
omitted a news story he would probably receive little
adverse criticism, but if he left out the astrology
column his phone would ring incessantly and his mail
would be filled with vituperative letters. This belief
in occultism is often given the phony dressings of
science, including attempts to fit data to the belief,
and not the other way around. Most often even the most
convincing evidence against some occult belief is
unacceptable, ignored, or openly repudiated by the
believers in the occult simply on the grounds that the
person finding the negative evidence was not a true
believer and therefore, of course, he cannot validate
the belief. Tirades against scientists from the
occultists tend to depend upon that kind of circular
argument to "prove" the unprovable.

Still another group attacks science and evolution
on the basis that either the record is incomplete or
definitive experiments cannot be accepted because they
are based partially on statistical reasoning. It is here
that the divergence between faith and scientific method
is possibly the greatest. The scientists reach their

conclusions based upon the best available evidence and they are constantly ready to alter their hypotheses if newer data shows them to be incorrect or incomplete. The occultist, or creationist, has one set of beliefs, and no evidence, no matter how compelling to others, can force them to alter their views. In simple language, their minds are closed to new ideas.

Basic to our form of democracy, set forth in the First Amendment of our Constitution, is the separation of church and state, for the mutual benefit of *both*. It is a fact that most of our educational system is state run in the broad sense. Thus the espousal of any set of beliefs on a religious basis by the state has been constantly disapproved by almost all lower courts, and repeatedly by the United States Supreme Court. Nonetheless, ever since the founding of the republic, probably, there have been those groups who have sought to impose their particular religious beliefs upon others by attempting to have them taught either exclusively or inclusively in the public schools. Each attempt has been ultimately unsuccessful, but apparently such groups are not convinced, or find new ways to try to force their will upon others through the school system. For example, the creationists now refer to their beliefs in the creation of the world on what they now call "creation science," and to demand equal time in schools to teach their beliefs under this name. The attacks upon evolution, however well meaning some of those carrying them out may be, are now based mainly on the right to have equal time in the teaching of biology for what cannot be taught in science, the acceptance of non-testable hypotheses. To call something a "science" does not make it such, particularly if acceptance of that subject must be based on non-scientific methods. There can be no objection to persons holding such ideas as a personal belief; that is equally granted to them in the constitution and its amendments; but to attempt to force others to accept such beliefs is clearly unconstitutional, and, at least to some observers, immoral. Such attempts to deny or forbid the teaching of evolutionary theory lead to the creation of a scientifically illiterate population. Even more damaging is the fact that publishers of biological texts, which they hope will be purchased in large numbers by state or local governing bodies, now shun the mention of evolution in many of their works. They fear that a careful and factually accurate discussion of the theory of evolution will lead to their largest purchasers shunning their works and turning to rival, more timorous publishers who simply omit the central theme of biology entirely from the books our children, through no choice of their own, are forced to use. To update a phrase of the late Nobel Laureate, H.J. Muller, "125 years of not teaching evolution is enough!"

Probably the most bitter attacks have come from the most misunderstood part of Darwinian theory. The creationists and their allies constantly mislead the public by stating that Darwin claimed that man descended from apes or monkeys. So firm are they in their misbeliefs that these people have never read Darwin, or attempted to understand him. At no time did he come close to making such a statement. What he carefully did was to point out the many similarities between man and the apes and suggest this might imply a *common ancestry* i.e., both man and the apes appear to have arisen from some ancient form and branched off in separate evolutionary pathways, one leading to the modern group of great apes and the other to man. Neither descended from the other. It is of further interest to point out that the modern science of chromosome analysis now shows the close relationship between these groups. Matching of the banding patterns of ape and human chromosomes indicate great similarity of structure, so great that it cannot be interpreted in any other way than that of close phylogenetic relationships between the groups. Indeed, it is likely that the groups may even share many genes between them. All of this is extremely hard, if not impossible, for a creationist to explain other than by the usual mystic ways such people rely upon. We are akin to the apes whether the fundamentalists like it or not. And philosophically, as Darwin concluded in his first edition of the *Origin of Species*, "There is a grandeur in it all."

Another disturbing aspect is the fact that regardless of past court decisions, the creationists are determined, and in some cases, well-financed groups. One legal defeat does not daunt them; they find technical means, such as mentioned above by renaming their beliefs as "creation-science," to attempt to find chinks in the law. The time, money, and bitterness engendered in such efforts is, frankly, disgraceful. One only need read some of the cases which follow to realize that the efforts of court after court, in state after state, and in the federal courts as well, certainly could be spent in different and more valuable endeavors.

There is probably little need to go into the most famous of all cases involving evolution, *Scopes v. State*, the so-called "monkey trial." The only thing which might be mentioned is the fact, often forgotten, that Scopes was found guilty of violation of the state law against teaching evolution in Tennessee in 1925, and his guilt was upheld by the Supreme Court of that state; however, the state court reversed the conviction on technical grounds concerning the power of a lower court judge to assess a fine of any amount over $50; amounts over this (Scopes was fined $100) had to be determined by jury. It was also noted that the Supreme Court suggested that Scopes not be retried. But the constitutionality of that state's law was not challenged by the decision itself.

It was not until 1960 that the constitutionality of any state law forbidding the teaching of evolution was decided by the United States Supreme Court. This case, arising in Arkansas, concerned that state specifically banning such teaching. In *Epperson v. Arkansas* the United States Supreme court upheld Epperson and found the state law clearly in violation of the First Amendment. The decision was based on the difference between secular and religious purposes for a statute and the understanding that government had no business meddling in strictly religious matters.

In 1972, A United States District Court, Southern District, Texas, again dealt with the problem of the teaching of evolution. The facts in *Wright v. Houston Independent School District* were somewhat different, although also quite similar. The similarity was the attempt to ban the teaching of evolution; the difference was that this was not by state law, but that suit was brought against a school district and the State Board of Education by a group of parents (the plaintiffs) in an attempt to ban both the teaching of evolution and the adoption of texts which presented such a theory. Based on *Epperson*, the parents sought injunctive relief against the action of the two governmental bodies. The lower court denied relief, and the appeal followed. The higher court upheld the claim for dismissal by the defendants. "Defendants, however, are not acting pursuant to either state law or school district regulation. Plaintiffs have not alleged that there exists even a school district policy regarding the theory of evolution. All that can be said is that certain textbooks selected by school officials present what Plaintiffs deem a biased view in support of the theory. The Court has been cited to no case in which so nebulous an intrusion upon the principle of religious neutrality has been condemned by the Supreme Court." The court could therefore find no reason whatsoever for a basis of this suit. "Plaintiffs' case must ultimately fail, then, because the proposed solutions are more onerous than the problem they purport to alleviate. For this Court to require the District to keep silent on the subject of evolution is to do that which the Supreme Court has declared the legislature is powerless to do. To insist upon the presentation of all theories of human origins is, on the other hand, to prescribe a remedy that is impractical, unworkable and ineffective." What the case, obviously, was about was the attempt by a group of parents to impose censorship upon the state and the school board because of their religious beliefs, and their apparent personal offense. The court also stated earlier in its opinion, "In the case at bar, the offending material is peripheral to the matter of religion. Science and religion necessarily deal with many of the same questions, and they frequently provide conflicting answers...Teachers of science in public schools should not be expected to avoid the discussion of every scientific issue on which some religion claims expertise."

Yet another method of attack by the fundamentalists and creationists arose in the 1980 case, *Crowley v. Smithsonian Institution.* Crowley brought suit both individually and as head of an organization known as "The National Foundation for Fairness in Education," and by another organization, "The National Bible Knowledge, Inc." against the Smithsonian for mounting an exhibition on the subject of evolution, claiming that this violated religious neutrality under the First Amendment, and promoted "Secular Humanism" as a religion. The case was decided by the United States Court of Appeals, D.C. Circuit. Both organizations claimed that they could "prove" that scientific evidence supported the proposition "that human and other forms of life were brought into existence in completed form, all at one time, by the Creator." The case was based upon the fact that the exhibit was supported by federal funds and the plaintiffs claim that this violated the freedom of religion clause of the First Amendment. The exhibit was entitled "The Emergence of Man," and a further exhibit planned was to "emphasize specimens from the Museum's collection depicting adaptations of plants and animals to their environment by such means as camouflage, the overproduction of offspring and other defense mechanisms. It was to include an introductory display of a variety of specimens such as trays of bird eggs, mammal skulls, and jars of amphibians. ...There were also to be displays on genetics, natural selection, and one showing differentiation of populations."

The crux of the matter appeared in the claim that

"Appellant's opposition to the summary judgment motion [the lower court had dismissed the case] was essentially a challenge to the concept of evolution. It questioned whether that concept is any more susceptible to scientific proof than appellants' concept of the supernatural origins of life, and characterized this issue as one of material fact. In support of this contention, they offered the affidavit of Dr. Richard B. Bliss, Director of Curriculum Development for the Institution of Creation Research, San Diego, California." This institution is named here, as it will appear again in further cases. The claim by Dr. Bliss was that evolution had to be a "faith" and not a scientific fact, as it could not be observed, nor subject to exact scientific tests. The Court dealt with this issue summarily, "Assuming *arguendo*, that, as asserted in Dr. Bliss' affidavit, the evolution theory cannot be proved "scientifically" in the laboratory and in that sense rests ultimately on "faith," such fact is not material because it would not establish as a matter of law that the exhibit in question establishea any religion such as Secular Humanism."

"The fact that religions involve acceptance of some tenets on faith without scientific proof obviously does not mean that all beliefs and all theories which rest in whole or in part on faith are therefore elements of a religion as that term is used in the first amendment. For example, appellees suggest that the theory of relativity defies absolute laboratory proof. Obviously the constitution would not interdict government development and diffusion of knowledge about relativity even if it were based on some hypotheses which are not susceptible to physically demonstrable proof." The opinion then points out that the government does not get involved in religious beliefs, citing the legality of such acts as birth control or abortion which are, to say the least, subjects of great religious controversy. The opinion continues to make the point: "Thus, the essential question posed by appellants has been resolved by authoritative decisions permitting public schools to teach the facts and theory of evolution to children who, unlike appellants, are compelled by law to come and look and listen." In other words, if the appellants do not like what they see, they are free to stay home and avoid exposing themselves to the exhibit. Finally the court, in deciding against the appellants, stated, "In view of the foregoing, it is unnecessary to labor and therefore we only note that we approve the District Court's application of the criteria. ...The solid secular purpose of the exhibits is apparent from their context and their elements. They did not materially advance the religious theory of Secular Humanism, or sufficiently impinge upon appellants' practice of theirs to justify inderdiction. Except insofar as appellants have themselves entangled religion in the exhibits, there is no religious involvement as that concept is used."

Again the court has denied that the teaching of the concepts of evolution is a violation of anyone's freedom of religion, and one would think that by now this would be accepted even by the creationists as the law of the land. Unfortunately, there appears to be no end to the fanaticism of this group, as the cases which follow indicate. To this observer, the refusal to admit the illegality of their position indicates the inability of the creationists and their supporters to understand the facts of democracy, and certainly an inability to live

within the constraints of legal jurisprudence. The
issues appear not only to remain with us, but in the
present social and moral climate to become ever more
embittered.

The classic modern case involving the attempt to
avoid the problems of the First Amendment and the
separation of church and state was *McLean et al v.
State of Arkansas*, decided in 1981 by the United States
Court of Appeals, Eighth Circuit. The state of Arkansas
had enacted a statute calling for the teaching of the
creationist theory on an equal basis with evolution. The
creationist argument in this case was a simple one.
Creationism, they claimed, was not a religion, but a
science, and therefore was exempt from the provisions of
the First Amendment. The case was also unusual in that
the plaintiffs were principally clergymen and religious
groups as well as members of the Arkansas Board of
Education and parents of children in the schools there.
Indeed, the list of plaintiffs occupies almost a page in
the opinion. Before the case was tried, an attempt was
made by the creationists to deny the right of many of
the plaintiffs to intervene, and this also was denied by
the court. The opinion involving the actual material of
the case is lengthy and strong. As I feel that the
judge writing the case intended his opinion to be a
final word on the subject, I shall quote from it
extensively. First, however, it should be emphasized
that the judge decided firmly that creationism was *not* a
science, and that the attempt to call it such was a
blatant attempt to circumvent the federal laws. The
entire basis of the case was again not the question of
First Amendments rights, but was whether creation
science was a religion.

The judge pointed out the difference as follows: "More precisely, the essential characteristics of science are: (1) It is guided by natural law; (2) It has to be explanatory by reference to natural law; (3) It is testable against the empirical world; (4) Its conclusions are tentative, i.e. are not necessarily the final word; and (5) It is falsifiable.

"Creation science as described...fails to meet these essential characteristics. First ...[it] asserts a sudden creation from 'nothing.' Such a concept is not science because it depends upon a supernatural intervention which is not guided by natural law. It is not explanatory by reference to the natural law." He then cites from the creationist literature: "We do not know how God created, what processes He used, for God used processes which are not now operating anywhere in the natural universe. This is why we refer to divine creation as special creation. We cannot discover by scientific investigation anything about the creative processes used by God." The Court then examined in detail most of the claims against evolution made by the creationists, and their own "scientific claims." He shows clearly that all of the so-called creation science claims are nothing more than a window dressing for an adherence to the book of Genesis, including the flood, and the belief that the events described there took place exactly as described. He adds, "Creation science as defined not only fails to follow the canons defining scientific theory. It also fails to fit the more general descriptions of 'what scientists think' and 'what scientists do.' The scientific community consists of individuals and groups, nationally and internationally, who work independently in such fields as biology,

paleontology, geology and astronomy. Their work is
published and subject to review and testing by their
peers. The journals for publication are both numerous
and varied. There is, however, not one recognized
scientific journal which has published an article
espousing the creation-science theory "...Some of
the State's witnesses suggested the scientific community
was 'close-minded' on the subject of creationism and
that explained the lack of acceptance of the creation
science arguments. Yet no witness produced a scientific
article for which publication had been refused. ... The
creationists have difficulty maintaining among their
ranks consistency in the claim that creationism is
science. The author of the act... said that neither
evolution nor creationism was science. He thinks both
are religion. ...The creationist's methods do not take
data, weigh it against the opposing scientific data, and
thereafter reach the conclusions stated. Instead, they
take the literal wording of the Book of Genesis and
attempt to find scientific support for it." Among other
citations to creationist literature in accord with this
view, the judge cites examples from the Creation
Research Society in which it is made plain that
membership depends upon undeviating subscription to the
belief that "The Book of Genesis is 'historically and
scientifically true in all of the original autographs.'

"While anybody is free to approach a scientific
inquiry in any fashion they (sic) choose, they cannot
properly describe the methodology used as scientific if
they start with a conclusion and refuse to change it
regardless of the evidence developed during the course
of the investigation." The judge continues to examine
the claims for scientific method in the creationists'

world, and finds not a single valid claim for their use
of the word "science." Further, teachers in the state,
under mandate to provide material for their classes
dealing with creation science, could not do so. A group
spent considerable time and effort in an attempt to see
whether there was any valid basis of knowledge so that
they could comply with the Arkansas law, and found none.
Indeed all of the literature they examined was so
permeated with religious beliefs and references as to be
unacceptable. Indeed the material prepared for
creationists by Dr. Bliss and his Institute, mentioned
above, was so full of factual errors as to be useless.
For example, one chart which purports to show the
evolutionists' views showed man evolving from a mammal,
itself believed true of course, but the mammal shown in
the chart was a rat! Another choice example from this
purported text stated, "Flowers and roots do not have a
mind to have purpose of their own; therefore, this
planning must have been done by the creator." Indeed
some of the literature examined by the committee
mentioned above carried the specific caveat that it was
not designed to be used in public schools. The court
sums it up best. "The conclusion that creation science
has no scientific merit or educational value as science
has legal significance in light of the Court's previous
conclusion that creation science has, as one major
effect, the advancement of religion...Since creation
science is not science, the conclusion is inescapable
that the only real effect of the act is the advancement
of religion."

"Implementation of the act will have serious and
untoward consequences for students, particularly those
planning to attend college. Evolution is the cornerstone

of modern biology, and many courses in public schools contain subject matter relating to such varied topics as the age of the earth, geology and relationships among living things. Any student who is deprived of instruction as to the prevailing scientific thought on these topics will be denied a significant part of science education. Such a deprivation through the high school level would undoubtedly have an impact upon the quality of education in the State's colleges and universities, especially including the pre-profession and professional programs in the health sciences." Further, "If creation science is, in fact, science and not religion, as the defendants claim, it is difficult to see how the teaching of such a science could 'neutralize the religious nature of evolution.'

"Assuming for the purposes of argument, however, that evolution is a religion or religious tenet, the remedy is to stop the teaching of evolution, not establish another religion in opposition to it. Yet it is clearly established in the case law, and also perhaps in common sense, that evolution is not a religion and that teaching of evolution does not violate the Establishment Clause. ...The application and content of First Amendment principles are not determined by public opinion polls or by a majority vote. Whether the proponents of the act constitute the majority or the minority is quite irrelevant under a constitutional system of government. No group, no matter how large or small, may use the organs of government, of which the public schools are the most conspicuous and influential, to foist its religious beliefs on others." The judge ends his opinion with the following words of Justice Frankfurter: "We renew our conviction that 'we have

staked the very existence of our country on the faith
that complete separation between the state and religion
is best for the state and best for religion.'...If
nowhere else, in the relation between church and state,
'good fences make good neighbors.'"

I have cited much from this opinion, as I am
certain it will be the most used one in all such cases
to come. A federal judge has decided, with vigor, that
creation science cannot be legally found to be true
science. All forthcoming cases, following this, will now
have the burden of finding other means to bring their
suits. Unfortunately, clever legal ways may be found in
an attempt to avoid this opinion, but the burden of
proof will, I hope, be a most heavy one.

There is a series of comments on this case in the
journal *Science* which bring out other facets I have
not dealt with. For example evidence is deduced that the
creationists themselves were not prepared for the act to
pass in the Arkansas legislature, and therefore not
prepared to defend the case strongly. Be that as it may,
the case has been decided. Upon consideration, the
creationist group decided not to appeal this particular
decision, basing their hopes on other pending cases, one
of which will be mentioned below.

Before coming to that case, reference must be made
to two other cases. The first, not reported in any form
in the legal journals, was described in an article in
Perspectives in Biology and Medicine, and concerned an
attempt in California to ban teaching of evolution by
the parents of a child as they believed it to be
contrary to his religious beliefs. Many biologists were
prepared to testify, and some did. At issue was the
syllabus of the State Education Department and the

previous claim that this established Secular Humanism as a state religion. Perhaps the reason the case was not more widely reported was the fact that the judge in the Superior Court of California delivered an unwritten opinion directly from the bench at the conclusion of the trial. The judge found no violation of religious rights and dismissed the case.

Another case involving this type of teaching arose in 1982 in South Dakota. The case, *Dale v. Board of Education*, was tried before the Supreme Court of that state. The issue here was complicated by problems of court interference with the freedom of actions by school boards, which normally is not done. However, the court could interfere in the case as their decision upheld the independence of the board.

Dale was a biology teacher in the school system for seventeen years when the school board decided not to renew his contract. Their reasons were simply that he had not complied with orders to follow the established curriculum in biology and instead spent the entire time in class teaching creationism. He was several times warned and ordered to cease this and to teach biology, warnings which he chose to ignore. The facts show that in one year he failed to cover, after three-quarters of instruction, the basic biology subjects appearing in chapters 4, 5, 7, 8 and 9 through 20 of the text he had been directed to teach. As most books at that level have seldom more than twenty chapters, it means that he had omitted almost the entire text. The record shows clearly that instead he had used the time to promote his beliefs in creationism. (One cannot help but wonder what took three-quarters of a year to teach.) Dale brought suit against the school board on the basis of the decision to

dismiss him as being arbitrary, capricious and an abuse
of discretion by the school board. In addition, as
pointed out by the court, there was some vague reference
to the principles of religious freedom, but none which
seemed well enough defined to be dealt with here. As
pointed out, much of the opinion dealt with the rights
of the board and the rights of the court to interfere at
all in such cases. But the court was clear that in
upholding the right of the school board to act as it had
they were not in essence dictating what should or should
not be taught in schools. As one concurring opinion put
it, "Essentially, Mr. Dale wanted to be a preacher, not
a teacher. This is intolerable in a classroom under our
state law, state constitution, and federal
constitution." However, one partially dissenting opinion
did show concern: "I dissent on the scope of review as I
fear that the South Dakota judiciary will go into
uncharted, unwanted, and unconstitutional waters
concerning the propriety of a school board's decision.
The majority opinion is radical departure from the
settled law of the state on teacher-school board
decision. It shatters previous precedent. I fear this
evolution of change, which seems spasmodic, in these
teacher-school board cases. When will we ever be able to
assure trial counsel, school boards, and teachers of
this state that, indeed, here is the law? I yearn for
stability in this area of the law. ...Above all, I fear
a judicial invasion into the doctrine of separation of
powers and the province of a school board to make a
decision, providing it is done in accordance with law. A
school board's good judgment is not in question. Only
the legality of the school board's decision is in
question. I would not permit a circuit court nor the

Supreme Court to run the school house." What the
dissenting judge is referring to was mentioned earlier,
i.e., the power of school boards to run their own
schools and not be interfered with unless they clearly
violate the law. Had the decision been made only on that
basis, there would have been no dissent; indeed, the
dissenting judge agreed to that part of the decision.
But he felt constrained to chastise his fellow judges
for delving into the facts underlying the dismissal
rather than just the dismissal itself.

The final case is one in Louisiana where the
creationists once again demanded equal time. The case
was to be defended against the creationists by a host of
organizations and was to be brought in a federal court.
However, when time came for trial, according to the New
York Times, the federal judge found that all state
remedies had not been exhausted and therefore the court
had no jurisdiction.

When the case then came to the Louisiana Supreme
Court, that court upheld the constitutionality of the
equal-time law, presumably under Louisiana's, not the
Federal, Constitution. The issue of the First Amendment
was not faced by the divided court and the case then
came before the federal court system on the basis of
violation by the state law of the First Amendment.

When the case come before the Federal District Court
as *Keith v. Louisiana Department of Education*, the main
issue did not deal with the facts of evolution. In fact,
the court specifically refused to deal with that issue
and decided the case only according to the technicality
of the laws under which it was brought. The whole matter
was based on whether the Federal Court, in this case the
U.S. District Court, M. D. Louisiana, had jurisdiction.

The suit was brought to have the court decide that Louisiana's *Balanced Treatment Act* which required equal time for the teaching of Creationism was constitutional or whether it violated various federal amendments. The attempt was obviously made as the Louisiana Courts held that the statute could not be enforced. In refusing to act, the District Court pointed out, "...To sanction suits for declaratory relief as within the jurisdiction of the District Courts merely because, in this case, artful pleading anticipates a defense based on federal law would contravene the whole trend of jurisdictional legislation by Congress, disregard the effective functioning of the federal judicial system and distort the limited procedural purpose of the Declamatory Judgment Act.

"Since the matter in controversy as to which plaintiffs asked for a declaratory judgment is not one that arises under the Constitution and Laws of the United States and there is no diversity jurisdiction present in this case, plaintiffs' suit must and shall be dismissed with prejudice at plaintiffs' cost." Thus, not only did the creation-science people lose their case, but they also had to pay for the considerable costs of it!

Nonetheless, the creationists were not satisfied by this decision obviously in opposition to their views. They somehow managed to have it appealed to the Fifth U.S. Circuit Court of Appeals. Again they were rebuked. At present there may be yet another attempt to have the case heard *en banc* before all members of the Fifth Circuit, but it is not yet clear whether this will be granted. However, The U.S. Supreme Court has recently granted *certiorari* and will hear the case during the

1986-1987 term.There seems to be no end to the amount of litigation the creationists cause; obviously they are well financed and entrenched in their inability to understand the United States First Amendment. The arguments cannot help but be the same, and, while it is dangerous to predict just what a court will decide, it would seem that the precedence of the Arkansas case would apply here. Somehow, it would seem, the issue should have been closed permanently by the earlier decision, but it has not been so, at least in Louisiana, or in the minds of the creationists.

Nor is the issue truly closed in other states. An attempt was made early in 1984 to have an identical bill introduced into the Arizona Legislature by a group of five legislators, and it is believed that similar attempts in other states will also be made. In view of the Arkansas and Louisiana case, such efforts would seem to be almost a deliberate waste of legislative time, and court efforts, but this does not seem to deter the creationists.

On the other hand, the evolutionists have made some progress in the state of Texas. Here the pressure has come from the educational system itself, and the Attorney General of that State. That official has ruled that the "state's textbook anti-evolution rules violate the First Amendment." The creationists will obviously attempt to fight that ruling, presumably again in appealing to the courts. If the Attorney General prevails, there cannot help but be a change in biology texts as Texas is the single largest purchaser of such materials. If the publishers can now include the usual discussions of evolution in Texas, then other states which also depend upon these publishers will be able to purchase biology texts which deal with evolution in an appropriate way. However, this ruling is also subject to even further legal action; one group in Texas has announced its intention to seek an injunction against the selection committee banning them from using books which do not deal with creationism and only deal with the facts of evolution.

It should also be noted that in New York State certain texts, which would be most suitable to the creationists in Texas, have been passed over on the grounds that they *do not* deal with evolution in an honest manner. However, the creationists will not accept legal decisions or opinions, and the struggle for honesty in science continues. Again, as stated in the text, this whole matter is a waste of valuable court time, and one wonders what steps can finally be taken to prevent the fanatical creationists from continuing to harass the courts, the public, and ultimately to damage the education of American children. These people simply do not understand the fundamental principles of the Bill of Rights, and if they should, by chance, understand them, they are against them!

I realize fully that I have stated strong and personal opinions throughout this chapter, opinions which will not make this a popular work in many states. However, this area, the teaching of biology, is my specialty, and in the areas concerned, evolutionary methods, I am more expert than most. If I did not speak out as I have, I would be in a sense betraying my own field of genetics. There is no doubt whatsoever in the minds of my colleagues that there are clear facts which cannot support any other hypothesis than that of organic evolution. We know, and readily admit, that we do not all agree on the exact processes of evolution, and that we are changing our views in this area as new data from all fields are introduced into scientific knowledge. But these new facts serve only to fortify the general principles of evolution and not to distort them. More importantly, these new data come under the rigid definition of science so well stated in the Arkansas case. Further, I would not be writing this book if evolution had not occurred.

ALPHABETICAL LIST OF CASES CITED

CASE	CITATION	YEAR
A Tale with too many Connections	215 Science 484	1982
Annas,Monkey Laws in the Courts	12 Hastings Cntr. Reports 221982	1982
Anti-evol.Rules are Unconstitutional	223 Science 1373	1984
Creationism law struck down	Knickerbock News (Albany, N.Y.)	7/9/85
Creationism on Trial in Arkansas	214 Science 1101	1981
Creationism on the Defensive	215 Science 33	1982
Creationist, ACLU to do Battle Again	222 Science 488	1984
Crowley v. Smithsonian Institute	636 F.2d 738	1980
Dale v. Board of Education, etc.	316 N.W.2D 108 (S.D.)	1982
Epperson v. State of Arkansas	393 U.S. 97	1968
Fed.Judge dismisses Creation Science Case	New York Times, 6/29 A-14	1982
Keith v. Louisiana Dept. of Educ.	553 F.Supp.295	1982
McLean v. State of Arkansas	663 F.2d 47	1981
McLean v. State of Arkansas	LR.C 81 322, (Slip)	1982
Seagraves v. State of California	25 Pers.in Biology & Medicine 207	1982
Wright v.Houston Independent School District	366 F.Supp.1208	1972

CHRONOLOGICAL LIST OF CASES CITED

YEAR	CASE	CITATION
1968	Epperson v. State of Arkansas	393 U.S. 97
1972	Wright v Houston Independent School District	366 F.Supp. 1208
1980	Crowley v. Smithsonian Institute	636 F.2d 738
1981	Creationism on Trial in Arkansas	214 Science 1101
1981	McLean v. State of Arkansas	663 F.2d 47
1982	A Tale with too many Connections	215 Science 484
1982	Annas,Monkey Laws in the Courts	12 Hastings Ctr Reports 22
1982	Creationism on the Defensive	215 Science 33
1982	Dale v. Board of Education, Etc.	316 N.W.2D 108(S.D.)
1982	Fed.Judge dismisses Creation Science Case	New York Times, 6/29 A-14
1982	Keith v. Louisiana Department of Education	553 F.Supp.295
1982	McLean v. State of Arkansas	LR.C 81 322 (Slip)
1982	Seagraves v. State of California	25 Perspectives in Biol. & Medicine 207
1984	Anti-evol.Rules are Unconstitutional	223 Science 1373
1984	Creationist, ACLU to do Battle Again	222 Science 488
1985	Creationism law struck down	Knickerbocker News (Alb.NY) July 9

The cases involving genetics and patent laws will
be arbitrarily divided just as the field of genetics now
is; i.e., into classical and molecular fields. The
former will deal with the use of genetics in the areas
of patent infringements and the complex cases of the
interplay between patent law, antitrust law, contract
liability, and implied warranties. The latter will deal
solely with the issues of whether microorganisms derived
by genetic means are to be considered under the laws
regulating the patentability of plant forms. In both
cases it will be necessary to quote at length from the
patent laws and the court opinions. Finally, some
comments will be made regarding the impact of these
decisions upon the science and upon the field of
genetics itself.

THE "CLASSICAL" CASES

These cases depend upon the wording of the plant
patent statutes, and it is necessary to quote parts of
them in order to understand the courts' decisions.
"Whoever invents or discovers and asexually reproduces
any distinct and new variety of plant, including
cultivated sports, mutants, hybrids, and newly found
seedlings, other than a tuberpropagated plant or a plant
found in an uncultivated state, may obtain a patent
therefor, subject to the conditions and requirements of
this title. The provisions of this title relating to
patents for inventions shall apply to patents for
plants, except as otherwise provided" (Chapter 15,
section 161, T. 35a U.S.C.A.-43. 1952). And, Chapter 57,
section 7, 2402, 1973, "(a) The breeder of any novel

variety of sexually reproduced plant (*other than fungi,*
bacteria, or first generation hybrids) who has so
reproduced the variety, or his successor in interest,
shall be entitled to plant variety protection therefor,
subject to the conditions and requirements of this title
unless one of the following bars exist" (Emphasis
added). The bars which follow, which will be referred to
in the second part of this chapter include the one that
the patent application be made within one year if,
"...the variety was (A) a public variety in this
country, or (B) effectively available to workers in this
country and adequately described by a publication
reasonably deemed a part of the public technical
knowledge in this country which description must include
a disclosure of the principal characteristics by which
the variety is distinguished... The change in the
reading of the plant patent acts was necessitated by
advances in genetic techniques which now made it
possible to propagate new varieties by sexual rather
than only asexual means. Also, one of the difficulties
in plant patents is the fact that often varieties are
extremely similar and only differ by slight changes in
color or aroma, and descriptions needed to secure a
valid patent are difficult to make. Nonetheless, as
anyone knows who has purchased a patented plant such as
a tea rose, plant patents for relatively minor genetic
changes are granted. The need to protect the
introduction of such varietal changes will be obvious in
the cases to be considered first, under the rubric of
the classical cases. What will also be apparent is the
need for the courts to understand genetics, and the
first case will quote at length from what is a patent
decision, not a quotation from a botany text nor from a

genetics book. In the 1976 case, *Yoder Bros., Inc. v. California-Florida Plant Corp*, The United States Court of Appeals, Fifth Circuit, found it necessary to state at the beginning of its opinion the following: "Chrysanthemums, in their natural state, blossom only during the fall. This is because they are photoperiodic in nature, meaning that their growth is affected by the relative lengths of lightness and darkness in the day. When days are long, the chrysanthemum plant remains in a vegetative state. As the nights become longer, the initiation process of the chrysanthemum bud begins. Thus, in early August, when the nights achieve a duration of nine and one-half continuous dark hours, the chrysanthemum plant in its natural state will begin the process of developing a flower. During the fall and early winter months, the mature flower appears. ... By applying black cloth shades over the chrysanthemums when dark hours were needed and applying artificial light when light hours were needed, it became possible to flower chrysanthemums on a year-round basis." The opinion goes on to mention the problems involved with excess heat which retarded bud initiation as well as loss of color. The court next goes on to discuss genetics *per se*: "New varieties of chrysanthemums may be developed in two major ways: by sexual reproduction and by mutagenic techniques. Sexual reproduction, the result of self or cross pollination, produces a genetically unique seedling, the characteristics of which are impossible to predict. Mutagenic techniques simply accelerate the natural rate of mutation in the chrysanthemum plant itself." (The effects of Xrays on mutagenesis resulted in the awarding of the Nobel Prize to Hermann J. Muller). A mutation was described...as a "change in the number of chromosomes or a change in the chromosome position or a specific change in the genes within those chromosomes. Technically, only those mutations that first express themselves as bud variations are properly called 'sports'; however, the word is used loosely in the industry as a general synonym for mutation, and we will so use it. Two kinds of sports can appear: spontaneous sports and radiation sports. The cells of all living organisms occasionally mutate, and spontaneous sports are simply the result of that process. Radiation sports, on the other hand, are induced artificially, through exposure to such things as gamma irradiation from radioactive cobalt and X-rays. These techniques do nothing that could not occur in nature apart from speeding up the natural mutation process. Although most of the mutations induced by radiation are not commercially usable plants, a skilled breeder will select for further development those that display such desirable characteristics as fast response time, temperature tolerance, durability, size, and vigor.

" After a breeder has successfully isolated a new variety, the only way he can preserve his creation is by means of asexual reproduction...the most common technique of asexual reproduction is the taking of cuttings from the stock plant."

The chrysanthemum industry is a multimillion-dollar- a-year business. The industry is complex. The first group, known as breeders, create the new varieties of the plant. At the next level are the propagator-distributors who expand the stock by using cuttings. The next level consists of the growers who develop the plants received from the propagator-distributor. Another

group, the self-propagators may combine the second two
functions. Finally either the self-propagators or the
propagator-distributor sells the flowers to the florist
for the retail trade. It is necessary to go into this
detail so that the Yoder case can be followed.

Yoder was the largest of the breeders, and also
functioned simultaneously as a large propagator-
distributor. California-Florida (henceforth Cal-Florida)
was solely a propagator-distributor, dealing mainly on
the west coast, but having a wholly owned subsidiary,
also a propagator-distributor, corporation in Florida.
They were at first in direct competition with Yoder.
Yoder, however, formed a corporation called BGA which
had the exclusive purpose of regulating the industry for
Yoder. Earlier, Yoder had formed another growers'
association, members of which, in return for receiving
new varieties from Yoder were also expressly prohibited
from selling, loaning, or otherwise disposing of
purchased cuttings. In effect this meant that the
members of this growers' group could not sell their
product to any propagator-distributor or self-
propagators. More importantly for this case, the members
of the association were also required to return to Yoder
any new sports which appeared in the cuttings originally
received from Yoder, in effect a monopoly against their
propagating new varieties on their own.

The BGA simply took over these restrictions. As
Yoder was by far the largest member of the BGA, it
controlled the industry. The first part of the long
opinion in this case dealt at length with these
agreements, their costs, and the fact that the United
States had obtained a consent degree from Yoder, under
the Sherman Antitrust Act, to disband the BGA. At the

same time Yoder was allowed to retain its patent rights for the varieties it had developed. As Yoder controlled most of the patents which were used for commercial production, and continued to patent new varieties which could only be grown by asexual reproduction and each plant so grown was subject to a royalty payment to Yoder, there appeared to be no real difference between the monopoly under the BGA and the present condition of almost complete control of patent rights. These then are the backgrounds for the case.

In the suit, it was alleged by Yoder that the Cal-Florida companies were infringing upon some twenty different Yoder patents. The Cal-Florida concern asserted the invalidity of twenty-two patents, and also that, as pointed out above, Yoder continued to avoid the antitrust agreement mentioned above. The lower court found in favor of Yoder on the patent argument and also asserted in their favor on the monopoly issue, but did find that Yoder had indulged in a group boycott which was illegal. The jury, however, found that Yoder had violated the Sherman Act and also found in favor of Cal-Florida regarding eight of the patent claims, infringement of patents for eight additional varieties, and was unable to decide on the remaining ones.

The United States Court of Appeals dealt with many issues and varying defenses. Among those raised were whether Yoder indeed had a monopoly such that the Antitrust laws applied (decided in the negative) and whether the statute of limitations applied (also negative) and the right of Cal-Florida to bring suit as an affected party (affirmatively decided). But the major reason for citing this case was the dispute over the patents, particularly the ones which the jury had found

were infringed by Cal-Florida. Again the court in these
claims referred frequently to the patent acts.
"Normally, the three requirements of patentability are
novelty, utility, and nonobviousness. ...For plant
patents, the requirement of distinctness replaces that
of utility, and the additional requirement of asexual
reproduction is introduced. ...As applied to plants, the
Patent Office Board of Appeals held that a 'new' plant
had to be one that literally had not existed before,
rather than one that had existed in nature but was newly
found, such as an exotic plant from a remote part of the
earth. ...The characteristics that may distinguish a new
variety would include, among others, those of habit;
immunity from disease; or soil conditions; color of
flower, leaf, fruit or stems; flavor; productivity,
including ever-bearing qualities in the case of fruits;
storage qualities; perfume; form; and ease of asexual
reproduction. Within any one of the above or other
classes of characteristics the differences which would
suffice to make the variety a distinct variety, will
necessarily be differences of degree." Further
discussion is given concerning the requirement of
obviousness which was one of the three requirements
mentioned above. The court continues to question the
meaning of the act: "...Rephrasing the ...tests for the
plant world, we might ask about (1) the characteristics
of prior plants of the same general type both patented
and non-patented, and (2) the differences between the
prior plants and the claims at issue. ...In the case of
plants, to develop or discover a new variety that
retains the desirable qualities of the parent stock and
adds significant improvements, and to preserve the new
specimen by asexually reproducing it constitute no small

feat. ...We suspect that part of our problem in applying plant patent concepts to the facts before us lies in the fact that we are dealing with ornamental plants. Beauty for its own sake is not often a goal of inventors-- indeed, even ornamental plant breeders might be more aptly described as seekers of beauty for profit."

Still another issue in deciding plant patent cases arose. What if the identical mutation occurs so that the originally patented plant is no longer unique? Would propagation of this independently derived plant be a violation of patent laws? A geneticist would deem this a most unlikely event; identical mutations would be extremely rare; what is quite possible is that another mutation at another chromosomal site might result in a plant of similar appearance. Only cross breeding tests could determine this, and no such genetic tests were carried out. Cal-Florida had raised this issue for one of the patent varieties, claiming it had arisen independently and there was no patent protection covering it. The court disposed of this issue by simply stating, "We therefore find that Congress did not intend to exclude the kind of sport that recurs from the Plant Patent Act. ...We do not think that sport recurrence would negate invention, however. An infinite number of a certain sized sport could appear on a plant, but until someone recognizes its uniqueness and found that the traits could be preserved by asexual reproduction in commercial quantities, no patentable plant would exist."

"The subtleties of the chrysanthemum business have given rise to a welter of legal issues in this case, both patent and antitrust. To summarize our holdings on the patent claims briefly, we have agreed with the lower court that evidence of sport recurrences is irrelevant to the patentability of plants, and that insufficient evidence was introduced to rebut the statutory presumption of patent validity." A partial dissent was based on other views: "It is enough for me to base this on my impressions. As I faced--in preparation for the oral arguments of a case all feared would produce an opus of the kind it did [41 pages of opinion] the complex briefs of these skilled advocates, I thought the whole thing was about chrysanthemums. ...In more traditional language this cryptic analysis bears fruit-- or more accurately--flowers." He sums up his feelings as follows, "Operating on what I hope is not a dubious notion that a Judge should have at least the common sense--although not encased in Seventh amendment armor-- of a jury I cannot escape the conviction that these competing factors call for fact-finding reasoning, not a deliverance of law from our non-horticulture hothouse. When one wants a Yellow Rose of Texas he is not satisfied with a Mrs. Miniver, no matter how cheap, available or beautiful in some other beholder's eye. A camellia for a hair dress offset to olive skin and a black gown is not filled by a carnation, or for that matter, a chrysanthemum.

"To each his own. And here David and Goliath are struggling over a single thing--chrysanthemums. Survival of one in this business depends on whether the other can be curbed. I respectfully dissent as to this feature of the Court's holding." Besides showing a flair for writing, the dissenting judge would also have probably upheld the antitrust violation. Considerable space has been devoted to this case as it epitomizes the problems with plant patent laws, concerning what is patentable, what must be shown to obtain the patent, and the pains a

holder of a valuable patent will take to protect his
assets.

The second case in the classical tradition, will
also involve chrysanthemums. (It is probably apparent
that cases involving other flowers or agriculturally
valuable patents have arisen, but as these two epitomize
the role of genetics in resolving such cases they form
the needed basis for most cases.) This case, *Pan-
American Plant Company v. Andy Matsui*, decided in 1977
by the United States District Court, N.D. California,
dealt again with the requirements for patentability. The
facts are, however, quite different in this first case.
The plaintiff had found a yellow sport of the flower and
applied for and received a plant patent for it.
Subsequently, the company found that the plant they had
patented was subject to a very high degree of
deformities, approximately 50-60% of the blossoms being
affected. Fortunately for the company, another extremely
similar variety from the same origin was found by an
independent producer of cut flowers. Both plants had
identical yellow blossoms, and Pan-Am grew them side by
side with their original mutant. When it was determined
that the mutant developed by the outside grower did not
show the same deformities, Pan-Am destroyed all of its
own stock and substituted the grower's plant for its own
patented material.

About the same time, the defendant, Matsui, had
discovered a similar yellow mutant from the same
original stock as was used by Pan-Am and had begun its
commercial exploitation. Pan-Am then brought suit
against him for patent violation; it should be clearly
kept in mind that the three yellow varieties, the
original Pan-Am patented variety, the substituted

variety used by Pan-Am and the Matsui variety all arose
from the same parent stock. In return Matsui raised
thirteen affirmitive defenses and counterclaimed against
Pan-Am alleging patent invalidity and violation of the
antitrust laws. Only the first issue, the validity of
the patent is of concern here.

It is obvious, as it was to the court, that the
plant originally patented was not a useful commercial
product due to its high rate of deformities. Pan-Am
claimed that it was not unusual for a breeder to
substitute other material under the same patent. The
court found, "To the extent that Danielson's patent
claim [the original patent for Pam-Am] describes his
plant material as 'producing very few culls' it is
clearly inaccurate. This inaccuracy is especially
troubling because before Plant Patent 3486 was ever
granted, Pan-American not only knew the Danielson plant
material produced 50 to 60% culls, but the company
actually destroyed all the plant material. Such a
knowing inaccuracy raises the spectre of fraud and
serious questions about the validity of the patent. ..
Failure to accurately describe in the patent claim the
Danielson plant material's known propensity to produce
culls also raises the question whether Pan-American's
description of the patented plant was 'as complete as is
reasonably possible' as required by 35 U.S.C. sections
1121 and 162." Although this quotation is from a
footnote and not from the body of the opinion, it would
seem that herein lies the heart of the case. In fact,
the Court found that the plant discovered by Matsui was
so sufficiently different from the original patent
granted to Pan-Am specifically on the basis of the fact
that Matsui's plant had no culls that Pan-Am had no

cause to claim patent violation. The important features of this case, and the reason for its selection here, are to underscore both the need for accuracy in patent applications and to point out that ease of asexual propagation is an essential part of establishing a plant patent.

To indicate that the issue of the use of classical genetics for obtaining new high yield strains of plants is still very much before us, one can cite the case, *Pioneer Hi-Bred International v. Holden's Foundation Seeds, et al.*, decided in 1985 by the U.S. District Court of Indiana, Hammond Division. While the decision dealt mainly with what records needed to be produced by the defendants, the basis for the suit was the purported use of a patented strain of corn from Pioneer as the origin of new strains of their own into which the genetic background of the Pioneer strain has been incorporated. Pioneer sought an injunction to recover from the defendants several inbred lines and to have them also "retain and dispose of all such material." In order to assure that this order would be obeyed, Pioneer demanded that almost all of the defendants' records over a great number of years be made known. Were this to occur, the defendants claimed they might then incur severe losses, not only at that time, but also in the future since anyone could then use the public records to duplicate t breeding techniques. The court here held that all the records need not be provided, and also that the expenses born by the defense concerning the cost of legal expenses involved in the dispute over the records might be recovered from the plaintiff. The decision does not in itself deal with the request for an injunction, but serves to protect the defendants against revealing trade secrets when the court deals with the requested injunction.

It can now be seen that plant patents are possible, provided the three characteristics mentioned in the first case are met. The classical genetic cases then depend upon finding a valuable mutant, sufficiently different from all others and not found in nature, and ease of asexual reproducibility of the new material, as well as the use of classical breeding techniques to obtain new, high-yielding plants. The molecular cases to be presented next will show a different type of patent claim, and in terms of modern biological advances, a far more important basis for high profits and beneficial uses.

PATENTS AND MOLECULAR GENETICS

These cases arise from the question whether bacteria or other microorganisms fall under the protection of the plant patent act. As pointed out above, the modified act of 1973 would seem to cover this point clearly; however the cases which arise will call into question the intent of the act in a specific manner. However, the first of the two cases dealing with this problem arose before the 1973 act and at that time it was clearly decided that bacteria *were not* to be considered as plants. It might be added that in modern taxonomic considerations, the bacteria are now considered by most experts to belong clearly to a kingdom of their own, neither plant nor animal, as the first case intimates.

This first case, *In re Arzberger* was decided in 1940 by the Court of Customs and Patent Appeals. The basic question then was whether Congress had intended to

include bacteria under the plant patent laws. The
applicant for the patent had found a new strain of
bacteria which were useful for the production of butyl
alcohol, acetone and ethyl alcohol. Reproduction was
asexual; at that time this was the only known method by
which bacteria could reproduce.

Most of the opinion, denying the right to patent,
is made up of a discussion of the biological taxonomic
position of the bacteria. There is expert testimony
against placing the bacteria in the plant kingdom. "In
holding that said bacteria are not plants within the
meaning of said provision the examiner held that the
scientific classification of bacteria in the living
kingdom is in doubt, stating that: ... 'The authorities
recognize that bacteria are not plants in the strict
sense. Applicant has submitted numerous references to
authorities on the subject of botany and bacteriology
which state that bacteria are classified as plants and
that plants include bacteria. It is not intended to take
issue with these authorities. Notwithstanding, however,
it is pointed out that a distinction is drawn between
this authoritative classification and the *fact* that
bacteria are midway between plants and animals, i. e,
possessing both plant and animal characteristics. This
classification is *optional* and based upon the
observations that bacteria have a preponderance of plant
characteristics [emphasis in original].'" Further the
court stated, "Certainly the House Committee on Patents
did not have bacteria in mind when it made the above
statement (the report of the Senate and House Committees
studying the plant patent laws). Further, pages 1 to 5
of each of the reports are replete with statements
indicating that bacteria were not to be included under

the Plant Patent Act. ... 'The exclusion of a chance find would indicate that bacteria are not included in the plant statute. A chance find is obviously something that can be seen by the naked eye. Bacteria are, of course, visible only upon isolation from their natural habitat and then only by means of a microscope. How, then, could bacteria be chance finds? If bacteria cannot possibly be chance finds, then what purpose would an exclusion clause serve?'" The court then goes on to examine the dictionary definition of plants, and to a study of the history of the then appropriate plant patent laws, and the purpose of these laws which state at one point, "Today plant breeding and research is dependent, in large part, upon Government funds to Government experiment stations, or the limited endeavors of the amateur breeder. It is hoped that the bill will afford a sound basis for investing capital in plant breeding and consequently stimulate plant development through private funds." The court adds, however, "Without quoting further from the report, it is sufficient to say that it fairly appears therefrom that the word 'plant' as used therein was used in its popular sense and not in its scientific sense, and that the bill was designed for the benefit of agriculturists and horticulturists.

"The word 'bacteria' was not mentioned in the reports. It is well known that bacteria occur in the human body, in plants, in air, in soil, and in water, and although they are classified as plants, we think that if Congress had intended that they should be included in the term 'plant' as used in the bill there would be some indication to that effect, either in the bill or the reports of the Committees.

" A drop of water may contain thousands of
bacteria, but outside of scientific circles a drop of
water would not be regarded as containing thousands of
plants." Further discussion of the issue concerning what
is meant goes on to distinguish between common usage of
words and scientific usages by citing a case in which
the courts had held that tomatoes are vegetables despite
their being technically fruits. The opinion concludes
therefore, "So here, we think that Congress in the use
of the word 'plant' was speaking 'in the common language
of the people' and did not use the word in its strict,
scientific sense." The conclusion, therefore, was that
even if bacteria were to be considered as plants, *sensu
strictu* they were not to be considered as plants under
the plant patent acts, and the application for appeal
was thereby denied.

The real issue arises now in the 1980's with the
vast advance in molecular genetics. In this age of
genetic engineering, gene splicing and cloning, it is
apparent that patentability will become a necessity for
ethical drug houses which wish to develop new medical
and other usages by these techniques. If their findings
are not protected by patents, then any new discovery can
be used by any competitor and the cost of scientific
research cannot be recovered nor is it possible for
industry to make a profit if there is no protection of
their product. A company might invest large amounts of
time and money to produce a new genetic form only then
to see the results of their findings become public
domain. Genetic research is costly and time consuming,
and none of the many gene-splicing firms which have
recently came into existence would be able to function
without some such protection. In addition, the now

widespread use of these techniques in agricultural research demands the same type of assurance that new discoveries can be utilized by them exclusively, or by licensing agreements. The case which made this possible is *Diamond v. Chakrabarty*, decided in 1980 by the United States Supreme Court in yet another of the recently common five to four splits.

The facts are these: Dr. Chakrabarty, then an employee of the General Electric Company at its Research and Development facility in Schenectady, New York, developed a strain of *Pseudomonas* bacterium which contained a set of plasmids which would allow the strain to be used to wipe out oil spills. The strain was thought to be able to utilize the oil as an energy source, and thus when applied to oil spills would metabolize the oil and, when the oil was depleted, be unable to undergo further growth and multiplication. (A plasmid is a section of DNA independent of the true bacterial molecular chromosome and capable of reproducing in synchrony with the chromosome so that each product of bacterial fission would continue to contain the same plasmid. The genes for the use of oil were contained in these plasmids.) Chakrabarty and the company applied for a patent on this new type of bacterium. The application contained three claims: first for the method of producing the bacterium, second for the inoculum used to grow the bacteria, and third for the bacteria themselves. The original patent examiner allowed the first two claims, but refused the third. The application was *not* made under the plant protection act, but rather under section 101 of the patent laws which allows for new inventions. The examiner decided "(1) that micro-organisms are 'products of nature,' and (2) that as living things they are not patentable subject matter under 35 U.S.C.Section 101." Chakrabarty appealed to the Patent Office Board of Appeals who upheld the finding, basing their conclusion on the Congressional exclusion of bacteria as noted previously and also that Section 101 was not designed to cover living material. The Court of Customs and Patent Appeals reversed the previous decision on the basis of a prior decision which held that "the fact that microorganisms...are alive...[is] without legal significance for the purpose of the patent law." After a certain amount of further legal procedures, the Commissioner of Patents and Trademarks, Diamond, applied for *certiorari* and the Supreme Court granted this and heard the case. The case had been joined to another, but that case became moot and only the case concerning Chakrabarty was heard.

The majority opinion, written by Chief Justice Burger, held that the patent for the bacteria should be granted. The decision was made by them wholly upon the basis of Section 101 which provides that "Whoever invents or discovers any new and useful process, machine, manufacture, or composition of matter, or any new and useful improvement thereof, may obtain a patent therefor; subject to the conditions and requirements of this title." The real issue as the majority saw it was whether the bacteria were a "manufacture" or "composition of matter" under the meaning of this part of the statute. The Supreme Court reviewed the history of the patent laws and came to the conclusion that Congress had intended "statutory subject matter to 'include anything under the sun that is made by man.' This is not to say that section 101 has no limits or that it embraces every discovery. The laws of nature, physical phenomena and abstract ideas have been held not patentable. ... Thus a new mineral discovered in the earth or a new plant found in the wild is not patentable subject matter. Likewise, Einstein could not have patented his celebrated law that E=mc2 nor could Newton have patented the law of gravity." The opinion contrasts this with an earlier case, *Funk v. Kalo*, where a refusal of patent was upheld on the basis that in that case there was no new single organism but only the use of multiple organisms to achieve a purpose. The court states, "Here, by contrast the patentee has produced a new bacterium with markedly different characteristics from any found in nature and one having the potential for significant utility. His discovery is not nature's handiwork, but his own; accordingly it is patentable subject matter under Section 101." The court then goes on to dispose of the issue raised by the defendant under sections of the Plant Patent Act, cited above, specifically denying rights to patent bacteria. They relied heavily upon the fact that in the 1970 act Congress had not suggested that 101 would not be applicable here. The majority pointed out that a patent for a bacterium had been granted to Louis Pasteur in 1873, and that prior to the passage of the Plant Protection Acts of 1967 and 1968 two patents of this kind had been granted. The majority opinion ends by saying, "We have emphasized in the recent past that [o]ur individual appraisal of a particular [legislative] course...is to be put aside in the process of interpreting a statute...' Our task, rather, is the narrow one of determining what Congress meant by the words it used in the statute; once that is done our powers are exhausted. Congress is free to amend 101 so

as to exclude from patent protection organisms produced
by genetic engineering...But, until Congress takes such
action, this Court must construe the language of 101 as
it is. The language of that section fairly embraces
respondent's invention."

The four dissenting judges felt strongly that this
decision in favor of Chakrabarty was a poor one. At one
stage, for example, the wording reads, "Because I believe
the Court has misread the applicable legislation, I
dissent." The dissenting opinion then goes on to stress
that the plant patent acts meant what they say about the
exclusion of bacteria from the realm of patentability.
The history of this act is followed closely, concluding
that "These acts strongly evidence a congressional
limitation that exclude bacteria from patentability.
...I cannot share the Court's implicit assumption that
Congress was engaged in either idle exercises or mere
correction of the public record when it enacted the 1930
and 1970 acts. And Congress certainly thought it was
doing something significant...Because Congress thought
it had to legislate in order to make agricultural
'human-made inventions' patentable and because the
legislation Congress enacted is limited, it follows that
the Congress never meant to make items outside of the
scope of the legislation patentable.

"Second, the 1970 Act clearly indicates that
Congress has included bacteria within the focus of its
legislative concern, but not within the scope of patent
protection. Congress specifically excluded bacteria from
the coverage of the 1970 Act...The Court's attempt to
supply explanations for this explicit exclusion ring
hollow. It is true that there is no mention in the
legislative history of the exclusion, but that does not

give us license to invent reasons. The truth is that
Congress, assuming that animate objects as to which it
had not specifically legislated could not be patented,
excluding bacteria from the set of patentable organisms.
But as I have shown, the Court's decision does not
follow the unavoidable implications of the statute.
Rather it extends the patent system to cover living
material even though Congress has legislated in the
belief that section 101 does not encompass living
organisms. It is the role of Congress, not this Court,
to broaden or narrow the reach of the patent laws. This
is especially true where, as here, the composition
sought to be patented uniquely implicates matters of
public concern."

The dichotomy in the court seems clear. The
majority based its opinion solely upon one section of
the patent laws and either read into, or denied, the
intent of another. The dissent clearly pointed out this
fact, and strongly stated that to ignore what Congress
had explicitly stated was not an act of legal wisdom, to
say the least. It is also interesting to note that the
majority of the court is now cast in an "activist" role
as compared to the "strict constructionist" views of the
minority. The split is also interesting as it underlines
the liberal as against the conservative viewpoints. The
liberals, as exemplifed by the minority, held that the
patent was invalid; the conservatives in the majority
opinion ally themselves with the interest of industry;
it is obviously in the latter's interest that patents
for genetic engineering be granted. It may be that the
general public will ultimately benefit from the use of
patents by the various industries, but the legal basis
for the decision seems to be have been made on other

grounds.

Another aspect of the patent problems has recently been discussed, that is the applicability of the one-year period for patent application after the facts relating to the application, or the material itself, have been made public. This is at present of great concern to several of the gene-G splicing companies and in particular to Stanford University. The University holds many of the patents which the companies are using in their development of new techniques for various methods of developing genetic strains useful in producing medical products. However, some of these patents may turn out to be invalid as their contents were given in public at meetings or in papers more than a year before the application was brought. As yet no suit has been entered, but it would appear inevitable that this will occur. A full review of this dilemma appears in the New York Times of August 29,1982. This can be a matter of much concern, since the financial reward to the university under license agreements is a very large sum. However, 1n 1984 Stanford did achieve a patent of "hybrid plasmids that are used to transfer genes from one organism to another. It complements and strengthens a broad patent award to the University in 1980 covering the basic techniques in gene splicing." The quotation is from 225 *Science* 1134 (1984). By this means Stanford may avoid further legal problems.

By far the most complex genetic case arose in 1982. *Novo Industries v. Travenol Laboratories, Inc.*, tried before the U.S. Court of Appeals, Seventh Circuit, dealt with a myriad of issues. Probably first and foremost from the genetic viewpoint was the issue of what is meant by a genetic species. The facts are complex, and

deal primarily with the use of fungi to produce
coagulating enzymes in the cheese-making process. Prior
to the discovery of fungal enzymes, the principle source
of such enzymes was from the fourth stomach of unweaned
calves. The major issue in the case at hand was the
discovery that certain fungi of the genus *Mucor* could
produce a similar type of enzyme. The case dealt with
the difficult questions whether Novo, a Danish concern,
had exclusive patent rights for a fungus, and whether
the defendant company, an American concern, had
infringed upon these rights intentionally. The issue
concerned dealt with whether the fungal species for
which Novo had filed a patent prior to a similar claim
by Travenol was a real species or whether it was only
a different strain of the species.

The Court found that it had to deal with the
complex genetic issue of what constituted a fungal
species. The first part of the opinion deals solely with
the genetic and biological classification of a true
species. In a footnote to the opinion, the court gave
a complete biological classification system,
surprisingly here, that of a Siamese cat! Further, the
Court had to deal with the problem whether a single gene
mutation could create a new species. In addition to the
biological questions involved, which included the above,
there were also major legal issues whether the
discovery and patent application of Novo was obvious or
not, the problem whether the patent had been misused by
Novo, and increased damages to Novo for a deliberate
infringement of their rights by Travenol.

The Court gave a detailed account of the process of
cheese making as a background for the opinion. A
complete history of the patents involved was given. The

basic genetic issue to be decided was whether the patent application by Novo was for the same species as the application by Travenol, and whether several misclassifications by various experts had a bearing upon the case. Basically, the early experts had mistakenly put two separate species into one, an error corrected by later type culture collections. Early descriptions of the fungus in question had categorized it as *Mucor pussilus* on the basis that it was a strain which was not capable of self-fertilization. A strain which was self-fertilizing was originally identified as a member of the same species; later this strain was correctly reclassified as not being of the species described but of a separate species, *Mucor miehei*. The research expert for Novo had isolated this species from natural sources and found that the strain he had was exceptionally useful for production of the cheese-making enzyme; the patent application was therefore for this now-recognized species of *M. miehei*.

Meanwhile a fungal specialist working for Travenol began a search for an organism to produce the enzyme and had learned about the possibility of the same species named in Novo's application as a useful source of the needed enzyme. He obtained the same strain named in the Novo patent application, and found the same utility for it; at this point Travenol applied for a patent for this same species as had been earlier claimed by Novo. The obvious conflict of applications was studied by the Board of Patent Applications and priority was awarded to Novo; the priority was upheld by Court of Customs and Patent Appeals, and the U. S. Supreme court denied *certiorari*.

Despite this Travenol marketed their enzyme from

Mucor miehei. The case then came before the Court of
Appeals over violation of patent rights, and for damages
against Travenol for deliberate violation of Novo's
patent claims.

The Court's opinion, as mentioned above, first
dealt with the question of what constituted a fungal
species. Having decided that *M. miehei* was indeed a
separate species from *M.pusillus* the court than had to
deal with the legal issues of obviousness of the claimed
invention. Travenol was put into the peculiar position
of claiming that the obviousness was such that no patent
should be granted, and at the same time applying for a
patent for the species! In their somewhat lengthy
opinion the Court dealt with that issue as well as
whether there should be damages against Travenol for its
use of the species in question. The Court decided easily
that Travenol was aware of the risk of selling the
enzyme produced by the species despite the prior patent
claim by Novo. Other legal issues were raised by
Travenol; the Court found all of them specious, and
damages were awarded against them for deliberate patent
infringements.

The court found also that Travenol's claims were
"...an-after-the fact argument designed primarily to
dazzle the district judge and take advantage of his
presumed lack of familiarity with the current state of
the art of biology and taxonomy. As judges we are not so
naive as to believe that such an approach is uncommon,
although we do not condone it. Neither are we so naive
as to think that 'obviousness' arguments are not derived
primarily from the inventiveness of patent lawyers....We
rely neither upon the esoteric nature of Travenol's
argument nor upon the obvious fact that the argument is

an artificial construct of counsel ... What we do rely upon is what underlies Travenol's argument, mischaracterization of the evidence and the deliberate avoidance of knowledge that might be disadvantageous to its position."

Novo thus won its case. There were other legal issues raised by both sides, but the obvious issues were whether the the fungal species was patentable by Novo and whether Travenol had illegally and knowingly violated the Novo patent. Again, it should be emphasized that the genetics involved were of utmost importance; the determination of whether Travenol had used the same species needed to be proved and therefore the genetic and legal definition of a species had to be made first. Once this was determined, it was apparent that Travenol had indeed infringed upon Novo's patent and the case was decided against Travenol upon that basis.

In these cases, the courts have been forced to deal with specific genetic issues concerning the meaning of species, and the novelty of usages of particular strains for specific industrial applications. In *Diamond v. Chakravarty*, the issue was whether the use of a microbe for industrial purposes was patentable; the second case dealt with the narrower issue of what a genetic species is. Both courts have had to delve into the genetics of speciation and both courts have dealt at some length with the latter issue. In the Travenol case, surprisingly, no mention was made of the conflicting clauses in the patent laws. Presumably after *Chakrabarty* this was no longer an issue in the Novo case, despite the specific prohibition of patenting either "bacteria or fungi."

Although no major new decisions in patent law

involving genetics have been published recently, as
predicted following *Chakrabarty*, the courts are becoming
cluttered with new patent suits involving gene splicing,
and even involving questions concerning the
patentability of a gene itself. Examples of cases now in
the courts include: *Genentech and Hoffman-Laroch v.
Biogen & Sherer-Plough* over rights to alpha-interferon,
an antiviral agent which might have some use in cancer
therapy (the case is already in the European courts);
Revlon v. Chiron and Genentech over monoclonal
antibodies, products useful for immunodiagnostic
purposes; claims by GCA/Precision Scientific Group for
patent rights on a cell fusion technique to yield new
genetic combinations possibly useful in a variety of
ways. A major issue here may well be the "failure to
properly describe the "invention," under the appropriate
clause of the patent laws. (One way might be to deposit
the new microbial form with the American Type Culture
Collection where it would be available to others.) So
profitable may be the field of patent law that one
chairman of a biotechnology firm has been quoted as
saying "If I had a child heading into a career now, I'd
want him to be a patent lawyer--preferably a
biotechnology patent lawyer."

Indeed in 1983 Stanford University and the
University of California attempted to devise a scheme to
resolve some of the difficulties by pooling their
patents in biotechnology and then turning them over to a
to-be-created foundation which would act as a clearing
house, sublicensing the patents with much of the
profits to flow back to the Universities. Whether such a
plan would succeed under anti-trust laws is not clear
nor is it even certain that all major patents would be

put into the pool. A similar venture has been begun
between the University of Maryland, Maryland's
Montgomery County where the research center will be
located, and the National Bureau of Standards. Again
there seems to be no certainty as to how patent rights
will be assigned for any biotechnological inventions.

A review of the various types of patents has been
published in 1984 by Sidney B. Wilson Jr. in which all
aspects of the problem of biotechnological plant patents
are studied, with particulor attention to the three
various U.S. Statutes involved, the Plant Variety
Protection Act, the Plant Patent Law and the General
Patent Law. Of particular interest will be the issue of
protection of *parts* of plants, for example recombinant
DNA in the nucleus of plant cells, the cells themselves,
plasmids (pieces of self-replicating DNA not associated
with chromosomes, but carrying genetic material
expressed by cells containing them). and even genes
themselves. None of the latter would apparently fall
under any of the three laws mentioned, and as the real
future of genetic engineering will almost certainly be
at this cellular or molecular level, the status of such
developments is quite uncertain. Still another
difficulty, not mentioned previously, may arise
particularly with plasmids. Following the recent
recognition of the early work of Barbara McClintock on
transposable elements in corn, it is now known that many
such "transposons" occur in nature and genetic material
may spontaneously rearrange itself without human
interference. Thus what might be a newly developed plant
plasmid may simply be one which could also occur in
nature, and it is not permitted to patent naturally
occurring plants. When, and if, such a patent

application for newly arranged genetic material is
sought, it will be interesting to see how the courts
deal with whether these are "inventions of man" or
simply discoveries of natural phenomena.

The dispute over genetic patents of microorganisms
is taking on world-wide aspects. European plant breeder
and European chemical companies are on sword's point
over who has the right to certain patents involving
molecular genetics. As most of the European firms also
either have American branches or may even be subsidiary
corporations of American firms, this issue is certain to
arise in the United States Courts within the near
future, and will then deal not only with patent laws,
but may take on an aspect of cases of international law
as well.

Another area in which patent law must be decided is
exemplified by the procedures used in the case of
surrogate embryo transfer. As will be recalled, this
technique involves using an ovum from a fertile,
unrelated female and fertilizing it *in vivo* with the
sperm of the husband of an infertile woman. The
developing embryo is then washed out of the donor's
uterus, recovered and implanted in the uterus of the
infertile wife where it may implant and produce a
foetus. Patents are now being sought not only for the
surgical instruments involved, but for the 5-step
procedure itself. Apparently the procedure has been
successful in a few cases, although it is not a simple
one. This is causing a furor in medical fields as well,
as any time a practitioner develops a new technique for
a medical procedure and receives a patent for it, no one
except those licensed by the patent holder can make use
of it, and it would certainly lead to the withholding of

information concerning new medical procedures if
patentability is to be permitted. This is the same type
of "secrecy" referred to in the main chapter of the
text.

Much of this has come about as the patent is sought
by *Fertility and Genetics Research, Inc*, a Chicago based
company which funded the research leading to this
procedure. There has been no decision whether this type
of patent will be allowed, but the threat of such
patents might do to medicine what the already mentioned
patent of microorganisms has done to general biological
research. (Other references to embryo transplants have
been made in the chapter on surrogate motherhood.)

Another example of the confusion brought to the law
by the introduction of molecular genetics concerns the
"ownership" of cells derived from a human and grown in
tissue culture. A Japanese investigator at a western
university brought with him cells from his mother who
suffered from cancer, apparently in the hope that they
might be useful in helping her. The cells were
cultivated at the university and studied intensively.
Some time later the Japanese scientist returned to his
home country, taking the cells with him without having
permission from his host laboratory to do so. He claimed
that they belonging to him by rights of inheritance!
There were inferences by some that he had simply
"stolen" the cells and it was further brought out that
his father operated a biotech firm in Japan. No court
case involving patent rights arose here, as there was an
amicable settlement whereby the Japanese investigator
retained rights for the cells in Japan, but not
elsewhere.

A somewhat similar case is also about to appear in

court. The facts here are somewhat different, but again illustrate the difficulties involved in patent law. A man suffering from a disease allowed cells to be taken from him and cultured. The cells were further manipulated in a research laboratory, and these may be used to produce products useful in treating disease. The original donor of the cells, claiming that he had given permission only for culturing of the cells, and had not waived his rights for their use, is about to bring suit against the laboratory for using his cells for commercial purposes without his permission. The case will not be heard for at least another year, and the ownership of these cells and the patent, if forthcoming, on their utility in producing a potential health cure will be a difficult one to decide.

A bitter dispute may come to court in the near future concerning the original patent for the AIDS virus. A group of French investigators are about to sue for priority of discovery and application for this virus; an American investigator is claiming priority and indeed a patent has been issued for his virus. Additionally a biotech firm is marketing a kit for the detection of the AIDS virus. The dispute here concerns whether the American virus and the French virus are one and the same or different; only detailed knowledge of microbiology and molecular genetics can settle this dispute.

Another matter of concern is the effect of these cases upon science itself. In biology, and in genetics, it has long been a tradition that all information, except perhaps that obtained by pharmaceutical concerns, is freely shared. Indeed, one of the major benefits of the annual meetings of the many societies is to allow

workers to share results, methods and ideas with one
another. From this stimulus many new ideas are
formulated. Also, if an investigator is having
particular problems in research, sharing of these
difficulties with others frequently leads to new ways to
approach a problem. Further, duplication of effort can
be avoided by open exchange of ideas; there is no sense
whatsoever in duplicating something which has already
been done, or in attempting to use methods which others
have tried and found to fail. Today these laudable
results of meetings are beginning to be suppressed.
Obviously, as scientists are as human as any other
persons, they may indeed be rather reluctant to share
results which might be extremely lucrative if
patentable. In this way, although there is still a great
deal of give and take in other areas, advances in
genetics, particularly in the fields of genetic
engineering, gene splicing, recombinant DNA or plasmid
genetics, to name a few, are now neither publicized nor
discussed until the material is patented. Another
negative feature may be the cessation of the sharing of
genetic strains which has always been a feature of
genetic research; stocks are exchanged freely between
investigators, but patent restrictions may bring an end
to this also. Further, as it is essential to maintain
secrecy concerning the methods by which new developments
are brought about, there will tend to be a monopoly by
large research organizations with large financing and
the effect upon smaller projects will be serious if not
stultifying. Many scientific developments come about not
from planning, but from an almost accidental discovery.
(Perhaps it is not necessary to mention penicillin
here.) If one is banned by secrecy from knowing and

sharing other workers' results, one cannot know which
area of a field even to begin working in. The
introduction of secrecy would seem to be the antithesis
of true science. Some method must be worked out also to
protect the rights of genetic research firms, whether it
be in agriculture or medicine, for an equitable return
for their research efforts and the rights of
investigators to share in the general progress of their
colleagues in their specialty. The present "wedding" of
large universities with large drug firms, taking place
at an accelerated pace, would seem to be moving in the
wrong direction; once a scientist is no longer free to
share or to publish his findings, science becomes the
right of a few and not the boon to society which it is.
This matter cries out for resolution, but as of this
writing no solution is either in sight or has been
proposed.

Nonetheless, the effects of these cases upon
biological research remains; patentability of organisms
either natural or "man-made" will certainly affect the
free interchange of information between scientists in
this area and these cases would seem to tend to suppress
freedom of formerly open and frank sharing of ideas. In
fact, at the time of this writing, it would appear that
many bacterial patents have now been approved. Many
biologists and geneticists are concerned over this
restriction upon the flow of information. If science is
to be conducted in secret, then science, and ultimately
the public, will suffer.

ALPHABETICAL LIST OF CASES CITED

Pioneer v. Holden's   (Slip) 81-60-E
                                   U.S.D. Ct. N. Ind.   1985
Protect.of Plant Varieties
              and Parts   225 Science 18   1984

Splitting Hairs over
     Splitting Genes   N. Y.Times 6/27 A16 1984
Surr.Embryo Trans. Perils   14 Hastings Center
           of Patenting   Reports 25   1984
Unus.Partners Launch
     Biotech Venture   223 Science 800   1984
Yoder Bros. v.Cal-Florida
           Plant Corp.   537 F.2d 1347   1976

CHRONOLOGICAL LIST OF CASES CITED

YEAR   CASE   CITATION

1930   35 U.S.C.   101
1940   In re Arzberger   112 F.2d 834
1948   Funk v. Kalo   333 U.S. 127
1962   35 U.S.C. 161
1970   7 U.S.C. 2402
1976   Yoder Bros. v.Cal-Florida
                  Plant Corp. 537 F.2d 1347
1977   Pan-American Plant Co.
                  v. Matsui   433 F.Supp. 693
1980   Diamond v. Chakrabarty   447 U.S. 303
1982   Novo Industries A/S
              v. Travenol Labs   677 F.2d 1202
1982   Patent Race in Gene-Splicing N.Y.T. 8/29 pg.F-4
1983   Onestop shop for gene-
              splicing patents   219 Science 1302
1984   Biogen will patent

	InterFeron product	N. Y.Times 2/23, D4
1984	Biotechnology's Patent War	N. Y.Times 3/19, D1
1984	Cohen-Boyer Patent Finally Issued	225 Science 1134
1984	Gene Splicers Square Off in Patent Court	224 Science 584
1984	Genetic Patent Disputes: Squabbles	Alb(NY)Times-Union 7/1 D-10
1984	Interferon Contest Kicks off in Europe	223 Science 1047
1984	Protect.of Plant Varities and Parts	225 Science 18
1984	Splitting Hairs over Splitting Genes	N Y.Times 6/27 A16
1984	Surr.Embryo Transplants Perils of Patenting	14 Hastings Center Reports 25
1984	Unus.Partners Launch Biotech Venture	223 Science 800
1985	AIDS Priority Fight Goes to Court	231 Science 11
1985	Commercial Giants Push for Patents	228 Science 1290
1985	Patent Dispute Divides AIDS Researchers	230 Science 640
1985	Pioneer v. Holden's	(Slip) 81-60-E U.S.D. Ct. N. Indiana

Probably the first geneticists, albeit unknowingly, were the primitive agricultural tribes. These people soon learned to save the best seed for next year's crops, or to select unusually strong or high yielding plants as the seed parents. Similarly the first groups to domesticate animals must have intuitively practiced genetic selection to improve their herds or perhaps even for aesthetic purposes, much as some pet fanciers do today. These untrained people also recognized "sports" and indeed one of Darwin's major sources in preparing for his monumental book was the observation of such sports in a variety of domestic animals. So, in a sense, genetics arose from agriculture and today agricultural genetics is a major industry.

The cases which follow involve a number of different types of interaction of genetics and the courts. Various fields of law are included: contract law, tax law, and tort law are all involved. The cases then present a varied approach to genetics in the courts but are grouped together here as they all share the common denominator of arising from agricultural cases. Of course, patent laws are also involved, but these have been treated in the previous chapter.

Probably the most frequent type of case arises when a genetically defined seed fails to produce the guaranteed results which the true type should. An early case exemplifying this occurred in 1966 in *Klein v. Asgrow Seed Company*. This case was based on the

*Title from the humorous book by E.P. Butler

failure of allegedly early maturing tomato seeds to
produce the promised results. Instead the seed was
mostly "rogue" and matured, if at all, either over a long
interval or too late for harvesting. Everyone got into
the act; the farmers who planted the seed sued the Farm
Supply Company; the Farm Supply Company in turn sued its
supplier of the seed who then sued his seed broker who
finally sued Asgrow, the original producer of the seed.
The lower court allowed the succession of claims, but
absolved Asgrow of direct liability. The District Court
of Appeals, Third District, California, upheld the
decision concerning the first set of claims and in
addition reversed the lower court's decision and held
Asgrow responsible. It became apparent during the trial
in the lower court that Asgrow, who had originally
obtained the parent material from the University of
California, was appraised by their plant breeder that
the seed was not pure and contained an unknown
percentage of "rogues." Nevertheless, Asgrow sold the
seed as a pure type. There were other damning statements
made during that trial which indicated that Asgrow knew
full well that there was a high risk that the seed was
impure, but had accumulated, without testing, so much
seed that not to sell it would have involved a financial
loss of a considerable amount.

The legal case centered upon the issue of sales law
and not of tort law. The question here is that of
express and implied warranties. "Any affirmation of fact
or any promise by the seller relating to the goods is an
express warranty if the natural tendency of such
affirmation or promise is to induce the buyer to
purchase the goods relying thereon." Also in the Civil
Code of California there appeared the statement, "Where

there is a contract to sell of a sales of goods by
description, there is an implied warranty that the goods
shall correspond with the description. ...Where the
goods are bought by description from a seller who deals
in goods of that description (whether he be the grower
or manufacturer or not), there is an implied warranty
that the goods shall be of merchantable quality." The
plaintiffs had sought to recover damages of about
$14,000 for the complete loss of their tomato crop; in
1966 this was a considerable sum. The defense had sought
to limit the amount of loss to the ultimate purchasers
on the basis that their liability was limited to only
the replacement cost of the seed. The legal points are
complex, based on many varied defenses by Asgrow on the
basis of warranties. The court carefully examined the
claims of Asgrow and found sound legal basis for
rejecting them and Asgrow was found liable for the total
amount of damages to the various parties. Asgrow was
given its choice of either making the award directly to
the injured party or to make it through the chain of
parties in the suit with the final benefit going to the
grower.

A somewhat similar case, *Pfizer Genetics v.
Williams Management*, arose in the Supreme Court of
Nebraska in 1979. Here the seed at issue was purebred
corn, and the case again depended upon the
interpretation of implied and express warranties. The
purchaser of the seed brought suit for damages due to
the failure of the corn to produce adequately. A test
plot yielded only 60 bushels per acre while other seeds
in similar plots were yielding up to 168 bushels per
acre. Prior to the purchase of the seed, the defendant
had been furnished a booklet entitled "Trojan, the Gold

Book," as well as the results of field tests on the disputed seed, and as a result of these materials claimed that express and implied warranties held. The defendant refused to pay for the defective seed. In the lower court it was found that the defendant must pay and also that the warranty claims by Pfizer excluded the the company from paying damages.The case also hinged upon the loss of one page of the sales agreement in which Pfizer claimed that direct denial of warranty was made by them and acknowledged by Williams. A purported copy of this missing page was introduced as evidence during the trial. Also during the trial, it was apparent that the trial judge either omitted or refused the request by Williams to instruct the jury properly concerning warranty claims. The jury had found for the plaintiff on all counts.

The Supreme Court reversed and remanded the case. They found that the furnishing of the copy of the missing page was not valid, and even if it had been, the disclaimer of warranty contained therein was not a valid one as the disclaimer, and also a similar disclaimer on the seed bags themselves, were furnished *after* the sale and were as such not part of the sales agreement. "Although this court has not specifically addressed the question, other jurisdictions have generally held that disclaimers of warranty made on or after delivery of the goods by means of an invoice, receipt, or similar note are ineffectual unless the buyer assents or is charged with knowledge as to the transaction." The court also ruled that the lower judge was in error for not instructing the jury properly in the matter of the law. The outcome was that Pfizer was held liable for the defective seed and the defendant could recover damages therefrom.

A somewhat similar case, again involving Pfizer, arose in Nebraska in the same court, *Pfizer Genetics, Inc. v Prochaska*. The case again involved allegedly defective corn seed. Some differences appeared, however. First, the seed was planted by the dealer himself and his claim was that there was only 50% germination. He also refused, therefore, to pay the amount agreed upon to Pfizer, who brought suit for the money due under the purchase agreement, plus accrued interest at the time of the suit. Pfizer also claimed that the seed had been sold for resale use and the dealer should not have planted it himself. The lower court held in favor of Pfizer on the amount due and against the counterclaim by Prochaska. The higher upper court maintained the requirement for payment with interest to Pfizer, but held that the counterclaim for damages should be allowed, and there was nothing in the purchase contract to forbid the dealer from selling the seed to himself and planting it himself.

A more complex case arose earlier, in 1977, in the case of *Agricultural Services Association, Inc. v. Ferry-Morse Seed Company, Inc.* Here the seed in question was okra, and the case came under federal jurisdiction as it involved actions in two states, California and Tennessee. This case involved the sale by Ferry-Morse [F-M] to the Agricultural Service Association [ASA] of 25,000 pounds of okra seed alleged by F-M to be type "C/S." The sale had not been direct but through a California intermediate, Waldo Rohnert Company [W-R]. F-M, from whom the 25,000 lbs. of seed was ordered, was unable at that time to furnish this amount, and purchased it from W-R. F-M then repackaged it in its own

bags and sent it to ASA with the representation that the
seed was indeed type "C/S". It was established that W-R
had not conducted effective tests on the seed prior to
sending it to F-M who had relied upon previous
experience with W-R as well as the label placed on the
seed by that company. The seed was ultimately used by
clients of ASA and found to be not true to type,
resulting in financial loss to ASA. In addition, ASA,
which was a newly formed corporation, sued for loss of
good will.

The United States Court of Appeals, Sixth Circuit,
dealt with these issues at length. The loss of good will
part of the suit was dismissed on the basis that there
was no proof that the purveying of the defective "C/S"
seed resulted in the loss of further sales from ASA. The
other counts were upheld, however and ASA was found to
be entitled to damages from W-R and F-M. The usual
claims concerning warranties were raised by the latter.
However, the court found that mislabelling of seed was a
violation of federal statutes as well as laws in both
the states involved. Damages here were not small,
amounting to over $75,000, plus 6% interest, to be
awarded to ASA. The warranty on the seed bags as sold to
ASA was plain: "Notice to buyer: We warrant that seeds
sold have been labelled as required under state and
federal seed laws and that they conform to label
description. We make no other or further warranty,
expressed or implied. No liability hereunder shall be
asserted unless the buyer or user reports to the
warrantor within a reasonable period after discovery
(not to exceed 30 days) any conditions that might lead
to a complaint. Our liability is limited in amount to
the purchase price of the seed." Obviously the labels on

the bags were wrong. The Court dealt also with the issue
of which state laws should take jurisdiction and agreed
that the suit by ASA should be properly brought against
F-M and W-R in Tennessee where the actual damages
occurred. This issue being disposed of, the Court went
on to find that W-R was liable to F-M whose claim that
F-M had failed to inspect the seed excluded damages
against W-R. The court did not agree, pointing out that
due to the required haste in supplying the seed to ASA
in time for the growing season, they had no time for
such tests, and it was actually the duty of W-R to have
done this prior to certifying the seed as being pure
"C/S". The upshot of the whole complex case was that ASA
was entitled to damages from loss of the okra crop and
that F-M was not liable for furnishing seed which they
had purchased under the normal assumption that the seed
was as represented when the purchase from W-R was made.
The ultimate responsibility lay with W-R for furnishing
the defective seed upon which this suit depended.

The case, as mentioned above, is not a simple one.
However the principle established by this and the other
seed cases is clear. Genetically labelled seed must
produce according to the claims made for it, and both
state and federal laws demand this. Although there
appears to be a good deal of "buck-passing" in these
cases, the producer of defective seed must ultimately be
held liable for damages occurring when an innocent
consumer suffers high losses caused by the failure of
the seed to live up to its claims, regardless of the
labels and disclaimers the producer or seller places on
the seed in an attempt to avoid just such a
responsibility.

A somewhat similar claim concerning warranties

arose in *Pennington Grain & Seed v. Tuten*, in the
District Court of Appeal of Florida, First District, in
1982. Here, the case dealt with the failure of 600
bushels of soy bean seed to germinate. In this case the
purchasers of the seed had performed careful control
germination tests, including replanting of the purchased
seed and further, the use of different seed in the same
area. The latter seed germinated perfectly well;
Pennington's seed twice failed to produce a crop. Part
of Pennington's warranty read "DO NOT PLANT THESE SEEDS
SHOULD THIS WARRANTY BE UNACCEPTABLE, BUT RATHER RETURN
THE SEED TO PENNINGTON WITHIN 10 DAYS OF PURCHASE FOR A
REFUND OF THE PURCHASE PRICE." Just how the buyer of the
seed could determine whether the seed was viable
without planting it, and thereby violating the warranty,
is difficult to determine. In any event, the guarantee
that the seed would give 80% germination was false. The
lower court decided that this type of warranty was
essentially ineffective, and awarded full damages to
Tuten. The District Court upheld the lower court and
determined that the defective seed had indeed allowed
the suit for negligence and breach of implied and
express warranty against the seller of the defective soy
bean seed.

While there is no expressed genetic issue cited
here, it is obvious that sellers of seed deemed to have
certain genetic traits, such as percentage viability and
yield, must be prepared to live up to the traits
claimed. Innocent purchasers of defective seed are
protected against loss when the seed does not meet
specifications, no matter how hard the sellers try to
exclude responsibility by writing warnings on seed bags.

A very special type of animal husbandry forms the

basis for cases involving interpretation of tax law.
These involve the large scale breeding of laboratory
animals for research purposes. The first case, *Charles
River Breeding Laboratories, Inc. v. State Tax
Commission* concerns the production of genetically pure
and relatively bacteria-free rats and mice, while the
second, *State Tax Commissioner of the Commonwealth of
Virginia v. Flow Research Animals, Inc.*, concerns the
raising of not only rodents, but also dogs, sheep and
goats. The claims in both cases are similar, namely the
facilities' claims that they are engaged in
manufacturing or processing and as such under the state
laws should be exempt from sales tax on the raw
materials such as feed and bedding which they purchase.
Both are large operations, employing several hundred or
more employees and both use specialized equipment in
their businesses.

The Charles River case was decided in 1978 by the
Supreme Judicial Court of Massachusetts, Suffolk. The
case came on appeal after the Appellate Tax Board had
decided against the laboratory breeders. Charles River
ships more than 6,000,000 animals a year to various
research organizations, each ordered by weight, sex, age
or strain. The present court described in some detail
the methods by which the animals are raised from the
time of Caesarean birth in bacteria-free environments
and subjected to special bacteria in order to stimulate
the ability to develop an immune system. All materials
used must be sterile, including bedding, food and
containers. Over 40 different breeding rooms, each
separated by germ-free barriers are maintained; no
personnel enter the rooms and the materials which enter
do so by means of a sealed pneumatic system. Large

amounts of equipment are required to maintain the
system, including back-up emergency diesel generators.
It is important to give these details, as it was the
claim by the laboratory that the maintenance and use of
this kind of equipment qualified them as falling under
the protection of industrial operations, and thereby
qualifying for the special sales tax exemption
Massachusetts had enacted to attract and maintain
manufacturers. The crux of the matter, however, is
whether a genetically purebred rodent can be considered
a manufactured product under the state tax laws. Upon
careful review of various other cases, the court ruled
otherwise, and decided that the laboratory did not
constitute a "manufacturing" business. "The statutory
goal was not to encourage and preserve all industry in
the Commonwealth but only that portion of industry
represented by manufacturing corporations. Whether such
a legislative distinction was and remains wise is beyond
our province." An important footnote in this case may
take on added importance in the years to come, and it is
apparent the court had an eye to the future. "We leave
to another day, if it comes, the question whether
processes which alter the genetic structure of animals
fall within the statutory concept of manufacturing."

With the present large companies involved in genetic
engineering, it would appear that the court might have
had this in mind. The opinion is not without a final
barb; in that same footnote, citing earlier cases, it is
stated, "The use of incubators did not make
manufacturers out of the owners of commercial chicken
hatcheries." Again this may be open to question as
certainly the purchase of young chicks, their being
raised in essentially a completely mechanical way and

sold in lots of thousands or more, may make this seem
more akin to the manufacturing of chickens rather than
farming. The footnote ends with a quote from an earlier
North Carolina case (*Master Hatcheries v. Cole*), "Every
layman of normal intelligence knows that a hatchery does
not 'manufacture' baby chicks, and the law does not
require that judges be more ignorant than other people."

The second tax case of this kind was decided by the
Supreme Court of Virginia in 1981. The facts are
essentially the same, except the animals raised are of a
greater variety. The lower court permitted the sales tax
exemption, but the higher court reversed the decision.
They pointed out along the way the difficulties inherent
in the business: "The animals themselves must meet very
definite specifications, both in breeding and raising,
according to international genetic standards and written
protocols and procedures either generated internally or
provided Flow Research by its customers." The court felt
certain, however, that this was not manufacturing and
decided the issue on the question of whether this could
be considered processing, defined as not just
"transformation of a raw material into an article of
substantially different character, but requires that
the product undergo a treatment rendering it more
marketable or useful. ... But even under our broad
interpretation of 'processing' it is contemplated that
the raw material will be treated in some manner, whether
by heat (as in pasteurization) or by blending (as in
making feed and fertilizer). In the present case, there
is no treatment of the breeding animals. They merely
provide a protected hygienic environment in which nature
takes its course. We do not consider this activity to be
processing of an industrial nature within the purview of

the exemption." Thus Flow Research was required to pay
sales taxes. A dissenting judge, however, had another
view. "It is true, of course, that the commodity the
taxpayer sells could not be produced unless 'nature
takes its course.' It is also true, however, that the
commodity is not the product of the laws of nature.
Rather, as an offspring of parents of different genetic
characteristics, it is a genetic amalgam with traits and
characteristics different from those of either parent.
The customary course of nature ends at birth.
Thereafter, the taxpayer applies a disciplined system of
scientific practices and procedures which manipulate
environmental and biological factors and divert growth
and development from the course nature normally takes.
The end result is a new product, wholly unlike the
unrefined raw materials from which it was derived, one
which meets the unique specifications of a market which,
though highly specialized, is international in scope."
Obviously this judge would have allowed the tax
exemption under the basis of processing. It might be
added, however, that the judge's explanation of genetics
could be considered in a different light, perhaps. We
are all an "amalgam" of traits; all of us differ from
our parents genetically, as do the offspring of every
sexually reproducing group of organisms. Perhaps, based
on this dissent, we ought each to seek to exempt
ourselves from sales tax for the special rearing we give
to our children, "amalgams" of our own lives!

An almost similar case arose in 1982 in New York,
In the Matter of the Petition of Blue Spruce Farms,
decided by the New York State Tax Commission. Blue
Spruce is a large breeder of laboratory rodents and
sought relief from sales tax covering the cost of feed,

claiming that the company was engaged in "farming." The
New York Law is explicit in defining farming as "raising
stock, dairy, poultry, or fur bearing animals, fruit and
truck farming, operating ranches, nurseries,
greenhouses, or other similar structures used primarily
for the raising of agricultural, horticultural or
floricultural commodities and operating orchards. ...
The breeding of dogs, cats and other pets or laboratory
animals is not farming. Despite the fact that rodents
are certainly fur bearing animals, the second part of
the definition dealing with laboratory animals was held
to exclude Blue Spruce's claim and no tax exemption was
allowed.

Be that as it may, the tax cases are of
considerable importance economically. They are not all
in favor of the state in regard to agriculture,
however, for in *Select Sires v. Lindley, Tax Commr.*, the
Supreme Court of Ohio decided a tax case in favor of the
plaintiff. Select Sires was a corporation engaged in the
furnishing of bull semen for artificial insemination. In
1979 the tax commissioner sought to impose personal
property taxes upon the equipment used in the production
and processing of semen. The Supreme Court to whom the
plaintiff appealed after an unfavorable review of their
claim for exemption by the Board of Tax Appeals, agreed
with the plaintiff and granted the exemption. The
opinion is of interest in the context of the detailed
description given by the court of the methods used by
the plaintiff as well as for the interpretation that the
laws of the State of Ohio were such that they could be
interpreted as applying to this type of business. As
many of the cases which follow will also deal with
artificial insemination, the description given by the

court here will also explain some of the genetic procedures in the cases to come. "...At this location the taxpayer raises and maintains bulls by a detailed 'progeny testing' procedure to have great 'genetic value,' in that the female offspring sired by these animals produce great amounts of milk compared to those sired by bulls of lesser 'genetic value.'

"Taxpayer collects bovine semen ...and tests this 'genetic material' for mobility and density of spermatozoa. The genetic solution is diluted, after controlled cooling, with an 'extender solution' consisting of double distilled water, sodium citrate, the liquid of egg yolks, glycerol and antibiotics. Premeasured amounts of the diluted bovine semen are injected into small plastic cylinders, referred to as 'straws,' and these straws are then sealed and placed into a freezer unit where, in a nitrogen vapor environment, straws are frozen. The straws are placed in a jug containing liquid nitrogen and are re-packaged in smaller, liquid-nitrogen filled containers for shipment for use in the artificial insemination of cattle." It was the described items here which the tax commissioner sought to have assessed as personal property.

The legal basis of the case was the interpretation of the pertinent Ohio laws regarding this case. The law reads, "'Business' includes all enterprises, except agriculture conducted for gain, profit, or income and extends to personal service occupations." No definition of agriculture is given in the statement of the law. The court, depending on earlier decisions, simply repeats at one point, "Appellant's business, in our opinion, is agriculture." The court pointed out that this statement made in 1969 had never been challenged by the legislature which could have reenacted the statute to cover such cases had it so desired; therefore the case at hand was decided in favor of the plaintiff and no tax for personal property was assessable.

Even in 1982, the issue of exactly what constituted agriculture and what constituted manufacturing was left unresolved by two contradictory opinions. The first, *McElhaney Cattle Company v. Smith*, arose before the Supreme Court of Arizona. Here the owners of a commercial cattle feed lot sued the county and state tax assessor and Department of Revenue for refund of taxes paid under protest and against a ruling that their operation was not agriculture, but business. They had at the time some 55,000 head of cattle on their feed lots and claimed exemption from taxation on the basis that they were wholesalers. The lower courts had upheld their claim but the Supreme Court reversed. The opinion is replete with genetic references, for example, "...significant technological and scientific developments in the areas of nutrition and genetics have affected the way cattle are raised for slaughter..." and "Genetic developments have allowed the feeder to select breeds of calf which will develop into cattle yielding the grade of meat the feeder desires." The court goes on to discuss the effects of castration and feeding of hormones to increase the yield further. However, the Supreme Court did not see fit to admit that this wholesale processing of calves into steers was covered in the statutes dealing with exemption for manufacturing or wholesaling and determined that McElhaney was liable for the usual taxes with no exemption allowed for being a "manufacturer" of beef.

The opposite was found in the case of *Bain v.*

Department of Revenue before the Supreme Court of
Oregon. Here the defendant, Aqua-Foods, a subsidiary of
Weyerhaeuser Corporation, was engaged in the raising of
salmon. More than twenty million fry were expected to be
raised in 1979. The company had PhDs in various fields
in their employ. Among other things each fish was
vaccinated before release and it was found that their
salmon not only had a higher return rate, but more
importantly they had a different bone structure and the
quality of the fish was far superior to those which
reproduced by natural means. The opinion goes into
considerable length concerning the entire processes
used, but the basic issue is whether this is or is not
manufacturing. The governmental agencies claimed that
the entire operation was akin to agriculture. The court
did not agree with that claim, stating, "We conclude
that it is not agriculture. The word 'agriculture' is
derived from two Latin words: 'ager' meaning 'field,'
and 'cultura' meaning 'cultivation.' [the court
continues for almost a paragraph to amplify this
definition]...Given the states of the art, or sciences
of agriculture in 1961, it is doubtful that the
legislature ...would have considered Aqua-Foods'
activities to be agriculture, or any other
nonmanufacturing use." The court also cites other cases,
not here fully described, involving the raising of
chickens. In the first case (*Perdue v. Dept. of
Assessment & Taxation*) it was decided that, as Perdue
was in the process of packaging 12,000 chickens per
hour, this was manufacturing, and in the other (*Master
Hatcheries v. Cole*) that a North Carolina chicken
operation producing 300,000 chickens per week was also a
"manufacturer." The court held that *Charles River*, cited

above, was not applicable here. The final statement of the opinion reads, "Based on our consideration of the legislative intent to provide industrial tax exemption in order to promote the economy of Oregon, coupled with the technological and scientific complexity of the operation involved here, we hold the Aqua-Foods' salmon hatchery is an income-producing industrial facility other than a nonmanufacturing facility and is entitled to the tax exemption."

The genetic purity of cattle strains is the main feature of many law suits. The difference between the sale price of a pure-bred and registered animal compared to an ordinary animal is marked, estimated in one case as approximately $750 for a breeding heifer as against only $125 for one sold for slaughter, a difference of $625. The cost differential between a bull for breeding and a bull for slaughter is approximately $375. Multiply these differences by the large number of cattle usually found in a herd and the genetic value of a pure-bred herd as against a beef herd becomes considerable. In addition, should genetic defects appear in a herd, the entire herd becomes suspect as being genetic carriers of a defective trait and its entire value as a breeding herd may be destroyed as no purchaser wishes to take the risk of having defective calves born to a possible carrier of the trait. Also included in the cases is another method of proving genetic purity, namely blood tests which can be used to prove the genetic background of the animal in question.

The first case of this type to be discussed deals not with the monetary value of cattle, but is a straight case in contracts. In *Meuse-Rhine-Ijssell Cattle Breeders of Canada v. Y-Tex Corporation*, decided by the

Supreme Court of Wyoming in 1979, the issue was whether a lower court judge had properly entered a summary judgment in favor of the defendant. The problem arose when Y-Tex, a distributor of semen for Meuse-Rhine wished to terminate their contract, but at the same time stated that they wished to receive 12,901 units from a particular bull, Mr. Image. Once this shipment was received the producers would be free to sell any additional units of this type of semen. Y-Tex noted that it had orders for the amount specified. The original termination of contract took place over the telephone, and a memorandum of the discussion was prepared by Y-Tex, sent to Meuse-Rhine with the request that if this was the correct agreement they should initial it and return a copy, which they did. Subsequently, Meuse-Rhine claimed that Y-Tex, which was leaving the business, refused to take delivery of the specified units. Y-Tex counterclaimed that the semen was not made available to them in time for the breeding season and would therefore not have been of use.

MRI claimed the agreement was a firm order, while Y-Tex held that this was not an order but only a reservation and it was not under obligation to buy any semen whatsoever. The lower court held that Y-Tex had indeed only placed a reservation, and granted summary judgment in their favor. The Supreme Court, however, did not agree. They held that there was sufficient basis for difference for the case to have been tried under contract law to determine whether the phone statements, followed by the letter setting forth the terms, constituted a genuine contract. The opinion often cites the Uniform Laws, Annotated, Uniform Commercial Code and its pertinent sections as a basis for its decision.

"(1) A term which measures the quantity by the output of the seller or the requirements of the buyer means such actual output or requirements as may occur in good faith, except that no quantity unreasonably disproportionate to any stated estimate or in the absence of a stated estimate to any normal or otherwise comparable prior output may be tendered or demanded.

"*(2) A lawful agreement by either the seller or the buyer for exclusive dealing in the kind of goods concerned imposes unless otherwise agreed an obligation by the seller to use best efforts to supply the goods and by the buyer to use best efforts to promote their sale*" (Emphasis in opinion). The critical factor here, which also had bothered the lower court, was the use of the specific number, 12,901, in the agreement. As there was considerable dispute over the meaning of the agreements, the court felt that these should have been resolved by trial and not by summary judgment. Indeed, it is not clear what was intended by the verbal agreement, and the court rightly held that this must be resolved.

In the case of *Roth v. Meeker*, the issues dealt more strictly with both genetics and contracts. This case, decided by The Appellate Court of Illinois in 1971 was most bitterly fought in the lower courts, so much so that the Appellate Court referred to it as, "...the Battle of Bull Run of Henderson County." Suit was brought alleging that the defendant had sold his property, refused to vacate it, and then intermingled his livestock with that of the plaintiff's to the damage of his herd. The actual land sale had been to a third party from whom the plaintiff rented the land for grazing purposes. Meeker had sold the entire property, with permission only to remain in a house thereon for a few months, and the rest of the property was to be turned over to the new owner immediately upon sale. Meeker also sold some of his cattle to Roth, and then refused to turn over to him the papers for the herd's registry as Angus cattle. (The advertisement of the sale of the cattle had mentioned registry.) The intermingling of the herd resulted in the interruption of the plaintiff's artificial insemination program, and it was also claimed that his cattle contracted the disease, pinkeye, as a result of the intermingling of the herds. Meeker's defense claimed that the lower court had improperly instructed the jury as to the awarding of penal damages as well as the direct loss to the plaintiffs. He also claimed that he had the right to continue to use certain lots and that the intermingling of the herds was due to the plaintiffs' negligence in maintaining the fencing in the pasture. There had been some wrangling concerning the judge of the lower court giving instructions to the jury, as mentioned. The Appellate Court dismissed this after summarizing the so-called instructions: "...we first entertain some doubt as to whether the trial judge's remarks should be categorized as instructions. We believe the remarks could be more properly classified as directives. We, however, see no merit in arguing semantics. The trial judge's remarks, whether they be termed instructions or directives, were permissible." Another point of law arose in that Meeker claimed he had oral approval, despite the terms of the contract to vacate all but the house, to retain possession of certain lots around the building. The plaintiffs' lawyers moved to strike this as it came under the *parol* evidence rule, a rule which

prohibits such oral agreements from evidence. The lower
court agreed, and the superior court found no error
here. They pointed out that almost all such agreements
are attempts "to find a way out of a bad bargain,
usually resulting from carelessness, no doubt, in
preparation of the written instrument." The fact that
the written contract omitted any mention of such lands,
and that it was apparent that the opportunity was there
showed it never had been intended for Meeker to keep the
lots in question (after all, the house was in the
contract). The defendant then fell back on the argument
that this was a case involving torts and not contracts,
in which case the *parol* rule would not apply. This too
was not considered favorably by either court.

The defense claim of contributory negligence by the
plaintiffs was also denied, as the records clearly
showed repeated inspection of the fencing and gates in
an effort to keep their own livestock confined. Later,
when the plaintiffs did release their livestock, the
intermingling of the herds had already occurred so that
the damage had been done by then. The defense then tried
to make a further claim that the plaintiffs had been
guilty of contributory negligence when they failed to
attempt to register the cattle purchased from Meeker,
specifically that they did not carry out the blood
typing required for registration. The courts found that
this could possibly argue for mitigating damages, but in
no way was negligence. "We are not impressed with
defendant's argument that the plaintiffs were under a
duty to register the offspring of the cattle which had
been intermingled and commingled by following the blood
typing procedure set forth by the Angus Association.
Even the defendant admitted that such a procedure would

be impractical. ...we merely make the observation that we harbor serious doubts if registration in the manner prescribed by the association would have been possible. ...The plaintiffs had a right to rely on the defendant's assurance that some registration papers would be provided for the livestock purchased by them. It appears from the evidence that the defendant did not have papers for all the livestock which he sold, but he admitted that he did have some papers and that he at all times refused to turn them over to the plaintiffs."

Still another legal point was raised by the defendant who now claimed there was no agreement to turn over the papers as there was no identification of the cattle. The court by now was becoming somewhat irate, to say the least. In good judicial prose they stated, "It was the defendant who advertised the existence of some registration papers and he cannot now hide behind a maze of confusion that he created. There is total lack of good faith on the part of the defendant as far as the delivery of registration papers was concerned." Further the court declared, "The damages incurred by the plaintiffs are not restricted only to the livestock purchased but in addition they suffered damages as to the offspring from the livestock. The plaintiffs were engaged in the cattle-breeding business and the conduct of the defendant prevented them from establishing blood lines and conducting an artificial insemination program. The record is replete with testimony concerning the value of registered cattle as opposed to the value of commercial cattle. Sufficient evidence was adduced upon which a jury could base an award of damages and hence the award in the instant case should stand."

The defense then further claimed that the jury had

been unduly influenced by stories in the press, another claim not deemed of consequence by the Appellate Court. Considerable details have been given here as this case illustrates the importance of the values of an artificial insemination program, and also shows the various different defenses, under different laws, which were attempted in order to avoid payment of damages by, to say the least, an outraged and imaginative defendant. Although one would think from all of this that vast sums of money were involved, the entire long original trial, the long appeal, leading to a twenty-three page opinion, amounted to the sum of $46,800!

Instructions to the jury also formed the basis, in part, of another case involving purebred cattle. This case, *Gyldenvand v. Schroeder*, was decided by the Supreme Court of Wisconsin in 1979. The facts are again complex, but it appears that the two parties entered into a partnership to raise purebred Simmental bulls, with Gyldenvand taking care of the paper work and finances and Schroeder actually raising the animals. It was decided to register the animals, and they went through the necessary steps, tagging, weighing and tattooing the animals. The application forms were filled out, and Gyldenvand took them with him. Schroeder understood the papers would be filed promptly. They were not filed at all, and Schroeder was not informed of this at the time. Subsequently Schroeder bought out Gyldenvand's interest, the purchase price being based on the assumption that the cattle were registered. When Schroeder asked for the papers he was told they were still being processed. Subsequently Schroeder had an offer to purchase eight cows and three bull calves. He went to Gyldenvand for the papers and was then informed there were none. A call by Schroeder to the American Simmental Association in Montana revealed that none had been filed, and that he was not even listed as a member of that group. Confronted with this, Gyldenvand then claimed that he had allowed the membership to lapse. Schroeder pointed out to him that once a breeder is a member, he is a life-time member and no further dues beyond the initial payment of $100 is required.

Schoeder then borrowed money to register the animals. Between the time that he had thought the registration was carried out by his former partner and the date of registration by Schroeder, the cost had increased from $354 to $1865. Next, one of the prospective buyers came back and offered to buy the whole herd for $30,000, with a $200 down payment. Schroeder's apparent gain, however, did not materialize as the buyer's corn crop failed and he could not pay the rest of the sum. Finally, the market for Simmental cattle collapsed; Schroeder was compelled to sell the entire herd for only $6500.

After a series of special instructions and special verdict questions, not objected to by the defendant's lawyer, the jury found for the plaintiff for $9,945, with Gyldenvand found 80% negligent and Schroeder 20% negligent. The appeal came on the basis that the award by the jury was not valid. "The general rule for appellate review of damage awards, as for other factual questions is that any credible evidence of the damage claimed is sufficient to sustain the jury's award." There are also other legal points regarding the case due to the fraudulent representation by the defendant when he agreed to register the cattle but did not do so. There is also a problem in law dealing with the "out-of-pocket" rule which suggests that "as a matter of logic, [it] is more consistent with the purpose of tort remedies, which is to compensate the plaintiff for loss sustained, rather than to give him the benefit of any contract bargain. "The treatise notes that it has been long suggested that the loss of bargain rule should be applied in cases of intentional misrepresentation, while the out-of-pocket rule should be applied when the misrepresentation is innocent. We hold that the out-of-pocket rule applies where negligent misrepresentation is found." (The treatise referred to is Prosser, *Law of Tort, 1971*.) On this basis the court remanded the case to the jury for reconsideration of the amount of the award to Schroeder. The court also dealt with the question of the special instructions and questions to the jury put by the trial judge, but decided that based on current laws these were proper and that issue should not be retried. This case again emphasizes the important point that purebred cattle are not the same as beef cattle in terms of market price. It might also point out the risks of raising some breeds as large losses may incur if the breed goes out of style or is supplanted by a better stock.

Another complex legal issue arose in 1980 in the case *Two Rivers Company v. Curtiss Breeding Service*. This case, decided by the United States Court of Appeals, Fifth Circuit, again involved artificial insemination. Two Rivers had purchased over one-hundred one-half blood heifers of the Chianina breed intending to develop a purebred herd by the process of backcrossing them to pure bulls of the same type. It is a genetic fact that at each successive backcross the amount of genetic material of the non-desired type will be halved, so that in a series of successive backcrosses to the pure type each generation will become more and more like the purebred. Curtiss Breeding Service is one of the world's largest furnishers of bull semen, marketing many different types. In 1972 Curtiss entered an agreement with a Canadian firm to market semen of the Chianina bull, Farro AC-35. As federal laws prohibit the import of some cattle, Curtiss maintains its bulls in other countries, but can legally export the semen to the United States. In any event, Two Rivers asked its agent to select proper semen and impregnate its heifers with that product. The agent selected the semen from the bull mentioned above and impregnated the cattle with it. Some time afterwards, Curtiss noted that the particular bull might be carrying a recessive gene for syndactylism and recalled all samples of the semen of Farro AC-35. By then Two Rivers already had sixty-four heifers impregnated with his semen. Of those, twenty-two produced live calves. Four calves were stillborn and showed the genetic defect of syndactylism. As the court pointed out, "Syndactylism is a genetic abnormality that can only appear when both the sire and the dam are carriers of the recessive gene. Therefore, Farro, as well as several of the heifers purchased by Two Rivers, were carriers. Syndactylism is exhibited by the fusion or nondivision of the functional digits of one or more feet of a cow. It is a hereditary genetic trait traced to the recessive gene. It is virtually impossible to detect the presence of a recessive genetic trait such as syndactylism until it is manifested by the union of two carriers of this recessive gene." Two Rivers brought suit for damages on two counts, the loss of the value of the stillborn calves, and more importantly, the loss of

the prospective value of the entire herd. The latter
claim is, of course, due to the fact that the appearance
of the defective calves announced to the world of cattle
breeders that the herd, at least the twenty-two
inseminated cows, were carriers of the recessive trait
and were to be avoided for future breeding use. The
lower court found in favor of Two Rivers claim for
$52,900 against Curtiss. The United States Court
reversed this decision on both counts. First, they
examined the loss pursuant to tort law. They found that
there were four distinct types of loss under Texas law
(where Two Rivers was established); the first was loss
due to personal injury, the second economic loss
resulting from a product with defective workmanship, the
third economic loss to the purchased product itself (the
typical case cited dealt with the loss of the value of
an airplane when a forced landing was required due to a
defect in the plane) and finally the fourth type of loss
involved physical harm to a plaintiff's other property
as well as to the product itself. The court here held
that none of these applied to the present case. They
dealt next with the question of strict liability, and
the two claims for loss as mentioned above. They
determined that under Texas law no case could be made
for the second claim: "After an examination of the
controlling Texas case law, it is clear that Two Rivers
has not stated a cause of action under strict liability
with respect to the second group of calves. If anything,
any damage incurred upon discovering and making publicly
known a latent physical defect in the herd of one-half
blood Chianina heifers purchased by Two Rivers
constitutes a loss governed by commercial law. A
plaintiff in Texas is precluded from recovering from

economic loss under strict liability." The loss was
again determined not to come under any of the stated
four categories. "The crux of Two Rivers complaint about
the non-Farro calves is that everyone knows about this
potentially deleterious gene and that since it is
impossible to distinguish the carrier from the
noncarrier calves, the value of all the calves is
reduced. This loss in market value due solely to the
stigma of an accidentally discovered defective gene is,
if anything, a commercial loss and that is not
cognizable in strict liability. ...The evidence showed
that all bull semen has the possibilities of carrying
recessive traits. But most importantly, the evidence
indicates that the 'ordinary' cattle breeder with the
ordinary knowledge common to the community of cattle
breeders expects that the bull semen will carry some
recessive genes. The ordinary custom is only to
guarantee that (1) the semen is from the correct bull,
(2) the bull is fertile, and (3) the semen will
impregnate a heifer. The custom is that the risk of
genetic defect falls on the owner of the herd. This is
especially true of exotic breeds such as the Chianina."
It is of interest, furthermore, to note the disclaimer
of warranty which Curtiss used, "Notwithstanding the
fact that artificial insemination is successfully
employed in the breeding of livestock on a worldwide
basis and Curtiss customers for many years have achieved
satisfactory results, Curtiss Breeding Service, a
division of Searle Agriculture, Inc., and its agents or
employees cannot guarantee the conception rate, quality
or productivity to be obtained in connection with the
use of its products or recommended techniques. Curtiss
does warrant that all semen sold will be processed

according to standards recommended by the American Veterinarian Medical Association and adopted by the National Association of Animal breeders. IT MAKES NO OTHER WARRANTY OF ANY KIND AND HEREBY DISCLAIMS ALL WARRANTIES, BOTH EXPRESS AND IMPLIED, OF MERCHANTABILITY AND FITNESS FOR A PARTICULAR PURPOSE. Further neither Curtiss Breeding Service, nor any of its agents or employees shall be responsible for any injury or damage caused to persons, animals or property through the negligent use or handling of its products, or for any damage or loss due to any defects in the containers used in connection therewith." (Large type is in the warranty.)

This statement, according to the court, would absolve Curtiss from any warranty claims such as were made here. The question whether this disclaimer could be extended to Curtiss through the agent who had purchased the semen was settled on the basis that the agent who purchased the semen for Two Rivers was not a seller of semen, but merely acting on behalf of Two Rivers itself and therefore the warranty applied. In other words, Two Rivers had no legal recourse under either tort or warranty laws. One judge did dissent, holding that Texas laws would permit recovery under strict liability, but the majority opinion left Two Rivers with no recourse for their alleged damage claims. It would thus appear that the purchaser of bull semen must accept the risk of spreading a recessive gene into his herd as part of the price he pays for using the service.

Another cattle case arose in *Morris Palmer Ranch Company v. Campesi*, decided by the United States Court of appeals in 1981. Here the issue was the failure of some 25% of the herd of 476 Maine-Anjou cattle purchased by

Campesi from the Canadian plaintiff to check out with breeding records by blood tests. Involved also was Campesi's failure to pay for the cattle. The issue was in part the Louisiana Redhibition law which is defined as, "the avoidance of a sale on account of some vice or defect in the thing sold, which renders it either absolutely useless, or its use so inconvenient and imperfect that it must be supposed that the buyer would not have purchased it, had he known of the vice." A seller who is unaware of the defect is bound to repair the defect, but if he is unable to repair the defect he is bound to reimburse the purchaser for any maintenance expenses occurred. Where the defects claimed do not rise to the level necessary to sustain redhibition, the buyer may seek a reduction of price.

The complexities in this case involve loss of markings by Campesi, questions of the validity of a random sample of the herd for blood tests, sudden cessation of calving records (ultimately proving to be because the herd had been in part sold, a damning factor for Campesi as the law also stated that for a claim to rescind a sale to be valid one must return the purchased item) and other claims. Much expert testimony concerning the possibility of other Maine-Anjou cattle, other than the designated sire, being the male parent, and testimony regarding the collapse of the market for this breed are part of the record. Campesi also claimed that only about sixty of the cattle involved were able to breed, but Palmer claimed that the breeding record was according to agreement. As pointed out it was difficult to prove Campesi's point as his breeding records stopped abruptly when half the herd was sold. Campesi also claimed that the failure of about one-quarter of the herd to pass the blood test cast doubt upon the value of the rest of the herd. All of these complicated issues were dealt with by the court which found that Campesi was not entitled to recision of the sale under Louisiana Law and also that he was not entitled to recover for lost profits.

The case is of interest primarily for the Louisiana laws, but also because it represents an example of the use of blood tests in determining the lineage of cattle. The tests at that time could not accurately determine just who the sire was, but could prove nonpaternity of a particular bull in question, in this case the supposed sire of the herd purchased by Campesi. Thus it would have been entirely possible that the cattle purchased by Campesi were sired by other Maine-Anjou bulls, but this would have required the impossible task of testing every bull as a possible sire. It is also of interest that the courts recognized a sampling of only fort-five animals of the herd, only about 10% of the animals, as a representative random sample of the herd. Also involved was the fact that even the ones who failed the test were recognized as at least 1/2 Maine-Anjou and the Association for the breed was willing even then to register these animals as representatives of the breed. On the basis of all these facts, Campesi was required to pay a reduced sum based on his debt and the lower value of the herd.

The dependency of modern methods of cattle breeding upon genetic techniques and their interactions with the law is found in *Pemberton v. OvaTech*, decided in 1982 by the United States Court of Appeals, Eighth District. The basis of this suit was the incorrect carrying out of superovulation and breeding back procedures. As

explained by the veterinarian involved, "The objective
of ova transfer, of which superovulation is part, is to
produce a number of offspring from an outstanding cow.
Through the process of superovulation a number of ...
eggs are released from a cow at one time; whereas,
normally in a cycle a cow would simply release one egg
at a time. We administer hormones which stimulate her
system to release a number of eggs. She is then
inseminated ... and about a week after she is
inseminated the fertilized eggs are flushed or collected
from her uterus.

"Once the ... eggs are recovered from the cow, the
material that is recovered is examined under the
microscope to identify the eggs that are found to
determine if they are indeed fertile and growing
normally.

"Those that are determined to be normally fertile
are then transferred into what we refer to as recipient
animals. These are simply animals that are well grown,
and can be of a variety of breeds. They make no genetic
contribution to the offspring that they are going to be
carrying. It is important that they be at the same stage
of the cycle as the animal from which the egg was
recovered...[T]he recipient animal ...merely acts as an
incubator or host for this developing embryo. ...

"In general, once an animal has been through the
superovulation process and ova transfer, if the owner of
the donor has obtained the [desired] number of
offspring...then in general...[the] animal [will be]
bred back...so that she can carry her own calf
naturally."

The actual case dealt with the desire of the
Pembertons to breed a particularly valuable Holstein

cow, Linda, who had difficulties in a previous number of
pregnancies. They selected sperm from a particular bull
and the superovulation process was attempted; however it
failed. Meanwhile, the Pembertons learned that the sperm
they had selected might have carried a recessive gene
for a defect and telephoned the veterinarian to
substitute another type of sperm for the next attempt.
He tried this and again failed. The third attempt was
successful; however in this case the veterinarian had
used still another type of semen and not the type
specifically requested by the Pembertons who had
requested that sperm for a different one of their cows,
Kyla. In any event, after the successful superovulation
the veterinarian collected seven fertilized ova and
implanted each egg into a recipient cow owned by
OvaTech.

The Pembertons then entered into a bull stud
contract with Select Sires for the sale of one bull calf
for $2000 and with another bull stud organization for a
bull for $1500 and with yet a third group for a cow for
$1400. When Mrs. Pemberton learned that the wrong sperm
had been used, she cancelled her check to the Ovatech.
Three weeks later a member of Ovatech went to the
Pembertons and attempted to remove their cows; he had
loaded several on a truck before he was stopped by a
representative of the local sheriff's office. The cows
calved, producing four bulls and three heifers. When the
three breeding associations learned that these cattle
were not from the union they had contracted for, they
cancelled their orders. The Pembertons ultimately sold
the four bull calves as beef cattle for about $1300 and
kept the heifers. Suit was then brought against Ovatech
and its veterinarian for damages from the loss of sales

and for loss of reputation. It appears that the Pembertons were attempting to establish themselves as breeders of Holsteins instead of just a dairy operation, and the sale of a bull or cow to the breeding associations would have been a large step in this direction.

The case took on federal jurisdiction as Ovatech is located in Wisconsin and the Pemberton farm is in Minnesota. The lower federal court found in favor of the Pembertons both for the loss in value of the seven calves and for the loss of the potential reputation they might have obtained had the proper semen been used. The Appeals Court reversed this decision concerning the loss of reputation. The first legal point to be considered was whether the Minnesota District Court had jurisdiction over the Wisconsin firm in this case under the so-called "long-arm" statutes of Minnesota. Ovatech had claimed that Minnesota had no jurisdiction over a Wisconsin case, but the court found ample reason to affirm its jurisdiction under that statute. The court then went on to deal with the substantive issue of the alleged loss of reputation and found that the damages were only speculative and should not be awarded. There were many points raised here, including the possibility of a national reputation for the Pembertons, but it was also pointed out that it would take more than five years before the bulls or cows could be tested and established as sperm or egg donors. The court could not find any real damage to the local reputation of the Pemberton operation and therefore declared that on neither basis could damages be awarded. The Pembertons were, however, allowed damages of $11,500 for the negligence and breach of contract claims which they had also made against

Ovatech.

Considerable detail has been given in the above cases, both to stress the importance of genetics and to illustrate the many different kinds of law which arise in genetic cases. Tort law, contract law, proper instruction of juries, jurisdictional disputes are just a few of the types of legal issues which ultimately become involved when genetic issues arise.

Still another phase of the law, environmental concerns, also arises in cases in which genetics may play a role. These are brought up here as the cases to be cited both arose due to agricultural methods. The disputes involve the use of pesticides or herbicides and their possible deleterious genetic effects upon other crops or upon persons exposed to them. The first case, *Riverside Citrus Farm, Inc. v. Louisiana Citrus*, decided in 1981 by the Court of Appeal of Louisiana, Fourth Circuit, dealt with the use by the defendant of the herbicide, phenoxy, on their lands. The lower court case produced a ten-volume record after an equal number of days of trial. Fortunately the Appellate Court opinion is brief. The main claim against the use of the herbicide was based on the testimony of a California allergist that "phenoxies cause flu-like illness and can even alter human genetic matter." The defendants had been temporarily enjoined from the use of the compound after testimony that several of the plaintiffs had suffered flu-like symptoms following an initial use of the phenoxy by the defendant. The appeal was based on the right of the lower court to impose the injunction against further spraying. The defendants claimed that their "more recent method of application of the herbicide, under the recorded wind conditions, could not

have allowed the herbicide to spread to plaintiffs'
property and therefore could not have caused the flu-
like symptoms some plaintiffs experienced or damaged
their citrus trees and vegetable plants." One cannot
help but wonder if the last clause was really the main
concern and the health claims brought in only to bolster
the seeking of the injunction. The Court refused to
intervene, saying that the decision to issue a
preliminary injunction was legally proper. "We conclude
that, on the question of sufficiency of proof for
preliminary injunction, allowing to the trial judge the
discretion that is recognized as his, the record
supports the preliminary injunction appealed from,
without prejudice to defendant's right to bring the
matter to full trial on the question whether a permanent
injunction should issue. There is arguable prima facie
proof of threat of irreparable harm to the health of
plaintiff that justifies the trial judge's exercise of
his discretion." One cannot help but wonder what the
length of that trial will be, given that the argument
over the granting of the injunction took 10 days and 10
volumes.

The second agricultural case is far more lengthy and
involves the spraying of strawberry fields where 10-and-
11- year-old children would be employed to harvest the
crop. This case, *Washington State Farm Bureau v. Ray
Marshall, Secretary of Labor*, obviously comes under
federal jurisdiction as it is a direct suit against an
officer of the United States. The United States Court of
Appeals, Ninth Circuit, dealt with the appeal from the
lower court which had held that the banning of the use
of the pesticides Captan and Benomyl was not
permissible. The problem was what the "Minimum Entry

Time" (MET) should be for these children. There are a
host of regulations concerning the use of the
pesticides, including those of EPA, OSHA, NIOSHA, and
other agencies dealing with whether the use of
pesticides will affect the health or wellbeing of minors
who enter the fields following their use. There were no
findings concerning the effect of these upon children
aged 10 or 11. Standards had first been promulgated for
these pesticides as requiring an MET of three days;
later in the same year's interval, Captan was removed
from the approved list for use on strawberries because
it and another chemical had been identified as suspected
carcinogens. Still another final rule shifted Benomyl
from the approved MET list to a list of pesticides which
had been reviewed and for which there did not seem to be
adequate data to establish an MET. The lower court found
that there was testimony by the plaintiffs' experts
which seemed far more credible than that of the
government and allowed waiver of the ban due to their
showing that the post-spraying interval before reentry
was not less than that required by the EPA and that
exposure would be trivial, and the Secretary failed to
perform a duty owed to plaintiffs when he denied them
waivers, and that these actions were "arbitrary and
capricious."

The Appellate Court disagreed. It cited a host of
various regulations from various agencies. The court
added, "This concern may or may not require absolute
assurance of zero risk to the childrens' health ...
However, we find it unnecessary to decide the
correctness of this suggestion and decline the parties'
invitation to express a contrary view for this circuit
and thereby to invite Supreme Court review. We hold only

that the district court erroneously required the
Secretary to rely on existing EPA standards for all
pesticides. We need go no further because we decide that
the agency's action in this case regarding Captan and
Benomyl were not arbitrary or capricious under either
the 'reasonable assurance' standard applied by the
agency or the 'absolute assurance' suggested by the D.C.
Circuit." The court also cites evidence that data
supporting the possible carcinogenic, mutagenic or
teratogenic effects of Captan have been established.
Thus the growers could not use these sprays on their
fields and expect young children to be employed as
pickers thereafter. Incidentally, the use of temporary
labor of children this age is in itself legal, and not a
violation of child labor laws.

This case illustrates the welter of regulations
which are promulgated by various governmental agencies.
The reading of all of these, and their proper
interpretation, is difficult, if not impossible, unless
one is a specialist in this field. The regulations are
voluminous, subject to many different interpretations,
and often contradictory. They are subject to constant
review and change, and what is permitted one day may be
forbidden the next as is shown in this case, and there
appears to be near chaos in this system. One need only
to turn to the opinion in this case, with its many
citations to various agencies and their rules to realize
that a layman, seeking to be law-abiding, may need an
expert lawyer simply to avoid criminal conduct.

Another case dealing with genetics, agriculture and
the law was decided in 1981 by The United States Court
of Appeals, Eighth Circuit. This is the case of *Select
Pork v. Babcock Swine*, the case which gives the title to

this chapter. The case might almost be humorous were it not for the obviously intentional deceit of the defendant corporation and the large amount involved in the suit, $552,781.29.

The facts are clear: "Babcock began selling boars and gilts as breeding stock in 1971. In 1973 Babcock began advertising for sale a feeder pig for breeding which it called the 'Midwestern Gilt.' As part of its national advertising campaign, the company prepared a brochure entitled 'Meet the Midwestern Gilt.' .... the pig was described as the 'gilt of the future' and 'not just another pig.' ...Other representations were that the gilts would be...as close to disease-free as possible; that the gilts would produce 825 pigs per month in a unit of 550 gilts; and that the defendants were experts in genetics and hog confinement management."

Based on these claims, the plaintiffs signed a purchase order for 550 Midwestern Gilts and twenty-five Meatline Boars. The usual warranty disclaimer were included in the contract. "At the time of the contracting, Babcock knew it did not have sufficient number of Midwestern Gilts to sell to the plaintiff or any customers. At the time set for delivery, a sufficient number of Midwestern Gilts was still not available. Babcock substituted pigs bought from a commercial dirt lot in Kansas. The hogs plaintiffs received were subsequently found to be diseased with atrophic rhinitis, myoplasma pneumonia and...swine mange. Instead of receiving three-way cross gilts and two-way cross boars bred to thrive in confinement, plaintiffs received a two-way-cross gilt and a purebred boar...Plaintiffs began experiencing trouble with the

pigs within a few months after delivery. Consultation
with veterinarians and medicine were required to treat
the various diseases. Production was lower than promised
or expected. After discovering the pigs were not
Midwestern Gilts and Meatline Boars, plaintiff filed suit"

One can hardly blame them. Instead of receiving
disease-free, fast-growing pigs, they received ill,
mangy, and slow-growing animals.

Babcock's defense was based solely upon their claim
of misinterpretation by the lower court of Iowa's
Uniform Commercial Code. They argued that there is no
basis on the facts given for nullifying the disclaimers
of warranty and limitation of remedies in the written
contract. Said the court, "We cannot agree." The court
went on to add, "The district court held that the
limitation-of-remedies clause failed of its essential
purpose because defendants did not deliver Midwestern
Gilts and Meatline Boars as described in the contract.
The court further held that the limitation on recovery
of consequential damages was unconscionable because the
defendants knew at the time of the contract they were
just entering the swine production field and that there
was a chance they would not be able to produce enough
pigs to meet contract requirements....We find no error
on this point. Had Babcock delivered the promised
Midwestern Gilts and Meatline Boars, then the clause
limiting damages to return of the purchase price would
have been reasonable. As events developed, however, the
very special pigs promised by Babcock were never
delivered. Babcock knew the pigs were not the types
promised; the plaintiffs did not....Having failed to
deliver the highly-touted special pigs, defendants may
not now assert a favorable clause to limit their

liability." A claim for fraud had also been made, but in light of the decision to allow full damages under the claims above, the court did not find it necessary to deal with that.

The case *Dempsey v. Rosenthal* illustrates the type of case which, perhaps, should not be cluttering legal volumes. In this case, before the Civil Court of the City of New York County in 1983, the problem involved a dog purchased for breeding purposes. This poodle, unfortunately, upon later examination had one undescended testicle. According to animal breeders this dog cannot therefore be registered for breeding purposes as the trait is know to be genetic. (The same apparently holds for horses or other livestock as well.) The plaintiff claims that the dog, a Mr. Dunphy, was sold to her as a breeder and that upon later examination by a veterinarian the lack of one testicle was discovered. The owner of Mr. Dunphy then sued to get her money back. The defendant claimed that the buyer should have been able to palpate the dog's scrotum and discover thereby that it was missing one testicle and that the dog could not be registered. A further claim was made that the cryptorchid condition would in no way impair the fertility of the dog, although the defendant does not seem to have discussed the fact the dog could not be registered. The court opinion in this major case occupied no less than nine pages, with a lengthy discussion on the meaning of merchantability and implied warranty under the N.Y. State laws. The judge awarded the plaintiff her money back. The entire sum, which occupied those 9 pages was $541.25! To make matters even more interesting, the case might have become moot, as, during the time it took to get the case tried, the

missing testicle did descend! While this case may seem
humorous, to say the least, it did deal with important
considerations of what a warranty really is; however, 9
pages of opinion seems a bit lengthy to decide the case.

Another type of case arose when one of the largest
rodent breeders for medical and other research found
that the animals it furnished were not genetically
homogeneous. This has been a near disaster for certain
research areas. Results cannot be duplicated as the
animals are different, and failure to achieve
replicability in scientific work casts a severe doubt on
the validity of the work. Many publications are called
into question. Months of time are wasted trying to
determine the failure of replicability when the real
cause is the accidental substitution of one strain of
rat or mouse for another. Young investigators whose
careers are just beginning are severely damaged, and
indeed their jobs may be in danger as their results are
called into question. It would have been surprising if
this did not provoke a lengthy and bitter law suit.
Based on the pig case, the expressed warranty clauses
may not be adequate defense if it were shown that the
rodent producer had knowledge that his animals were
impure. Indeed, the case did go to court. According to
the journal, *Science*, a Miss Kahan brought suit against
Charles River Breeding Laboratories, mentioned above,
for just such damages. The plaintiff asked at least
$200,000 for compensatory and punitive damages, claiming that
her ability to attract research support and/or promotion
at the University of Wisconsin had been impaired as the
use of genetically impure strains of mice has rendered
her research results invalid. The case was to have come
before a Federal Court under the product liability law.

The case was entitled *Kahan and the University of Wisconsin v. Charles River Breeding Laboratories*, and involved suit by the researcher and her employing institution against the breeding company for supplying her with improper genetic strains of rats, thereby rendering her work unrepeatable, and her position at the University no longer assured. However, prior to trial the case was settled out of court by Charles River awarding a sum of $40,000 for a research fund, two-thirds of which is to be used to support Kahan's further research with the remainder to support general zoological research at the University. No terms of the personal settlement between Kahan and the breeder have been released.

A basic issue which might also be covered under environmental law or under agricultural developments arose when a suit was brought by an interested third party to block a genetic experiment. In this case a University of California at Berkeley scientist was about to release a new genetic strain of bacteria into field conditions. This strain would allow the recipient plants to have a higher tolerance towards frost damage, thus extending the growing season and increasing yield. The suit was brought against the National Institutes of Health to block the experiments. Similar suits were also brought against Cetus Madison Corporation and BioTechnica International, both of which had planned similar experiments. The basis of the suit was the safety of releasing new strains of bacteria to replace wild strains in the environment, with the claim being made that adequate safety tests had not been carried out. The Federal Judge involved (Judge Sirica, of Nixon fame) issued an injunction against carrying out these tests until the case was decided, and there was a strong suggestion that the judge will decide against the tests. There was a question whether the decision would deal with the safety of the experiment or whether it would be on the basis that the N. I. H. simply had not followed proper procedures in approving the tests. Nevertheless, this third party suit indicates the legal confusion about such procedures, and brings the important issue of safety testing and proper procedures to the courts. Ultimately Judge Sirica did decide the case against the N.I.H on the narrow basis of its failure to carry out adequate environmental studies, and the tests were forbidden until proper consideration of this issue was made.

According to the New York Times, the first successful cloning of a cow has taken place, in this instance by taking an early developing egg and splitting it into two parts, each of which was then implanted into different recipient cows. (This was done as usually only one calf is produced by a cow, and twinning is rare in cattle.) The resulting calves are, of course, genetically identical, and it would seem to raise the spectrum of possibly similar means to be used in other organisms, including humans, to provide identical twins. While there would appear to be no legal issue in this case involving cattle, there certainly may be some if this procedure is attempted in humans, for example if identical-birth defect children resulted. How far this method can be carried is not clear now, but with the many successful Sin vitro with humans, the implications may be more serious than just having identical purebred cattle.

These, then, are a few of the cases which involve

Genetics in the Courts in various fields of agriculture and animal husbandry. It is certain, again, that many more will come. Every large seed company, or plant grower, or animal producer, carries on genetic research. In these days of gene splicing, recombinant DNA, cloning, and other new genetic techniques, no company can hope to remain competitive without such research. And, of course, as more genetic research is done, more cases involving liability and related issues will arise.

Another area where suits will proliferate certainly will be that of environmental law and the use of various chemicals in agriculture, as well as hormones in animal feed. As pointed out earlier the almost inscrutable welter of regulations would seem to openly invite such suits. While genetic damage may be difficult to prove, as often such effects do not occur for many years after exposure, there will be attempts to do so. Another feature will be the fact that the damage may not be caused to individuals who were directly exposed to the mutagen, but to their offspring, and the cases dealing with radiation and other mutagenic agencies dealt with in another chapter will be applicable to the harmful effects of agricultural use of chemicals as well. Indeed, it appears that the courts will be cluttered with such suits, and one can predict that there is a rich field for the legal profession in these areas.

ALPHABETICAL LIST OF CASES CITED

CASE	CITATION	YEAR
Agr.Svs.Assn's. v. Ferry Morse Seed Co.	551 F.2d 1057	1977
Bain v. Dept. of Revenue	645 P.2d 12 (Ore)	1982
Charles River, etc. v. State Tax Comm.	372 N.E.2d 768 (Mass)	1978
Cloning of Bannermaid, a cow	N.Y. Times, 3/25	1984
Dempsey v. Rosenthal	121 Misc 2d 612	1983
Gyldenvand v. Schroeder	280 N.W.2d 253(Wis)	1979
Klein v. Asgrow Seed Company	54 Cal.Rptr. 609	1966
Master Hatcheries v. Cole	212 S.E.2d 150(N.C.)	1975
Matter of Petition of Blue Spruce Farms	NY State Tax. Comm (Slip)	1982
McElhaney Cattle v. Smith	645 P.2d 801 (Ariz)	1982
Meuse-Rhine Cattle Breeders v. Y-Tex Corp.	590 P.2d 1306 (Wyo)	1979
Morris Palmer Ranch Co. v. Campesi	487 F.Suppl.1062	1980
Pemberton v. OvaTech, Inc.	699 F.2d 533	1982
Pennington Grain & Seed Co. v. Tuten	422 S.2d 984 (Fla)	1982
Perdue Foods v. Dept.Assessment & Taxation	288 A.2d 170 (Md)	1972
Pfeizer Genetics v. Prohaska	316 N.W.2d 777(Neb)	1982
Pfeizer Genetics v.Williams Management Co.	281 N.W.2d 536(Neb)	1979
Riverside Citrus v. Louisiana Citrus Lands	400 So.2d 263 (La)	1981
Roth v. Meeker	389 N.E.2d 1248(Ill)	1979
Select Pork v. Babcock Swine, Inc.	640 F.2d 147	1981

Select Sires v. Lindley   389 N.E.2d 1135(Oh) 1979
State Tax Com'r v.Flo
Research Animals   273 S.E.2d 811 (Va) 1981
Suit over genetically
   impure mice settled   224 Science 969   1984
Two Rivers Co. v. Curtis
       Breeding Service   624 F.2d 1242   1980
Unif.Comm.Code   34-21-101 to
                       34-21-1002   1977

Washington State Farm
   Bureau v. Marshall   625 F.2d 296   1980

CHRONOLOGICAL LIST OF CASES CITED

	v. Campesi	487 F.Suppl.1062
1980	Two Rivers Co v. Curtis	
	Breeding Service	624 F.2d 1242
1980	Washington State Farm	
	Bureau v. Marshall	625 F.2d 296
1981	Riverside Citrus v.	
	Louisiana Citrus Lands	400 So.2d 263(La)
1981	Select Pork v. Babcock	
	Swine, Inc.	640 F.2d 147
1981	State Tax Com'r v. Flow	
	Research Animals	273 S.E.2d 811(Va)
1982	Bain v. Dept. of Revenue	645 P.2d 12(Ore)
1982	Matter of Petition of Blue	NY State Tax. Comm
	Spruce Farms	(Slip)
1982	McElhaney Cattle v. Smith	645 P.2d 801(Ariz)
1982	Pemberton v. OvaTech, Inc.	699 F.2d 533
1982	Pennington Grain & Seed Co.	
	v. Tuten	422 S.2d 984(Fla)
1982	Pfeizer Genetics v. Prohaska	316 N.W.2d 777(Neb)
1983	Dempsey v. Rosenthal	121 Misc 2d 612
1984	Cloning of Bannermaid, a cow	N.Y. Times, 3/25
1984	Suit over genetically	
	impure mice settled	224 Science 969

Chapter 10
"Murder will out"
THE CRIMINAL CASES

The use of genetic evidence in solving felonies is becoming widespread. Once again, the major issue is the admissibility of evidence. But it must be remembered, as pointed out before, that the rules of evidence in criminal cases differ from those previously noted in paternity cases. There, only evidence of innocence of paternity was admissible until recently. In criminal cases, evidence tending towards proving either innocence or guilt has always been allowed, provided that expert testimony was given and that the criterion of the *Frye* test was followed. Thus the crux of a criminal trial is frequently the *weight* of the evidence as determined by a jury rather than the admissibility of that evidence. Also, unlike paternity cases, criminal cases require the finding of guilty "without a reasonable doubt," a stricture not found in paternity cases.

The criminal cases are of great interest in a study of the admissibility of genetic evidence. Unlike the paternity cases where the issue is usually the same, namely who was the father, criminal cases deal with a wide variety of issues. Before examining the cases dealing specifically with genetic evidence, it is necessary to review in general the types of evidence admitted, with specific attention to the admissibility of novel means of detection and solution of crimes, and then to compare these to the use of genetic evidence. It will be obvious that the use of evidence from physical and chemical sciences appears to be more readily admissible than the evidence from biological sciences. Therefore, this chapter begins with a series of cases decided on the basis of evidence from the former two sciences before turning directly to the cases involving "genetics in the courts."

A major case, not involving blood types, but showing the use of scientific evidence by the courts, arose in *Coppolino v. State of Florida*, before the District Court of Appeal of Florida, Second District, in 1969. Here in a case remarkably similar to a more recent one, a physician allegedly killed his wife by a deliberate injection of the lethal agent succinylcholine chloride. (The more recent one dealt with the accusation of attempted murder by injection of insulin; the defendant in this case was ultimately found innocent.) The defendant here had been found guilty and appealed, obviously. His medical knowledge had led him to the then correct conclusion that this compound could not be detected in the corpse, and the medical cause of death would be undetermined. Two errors led to his undoing. First, there was a discovery of a hypodermic wound in the left buttock of the deceased as well as the track of a needle in her subcutaneous fat. Paradoxically, it was pointed out that she was right- handed, and would hardly have injected herself on the left side of her anatomy had she wished to do herself in. This is a detail, and the major point of the case was the introduction as evidence by a scientist of a new method, especially developed for this case, for the detection of this toxin in the deceased. As this method was specifically developed for the case, it certainly should not have met the *Frye* standards. Nonetheless, the court ruled that this novel method was admissible. Many other medical experts were called in during the trial, and all testified that the lack of finding of any other known cause of death, coupled with the positive finding of the original expert, led them to believe that indeed the woman had died from the toxic injection.

The court dealt with this by citing various rules of evidence, one of which said,0 "Where the evidence is based solely upon scientific tests and experiments, it is essential that the reliability of the tests and results thereof shall be recognized and accepted by scientists or that the demonstration shall have passed from the stage of experimentation and uncertainty to that of reasonable demonstrability." As this was the first time this technique had been used the *Frye* rule seems not to have been applied in Florida in 1969. The case was also decided on other legal grounds, and there was a good deal of legal wrangling in the appeal as to the nature of charges to the jury on legal points. For example the defendant was found guilty of murder in the second degree. Only murder in the first degree requires proof of premeditation. As he could hardly have given his wife the lethal dose without premeditation, he claimed he could not be guilty of murder in the second degree. He also claimed that the jury was improperly instructed on other matters, but his claims were inconsequential compared to the evidence concerning the detection of the proven use of succinylcholine chloride which was obviously important in persuading the jury of his guilt.

A second case involving an attempt to introduce a new type of physical evidence arose in *State v. Stout* before the Supreme Court of Missouri in 1972. In this murder case the principle evidence against the defendant was the use of neutron activation analysis of blood, comparing blood found on the floor of the accused's car to that found on his tee-shirt and on a blood stained

sheet which covered the corpses. A professor from the
University of Missouri had used their nuclear facility
to compare the three bloodstains and concluded that the
three were identical. The court here found the evidence,
allowed by the lower court, to be inadmissible as the
testimony showed that he was the only person who could
carry out these tests and their validity, under the *Frye*
rule, could not be established. There was a general
agreement that "neutron activation analysis is generally
accepted as a 'scientific technique of chemical
analysis.' The State argues that this is sufficient to
meet the *Frye* test. We do not agree. The issue must be
narrowed to whether the application of the technique to
blood samples has 'gained general acceptance in the
particular field in which it belongs.' In this case, on
the record presented, we conclude it has not. 'It may
well be, of course, that some day in the near or distant
future the * * * [technique] will achieve the same
degree of acceptance as the present standard blood
tests. In that event the courts will welcome * * * [its]
use as an aid in our never-ending pursuit of truth. But
until that day we must continue to * * * [protect] both
litigants and jurors against the misleading aura of
certainty which often envelops a new scientific process,
obscuring its currently experimental nature.'" The court
then went on to discuss the use of similar techniques of
neutron activation for analysis of hair which had
originally been ruled inadmissible, but presently was
generally accepted. They pointed out that the use of
neutron activation tests for blood might well become
equally established in the future. Without any attempt
to judge the guilt of the accused, the case was remanded
for retrial without the use of the neutron activation

evidence upon which the state had based its case.

Again, showing that physical evidence is more acceptable, is illustrated in the case, *United States v. Stifell, II.*, in the U.S. Court of Appeals, Sixth Circuit, 1970. In this murder case, a young man was murdered by a bomb sent through the mail which exploded when opened. The data showed that the defendant had worked at a place where similar material, as demonstrated by neutron activation, was available and was identical to that found in the remains of the "infernal machine" opened by the deceased. The claim was made by the defense that the neutron activation analysis was too new and unproven and should not have been admitted as evidence. The court, obviously shocked and upset by the crime, examined all previous cases, examining some in detail, and concluded that on the basis of those previous cases where neutron activation techniques had been allowed here too the evidence was correctly allowed and the conviction of the defendant was upheld. There was a rather full description of the methods and technical processes used for this type of analysis, and that part of the opinion concluded by saying, "On this record and on the authorities cited to and found by us, we discover no basis for holding that the test results based on neutron activation analysis are inadmissible as a matter of law or that the District Judge abused his discretion in admitting the expert witness testimony... While we believe that the neutron activation analysis evidence meets the test of admissibility in this case, we also note that like any other scientific evidence, this method can be subjected to abuse. In particular, if the government sees fit to use this time consuming, expensive means of factfinding, it must both allow time for a defendant to make similar tests, and in the instance of an indigent defendant, a means to provide for payment of the same. We do not enumerate these problems because we think they are the only ones which may develop, but simply because we think they are the examples which come first to mind." It might be added that there was also considerable other evidence of the guilt of the accused, namely the finding of a number of old newspapers in the defendant's living quarters, all dealing with this particular crime. All of the other evidence, was of course circumstantial, but the defendant's claim that the neutron activation test was improper evidence under the *Frye* test was denied.

One cannot help but wonder whether the judges were so shocked by the this type of crime that the decision was so easily made in favor of the admissibility of the evidence.

Yet another case allowing the admission of a new type of scientific evidence arose in *People v. Palmer*, decided by the Court of Appeal, First District, in California in 1978. Again this was a murder case, and the evidence was that of proof of having used a scanning electron microscope to determine the presence of chemical elements which were found on the accused's right hand and were determined to have come from the ammunition used in the lethal weapon; in this case a pistol was fired at almost point blank range into the head of the deceased. This was the first use of this type of evidence and the judge defined the instrument. "The scanning electron microscope is somewhat similar to a light microscope. What little knowledge is necessary to run the device can be acquired in a short time." It must be pointed out that my colleagues in the field of electron microscopy would certainly not agree with the judge's estimate of the ease of use of such an instrument, and indeed would take considerable umbrage over such a suggestion. Nevertheless, the evidence was found admissible, and this, along with other evidence was sufficient to convict the defendant of murdering her husband. The judge examined the applicability of the *Frye* test once again, and, although this was the first case of its kind, found enough data in the scientific literature to decide that the expert witness for the prosecution was fully qualified in this field. It was determined that although this was the first time such scanning electron microscope evidence was used, it was up to the jury to decide on the *weight* of the evidence rather than it being an error by the judge in admitting the evidence. Again, while other legal issues were also involved, the major point is that a new use of an established scientific method was found to be admissible under the *Frye* doctrine.

As was pointed out earlier, these cases just cited involve the use of physical means to determine guilt, rather than to assay biological data for such a purpose. But they are necessary to point out that in many instances the mere fact that data are novel or that a case of the first instance may be before a court is not in itself a compelling reason, under *Frye*, to deny the use of such evidence in criminal trials.

The cases involving genetic evidence are numerous and again an attempt will be made to select those which show some novel usage or are cases of first decision involving the use of evidence.

One of the earliest cases involving biological data

was that of *State v. Knight* before the Supreme Judicial Court for the Eastern District of Maine, in 1857. The case involved the use of blood tests in determining guilt in a murder case. As, of course, no blood types were discovered until the early 1900's, this may seem peculiar, to say the least. However, the case is real, and the tests were real. But they were not for blood types, but for the source of the blood found on a knife with which the defendant was accused of slitting his wife's throat, or as the court put it in the language of that day "...with force and arms, at said Poland, in said county of Androscoggin, in and upon one Mary Knight, of said Poland, she, the said Mary Knight, then and there being a human being, and she then and there being in the peace of said State, feloniously, wilfully, and of his express malice aforethought, did make an assault; and that he then and there in his right hand had and held her, the said Mary Knight, in and upon the throat of her, the said Mary Knight, then and there feloniously, wilfully, and of his express malice aforethought, did strike, cut, stab and thrust, giving to the said Mary Knight, then and there, with the knife aforesaid, in and upon the throat of her, the said Mary Knight, one mortal wound of the length of five inches, and of the depth of three inches;--of which said mortal wound, the said Mary Knight then and there instantly died." In brief, he grabbed her and slashed her throat. (It would be of some interest to muse upon what legal records would look like if they had continued to be written in this style! The main body of the decision does not come until page 108 of the opinion.)

What is of interest, however, is not the language of the case, but the claim by the defendant that the knife in question, while indeed covered with blood, had become in that condition as he had slaughtered a sheep with it on the day of the purported murder. The state called a medical expert who gave lengthy testimony on the method whereby he obtained blood from the knife, carefully reconstituted it with the appropriate salt solution and then measured the diameter of the red blood cells found therein. He could testify without hesitancy these could not have come from a sheep as the diameter of the red blood cells was far too large to have been that of the sheep. "My opinion, on a careful microscopic examination is that no part of the blood upon that knife flowed from the blood vessels of a sheep," the doctor testified. The good doctor was thus instrumental in convincing the jury that murder most foul had indeed been conducted by the "said George Knight."

The defense claimed also that the deceased had slit her own throat while sharing a bed with his mother, but it was pointed out that it was most peculiar that the elderly lady had no blood on her clothing even though in close proximity of what must at best have been a gory mess. But there is also real irony to this case. Had said Knight claimed to have slaughtered either a rabbit or a steer, the expert witness could not have testified as he did; the diameter of red cells of these common farm animals is roughly that of the human erythrocyte and no distinguishment could have been made. Thus, while not feeling great sympathy for the defendant, one wonders what other evidence could have been brought had he not claimed a sheep rather than other animals as the source of blood.

There are countless cases involving rape and/or

murder as well as other felonies whose successful
solution rests upon the use of blood tests. A few will
be cited so that the general picture is clear, but this
is by no means an inclusive list.

First, another genetic factor has to be explained.
Certain people are genetically "secretors;" i.e., the
antigens normally found only in red blood cells are
expressed in various body secretions such as seminal
fluid, vaginal fluid or saliva. In criminal cases
involving rape or assault, these fluids may be subjected
to analysis just as would be the use of normal blood
tests, and the establishment of the blood type of the
person secreting these fluids is readily and routinely
performed. For example if ejaculation has occurred
during a rape, the seminal fluid can be examined, and if
the rapist was a secretor, then his blood type can be
determined using the standard ABO and rH series; this
can be compared to other vaginal secretions by the
victim, and if a discrepancy occurs, then the accused
belongs to a class of males who could be the possible
rapist. Such cases, as well as others involving standard
blood tests, will now be discussed.

An example of the use of blood-stains found upon
clothing following an alleged rape is found in *Shanks v.
State,* decided in 1945 by the Court of Appeals of
Maryland. The accused had been found guilty and
sentenced to hang. He appealed on the basis that the
blood type evidence, admitted as part of the chain of
evidence leading to his conviction, should have been
barred as its "scientific" nature would have an undue
effect by prejudicing the jury against him. The court
made a lengthy study of the use of blood types in cases
prior to this, finding that they had been used in

various European courts since 1924, albeit mainly in
paternity cases, but that by the time of this case many
thousands of cases existed overseas, in Germany,
Austria, and Great Britain. The court also recognized
the decision in *State v. Damm*, mentioned in the chapter
on paternity, and that the use of blood typing was
legally justified by that decision. The court also cited
numerous other paternity cases to establish the validity
of the use of blood tests in the legal system.

The facts here were genetically simple. Following
the crime an overcoat belonging to the accused was found
hidden in a closet. The coat was found to have blood
stains upon it. The accused claimed these had been the
result of a fight with another woman, and had nothing to
do with the case at hand. Unfortunately for his defense,
the blood type tests were carried out on the victim of
the rape, the girl mentioned, and on the stains found on
the coat. The victim was found to be type O as were the
stains on the coat. The other woman, to whom the
defendant attributed the blood-stains, was found to be
type A, thereby eliminating her as the source of the
blood-stains. In addition, blood stains found in the
snow at the place of the alleged rape were also found to
be of type O. Surprisingly, nowhere in the opinion is
the blood type of the accused mentioned. The opinion
does point out that, in as much as 45% of a population
may be type O, this simply put the accused in that
group, and nowhere does it point out that had the
accused been of type A the stains on his coat could have
come from him and the blood in the snow could not have
come from him. However, his alibi for the blood-stains
on the coat was demonstrably false as they were not of
type A as they must have been had the blood come from

the woman to whom he attributed the stains.

There was much other evidence also to connect the accused with the crime: he had been seen in the vicinity, and he had been observed on a street car covered with blood soon after the time of the rape; the victim was bleeding profusely when found, and the blood in the snow where the rape was committed came from her. Thus, the use of the blood tests in this case was only one of a chain of circumstances leading to the conviction, and of course was only of real value in eliminating the defendant's alibi. Taking all of these into consideration, and ruling that no rights of the defendant had been violated by the admission of the blood tests as evidence, the court upheld the conviction. The court concluded its opinion in part, "We see no valid objection in the idea that the jury (or the Court in this case) might attach too much importance to the scientific evidence. Recently, during World War II, blood banks, as they were called, were accumulated from millions of people...There could be few people in the country who failed to know of this, and who did not also understand that when a transfusion was made, it had to be of blood of the same type as the patient. Blood types, therefore, are now matters of common or ordinary knowledge...Judges and juries must be presumed to have average intelligence at least, and no assumption to the contrary can be made for the purpose of excluding otherwise admissible testimony."

Blood tests are also used in solving burglary cases. In *People v. Gillespie*, in 1974, the Appellate Court of Illinois was faced with such a case. In this case, two men were accused of smashing a plate glass window in order to gain access to a television store. A

witness had seen a car parked nearby and given the
police its license number which was traced, and the
police went to the accused's address where they found
blood-stains in front of the door and the accused with a
primitive bandage covering much of his arm. Blood stains
were also found on a rag and on the tags of several
television sets found later to belong to the robbed
store. The police also had found many blood-stains
within the store, presumably from the burglar at the
time the window was broken. All of these stains were
type A and rH positive, as was the blood of the accused.
By combinational analysis, such persons would exist in
about 2.7% of the black population, of which the accused
was a member.

An interesting legal issue arose here, however. At
the time of the original trial, the defense did not
object to the admission of the blood type evidence, and
the superior court noted that this precluded him from
asserting that this evidence was an error during the
appeal. The court, nonetheless, saw fit, as this was a
case of first impression in Illinois, to deal with the
admissibility of blood tests in criminal cases. They
examined other cases and concluded that "Accordingly, it
is appropriate that the court take judicial notice of
the scientific reliability of the ABO system of blood
grouping, if predicated upon a sufficient foundation in
conducting the tests." The court noted the fact that at
that time the use of blood tests to prove paternity was
not allowed, but pointed out that there were social
reasons for such laws which did not apply to criminal
cases. The court also dealt with the issue of
probability, and was not particularly pleased by the way
it was presented in this case (the probability mentioned

above had come from a publication by a pharmaceutical house). While admittedly this was heresay at least "to the second or third degree," they pointed out that such publications are generally held admissible by an exception to the heresay rule under most laws of evidence. Referring to another case, they quoted, "In our opinion, expert opinion will be a more effective tool in the attainment of justice if cross examination is permitted as to the views of recognized authorities, expressed in treatises or periodicals written for professional colleagues. The author's competence is established if the judge takes judicial notice of it, or if it is established by a witness expert in the subject." As such examination and objection to the statistical nature of the evidence had been brought during the trial, they concluded, "In accordance with the foregoing, the statistical information relating to percentages of blood types in the Caucasian and Negro populations and population as a whole was properly admitted for consideration by the jury." It is of interest here that this admission of statistical evidence which plays so major a role in paternity cases involving the HLA antigen system was not more widely cited by other courts.

A case of felony-murder and rape was tried before the Court of Appeals of Michigan in 1980. In *People v. Horton*, the evidence centered on the issue that the accused was a nonsecretor (only 20% of the population is in this class) and was of blood type O. The victim was of type B. Blood found on the accused's trousers and shoes, as well as blood found on the sheets was, type B. Seminal fluid found on the same sheets was from a nonsecretor. The defendant claimed that all of

the bloody mess found was due to the fact that he had
been in a fight and the blood found upon him came from
two causes- first, the fight, and secondly th fact that
the deceased had a severe nosebleed thereby causing the
blood on his clothing and on the bed sheets. He had no
explanation for the fact that the seminal fluid found in
the vaginal tract of the dead woman was from a
nonsecretor. The case was compared by the court to the
one just cited above where admission of a figure of 2.7%
for the frequency of a blood type was used. Here the
figure for nonsecretors is 20%. The court dealt with
this by more or less ignoring the difference in saying,
"If established data such as this is to be used at all,
we believe that the statistics themselves are of no
significance. As the population group connected with a
crime grows larger, the probative force of that
connection will decrease accordingly. As observed [in an
earlier case] 'To exclude evidence merely because it
tends to establish a possibility, rather than a
probability, would produce curious results not
heretofore thought of.'

"We conclude that deposits of blood and other
identifiable substances do not differ from other pieces
of physical evidence such as clothing, hair styles,
physical stature, or other observations that show
possible connections between defendants and criminal
acts. The weight of such evidence is for the jury's
determination." The use of the blood and seminal fluid
evidence is thereby established. However, it should be
noted that there were other confusing legal issues in
regard to the way the case was brought and the case was
remanded for rehearing, but with the evidence concerning
the seminal fluid and blood typing to be admissible in

the retrial.

The value of semen samples recovered after a sexual assault on a female is shown clearly in *People v. Nation*, decided in 1980 by the Supreme Court of California. In this case a man was convicted of lewd and lascivious assault on a child under the age of fourteen. While rape was attempted, no penetration occurred so that the lesser charges were brought. Semen was, however, found in the region of the victim's vagina. In this case no tests were made of the semen, but it was simply retained by the police on a smear slide. When the defense requested the slide some time later, after the initial trial, some type B blood group activity was determined. However, a more extensive analysis could not be made as the slide had not been properly preserved. No blood tests were made of the accused at the time of the original trial, and it was not determined whether this was the type of the accused or of the victim. This was the major basis of the appeal as well as a somewhat irregular identification procedure of the defendant. The Supreme Court stated, "We conclude that if the state recovers a semen sample of one who has made a sexual assault, it has a duty to take reasonable steps to preserve that evidence and to make it available to the defense. However, in the circumstances of this case we hold the defendant has not demonstrated that the people failed in their responsibility to adequately preserve this evidence."

The court goes on to determine the value of semen evidence: "While there are many possible analyses that may be performed on semen to identify the donor, and, by corollary, to eliminate others from the class of possible donors, the two analyses deemed most commonly feasible are ABO blood typing and identification of the genetic marker phosphoglucomutase (PGM). A recent discrimination probability study of white males in California indicate that an analysis of ABO type and PGM would eliminate approximately 80% of such males as the donor of a particular semen sample.

"Thus, an analysis of the semen sample in the present case might have not only impeached the credibility of the prosecution's witnesses, but also might have completely exonerated the defendant. Whether or not the police deem the crime sufficiently important to warrant such an identification analysis, they cannot make the decision for the defendant. Accordingly, when a woman has been the victim of an attempted or actual rape and the police recover a semen sample of the assailant, the authorities must take reasonable measures to adequately preserve this evidence." While this might seem a favorable opinion for the defense, the court went on to point out that the defense had also been lax. "As noted above, the state delivered the semen sample to the defense in response to a pretrial discovery request. At that time the defense counsel neither submitted the slide to a laboratory for analysis nor took any precautions to assure its preservation. Instead, it appears from the record that counsel gave the slide to the defendant's father, and only after the defendant was tried and convicted was the semen sample retrieved and sent to a laboratory for analysis. Even at this late date the laboratory was still able to identify the blood type of the sample. While it is theoretically possible that a more complete analysis might have been successfully undertaken had the state refrigerated the evidence, under these circumstances we cannot say that the deterioration of the sample was caused by the state's inaction. Certainly the prosecution had no duty to preserve the evidence once it was in the hands of the defense." Thus, this invaluable evidence was of little use in the trial. However, the case was reversed on the basis of the identification evidence referred to previously, and one will never be certain whether the accused was or was not guilty of the offense charged. The important point, in terms of genetics, however, was the recognition by the court that had the evidence been properly preserved and used correctly in the original trial, it would have been invaluable in deciding the guilt or innocence of the accused.

An almost similar case arose in 1982 in *People v. Newsome*, before the Court of Appeals, First District, again in California. Here following an alleged rape, proper vaginal swabs were made and the samples frozen. However, no tests for PGM were made as it was the custom not to do this test until other blood tests of the accused were carried out, tests which were never made. At a later date when PGM tests were done, there were inconclusive results. The original trial of the accused resulted in a hung jury, and the suspect then moved to have the charges dismissed as a result of the failure of the prosecution to preserve the samples properly. A lower court upheld the defendant; however, the appellate court overruled that verdict on the basis that the evidence had been properly obtained, examined and stored, and that the defense knew such evidence existed, but made no attempt to have the necessary testing carried out. Although the prosecution may have also been lax, the court found that, like the previous case, there had been no negligence by the prosecution.

Not only semen samples must be carefully preserved, but also any blood samples should be similarly treated. This issue arose in the Supreme Court of New York, Queens County, in 1982. Here the case, *People v. McCann*, involved sodomy, robbery, sexual abuse and possession of a weapon. The major issue once again dealt with the failure of the police to preserve evidence. Blood stains had been found both on the victim's pants and on the wall and floor where the alleged crimes were committed. Acting on poor advice, the investigating officers discarded the floor and wall samples and actually sent the woman complainant home wearing the blood stained pants. Efforts to obtain the pants later were cursory. The court ruled that the police had been negligent in their actions, as these stains might have exonerated the accused, both by the usual ABO and the PGM tests, particularly as the only other evidence against the accused was one witness to a crime purported to have been committed three years ago. The court ruled that the case be dismissed. One of the more literate judges pointed out, "Unlike Lady Macbeth in her unsuccessful effort to wash away the blood stains of guilt, defendant McCann seeks out the perpetrator's blood stains, found by the police at the crime scene and on the victim's pants." More importantly the Court stated plainly, "The court is of the opinion that the failure of an investigative agency to preserve potentially exculpatory evidence, when it should have been reasonably foreseen that such evidence was material on the issue of guilt, violates the plaintiff's due process rights regardless of whether he was a suspect at the time."

An unusual use of expert testimony arose in another case where the accused was found to have blood-stains on his clothing which arose from the fatal bleeding of the deceased. In *State v. Hall*, tried in the Supreme Court of Iowa in 1980, the defendant did not deny the presence of the deceased's blood on his clothing, but claimed that this resulted from his coming across her body and his dragging the body to get help. The state introduced a criminologist who testified that the "blood patterns on the defendant's clothing could only have been produced by being in the immediate vicinity where blood was spattered at a great velocity, as in a stabbing or beating and that it could not have resulted from the mere contact with the body. He also testified that the blood patterns on the defendant's pants were consistent with wiping a bloody knife like the murder weapon." The witness demonstrated that he was an expert on the flight characteristics of blood and had testified previously "on the geometric interpretation of bloodstain patterns on forty-two different occasions." Although he was one of the few people in the country in this field, he noted his experience and also that similar work on blood spattering was being studied in another place. The defense contested the admissibility of this type of evidence under the *Frye* rule. The court disagreed: "We believe that the rationale of *Frye* should apply insofar as it bears upon the reliability of the proferred evidence. Accordingly, we do not believe that 'general scientific acceptance' is a prerequisite to admission of evidence, scientific or otherwise, if the reliability of the evidence is otherwise established.

" Later in the opinion the court adds, "This evidence need not wait an assessment by the scientific community; the foundation evidence of reliability and the inherent understandability of the evidence provided

sufficient bases for its admission." A special
concurrence on most points of law, of which there were
many others besides the admissibility of the physics of
blood splattering, did not agree with the use of this
evidence. "I am unable to agree that scientific evidence
cannot be sufficiently identified, that admissibility of
scientific evidence should be determined on an *ad hoc*
basis, that reliability of a novel scientific technique
can be established solely on the basis of the success of
its leading proponent in peddling his wares to
consumers, that the general acceptance standard is
inconsistent with Federal Rules of Evidence, or that the
inference sought to be established in the present case
was not of great breadth, sensitivity and importance.
However, I concur in this result because I am willing to
take judicial notice that the principles of physics upon
which the blood spatter analysis depended in the present
case are well enough established to be subject to
judicial notice." Another complete dissent was brief. "I
believe the admission of blood dynamics evidence was
subject to the infirmities listed in the special
concurrence, but I am unable to agree that the error is
cured by taking judicial notices of 'certain principles
of physics.'" Despite the dissent, however, the guilt of
the accused stands. Although there is little of interest
from the genetic point in this case (except of course
the identification of the blood stains as being from the
stabbed victim), the point made earlier, that physical
and chemical evidence apparently is more readily
admissible than biological evidence, is the reason for
citing this case in detail.

Another case involving both intercourse and
fellatio arose in Michigan in 1981 before the Court of

Appeals. The evidence in *People v. Camon* was based upon
blood type evidence in part as well as other factors,
including visual identification of the defendant by the
victim during the course of the crimes. The defendant
was found guilty by the lower court and appealed on the
basis that evidence based on analysis of seminal fluids
was improperly admitted during his trial. Fluids found
upon the plaintiff's panties were found to come from
secretors of type A and type O blood. Both plaintiff and
defendant were secretors. The plaintiff was type A while
the defendant was type O. Thus the seminal fluid could
have come from the defendant. It was pointed out that
probability analysis showed that 36% of the population
would be type O secretors. The defense claimed that
admission of such evidence was improper, based on
previous Michigan cases. However, in this case, the
court ruled that the weight of other evidence was such
that the introduction of the semen analysis was not of
importance as "The test of relevancy is whether
evidence has any tendency to make the existence of any
[material] fact *** more probable or less probable. The
blood type evidence admitted at trial provided one
additional circumstance contributing to the
identification of the defendant. The objections of
remoteness goes to weight and is more appropriately a
matter for argument before the jury."

What made the blood type evidence less important
was the very positive identification of the accused by
the victim, including description of clothing and foot-
wear which were conclusively linked to the accused. It
is difficult to see how, from this decision, the court
would have treated the semen analysis data had they been
the sole means of identification, but the implication in

this decision is that this type of evidence alone would not have been sufficient for conviction. Although other legal issues arose, the primary interest in this case is the fact that admission of semen analysis was not ruled out by the court where other evidence was sufficient to insure proper identification of the defendant.

A much more sophisticated analysis was used in the case *State v. Anderson*, before the Supreme Court of Iowa in 1981. Here the charge against the defendant was first-degree sexual abuse against an 85 year old woman who was first beaten and then sexually abused. As the alleged perpetrator of the crime was well known to her she immediately called the police who arrested the suspect later the same day. The evidence used was that he had broken into her home by smashing a window in her side door, thereby cutting his hand, and bleeding onto the victim's bedding during the course of the crime. The blood was analysed and found, when compared to the defendant's blood, to bear four identical genetic factors. This resulted in a probability of less than two in one hundred for a person who was of Anderson's race, which was black. Another legal issue besides the admissibility of this evidence also arose. Evidence was found that the defendant might have claimed a defense of "diminished responsibility" at the time of the trial. However, the defense counsel faced a genuine dilemma. If he filed a timely notice of appeal, he would protect the right to appeal. If he did so, however, it would have precluded his right to file for a new trial. Thus there was a real issue of whether filing for the appeal, which he did, thereby excluding the application for new trial, was in fact a real conflict in the laws of the state of Iowa. This legal mixup was perhaps the main issue of the case, and the court had great difficulty with it. The court wrestled with this issue, and decided that the defendant could have done both, with appeals from both, i.e. he could have filed for a new trial and appealed that if it were not granted, and also appealed, as he did here from the original lower court finding of guilty in the first trial. Still another legal issue was involved which dealt with the definition of serious injury to the victim. The 85-year-old woman was described as "frail" and suffering from arthritis and hardening of the arteries. During the course of the assault she suffered two broken ribs and bruises to her head, as well as lacerations which caused her to be hospitalized for ten days. Apparently she was tougher than it would have appeared, as she recovered despite the apparent threat to her life. It is of interest that the court apparently accepted the results of the blood tests without any discussion of them in the opinion and simply dealt with the other issues described. It would therefore appear that the court accepted this evidence as valid, and that what was of importance here was not the guilt of the accused, but the various legal implications surrounding the trial. Therefore, the blood type evidence was admissible, and did not need to be discussed further.

A few 1982 cases need to be discussed in light of the above trials. The first, *People v. Wilson*, was decided by the Court of Appeal, First District, in California. The defendant was convicted of rape, robbery, and burglary. The appeal was based on the fact that the defendant's counsel was inadequate in his defense. The victim was taken to a hospital following her attack where two vaginal swabs were taken and one of

them examined. The presence of sperm and of an
aggravated level of acid phosphatase, indicating the
presence of semen, was detected. However, the expert
testifying in the case stated that there was only a
borderline level for determining the level of the enzyme
and therefore she did not conduct any blood tests of the
defendant for enzyme groups. As the general blood group
of both the victim and the accused was type O, no
inference as to guilt or innocence could be made on that
basis. The defendant claimed that, "his counsel was at
fault for failing to have the semen sample analysed to
determine the presence of various genetic markers,
including peptidase A which he alleges is particularly
useful in determining the identity of a semen donor when
a black individual is the suspect, as well as PGI
(phosphoglucose isomerase) and PGM. He asserts, on
information and belief, that 0.04 microliters of semen
[less than the 1.3 microliters found, but an amount
claimed too low for analysis by the expert witness] is
sufficient to determine the presence of peptidase A, .01
microliters to determine the presence of PGI and 0.10
microliters to determine the presence of PGM. The
defendant's claim of the inadequacy of counsel is thus
based on the failure of this attorney to request the
genetic analysis which the defendant stated could have
been carried out. The court thus had to deal with two
issues: the defendant's right to adequate counsel, and
the probative values of semen analysis. The court dealt
with both issues in determining to affirm the
conviction. First, they pointed out that "Routine
genetic typing of semen from post-coital vaginal swabs
appears to be a relatively recent phenomenon in forensic
science." They cited *People v. Nation*, discussed

previously where the court had stated, "While there are many possible analyses that may be performed on semen to identify the donor, ...the two analyses deemed most commonly feasible are ABO blood typing and identification of the genetic marker phosphoglucomutase (PGM). The Supreme Court did not mention peptidase A, which constitutes the principal focus of Wilson's appellate attack. While it may be that reasonable inquiry by Wilson's trial counsel would have led him to peptidase A, that is by no means apparent on this record. ...We could not expect Wilson to demonstrate that the peptidase A test, if performed, would have exonerated him. ...What we have is a declaration to the effect that the amount of semen present 'may have been sufficient' to test for peptidase A, with no consideration or explanation of the factors which might have bearing upon the degree of probability that this would be so." In a footnote to this the court adds that the data upon which Wilson bases his case were from semen obtained from volunteers by masturbation, liquefied at room temperature for 30 minutes and then treated or stored at low temperatures. "The samples in the case at bar were obtained by vaginal swabbing some time after the rape. The literature indicates that protein degradation and loss of enzyme activity probably occurred in the interim." Thus the court held that legal counsel who did not request the tests now cited by Wilson as possibly acquitting him was not remiss in his defense actions, and also that the peptidase A evidence was of little apparent use in the case, as both the evidence for its use and the probability of finding such an enzyme in a vaginal swab were uncertain.

A similar case involving loss of a semen sample

arose in *People v. Jackson,* the Court of Appeal, First District, of California, in 1980. Here the semen samples taken were simply lost by the police. "Appellant had moved for production of the semen samples, but the district attorney advised the court that the evidence no longer existed." The appellant claimed that the charges against him should have been dismissed because "Comparison of genetic markers found in the semen specimen with those in an innocent man's blood might have afforded an innocent individual a high probability of scientific exclusion." The court, relying on an earlier decision, decided that there had been no "bad faith" on the part of the prosecution and denied this claim. The final decision, upholding the defendant's guilt was then based on other available evidence which was found more than sufficient; the two separate events of rape and oral copulation had sufficient commonality of events to make certain that the same person had been the perpetrator of both and the evidence from the semen samples was deemed unnecessary.

Evidence obtained from seminal fluid following a rape-murder case was found admissible by the United States Court of Appeal for the Fourth Circuit in 1982. This case, *Anderson v. Warden,* also involved the finding of the defendant's fingerprints at the scene of the crime. But most telling was the presence of seminal secretions on his pants which contained upon analysis proof of both the blood type of the victim and the accused, as the deceased was type B, a less common type found in only 10% of the population. In addition pubic hairs demonstrated that it was possible for him to have been the assailant. The complete set of evidence was judged sufficient to convict him of the crime. Anderson also based his appeal on the issue of the use of the written confession which he had signed. As he was illiterate, this confession was taken orally and then transcribed for his signature. His claim that due to his inability to read he could not tell what he was signing was also part of the appeal. However, the court upheld the guilty verdict on the basis of the evidence presented.

While, as was pointed out previously, the courts are frequently slow to accept new genetic evidence, defense counsels are not. During the early and mid 1970's, a new legal defense of insanity due to genetic causes arose. During the previous decade, evidence had accumulated, mainly from penal institutions abroad, that comparison of violent prisoners as against more tractable ones showed a much higher preponderance of a chromosome abnormality in males, namely the presence of an extra Y chromosome. This anomaly arises during the formation of the sperm when instead of each spermatazoon containing either an X or a Y chromosome, a rare sperm will contain no X chromosome and two Y chromosomes. As every normal egg contains a single X chromosome, union of this rare sperm with the usual egg will yield an individual with an extra Y chromosome, the XYY condition. Due to the presence of the Y chromosome the individual will be male. Most such persons containing the extra Y turn out to be taller than the average, but otherwise appear normal.

At the time of the original study, much was being made of the possibility of a genetic cause for criminality. Also, of course, if a crime were committed by an XYY individual, he would then be able to plead innocent by reason of genetically-caused insanity, and such a defense, if proven, would offer the possibility of a not guilty decision by reason of insanity. The various rules under which a court operates in determining insanity are complex, and would, in themselves, constitute a different type of book than this and for this reason, cannot be discussed in detail here. The reader may, if interested in this complex area, consult any of the books on rules of evidence and of insanity.

The first case dealing with this novel defense was in 1970, before the Court of Special Appeals of Maryland. This murder case, *Millard v. State*, did not essentially involve the guilt of the accused, but whether he was mentally responsible for his actions. Two expert witnesses testified that in their opinion the finding that the defendant was of the XYY type would be exonerating evidence. As one put it, "That the presence of this extra chromosome constituted a 'basic defect in the genetic complement of the cell' affecting not only the cell....but also the physical growth of the body itself; that the presence of the extra Y chromosome caused 'marked physical and mental problems' affecting the manner in which persons possessing the extra Y chromosome will respond to certain stimulus; certain physiological problems; certain behavioral characteristics." He went on to testify concerning past data on XYY persons as having, "marked antisocial, aggressive and schizoid reactions and were in continual conflict with the law." The second expert essentially agreed with these statements. It was pointed out that the defendant was an unruly prisoner and had also attempted suicide and his overall actions were "not consistent with sanity." Nonetheless, the court was not impressed with the testimony of these experts and the conviction stood.

The next case arose in California later in 1970. Here, in *People v. Tanner*, the Court of Appeal, Second District, was asked to rule whether the defendant could change his plea after being charged with kidnapping, forcible rape, and assault with intent to commit murder. Tanner pled guilty to the assault charge and the others were then dropped. The court had him committed to a state hospital for examination as a "possibly" mentally disordered sex offender; after this was determined to be the case he stayed at the institution for 6 months at which time it was determined that he did not respond to treatment and was still a menace to society. He was then returned to the trial court and the criminal proceedings against him reinstituted. It was at this time that he sought to change his plea.

During his stay at the institution, it had been determined that he was an XYY person. His entire defense was to be based on this; thus the change of plea sought by him. The court spent considerable time carrying out research on the XYY situation and noted that this defense had been upheld in early cases in Australia and France, but pointed out that the "genetic criminal" was unknown in the United States. They further noted that a more recent article, challenging the concept of inherited criminal tendencies in XYY individuals had been published. The testimony in the case was apparently long and detailed. The court noted this by saying, "The testimony and documentary evidence on this point was voluminous and complex. Certain facts of special importance are apparent, however. The studies of the '47 XYY individuals' undertaken to this time are few, they are rudimentary in scope, and their results are at best inconclusive. "... The evidence collected by these experts does not suggest that all XYY individuals are by nature involuntarily aggressive. Some identified XYY individuals have not exhibited such behavior." The court therefore concluded that the defendant was guilty as charged, and that the presence of the extra Y chromosome was not sufficiently demonstrable evidence for a change of plea to legal insanity.

In *Knight v. Texas*, heard by the Court of Criminal Appeals of Texas in 1976, the issue in this murder trial was whether the defendant's request for a chromosome analysis to determine if he were XYY, and the failure to grant such a request, were grounds for determining that he had not received a fair trial. There was no question of guilt (the defendant had choked the female to death with a strangle hold he had learned while serving in the U.S. Marines), but the issue was stated by the court quite clearly. "Appellant contends the court erred in refusing to grant his motion for a medical examination to determine if he had a chromosomal abnormality.

"Appellant argues that recently developed medical information indicates that an individual with the 'XYY Syndrome' may, because of his chromosome abnormality, be more aggressive and violent than an individual with a normal chromosomal makeup." The appellant had also requested that the approximate cost of $1000 for the tests be borne by the county as he was unable to raise such a sum. The lower court heard one expert witness who simply testified that in his opinion the defendant did not have "a premeditated plan" to take a human life at the time in question." The appellate court found no reason to order the chromosome test and that failure to do so had not denied the accused a fair trial.

The XYY defense reached New York State in 1975 in the case of *People v. Yukl*, tried before the Supreme Court, Trial Term, in New York County. Yukl had a previous record of a brutal murder for which he was convicted, served 7 years and then was released on parole. Yukl was again charged with a similarly brutal murder and brought to trial. There was considerable legal wrangling over his motion to separate the trial into two issues, one of guilt, and one of insanity. The court did not see fit to do so as it felt the two issues were so intertwined that they could not be separated. The court then dealt most specifically with the XYY issue. The judge cited the previous cases and turned to the insanity defense rules in New York. "...an insanity defense based on chromosome abnormality should be possible only if one establishes with a high degree of medical certainty an etiological relationship between the defendant's mental capacity and the genetic syndrome. Further, the genetic imbalance must have so affected the thought processes so as to interfere substantially with the defendant's cognitive capacity or with his ability to understand or appreciate the basic moral code of his society." Apparently an earlier case had dealt with this same plea in New York, namely *People v. Farley*, not cited here. The court here was consistent with that case in denying the XYY condition as a basis for an insanity plea.

The Court of Appeals in the State of Washington, Third Division, came to a similar conclusion in 1976 in *State v. Roberts*. The accusation here was less serious, namely grand larceny and taking and riding in a motor vehicle without permission of the owner. The lower court

had refused to grant continuance to the defendant so
that an XYY test could be conducted. The defendant,
incidentally, was already on parole for similar crimes,
and the defense attorney had suggested that his client
"has characteristics of the 47 XYY syndrome, affecting
mental competence." The court, as in the previous cases,
reviewed the opinions concerning this genetic condition,
and again refused the plea for a chromosome test. "The
XYY individual is said to often be extremely tall, with
long limbs, facial acne, an aggressive behavior often
directed toward property rather than persons, with no
significant family history of mental illness or criminal
activity. The XYY criminal also tends to be more
resistant to conventional corrective training." The
court went on to cite medical evidence, "But ...
presently available evidence is unable to establish a
reasonably certain casual connection between the XYY
defect and criminal conduct. ...In the absence of sound
medical support of the XYY defense, the trial court did
not abuse its discretion in denying defendant's motion
for a continuance."

An important change in emphasis is seen in this
case. It should be noted that the XYY individual is now
deemed to be more likely to commit crimes against
property, and not against persons, something which
medical evidence was beginning to reveal. More
importantly, this is probably the last case in which
this defense was attempted, and justifiably so. The
courts, in refusing to leap on the XYY defense
bandwagon, were in this case completely justified in
their conservatism. It is now believed that the earlier
data, indicating the violence and criminality of the XYY
individual, may have been based on a particularly biased

sample, namely institutionalized persons who certainly tend to be more violent as a population. The finding of a disproportionate number of XYY cases in this restricted, and certainly not general, population appears now to be not a valid genetically controlled characteristic but rather simply a sampling error. Further, as most of these men are quite tall, it is equally possible that they may be viewed as more threatening to the normal person, and therefore attributes of violence are easily given to them. Perhaps a study of professional basketball players should be carried out. Certainly these are all tall men, and some of the actions on that court could readily be interpreted as rather violent!

Competence to stand trial may also be a question in cases involving defendants with low I.Q. or with inherited mental disease. While little is known concerning the inheritance of mental abilities, particularly in regard to the I.Q. tests, it is now strongly believed that some mental diseases, such as schizophrenia, have a high genetic component. Two cases, chosen from others, illustrate these two points.

First, in *United States v. Wenzel*, decided in 1980 by the United States District Court, D. Nevada, the question of the mental ability of a defendant in a case involving the sale of drugs was a basic issue. It was found that the adult defendant had only a third or fourth grade ability to read, write and do arithmetic. Testimony of an examining psychiatrist indicated that there was "a belief on his part that the defendant suffers a disability in the left hemisphere of his brain and that this is due to a genetic or biological developmental defect. In other words, in his view it is not due to any mental disorder, but rather it is a physical defect... not mentally retarded in a general sense, but only in the specific areas of reading, writing and arithmetic and finds him to be in the range of mild retardation. Defendant's counsel therefore argues that the defendant is for these reasons mentally incompetent and unable to understand the proceedings against him or properly assist in preparing his defense." One must confess some difficulty in understanding why a physical brain defect is different in its effect from an inherited mental condition, but apparently the court did not find the two similar. The court disallowed the claim of incompetence and stated in part, perhaps somewhat sarcastically, "No doubt there is a level of mental retardation which would make a defendant so mentally incompetent that it could be assumed he would be unable to understand the proceedings against him or properly assist in his own defense. But based on this record it must be concluded that the extent to which defendant suffers mental retardation does not reach this level. "...It simply does not follow that because a person has only a fourth grade ability in reading, writing and arithmetic, or that he is unable to think abstractly when faced with proverbs, that he will be unable to understand the nature of the proceedings against him or to assist in his own defense." In addition to the defendant's own inabilities it was suggested that his parents also had similar difficulties with the basics of education, but the possibility of the defendant's suffering a genetic defect was not seriously considered. Thus the court concluded that knowledge of the 3 r's at the fourth grade level was adequate.

The opposite results were found in *People v. Samuel*

by the Supreme Court of California in 1981. Here the issue was the mental condition of a person convicted of first degree murder during a holdup of a gas station. The defendant's counsel, prior to the initial trial, suspected that the accused was mentally ill and therefore incompetent to stand trial. All of the 5 court-appointed psychiatrists, 3 psychologists, a medical doctor, a nurse, and 3 psychiatric technicians testified on behalf of the defendant. Each and every report stated that the defendant was certainly incompetent to stand trial. The prosecution offered only two lay witnesses, neither of whom contradicted the above reports. Yet, for some reason he was deemed capable to stand trial and was convicted.

The Supreme Court examined the well-defined record and noted that the accused had shown signs of mental illness since the age of 8, experiencing catatonic episodes, hearing disembodied voices, and in general showing all of the clear signs of schizophrenia. In addition, there was apparently a strong genetic basis for this as his sister also suffered hallucinations. As he grew older the symptoms of the disease worsened and much of his life was spent in jails or a mental institution, from which he had escaped prior to the murder. After being captured and placed again in jail he became even worse and had to be put into a padded cell lest he injure himself. Treatment with vast dosages of medication, roughly 100 times the amount a normal person could tolerate, showed that he could not be feigning the disease. The consensus was that the defendant suffered from not one, but three mental diseases, chronic schizophrenia, mental retardation, and some organic disfunction.

Obviously the higher court, while taking time in its opinion to discuss the insanity defense, ruled that the defendant was completely unable to understand or participate in his defense and the failure of the lower court to determine his incompetency was reversed. It is quite unclear how the lower court could possibly have found him competent. However, the point to be made here is the quasi-recognition that schizophrenia may be an inherited disease and therefore this is a genetic case. Unlike the XYY cases, these two cases illustrate the proper use of genetic evidence and its availability for decision making.

The use of additional-blood test evidence besides the classical ABO system and of new techniques for their determination are now well established in the courts. The principle now used is the technique of electrophoresis. In method, the PGM and the EAP system of blood enzymes is used. As the judge in the case *State v. Rolls*, tried before the Supreme Judicial Court of Maine in 1978, stated, "The erythrocyte acid phosphotase (EAP) and phosphoglucomutase (PGM) blood classifications are based on the presence of enzymes in the blood. These systems are genetically controlled, as in the ABO classification, and all three are independently inherited, such that the blood type in one system does not depend upon the type in either of the others.

"The theory of the PGM system is that there are slight differences between enzyme molecules in the blood. Under certain conditions, these different molecules separate and migrate and can be visually classified by the formation of 'banding patterns' in reaction to charging by an electrical field (electrophoresis) in a starch gel solution.

"The theory of the EAP system is similar except that it involves the use of ultraviolet light rather than an electrical charge. The end result which is subject to typing is again a banding pattern." The judge then goes on to discuss the three classifications in the PGM system and the six possible classes in the EAP system.

The case involved the rape of a 14-year-old girl, and the finding of blood stains on the accused's pants afterward. The initial blood typing proved of no value as both victim and suspect were type A-1 in the ABO system. The opinion includes a table which is worth showing here as it demonstrates the type of analysis in the case:

	ABO	EAP	PGM
Pants	A-1	BA	2-1
Victim	A-1	BA	2-1
Defendant	A-1	A	1-1

Thus the double finding of the EAP type and the PGM type on the accused's pants shows that the blood stains could not have come from him. Furthermore, the expert witness for the prosecution testified that the combination of the three types gave a probability indicating that "approximately 5% of the population, or one person in twenty, would possess all three blood types which the victim possessed and which were found on the defendant's pants."

Needless to say, there was considerable argument about both the admission of the evidence gained in this novel way, and also the properness of allowing the testimony concerning the 5% probability. The defendant claimed that the blood stains had come from an event some months ago when his girl friend's daughter had cut herself in his presence. But he did not do his case any good when he admitted that he knew that blood tests would be used as evidence and had not bothered to obtain a sample test of the proposed source of the blood. The Supreme Court decided that both types of evidence were admissible, and went to some length to discuss the use of the evidence, stating in part, "The physician makes life and death decisions in reliance upon ...[matters which would not necessarily be admissible, at least without time-consuming authentication]. His validation, expertly performed and subject to cross-examination, ought to suffice for judicial purposes."

Again citing an earlier case, the court declared that the expert "is fully capable of judging for himself what is, or is not, a reliable basis for his opinion. ...Experience teaches the expert to separate the wheat from the chaff and to use only those sources and kinds of information which are of a type reasonably relied upon by similar experts in arriving at sound opinions on the subject. The court also, in a foot note, looked at the confidence limits of the 5% statistic, and noted that had there been some error in the statistics for each blood typing used, the chances would still be, at the best, that 1/10 rather than 1/20 of the population would be of the type of the victim. This court then found no error in either the use of electrophoresis or the use of probability as evidence.

It should also again be noted that this was only part of the evidence presented in the case, and, despite the sophisticated blood tests, the court still found the defendant guilty. There were other, obviously more seriously considered reasons for affirming his guilt. Identification, while not positive, was fairly accurate (the victim, without her glasses, could only describe the approximate height, build and weight of her assailant) in that the accused fit this description and his alibi for his activities during the time of the crime was totally without substantiation. In addition, the claim that the blood-stains came from the girl friend's child as he claimed were patently false as they were still found to be damp some weeks after the incident the defendant said had caused them. For all these reasons, his guilt was affirmed and the use of blood tests of a novel type while now established as admissible were not sufficient corroborative evidence in his claim of innocence.

Robinson v. State is another case in which electrophoresis was employed to help establish the guilt of the accused. In this murder case, tried before the Court of Special Appeals of Maryland in 1981, there was also additional evidence, but the critical issue was the admissibility of the identification of the blood found on the accused's knife.

Maryland is an interesting state in terms of the interpretation of the *Frye* Doctrine. In a classical case involving the use of evidence in criminal cases *Reed v. Maryland*, the opinion of the Court of Appeals in 1978 was in reality a "law review" of more than sixty-five pages. The opinion dealt with every form of evidence, voiceprints (the basis for the appeal), fingerprints, ballistics, blood tests, intoxication tests, and other scientific evidence. The bulk of the opinion, it should be pointed, out is a scholarly dissent, and a thorough review of the *Frye* standards, including the introduction of new types of scientific evidence and the concept of relative acceptance. The two views on acceptability of evidence, *McCormick* and *Frye*, are compared and contrasted in detail, and the reader interested in the use of scientific evidence is strongly urged to examine this opinion. In general, Maryland seems to adopt a position between the two opposing views, and certainly is not bound by *Frye*, which their courts have not fully endorsed.

To return to the case at hand, the defendant had been convicted of first-degree murder, rape and aggravated burglary. The victim had been stabbed no less than eighteen times, and there was no question that the assailant would be blood-stained. The defendant had appeared at a local hospital to be treated for stab wounds, and was arrested on the basis of an outstanding bench warrant for a traffic offense. Subsequently after being fingerprinted he was released. However, his fingerprints were found at the scene of the murder and examination of pubic hairs found on the victim established no less than 21 similar microscopic features, not enough in itself, given the state of the art of identification of such hairs, to prove guilt. However, blood samples from him were found to match those at the scene of the crime as well. Here no less than 8 tests were used, the standard ABO as well as 7 blood proteins detected by electrophoresis.

The expert for the prosecution, examining the blood of the victim and of the accused could state confidently that less than 3.1% of the population could have contributed the blood found at the scene of the

crime. In addition, semen samples showed that the
rapist-murderer was a nonsecretor, a group, as pointed
out earlier, comprising only 20% of the population.
Adding this onto the various blood type tests, it was
determined that only 6/10 of 1% of the population would
match all these characteristics, or to state it
conversely, 99.4% of the population would not match
these types. This was the first use in court of what has
come to be called "The multisystem analysis" and the
expert testified that this is now widely used in
forensic medicine. Testifying against the use of the
tests was another research biochemist who claimed the
method to be unreliable on several counts, most
specifically that the type of test used could not be
accurately carried out on dried blood. To rebut this,
the state brought in a forensic chemist who had been a
coworker with the defense witness, but had dismissed the
latter's work as being so unreliable and poorly
performed that his results were deleted from a joint
report done by this expert and yet another coworker.

The prosecuting witness testified that the
multiple enzyme system was being used in over a hundred
crime laboratories in the U.S. and Canada, and also that
at this time the F.B.I. had now begun to use it
routinely. The defense based its claim on this issue on
Frye and also that "there is a difference between the
field of forensic science and the field of biochemistry,
which renders a forensic expert incompetent to testify
on the reliability of blood analysis tests, in
contradiction to the testimony of a biochemist." It is
obvious that the defense witness was a biochemist and
the prosecutions experts were forensic scientists. There
was a further challenge to the ability of the forensic

scientists in this case, but their credentials for the type of work in this case were impeccable. Every bit of evidence, from the data from the blood tests and from the probability analysis was bitterly challenged by the defense. There was in addition a challenge by the defense fact that the victim of the crime did not show injuries to her sexual parts, with the claim that all such victims received at least minor laceration, and none were shown by the deceased. Rebuttal evidence from yet another expert showed that in her examination of 160 rape victims, only 6 showed such injury.

The court, faced with all of these bitterly contested scientific statements decided, against the defendant. They held that all of the evidence was admissible and that the issue again was the weight of the evidence, something which was a jury, not a judge's decision. The evidence on physical injury was perhaps not given in the best manner, as the witness who testified was herself an assistant district attorney, and was both not qualified as a medical expert and possibly somewhat on the side of the prosecution. Indeed, one judge did suggest that her testimony should not have been admitted, but agreed that the bulk of the other evidence was sufficient so that even without that testimony the accused was guilty. But the real heart of the case, for this book, is the acceptance of the multi-enzyme blood tests, and the allowance of probability results to be used as evidence, and their obvious acceptance, despite the bitter conflict about their validity, by the jury. General acceptance of this method under *Frye* might not be firmly established, but the fact that the novel technique was now used in a widespread manner was sufficient to gain its admissibility in

Maryland.

In another rape, kidnapping and armed robbery case in Florida in 1981, *Jenkins v. State*, the Court of Appeals of Georgia also accepted data from electrophoresis. There were, as always in criminal cases, other issues raised in the appeal, but without going into detail, the court stated, in regard to the tests, "An expert witness testified concerning the identification of certain blood samples. His testimony was based on the results of a procedure known as electrophoresis. Appellant moved to have the witness's testimony stricken from the record, basing his motion on a contention that the reliability of the procedure and the acceptance of the procedure by the scientific community had not been shown. We find no error in the trial court's refusal to strike the testimony. The opinion of an expert on any question of science is always admissible....We do not find that the fact that the procedure is relatively new requires that the testimony be excluded. The witness explained the procedure, thereby giving the facts upon which his opinion was based. The question, then, is not of admissibility but of the weight to be given the evidence by the jury."

A further issue, not commented upon in detail previously, was raised, namely the "chain of custody of the blood samples used by the witness." However, as in many other cases this defense was not raised at the original trial, and therefore could not be raised here. However, it should be pointed out that the issue is not a moot one. In order for such evidence to be used, it must be conclusively shown that the evidence was properly obtained, and identified, and that the samples were indeed given proper care. No details of the defense claim were given, but if it had been shown at an earlier stage that there was some possibility that the sample analysed by the expert had not been given the care mentioned, the results would have either been inadmissible, or discredited by the jury.

The same issue of the validity of electrophoretic evidence arose in *People v. Young* before the Court of Appeals of Michigan in 1981. This felony-murder case involved much of the same type of evidence as before, with a probability being shown that the blood factors found on the victim's porch were found in only 1.3% of the population. The defendant's blood matched these factors completely. The admissibility of the evidence on the grounds that it did not prove guilt but only indicated that the defendant could have been guilty was the major issue. The lower court heard the expert witness and admitted the testimony. Further refinement of the data indeed showed that only 1/2 of 1% of the population would have the blood factors found here, and indeed they did match the accused's blood. Here, there is an interesting history in Michigan. Earlier cases, not cited, had refused to allow admission of evidence which put a defendant in a class where 20% of the population would be found (nonsecretors) but had allowed evidence when the tests limited the class to 2.7% of the population (c.f. Horton, above). The present court appeared to straddle the issue of the size of the class by stating at the end of the opinion "...the Court went further and held that the admission of blood analysis results including a defendant in a class of possible defendants was not restricted by the size of the group but instead should be permitted in accordance with the

rules for admission of other physical evidence and that
the weight of such evidence was for the jury to
determine." This, of course, is a paraphrase of Horton.

Electrophoresis has also been recognized in 1982
both in New York and in Kentucky. The former case,
People v. Borcsok before the Supreme Court, Westchester
County, recognized the use of electrophoresis and
probability in a murder trial and the latter case, *Brown
v. Commonwealth of Kentucky* also involved
electrophoresis data and conviction of murder.

Once again, one cannot help but be impressed with
the ease with which courts accepted electrophoresis as a
valid scientific technique, based on physical as well as
biological principles. In contrast to the courts'
refusal to accept the XYY evidence based solely on what
were once thought to be biological data, there seems to
have been only limited debate over the acceptance of
this novel method. Perhaps it should also, however, be
stressed that the biological basis was also accepted
simultaneously with the physical principle. The various
blood enzymes and factors found were accepted as
genetically based realities almost at once. This may
well be because there was at no time any group of
medical or biological scientists opposing the reality of
the genetic basis of blood typing, while there always
was a medical, and biological, basis for doubting the
XYY. In any case both decisions, refusal to accept XYY
and admission of electrophoresis is now known to be
biologically sound.

The use of *dried* blood in electrophoresis tests
continued to arise in criminal cases. While there now is
no question that evidence using this technique is
acceptable for *fresh* blood, there still may not be

sufficient data for the use of dried blood to satisfy
the *Frye* rule. The new reference case is *Graham v. The
State*, before the Court of Appeals of Georgia in 1983.
Interestingly, although the opinion discusses *Frye* in
detail, Georgia has never accepted it as binding on
their courts. Citing an earlier case, the judge here
states, "[W]e conclude that the *Frye* rule of 'counting
heads' in the scientific community is *not* an appropriate
way to determine the admissibility of a scientific
procedure in evidence." There is also criticism that the
main proponents of the use in evidence of dried blood
with electrophoretic techniques are all forensic experts
associated with the prosecution, and there remains a
need for impartial examination of this method by
investigators not involved in actual criminal cases.

Nonetheless, the court was willing to find the
evidence from this technique admissible, but drew a
strong line against the use of implications obtained by
statistical analysis. The court cited an earlier case in
stating, "We hold that mathematical odds are not
admissible as evidence to identify a defendant in a
criminal proceeding *so long as the odds are based on
estimates, the validity of which have not been
demonstrated*." (Emphasis supplied in opinion.) Another
example which stresses the point for the need for
understanding of genetics and probability by courts is
clearly shown in this opinion. Citing from an earlier
case, this time in Maryland, the opinion states, "On
page 28 Judge Ross is quoted: 'The statisticians have
explained it to me at least 50 times, and I always
understand it when they do. Once the explanation is
over, however, I can't explain it to anyone else.' If
Judge Ross can understand it, then the jury of non-

experts should also be able to handle it." However, the judge, still worried about the use of probability, managed to find sufficient other evidence in this rape case to uphold the lower court's verdict of guilty.

Michigan does adhere to the *Frye* rule. In *People v. Young*, before the Supreme Court of Michigan in 1983, the court dealt with the admissibility of electrophoresis. The problem here was that the lower court had admitted the evidence without proper establishment of its reliability. The court quotes an earlier opinion by Justice Butzel which stated: "The tremendous weight which such tests would necessarily carry in the minds of a jury requires us to be most careful regarding their admission into evidence and we should not do so before its accuracy and general scientific acceptance and standardization are clearly shown." Based on this, and other cited cases, the court remanded this felony-murder case so that the admissibility of evidence from the electrophoresis tests could be determined before trial. There was also a long detailed argument concerning the correct interpretation of Michigan's statutes regarding burglary, but this is of secondary interest.

Blood enzymes and blood types also served as probitive evidence in the state of Montana. In this case, *State v. Bauer*, on appeal before Supreme Court of Montana in 1984, the charge was again rape, and the charge was strengthened by evaluation of semen and vaginal fluids. In this instance, an expert testified that "...the intercourse had been with a male secretor of blood type O, and PGM subtypes 1 + 1 +. Bauer is a secretor, and has PGM subtype 1 + 1 +. The percentage of males with these genetic markers is 7.15 percent. ...Furthermore D.K.'s husband, who had been at work all day January 26, the time of the rape, has blood type A, is a non-secretor, and has a PGM subtype of 1 + 1 -, different from Bauer in all respects."

Bauer objected on the basis that the evidence only showed a possibility that he was guilty and certainly other men could have been involved, and the introduction of these data should not have been permitted. However, his case was considerably weakened as a a very positive identification of him was made by the rape victim, and also samples of pubic and head hair obtained following the rape agreed with his to the extent that another expert in that area testified that the "chances of another person having the same type of pubic and head hairs were one in ten thousand." The court found no error in admitting the blood type evidence and based on it plus the other factors sustained the lower court's verdict of guilty.

Another case involving the use of blood-stains depended not only upon the identification of the blood types, but the question whether in obtaining that information the use of the complete amount of material available for analysis denied the defendant an opportunity to have an independent expert repeat the tests. This murder case, *State v. Carlson*, was before the Supreme Court of Minnesota in 1978. Here a 12-year-old had been assaulted sexually and then murdered. The corpse was badly battered but the police were able to obtain samples of pubic and head hairs. In addition there were apparent vaseline stains in the groin area of the victim. The accused, who had been in the company of the deceased, was found to have blood-stains on his jacket. At first he claimed they were caused by ketchup, but when informed they were blood-stains he then

maintained that his dog had been killed and he had
carried it, spilling blood upon himself. When it was
pointed out the dog had been seen alive, he then again
changed his story, saying the dog had only been injured.

The stains were from human blood, and showed ABO,
PGM, and EAP characteristics identical to that of the
victim. In order to be certain, the tests had been run
twice, and as a result the entire amount of stains had
been depleted by the two tests. In addition the hair
matches from the victim and the accused were identical,
and an empty petroleum jelly jar was found in the
accused's room. The only real issue was the use of the
blood evidence, and the destruction of the blood spots.
The court had to weigh the loss of the possibility of
further tests and the effect this might have on the
outcome of the case. They examined other cases where
evidence had been destroyed prior to trial, and
concluded in this instance, a case of first impression
in their state, that the prosecution had not
deliberately destroyed the evidence, and citing an
Alaskan case (*Kauderdale v. State*) stated, "We find that
due process of law does not require that the defendant
be permitted independent expert examination of evidence
in possession of the prosecution before such evidence is
introduced at the trial. * * * In those cases where
expert analysis exhausts the substance there is clearly
no error in the admission of evidence regarding the
analysis in the absence of allegations and proof of
deliberate destruction, or deliberate attempts to avoid
discovery of evidence beneficial to the defense."

There was also an objection made to the
admissibility of statistical probability that indicated
that "only .85 percent of the population would have

blood with the same combination of ABO, PGM, and EAP as the matching bloodstain found..." Here, interestingly, is a reversal of the usual case; most frequently it is the blood type of the accused, found at the scene of the crime or upon the victim, which is damning; here the blood type of the victim proved to be a major factor in the finding of murder. It should also be noted that the both the pubic and the head hair evidence, found on the victim gave the accused less than a 1-in-4,500 chance of being innocent.

The same issues of probability arose in the Appellate Court of Illinois in another murder case, *People v. Harbold*. In this 1984 case the blood samples in question had been analysed for ten factors and gave a probability "of an accidental match of less than one in 500." The defendant was of type O. There was a problem in that the victim had received several pints of type O blood in a futile effort to save his life, and the sample taken from him prior to his death had arrived at the crime laboratory in a broken vial. It was "assumed" that stains found upon the victim's clothing were her own blood; these stains were of type A, not that of the accused. The latter was found to have both type O and type A stains on his sport coat, (and surprisingly, upon a pair of panty-hose he was presumed to be wearing at the time of the crime), as well as type O on a pair of gloves apparently worn during the murder, and type A on the knife identified as the murder weapon. The accused claimed that the blood came from his injury while using a power saw; no reason for the presence of type A on his clothing was given. In addition a cut in the gloves coincided exactly with the position of the cut in the accused's hands, making it unlikely that the blood had come from an accident with a power saw.

Despite this, and other evidence linking the defendant to the crime, the Superior Court felt it necessary to reverse the lower court and remand the case for retrial. This was because the prosecuting attorney, in what the court deemed to have been an excessively prejudiced manner, had so acted to interfere with the presentation of evidence as to make a fair trial impossible. The court had no difficulty with accepting the blood evidence as relevant, but both the way in which it, and other evidence were presented was prejudiced. "We believe that defendant's right to a fair trial was consistently compromised by irrelevancy and error which enhanced the State's case." The irrelevancy was an attempt to introduce a motive for the crime which was clearly not proven, and by the way in which the prosecutor made his final plea to the jury. What will happen to the case upon retrial is unclear, of course, but the blood evidence if properly presented and if a fair trial is held may be sufficiently weighty that a jury can find guilt even in an honest trial.

Blood enzymes were also found to be admissible in 1985 by the Supreme Court of South Dakota. In this third-degree burglary case, *State v. Dirk*, samples of dried blood taken from the scene of the alleged crime and from clothing as well as fresh blood from the accused all agreed and were of type A and also matched in terms of esterase and PGM enzymes. This court also had cause to examine *Frye* and the contrasting views of *McCormick* and found that while the *Frye* standard seemed to be met here, perhaps it might be time to reexamine that decision in light of new scientific advances.

A Minnesota case, *State v. Boyd*, before that

state's Supreme Court in 1983, further illustrates complexities in the admissibility of evidence. In this instance the charge was of criminal sexual conduct. A man was charged with being the father of a child born to a 14-year-old girl. The issue here was that "sexual penetration" must be proved before conviction on this charge, and the state attempted to use blood type evidence to prove that the child was indeed his and obviously penetration must therefore have occurred. The sticking point here was the use of statistical evidence to prove paternity. The lower court ruled the evidence inadmissible, and the prosecution appealed to allow this evidence. An important part of the lower court's decision may be the fact that the expert assumed that penetration had occurred and then applied statistical evidence of paternity on this assumption, a type of circular analysis, similar to the one causing difficulties in the use of HLA tests mentioned earlier. The higher court, however, remanded the case to allow the use of the blood type evidence, but limited it solely to a discussion of blood type testing and expressly stated that probability could *not* be used. The expert could testify that none of the fifteen blood tests used excluded the man from paternity, but could not give an index of the probability of paternity. Here we see the opposite of the paternity cases where the probability of paternity has recently been admitted as evidence, when earlier it was inadmissible. Now in a criminal case, where admissibility standards are much looser, probability seems to be disallowed!

Another facet of the use of HLA tests in criminal cases was discussed previously in the chapter dealing with paternity. This 1985 case, *Davis v. State of*

Indiana, dealt with the establishment of paternity and therefore of child neglect by parents who had abandoned their new born in a gravel pit.

In *Commonwealth v. Jewett*, before the Appeals Court of Massachusetts, Middlesex, in 1984, the issues in a rape case were both of mistaken identity and of blood typing. The blood type of the accused (only the ABO system was used) did not exclude him from the ranks of possible guilty persons. PGM tests were also attempted on the semen found, but the sample was too old to give any results, and the defendant claimed that this was caused by a failure of the state to take proper precautions to assure that such a test could be carried out. The court rejected this claim, but remanded the case due to other factors, particularly the strong likelihood of misidentification by the victim. (The defendant had once before been accused of rape; the charges then were dismissed when it was found that the defendant was incarcerated at the time of that purported event.) Obviously the defendant, if freed in this case, should take immediate steps to change his appearance.

In another rape-murder case, this time of an elderly woman, semen analysis was admissible. This case, *State v. Abbott*, decided by the Missouri Court of Appeals, Southern District, Division Three, in 1983, was a particularly gruesome rape and murder of an elderly woman. The evidence showed that she had type A blood. Evidence linking the accused to the crime was found when it was determined that he was a secretor and of type B. Semen analysis showed that the semen found was type B and this evidence was part of the chain which convicted him. There was sufficient other evidence for conviction as the accused had possession of the decedent's automobile, there was evidence of a struggle involving a bloody foot- print on a bathtub, the accused lived within a few blocks, and forced entry into the house had taken place. The semen analysis was simply one link in the chain, and even had it not been available there is little doubt that the accused would have been convicted.

Not all new genetic evidence is admitted as readily as was electrophoresis. In *People v. Alston*, decided by a New York Supreme Court in 1974, the evidence which the state wished to use was the identification of blood stains on the accused man's coat, the sample now being 22 months old. As both he and the victim were of type O other evidence was needed. The state claimed that the development of modern techniques allowed one to ascertain the sex of the donor of the blood. As these cases will come up again some day in the future, it is worth quoting the state's expert witness at length to establish the genetic facts. He testified that "...such discriminations are accepted in the scientific community and the three techniques utilized are: (1) The F or Y body test (2)the drumstick test and (3)the Barr body test.... in the F or Y body test, developed and accepted in 1967 or 1968, the male Y chromosome fluoresces after being stained and that this Y chromosome shines like an 'eastern star', indicating male origin. ....the drumstick tests seeks to identify in mature white blood cells called polymorpho-nucleated leukocytes, a nuclear appendage on the X chromosome of a definite shape within a certain micron range. This is called a drumstick and indicates female origin. ....the Barr body test, recognized and accepted scientifically since 1949, seeks to identify the Barr body in tissues other than blood. It is a hollow chromatin mass of a certain micron range located on the X chromosome near the perimeter of the nucleus, again indicating female origin." The expert examined some 446 cells from the coat and found no Y chromosome in any of them. He also claimed to have found at least two definite drumsticks, within the range expected for such a sample (the drumsticks do not appear in all cells from a female's white blood cells of the type mentioned, but finding any such is in the laboratory conclusive evidence of female origin). He also identified cells on the coat as being squamous epithelial cells from the mouth of a female. It should be pointed out that the presence of Barr bodies is not limited to blood cells, but they may be found in all tissues of female origin.

The issue of the admissibility of the evidence was argued without the presence of the jury, and the issue depended upon whether cells of the age of those on the coat could validly be used for such testing, as there was no way to have an adequate control (the best evidence was from a slide only 1 year old). A second prosecution expert validated the results of the first, but not for the Barr body evidence which he believed could not be accurately relied upon. Such things as the fabric of the coat and the chemistry of such fabric might in itself cause false readings, but nonetheless he felt that the other results certainly pointed toward a female origin of the cells. The defense called its own blood experts who completely denied that the use of blood samples of this age could be considered a valid scientific procedure. A second defense expert agreed with this as did still a third expert called by the defense. Thus the matter rested with the court to decide upon the admissibility of the prosecution's expert

testimony. Citing *Frye* once again, and other New York
cases, the judge declined to allow the evidence from the
prosecution to be admitted. As the entire case was based
upon circumstantial evidence, the court felt that
admission of this dubious chain of proof that the aged
blood sample forged would be the cause of undue
prejudice to the defendant. "The admission of this
testimony, with a seeming scientific imprimatur,
pointing to female blood on the defendant's jacket in
contradiction to defendant's claim that the blood was
his blood, would in the Court's view be highly
prejudiced to defendant, far outweighing its probative
value."

As one reads this case, it appears striking that
the major issue was the use of the Barr body and the
drumsticks, but no real contest over the Y chromosome
seems to have been fully explored. Presumably if the
first two tests might be invalid, the judge assumed that
the third also should be excluded. Had the blood been
fresh, the case might have been differently decided; the
three tests noted above are standard tests now, and can
be carried out by almost any blood analysis laboratory.
It would seem obvious that in addition to the other
blood tests previously discussed both here and in
paternity cases, these will also become standard parts
of forensic medicine in the near future.

A novel use of blood tests arose in the case *State
v. Lawson*, before the Supreme Court of Appeals of West
Virginia, in 1980. In this statutory rape case, the
defendant denied all charges. The woman swore she had
intercourse only once, and that was with the defendant.
As luck would have it, she also became pregnant. The
defendant asked for and received a continuance of the

case until such time as the child was born, and appropriate blood tests could be made. There was also considerable disagreement concerning the events involved in the alleged rape, and the court felt the evidence to be sufficiently clouded as to make the decision for a continuance. They decided that should the blood tests exclude the defendant, he would obviously be entitled to a new trial, but if they did not do so, his conviction would stand. No further record of this case seems to exist, and it is regrettable that it has not been possible to find what the blood tests revealed; if they exonerated the accused, then this certainly was a wise decision. Even if they did not, at least the opportunity was given him to establish his case and justice was done.

By far one of the most complex criminal case involving paternity determination, wrongful life, child custody, seduction, breach of promise, assault, invasion of privacy, and the proper state in which to try this variety of issues is *Slawek v. Stroh*. The case was dealt with in 1974 by the Supreme Court of Wisconsin. The facts cannot be summarized briefly. Slawek was a Philadelphia physician who claimed to be unmarried when in fact he had a wife and children. Representing himself as single he convinced Stroh that he loved her and they had intercourse many times, both in Pennsylvania and in New Jersey (the seduction). The first child resulting from their union was placed for adoption. When Stroh become pregnant the second time, the physician gave her an injection supposedly to help maintain the pregnancy, but in actuality he injected an abortion-inducing drug and she had a miscarriage (the assault). The third pregnancy resulted in the birth of a daughter. Just to make the

case easier to read, the mother gave the daughter her
own name. Stroh then moved to Wisconsin with the child.
Slawek wished to obtain custody of the child, and
requested the district attorney of that state to
initiate paternity proceedings, which he did not do.
Slawek also apparently continued to harass her by phone
calls at all hours of the night (invasion of privacy).
In addition, the daughter underwent the stigma of
bastardy and claimed that she had a suit for wrongful
life. The father further contended that the mother could
not support the child adequately and sought to have his
paternity established so that he could take custody of
the young girl. There was still a further issue involved
because of the seduction charge being invalid due to the
statute of limitations.

As the case was brought in Wisconsin, the court
there first had to determine whether they had
jurisdiction; after citing many laws of various states
they agreed that indeed they did. They then found the
mother's claim for seduction to be valid as well as the
assault charges being legally brought. They immediately
dismissed the wrongful life claim as being invalid under
the law at that time. They held that there was no legal
basis for breach of promise. What the court really
seemed to do here is to reaffirm the claims, and decide
that despite all of the father's wrongdoing, he had
legal rights to establish paternity and the district
attorney should have brought the paternity case as
requested. This led one of the dissenting judges to
suggest that a rapist would have the same rights! In any
event, this is a criminal case, and of course the
determination of paternity would involve genetic
evidence when, and if, it was tried.

A different type of genetic evidence was brought in a double murder and brutal rape case heard before the United States Court of Appeals in 1985. This case, originating in South Carolina, *Roach v. Martin*, was an appeal against the imposition of a capital sentence upon him, based on an attempt for him to obtain a writ of *habeas corpus* so that new evidence could be brought forward which might have the effect of lowering his death sentence to a lesser sentence. His claim was that he had found out subsequent to his trial and sentencing that he might be suffering from incipient Huntington's Chorea and may have also been under the influence of involuntary drug-induced psychosis at the time the crimes were committed. He claimed that he now could come forth with a material witness not available at the earlier time, and that the neurological expert assigned to examine him prior to sentencing was not accurate. He further claimed that he had mistakenly given that expert the name of the wrong drug, and that the drug he was influenced by might have made his actions such that he could not be held accountable for them.

However that may be, the genetic issue of the claim of incipient Huntington's disease is the one pertinent here. It should be noted that this invariably fatal disease is a progressive one, resulting in the gradual deterioration of mental and physical abilities. (The most famous case is that of the late Woody Guthrie.) The judges made some examination of the facts of this hereditary disease, but realized that the age of the defendant, then a young man, was such that symptoms of the disease are extremely unlikely; the onset of the disease is usually not until the age of 35 or 40. Even if the claims by the defendant that a new method,

recently developed, could be used to determine whether
he might eventually have this disease, were correct,
there appeared to be no reason to think that at his age
it could have affected his behavior in any way. As a
result, the federal court found no basis for his appeal.
There was much other legal wrangling over various
aspects of the case, but no errors were found here
either, and the genetic defense was found to be invalid
also.

Other new genetic tests and biological factors are
entering the field of criminal justice. While still not
firmly established, these should be mentioned as they
will become more widespread in the future. The first is
the use of fingernail clippings as positive
identification. This arose in 1981 in *People v. Wesley*,
before the Court of Appeals of Michigan. This felony-
murder and kidnapping case involved an attempt by the
prosecution to introduce finger nail analysis as part of
the evidence against the accused. The expert in this
case did not, it seems, make an impressive witness, as
he said that his experience was extremely limited (he
had examined his and his boss's clippings only). Despite
his claims that at a recent meeting much had been made
of the possibility of such evidence being used for
identification, the court had no option but to deny the
use of this type of evidence under *Frye*. Not only was
there no consensus of opinion by any scientific group,
but also there was almost no scientific literature on
the subject. However, the possibility remains that this
type of evidence will be admitted in the future if
further developed. Fingernails are, of course,
biochemical compounds and if it can be established that
the type of compound found here is controlled by

hereditary factors, as are the blood proteins, then such evidence, with proper analyses, will certainly play a role in the future, as can be seen in the following recent case.

Another as yet incompletely proven type of evidence is the use of hair samples as identification. As was pointed out earlier, there is an inherited genetic difference between the structure of hair in the black and white populations. Whether further genetic distinctions within these groups can be made is not yet certain. In *People v. Watkins*, also tried before the Court of Appeals in Michigan, the court examined in 1977 the admissibility of the use of microscopic analysis and comparison of hair samples taken from an accused first-degree murderer and upon the deceased victim. They found that microscopic examination by an expert who found fifteen similarities between the two, and who also testified that a single disparity would have been enough to exclude the defendant, was admissible evidence. Again, while this is not strictly, as yet, genetically understood, if further evidence shows this also to be the case, then this test will join the battery of genetic tests available to the courts for identification of a criminal.

As pointed out earlier, neutron activation methods are in use for identification in criminal cases. *State v. Stevens*, tried as early as 1971 in the Supreme Court of Missouri, involved the use of such evidence in a second degree murder case. While much of the evidence dealt with threads, there was also as an analysis of the hair of the victim and hair found on a shirt belonging to the accused. In addition, the victim had a different blood type from the accused and bloodstains matching her

type were also found on his clothing. The court cited a
list of seventy-five cases where neutron analysis had
been used previously and admitted the evidence. The
accused also raised the by now customary constitutional
issue concerning the withdrawal of blood, fingernail and
shoe scrapings, etc. The court, probably somewhat
wearily, referred to *Schmerber* and dismissed these
claims. Another plea which the defendant raised was that
there was some indication that his crime had been
planned; therefore he could not be guilty of second-
degree murder, but must be guilty of first-degree murder
or nothing else. The court also gave this short shrift.

Still another murder case involving genetics, and
similar to one in Indiana discussed previously in the
"Baby Doe" case, arose in England when a physician,
acting upon the advice of parents, withheld nourishment
and injected a Down Syndrome child with a drug which
would ease his pain, but also might ease him out of this
world. The British courts had to deal with this when an
anonymous informer reported this to the police. There
was some dispute about the use of the injections, but
the British jury found that the accused physician was
innocent. There was no question that the parents did not
want the child to live, and the only issue was whether
the injections were to ease the pain of the newborn or
were intended to be lethal.

Yet another facet of genetics, race and crime arose
in a classic case, *Loving v. Virginia* before the United
States Supreme Court in 1966. Here the issue was the
State of Virginia's miscegenation laws. A black woman
and a white man were married in the District of Columbia
and returned to Virginia and established a home there.
They were arrested and they pled guilty to the charge of

violating Virginia's miscegenation laws. They were
sentenced to one year in prison, but the trial judge
ruled that the sentence would be suspended for twenty-
five years if they moved out of the State of Virginia.
The flavor of his opinion accurately reflects the tenor
of those times: " Almighty God created the races white,
black, yellow, malay (sic) and red, and he placed them
on separate continents. And but for the interference
with this arrangement there would be no cause for such
marriages. The fact that he separated the races shows
that he did not intend for the races to mix."

The Lovings moved to the District of Columbia and
began the suit which eventually reached the U. S.
Supreme Court. The Warren Court was unanimous in
declaring the Virginia laws unconstitutional, thereby
also striking down the laws in fifteen other states
similar to this one. "There can be no question but that
Virginia's miscegenation statutes rest solely upon
distinctions drawn upon race. The statutes proscribe
generally accepted conduct if engaged in by members of
different races. Over the years this Court has
consistently repudiated 'distinctions between citizens
solely because of their ancestry' as being 'odius to a
free people whose institutions are founded upon the
doctrine of equality.'"

Further they pointed out that "These statutes also
deprive the Lovings of liberty without due process in
violation of the Due Process Clause of the Fourteenth
Amendment. The freedom to marry has long been recognized
as one of the vital personal rights essential to the
orderly pursuit of happiness by free men." A brief
concurrence stated, "I have previously expressed the
belief that 'it is simply not possible for a state law

to be valid under our Constitution which makes the
criminality of an act depend upon the race of the
actor.'"

An unusual case which depends upon the fact that
there are genetic differences between the sexes arose in
1976. This case, *Sumpter v. State*, came before the
Supreme Court of Indiana. Here, Johnnie Marie Sumpter
was convicted of the crime of living in a house of ill
fame. Her defense was that no one had definitively
established her sex! Obviously, Indiana found it no
crime for a male to live in such an establishment, and
therefore unless it could be proven that she was a *bona
fide* female, she could not be convicted. She also
claimed that remanding the case for sexual determination
would then put her in double jeopardy. The court did not
seem overly impressed with the latter claim, but did
remand on the basis it had not been legally established
that she was female. Ridiculous as this case may seem,
it should be remembered that in contemporary Olympic
competition sex chromatin tests (Barr bodies,
drumsticks) are required of all female athletes. In
addition, today there might be a federal case in that
the Indiana laws discriminated against females and could
convict them of a crime which was not a criminal act
when performed by a male. In the time of the case here,
such tests were not, of course, well-known, but the case
shows the extent to which a defendant will go, using
genetic claims, in order to defend herself.

As an example of the type of further complexity
which may face the courts, the journal *Science* recently
carried an article entitled "Genetic Influences in
Criminal Convictions: Evidence from an Adoption Cohort."
This article attempts by a study of adopted children to

show that there is a strong correlation in criminal behavior between the adopted children and their natural fathers; in contrast no correlation between the rate of criminal behavior of adoptees and their adoptive parents was found. The article is based on a large study, but it would seem that in a matter as controversial as this further information is needed.

Should this type of evidence be admitted, then there might well be a "genetically-caused" defense similar to an insanity plea. There is, however, some question of this finding. Indeed a refutation of this was published by another author in a succeeding issue of *Science*. It remains to be seen whether this new claim will become established, or like the XYY cases discussed earlier, will find its way into scientific oblivion.

As pointed out in the introduction to this chapter, there are countless cases involving the use of genetics to solve criminal cases, and their numbers will undoubtedly increase. As more refined genetic means of identification are discovered, just as was the case with the paternity suits, judges will perforce have to deal with genetic issues. Perhaps someday a short introductory course in human genetics will be part of law-school training, and will be an absolute prerequisite to election or appointment to a judicial post. As molecular genetics becomes increasingly more sophisticated, and the applications of this field become widespread in the law, lawyers and their clients, as well as judges, will be required to prepare cases and opinions based on an understanding of this field. There will be constant new applications of genetics to forensic practice. Indeed the use of minute amounts of blood for a variation on electrophoresis, isoelectric focussing, has recently been reported by the press. It is clear that many challenges to the *Frye* doctrine will be forthcoming as each new method is introduced and more sophisticated genetic techniques are introduced. The need for "general acceptance" under that doctrine will have lead to an investigation of the the validity of new genetic discoveries, particularly in cases of first impression, as it is apparent that discoveries now move quickly from the laboratory to the courtrooms.

An example typifying this statement is the recently published finding that in rape cases it is possible to separate the DNA found in sperm from that found from cells in the vaginal tract, and that this can be done on material from semen or blood-stains more than 4-years old. By further comparisons of the DNA of the accused and the victim of the purported rape, positive determination, or positive disproof, of the guilt of the accused can now be obtained. It remains to be seen how long it will take for this type of evidence to be admissible under the *Frye* standards.

ALPHABETICAL LIST OF CASES CITED

<u>CASE</u>	<u>CITATION</u>	<u>YEAR</u>
Amissibility of Novel		
Scientific Evidence	80 Col.L.Review 1197	1980
Anderson v. Warden,		
Md.Penetentiary,	81-6626 U.S.Ct App.4	C1981
Brown v. Commonwealth		
of Kentucky	639 S.W.2d 758(Ky)	1982
Commonwealth v. Jewett	458 N.E.2d 769 (Mass)	1984
Coppolino v. State	233 SO.2d 68(Fla)	1969
Criminality and Adoption	227 Science 983	1985
Forensic Application of		
DNA "Fingerprints"	318 Nature 577	1985
Frye v. United States	293 F. 1013	1923
Genetic Influences in		
Criminal Convictions	224 Science 891	1984
Graham v. State	308 S.E.2d 413 (Ga)	1983
Jenkins v. State	274 S.E.2d 618(Ga)	1980
Knight v. State	538 S.W.2d 101(Tex)	1975
Loving v. Virginia	388 U.S.1	1967
Millard v. State	8 Md.App. 119	1970
People v. Alston	N.Y.Supp.2d 356	1974
People v. Borcsok	114 Misc 2d 810	1982
People v. Camon	313 N.W.2d 322(Mich)	1981
People v. Gillespie	321 N.E.2d 398(Ill)	1974
People v. Harbold	464 N.E.2d 734 (Ill)	1984
People v. Horton	297 N.W.2d 857(Mich)	1980
People v. Jackson	162 Cal.Rptr. 574	1980
People v. McCann	115 Misc 2d 1025	1982
People v. Newsome	186 Cal.Rptr.676	1982
People v. Palmer	145 Cal.Reptr. 466	1978
People v. Samuel	174 Cal.Rptr. 684	1981

People v. Tanner	91 Cal.Rptr. 656	1970
People v. Watkins	259 N.W.2d 381(Mich)	1977
People v. Wesley	303 N.W.2d 194(Mich)	1981
People v. Wilson	179 Cal.Rptr 898	1982
People v. Young	308 N.W.2d 194(Mich)	1981
People v. Young	340 N.W.2d 805(Mich)	1983
People v. Yukl	83 Misc 2d 364	1975
People v.Nation	161 Cal.Rptr. 299	1980
R. v. Arthur	283 Brit. Med.J.1340	1981
Reed v. State	391 A.2d 364(Md)	1978
Roach v. Martin	757 F.2d 1463	1985
Robinson v. State	425 A.2d 211(Md)	1981
Shanks v. State	45 A.2d 85(Md)	1945
Slawek v. Stroh	219 N.W.2d 9(Wis)	1974
State v. Abbott	654 S.W.2d 260 (Mo)	1983
State v. Anderson	308 N.W.2d 42(Iowa)	1981
State v. Bauer	683 P.2d 946 (Mont)	1984
State v. Boyd	341 N.W.2d 805 (Minn	1983
State v. Carlson	267 N. 170 (Minn)	1978
State v. Dirk	364 N.W.2d 117(S.D.)1985	
State v. Hall	297 N.W.2d 80(Iowa)	1980
State v. Knight	Sup.Jud.Ct.E.Div.(Me)1857	
State v. Lawson	267 S.E.2d 438(W.Va)	1980
State v. Roberts	544 P.2d 754(Wash)	1976
State v. Rolls	389 A.2d 824(Me)	1978
State v. Stevens	467 S.W.2d 10(Mo)	1971
State v. Stout	478 S.W.2d 368(Mo)	1972
State v. Washington	622 P.2d 986(Kan)	1981
Sumpter v. State	340 N.E.2d 764(Ind)	1976
United States v. Stifel	433 F.2d 431	1970
United States v. Wenzel	497 F.Supp. 489	1980

CHRONOLOGICAL LIST OF CASES CITED

YEAR	CASE	CITATION
1857	State v. Knight	Sup.Jud.Ct.E.Div.(Me)
1923	Frye v. United States	293 F. 1013
1945	Shanks v. State	45 A.2d 85(Md)
1967	Loving v. Virginia	388 U.S.1
1969	Coppolino v. State	233 SO.2d 68(Fla)
1970	Millard v. State	8 Md.App. 119
1970	People v. Tanner	91 Cal.Rptr. 656
1970	United States v. Stifel	433 F.2d 431
1971	State v. Stevens	467 S.W.2d 10(Mo)
1972	State v. Stout	478 S.W.2d 368(Mo)
1974	People v. Alston	362 N.Y.Supp.2d 356
1974	People v. Gillespie	321 N.E.2d 398(Ill)
1974	Slawek v. Stroh	219 N.W.2d 9(Wis)
1975	Knight v. State	538 S.W.2d 101(Tex)
1975	People v. Yukl	83 Misc 2d 364
1976	State v. Roberts	544 P.2d 754(Wash)
1976	Sumpter v. State	340 N.E.2d 764(Ind)
1977	People v. Watkins	259 N.W.2d 381(Mich)
1978	People v. Palmer	145 Cal.Reptr. 466
1978	Reed v. State	391 A.2d 364(Md)
1978	State v. Rolls	389 A.2d 824(Me)
1978	State v. Carlson	267 N. 170 (Minn)
1980	Admissibility of Novel Scientific Evidence	80 Col.L.Review 1197
1980	Jenkins v. State	274 S.E.2d 618(Ga)
1980	People v. Horton	297 N.W.2d 857(Mich)
1980	People v. Jackson	162 Cal.Rptr. 574
1980	People v. Nation	161 Cal.Rptr. 299
1980	State v. Hall	297 N.W.2d 80(Iowa)
1980	State v. Lawson	267 S.E.2d 438(W.Va)

1980	United States v. Wenzel	497 F.Supp. 489
1981	Anderson v. Warden, Md.	81-6626
	Penitentiary	U.S.Ct App.4 C
1981	People v. Camon	313 N.W.2d 322(Mich)
1981	People v. Samuel	174 Cal.Rptr. 684
1981	People v. Wesley	303 N.W.2d 194(Mich)
1981	People v. Young	308 N.W.2d 194(Mich)
1981	R. v. Arthur	283 Brit. Med. J. 1340
1981	Robinson v. State	425 A.2d 211(Md)
1981	State v. Anderson	308 N.W.2d 42(Iowa)
1981	State v. Washington	622 P.2d 986(Kan)
1982	Brown v. Commonwealth of	
	Kentucky	639 S.W.2d 758(Ky)
1982	People v. Borcsok	114 Misc 2d 810
1982	People v. McCann	115 Misc 2d 1025
1982	People v. Newsome	186 Cal.Rptr.676
1982	People v. Wilson	179 Cal.Rptr 898
1983	Graham v. State	308 S.E.2d 413 (Ga)
1983	People v. Young	340 N.W.2d 805 (Mich)
1983	State v. Abbott	654 S.W.2d 260 (Mo)
1983	State v. Boyd	341 N.W.2d 805 (Minn)
1984	Commonwealth v. Jewett	458 N.E.2d 769 (Mass)
1984	Genetic Influences in	
	Criminal Convictions	224 Science 891
1984	People v. Harbold	464 N.E.2d 734 (Ill)
1984	State v. Bauer	683 P.2d 946 (Montana)
1985	Criminality and Adoption	227 Science 983
1985	Forensic Application of	
	DNA "Fingerprints"	318 Nature 318
1985	Roach v. Martin	757 F.2d 1463
1985	State v. Dirk	364 N.W.2d 117 (S.D.)

Originally the cases involving genetics and
environmental law were to have been included as a small
part of the chapter dealing with miscellaneous types of
genetic cases, but there are now good reasons to reserve
a special section for discussion of the environmental
laws and their dependence upon genetic issues.

Perhaps no other types of decisions by the courts
have increased in number as have the suits brought
dealing with the regulations of the various agencies
responsible for enforcing environmental protection.
There is a vast multiplicity of agencies, both federal
and state, and there is a myriad of regulations issued
by each of these regulatory agencies. As a result,
almost every case is both complex and lengthy, and the
judges have had to labor long and hard in the
environmental vineyard.

Most of these regulations deal with potential
carcinogenic effects or other possible toxic effects of
various compounds in either the work place or in the
general environment either by contamination of the air,
the water, or in the case of chemical or radioactive
dumps or landfills, contamination of homes and damage to
persons living near such areas. The number of the
chemicals in use today is in itself almost overwhelming;
estimates may run as high as 66,000 such compounds of
which most have been classified as potential if not
actual health hazards (*Time Magazine*, Oct.14, 1985).
Obviously, there is no way any agency can test each and
every one of these. At present at least 685 compounds
are classified as known carcinogens or mutagens and
strict compliance with health and safety standards under

a variety of federal and state laws is required for this relatively small, but extremely toxic, number of compounds.

The genetic issue is also clear. It is generally considered that any carcinogen is a likely mutagen, and vice versa. In fact, the standard test for new compounds is the Ames *Salmonella* test. In this method, special bacteria of that genus are exposed to a suspected compound, and an increase in mutation rate found with some of these is in itself sufficient to deem the compound a carcinogen. This is a convenient, rapid (over-night) test, but the prospects of testing 60,000 or more compounds are slim. Even when a compound is identified as a potential carcinogen by this test, it then usually must be further tested on other organisms, fruit flies, rats, etc., for definitive proof of its harm to higher organisms. Additionally, there is no certainty that a negative mutagen test in *bacteria* would hold for man. We are in a real dilemma here, and there seems little chance of resolving the problem of the introduction of these compounds into our environment.

There is an added difficulty as many untested compounds, if dangerous, will express their effects only after a lengthy period of time, and by then it may be difficult to trace the source of an injury caused by a toxic agent many years ago. In addition, if there is a delayed effect, in most instances the statute of limitations will have run out, and an injured person may be barred from bringing suit.

For this reason, the government as mentioned above has introduced a variety of acts dealing with this in an attempt to protect its citizenry. The Atomic Energy Act of 1954 dealt with the problem of radiation release, and with the problem of how to handle spent, but extremely radioactive, waste products from the generation of atomic energy and from medical or other research uses of isotopes, an ever-increasing problem today. The National Environmental Policy Act of 1969 (NEPA), the Occupational Safety and Health Act of 1976 (OSHA), the Clean Air Amendments of 1970, the Water Pollution Control Act of 1972, the Resource Conservation Act, and the Hazardous Materials Transport Act (HMTA), the latter in 1981, are a few dealing specifically with attempts to preserve the quality of the environment.

Separate acts are in place to preserve endangered species of animals and plants, the Endangered Species Act of 1973 (ESA) and the Marine Mammal Species Protection Act (MMSP). Of course any act dealing with species preservation is in itself dealing with genetics and, similar to some of the patent cases, courts find themselves having to study genetics in order to determine what is meant by the genetic term, species.

Further difficulties arise when one begins to study these cases. Each agency in charge of these acts issues regulations, often in large numbers, and in some instances the regulations may appear contradictory. Almost all of the court decisions are lengthy, and the judges have had to deal in considerable detail with complex issues raised by these regulations and whether the various agencies involved have acted properly to enforce them. At times it would seem that various organizations wishing to obey the regulations are not even sure which set of rules covers their operations; at other times some have flagrantly, and perhaps deliberately chosen to disobey them and take their chances with the courts.

As mentioned earlier, many states have their own environmental statutes, often more strict than the federal regulations. As a result a company or research laboratory must often fulfill two sets of standards, one to comply with state regulations, and one with federal dicta. It is perhaps no wonder that the field of environmental law is one which will occupy more legal talent than most.

What follows is an attempt to demonstrate the difficulties with environmental laws and their enforcement as they relate to genetic issues. Of course other earlier cases in the chapter dealing with agriculture might fit into the rubric of environmental cases, for example *Washington State Farm Bureau v. Marshall* dealing with regulations on the spraying of strawberry fields, or *Riverside Citrus v. Louisiana Citrus Lands* which dealt with spraying of various chemicals and the effect on neighboring lands. Also, the cases dealing with radiation and the effects of mill-tailings on the use of land or homes could fit in here.

It is easier to turn first to the issues involved in the protection of species, and then deal with the more numerous cases involving direct environmental protection. The major regulatory statutes in these cases are the MMPA as well as those governing other agencies dealing with survival of non-mammalian species, such as the Fish and Wildlife Service. The laws deal with endangered species, i.e. species which either have, or are about to have, become sufficiently rare that the survival of the species is in doubt unless special care is taken not to further deplete their numbers. In addition, as will be seen, many of these species live only in a very small area or range, and if that area is

disturbed there is no other place on earth where that
species can either be moved to or be found in nature.

A typical protection case may be *Conner v. Andrus*,.
dealing with hunting regulations imposed by the United
States Fish and Wildlife Service of the Department of
the Interior. The case, heard by the United States
District Court, W. D. Texas, El Paso Division, in 1978,
concerned the banning of hunting in a region where a
species of ducks, believed to be endangered, was found
to be prevalent only in areas previously regularly
hunted. The Bureau had defined a species as follows: "a)
they have a common genetic pool; b) they interbreed
within the genetic pool; c) they produce viable
offspring (i.e. offspring capable of reproducing) and;
(d) they are genetically isolated from any other species
(i.e. they cannot interbreed with any other species)."
The bird in question here was the Mexican Duck.
Considerable argument was made whether the change of
environment in the area or the pressure of hunting was
the cause of the birds' near extinction. Opposition
arguments were made that cessation of hunting might even
allow the population of the birds to *increase*
sufficiently so as to permanently damage the
environment, a result found sometimes when an overly
large number of animals destroy the environment they
need by so depleting the food resources that starvation
may lead to the death of all of the species. At the time
there were some 300 Mexican ducks in the area and a
reserve population estimated to be from 20,000 to 40,000
in Mexico. This is not a large number of organisms for a
population to maintain itself, however. There was also
considerable argument whether this bird was indeed a
true species under the definition given above. One

expert felt that this bird was simply a subspecies of
the common mallard. The court, wrestling with these
arguments, about population genetics and the meaning of
species, concluded that the federal regulations had
overstepped the bounds and allowed the hunting area to
be reopened, as there was insufficient data to show that
hunting would cause the ultimate demise of the Mexican
duck.

The case, *Balelo v. Baldridge*, decided early in
1984 by the U.S. Court of Appeals, Ninth District, *En
Banc*, is more complex. Here the genetic issue is the
survival of a species, namely porpoises; the legal issue
is the applicability of the Fourth Amendment to this
case. Congress passed the Marine Mammal Protection Act
(MMPA) in 1972 specifically dealing with the type of
species involved in this case. The tuna fishing
industry used nets to capture their fish, but at the
same time could not avoid taking porpoises in these
nets. Once porpoises are captured in the fish nets there
is no way to free those mammals from the nets, and their
capture results in their death. Prior to the passage of
the MMPA, the sighting of porpoises feeding on tuna was
a useful method of obtaining large hauls of the fish.
The MMPA mandated that the tuna industry change their
methods so as to avoid capturing the porpoises.

Included in the MMPA was the authorization of the
positioning of observers on fishing boats to collect
data, and if necessary to use these data as evidence of
violation of the laws. The captains of the fishing fleet
brought this action under the Fourth Amendment, claiming
that this was an illegal "search" as at the time of the
placing of the observer aboard the fishing vessel no
violation of any law had occurred. The captains did not

object to the placing of observers, but to the use of
the data collected by them as possible evidence of
illegal capture of the porpoises, or in their claim what
amounted to an unconstitutional "warrantless search."
The court dealt at length with the meaning of the MMPA
and with the history of the use of "warrantless
searches" where no other means of obtaining information
was possible. In fact, it would appear that other cases
where government observers are stationed, such as in
meat packing, OSHA inspections, etc., similar
positioning of observers is not questioned. The court
felt that the issue under the Fourth Amendment is of
paramount importance and two concurring opinions were
written, both explaining and upholding the placement of
the observers. One such opinion contained the following
phrases: "If hard cases make bad law, I fear the result
of cases such as this. I write to reveal the
extraordinary difficulties I find in this case, and to
explain its limited applicability." This concurrence is
based on the assumption that a commercial fishing vessel
is a "workplace." By finding the boat to be a
"workplace" the judge could then find sufficient reason
to allow the placement of the observers. He was
extremely careful, however, to point out that this case
should in no way be used as precedence for general
violations of the Fourth Amendment, but was a special
case outside of that protection.

The judge was indeed moved by the environmental and
genetic factors, saying, "Second I write to emphasize
the magnitude of the governmental interest involved in
this case. If the world loses genetic diversity, it has
suffered irreparable harm. Marine mammals have long been
threatened by the onslaught of technology; if we must

take drastic steps to avoid further encroachment, so be it." The decision was 3-2; two judges entered a dissent on the basis that there was a serious violation of the Fourth Amendment here, but the placement and use of observers was upheld by the majority. The opinion with its two concurrences and its dissent is lengthy, but is a reference point for those interested in applications of the MMPA and other environmental acts passed by Congress. It should again be stressed that the decision by the majority was applicable only to this case and was not designed to set a precedence against the Fourth Amendment's "unlawful search and seizure" clause.

However, in another case, *Cabinet Mountain Wilderness v. Peterson*, the claimed endangered species, in this case grizzly bears, did not fare as well. The plaintiffs owned a 94,247-acre tract which was one of the few areas claimed to be able to support this large mammal. The defendant, head of the U.S. Forest Service, joined, with court permission, by an oil prospecting firm, had granted a prospecting company the right to continue drilling explorations in this area, and suit was brought to block this action claiming the permission granted continuation of previous drillings violated both the ESA and the NEPA. The case was an appeal from a summary dismissal in favor of the defendant by a lower court, the U.S. District Court for the District of D.C., and was heard by the U.S. Circuit Court of Appeals, District of Columbia, in 1982. The claim was made that no EPIS had been made, and that the Forest Service had not explored thoroughly the damage to the grizzlies' habitat which would ensue, and that the drillings would endanger the bears' habitat. However, the prospecting company had taken specific steps, ordered by the Forest Service,

upon consultation with the Fish and Wildlife Service, to
avoid as much as possible any destruction of the bears'
surroundings and to be certain that they were not
endangered. It might also be pointed out that although
probably no more than about a dozen of these animals
lived in the area concerned, "the area has been
recognized as having a high potential for grizzly bear
management." The Forest Service decided that no EIP
statement was needed as the company had fully complied
with the restrictions the service had put upon them, and
that under the circumstances no real harm would accrue.
The Appellate Court took the latter claim into full
account and ruled that there had been no violation of
either ESA or NEPA and upheld the lower court's summary
dismissal of the plaintiff's claims, agreeing that there
was no reason to believe that damage to the environment,
and to the bears, would occur. In a sense, it might seem
here, the court seems to have evaluated the
environmental claims much as if it were the EPA.

In *Commonwealth of Massachusetts v. Clark*, the
issue was the potential destruction of the Georges Bank
fishing grounds off the coast of Massachusetts. The
Department of the Interior had sought to open bidding
for the lease of twenty-five million acres of ocean bed
for prospecting including the "most productive fishing
area on the eastern coast of the United States and one
of the most fertile fishing areas in the world." In
addition, there was evidence that the fishing grounds,
which supported over two billion dollars of economic
activity in 1983 may already have been in a state of
decline, and particular concern was felt over the
dwindling lobster populations. The case is complex, but
basically the court held that the Department of the

Interior, of which Clark was the head, had failed, in
violation of NEPA, to prepare a proper EIP, and in
particular had not taken into account suggested
alternatives which would have exempted most of the
fishing grounds. The plaintiffs sought an injunction to
prevent the sale of the oil and gas leases, and the U.S.
District Court, D. Massachusetts found that there would
be no "irreparable harm" done to the defendant if the
sale were postponed until a proper EIP had been
prepared. More importantly, the court found that
"irreparable harm" might be done to the plaintiff and
that the requested preliminary injunction would prevail
upon a full hearing. As a result the government was
blocked from conducting the sale. The case is by no
means concluded, and it is expected that further rulings
concerning a modification of the sale, or an appeal of
the court's granting of the preliminary injunction will
be carried out. However, as of the time of the decision
government was enjoined from proceeding with the sale
until such time as it could be proven that no harm would
result from drilling explorations in the area concerned.

Another example of endangered species is found in
the long controversy between the U.S. Environmental
Protection Act and five New York Power companies
concerning the development of atomic power plants on the
Hudson River and the possible impact of these plants
upon the fish species, striped bass, as well as all
other animal populations. The problem is two-fold:
first, it is impossible not to draw small fish into the
water intakes needed to cool such plants, thereby
killing them, and second the heating of the river by the
water discharge from these plants in itself changes the
environment and may make it less favorable to local

animal populations within the region. A compromise seems to have been effected so that the redesigning of the water systems for the plants will alleviate much of the problem and supposedly not interfere with the many species otherwise endangered by the building of the power sources.

The striped bass also were the foremost actors in the long suits involving the building of the super-highway and surrounding development of the Westway Project in New York City. After voluminous testimony and many court decisions, it has finally been decided that the building of that project would interfere with the habitat of the fish, and the project has been permanently halted. Fortunately for New York City the large funds from the Federal Government for that project may be "traded in" for other transportation improvements, such as the subways, and the funds have not been irretrievably lost. It might be recalled that the endangerment to another small fish, the snail darter, held up construction of a major dam in 1978. The issue may now be moot as other small populations of the same species have been found in different locations. But the courts are dealing with problems of genetics, of the meaning of species, and the value of their survival as well as the interpretation of the special language of regulatory acts such as the MMPA or ESA.

Yet another example of consideration of the survival of genetic species is found in the on-going controversy over Japanese fishermen continuing certain whaling activities despite a decision by the International Whaling Commission to declare a moratorium on the hunting of the sperm whale lest this species be pushed to extinction. The Reagan Administration had

decided not to certify that the Japanese were in
violation of the Commission's moratorium and not to
invoke the required sanctions against that nation.
However, the District Court of D.C. has decided that
Japanese fishing rights in U.S. waters for other sought-
after species must be revoked under United States
Congressional Acts (the Pelley Amendment and the
Packwood-Magnuson Amendment) unless they agree to the
international terms for preservation of this endangered
whale species. As a result, the Japanese will lose 50%
of their fishing quotas within the U.S. 200-mile
economic zone in the first year and 100% in the
following year unless they stop taking sperm whales. The
government did appeal this decision but the U.S. Court of
Appeals affirmed the lower court's judgment and has
instructed the Secretary of Commerce to certify Japan as
being in violation of the treaty and invoke the
sanctions or appeal within ninety days. The Japanese
government would certainly like to see such an appeal to
the United States Supreme Court carried out, but it is
not certain whether this will occur. If not, the
Japanese will have their choice of giving up a
$50,000,000 yearly income from whaling or suffering the
loss of a $500,000,000 income from other fisheries.

The possible impact of a newly introduced organism
upon the environment is the basis for *Foundation on
Economic Trends v. Heckler*. Before discussing
this suit it is necessary to mention briefly both some
of the general ecological facets as well as some of the
molecular biology and economics of this type of case.

The effect of an introduction of a new species of
organism into an environment is difficult to assess. The
organism may spread, eliminating native populations and

becoming a pest in itself, and, as its parasites or predators or other control mechanisms are usually not introduced with it, the spread of the organism may be impossible to control. If the new organisms are more reproductively efficient, or can compete more effectively for a limited food supply, they may well displace native species. One need only think of the English sparrow, the starling, or more recently the almost explosive spread of the house finch in the eastern United States. (Paradoxically, this bird formerly lived only on the west coast, and if the facts leading to its establishment on the east coast were not known its presence here would confront ornithologists with a most difficult problem. The birds arrived at their present habitat due to a law the State of New York passed some fifty or more years ago forbidding the keeping of wild birds for sale in pet shops. The house finch was sold by pet shops in New York City as a cage bird. To comply with the law, they were simply set free, probably in the expectation they could not survive; surprisingly they not only survived but are extending their range throughout the eastern regions.)

In addition, economic considerations of the effect of newly, and often accidentally, introduced species must be considered. One need only consider Dutch Elm disease to realize the potential explosion of a new organism when all natural controls are removed. More recently there have been other events demonstrating the dangers of the introduction of foreign organisms into the United States, namely the large snail in California which, if left unchecked, might be a potential disaster to crops there, and, of course, the spread of the "killer bee" originally introduced in South America and now apparently spreading through Central America and into the U.S. by outbreeding the less vicious species. As the use of honey bees for pollination is a large commercial industry in agriculture, it is obvious that bees which are more dangerous than those used for this needed purpose are certainly undesirable.

Molecular biologists have for the past years been attempting to turn their science into practical results which would be economically beneficial. For example the genes of grain crops might be altered in such a way that these plants could support the nitrogen-fixing bacteria found in association with the roots of legumes, and there would be no need for the extensive use of fertilizer on grain crops. Much research is being carried out in attempts to do just this, although no real success has as yet been accomplished. Other uses of genetically-changed organisms are being carried out, for example the commercial production of insulin by a common bacterium which normally does not carry the genetic information to make this hormone. The prospects for agriculture and industry may be almost staggering to the imagination, and the entire biotechnology industry is built on this type of expectation.

The case at hand deals with just such a bacterial product. It involves the deliberate release into the environment of a new type of bacterium, formed by the process of recombinant DNA. Originally, under the guidelines formed by the National Institutes of Health (NIH), there were restrictions upon recombinant DNA experiments for all recipients of grants from this agency, restrictions agreed to and based on recommendations of the foremost investigators in this field. It is noteworthy that the geneticists working in this field were themselves the first to point out possible dangers of this technique and had agreed to a "moratorium" upon certain types of experiments as well as to the prohibition of the release of any recombinant forms into the environment. It should additionally be made clear that the restrictions put forth by the NIH only apply to investigators working under grants or other funds from this agency and do not pertain to industrially sponsored research.

Foremost for this case was the absolute prohibition of the deliberate release into the environment of any recombinant organism, and indeed many of the voluminous rules governing recombinant DNA experiments were to prevent the accidental release of any organisms. Several times subsequently, the NIH has revised its guidelines, finally leaving it up to the director, with possible referral to and advice of a committee, to determine whether planned release of some recombinants might be permitted.

The case here is based, however, not on the intricate genetics of recombinant DNA, but simply on the action of the NIH in approving the deliberate use of recombinant organisms in field trials without preparation of an EIP. The plaintiffs sought a preliminary injunction against the experiments involved here, the introduction of "genetically altered bacteria onto a row of potatoes in Northern California." The hope was that the presence of these bacteria would reduce frost damage, thereby allowing a longer growing season and a higher crop yield. Further the plaintiffs sought an injunction against all further types of experiments.

The case was first decided by the Circuit Court,

District of Columbia, in 1984. The court recognized the problem clearly, stating, "The issues that this Court must confront are narrow, legal questions. Accordingly, while speculation on the possible benefits and hazards presented by the emerging technology of genetic engineering may intrigue the parties to this litigation, the Court has no desire, authority, or competence to so speculate. The sole task is to review whether the federal defendants should have issued an environmental impact statement under the circumstances of this case." The judge then spent considerable pages discussing the history of the NIH regulations, and the use of their Recombinant DNA Advisory Committee as their reference group, and also of the changes which empowered the Director to issue permission for this type of experiment, and the lack of the preparation of an adequate EIP. As a result of his inquiry into the history of the case, the judge decided that a preliminary injunction would be granted, on the basis that the plaintiffs would be likely to succeed in obtaining a permanent injunction, and that irreparable harm might be done to their cause, and none to the defense, if the injunction were not immediately granted. Indeed the judge agreed not only with the injunction against this release, but also granted a preliminary injunction against all experiments of this type. As a side note to this case, the Regents of the University of California had voluntarily joined as defendants; as they were to be the immediate sponsor of the research, they wished to be legally certain of their position.

The defendants appealed this decision to the U.S.Court of Appeals, D.C. Circuit. This case, also *Foundations on Economic Trends v. Heckler*, was decided

early in 1985. The higher court sustained the granting
of the injunction against the potato experiment on the
same basis as had the lower court, the failure of the
NIH to issue an EIP, but did reverse that court in
regard to future experiments. The question was again
posed, "For this appeal presents an important question
at the dawn of the genetic engineering age: what is the
appropriate level of environmental review required
before the National Institutes of Health (NIH) before it
approves the deliberate release of genetically
engineered, recombinant-DNA-containing organisms into
the open environment? More precisely, in the context of
this case, the question is whether to affirm an
injunction temporarily enjoining NIH from approving
deliberate release experiments without a greater level
of environmental concern than the agency has shown thus
far. ...We emphatically agree with the District Court's
conclusion that the NIH has not yet displayed the
rigorous attention to environmental concerns demanded by
law."

The two opinions by the different courts thus reach
the same decision regarding the present experiments, but
differ concerning the possibility of future attempts of
this kind, the higher court finding the injunction too
broad. Both opinions are lengthy, cover much of the same
historical ground concerning NEPA and its applicability
here, as well as the history of the NIH guidelines and
the decision by the latter that no EIP was needed. The
second opinion perhaps goes further in delineating the
problem by quoting from a House of Representatives
Report, "The potential environmental risks associated
with the deliberate release of genetically engineered
organisms or the translocation of any new organism into

an ecosystem are best described as 'low probability, high consequence risk;' that is, while there is only a small possibility that damage could occur, the damage that could occur is great."

That this will not be an isolated event is demonstrated by the present plans of Monsanto Company to test a pesticidal bacterium sometime in 1986, using a genetically altered bacterium to coat corn which will result in plants carrying pesticidal abilities and to plant them in carefully isolated plots. The Company has prepared an 800 page report to present to the EPA. The fate of this proposed research has not been settled as yet.

Still another example of the use of foreign organisms introduced into the environment is the recent attempt to use the bacterium, *Pseudomonas syringae*, as a means of making artificial snow for ski slopes. The bacteria are being produced commercially and released after they have been killed. It is hoped that an increased efficiency of from 20 to 80% will result, and that the snow making may take place at warmer temperatures. The bacteria are not the result of recombinant DNA and are found naturally on several species of plants, and are thought to be harmless to humans. The problem here which concerns some scientists is that the bacteria must be used in very high concentrations if they are to serve as nuclei for snow crystals, and there is a chance that under certain wind conditions they could be carried great distances. The applications would occur at the seasons when crop plants are most susceptible to frost, in early spring and fall. If these bacteria were to act as they are claimed to on the slopes, then they might induce similar results and cause great crop damage. Surprisingly, there seems to be no legal suit brought, and no claims for the need of an EIP have been made. One scientist attributes this to the great attention given to recombinant forms, and the lack of interest in natural forms.

Turning now to more general environmental problems, a major issue has been the protection of the environment by prevention of dumping dangerous wastes either on land or in places where the water supply can be contaminated. This type of case, exemplified by an early case in 1979, *Bokum Resources Corporation v. New Mexico Water Quality Control*, is now common as will be seen. In this instance, the issue dealt with the vagueness of the regulations put forth for water control in New Mexico. The Supreme Court of that State in 1979 dealt with the issue of what was a "toxic pollutant." The State regulations specifically included in that category the statement, "Those water contaminants, or combination of water contaminants present in concentrations which, upon exposure, ingestion, inhalation or assimilation into humans or other organisms of direct or indirect commercial, recreational or esthetic value, either directly from the environment or indirectly by ingestion through food chains, will, on the basis of information available to the Director or the Commission, cause death, disease, behavioral abnormalities, *genetic mutation*, physiological malfunctions or physical deformations in such organisms of their offspring." Let one hasten to add that the grammar here is that of the regulatory agent, not the court! Nonetheless, it is the misfortune of the courts to have to deal with this kind of prose and it has been cited here to give some flavor of how these regulations are written. In the present case, the court decided that the regulations as promulgated (of which the above quotation is only a small part) were so vague as to have little validity. While the purpose of such regulations is admirable, one wishes they might be written in English!

The main issue in *Save Our Ecosystems v. Watt* before the U.S. District Court for the District of Oregon, in 1983, was whether the Department of the Interior had acted properly in approving a large herbicidal-spraying program on public lands. Some 6000-7000 acres were to be sprayed with a variety of herbicides, including 2,4-D and Tordon, a compound which contains the chemical picloram. The plaintiffs question the preparation of the necessary "worst case analysis" by the government (in attempting to show that a procedure is harmless, the applicant must demonstrate what the effect of the most dangerous situation, given all factors, might be). Although analysis showed "no observed effect [for] the dosage levels for each of the herbicides proposed for use, these are levels at which the herbicides have not appeared to have any toxic effects on test subjects. The worst case analysis treats them as the highest dosage levels which are safe for human beings exposed to the herbicides. By the BLM's [Bureau of Land Management] own admission, however, the 'no observed effect' levels have not been proven to be safe dosage levels from the point of view of the potential, if any, of these herbicides to cause cancer or to cause mutations of the genes." The judge adds, "This worst case analysis is deficient in that it at no point assumes for purposes of discussion that these herbicides are carcinogens and mutagens." The judge, for this reason, then enjoined the spraying of the lands until such time as the BML prepared a proper worst case analysis including the possibility of carcinogenic and mutagenic effects of the compounds.

Examples of suits brought against the EPA for allegedly not carrying out its proper functions are not rare. In 1982, the case *Sierra Club v. Gorsuch* was heard by the U.S. District Court, N.D. California. Gorsuch was then administrator of the EPA and the Sierra Club brought suit against her and the agency for failing to establish national standards for hazardous air pollutants. The EPA had simply not proposed standards within the 180 days required by law and obviously had not then followed this initial period and adopted standards, after allowing for hearings, within a subsequent 180 days. The pollutants in this case were radionuclide emissions which had clearly been identified as being in the category of hazardous emissions. The EPA claimed that it would need no less than nine more years to adopt such regulations. While there is power within the court system to extend the legally mandated time periods, courts can only do so if they find "good faith" on the part of the EPA or that the time limits could not be met for reasons of limited staff or budget, or if there were clear-cut needs for further study. The court noted that a similar earlier case, *Illinois v. Gorsuch*, had occurred and had decided that the five-year delay in that case was not excused by the reasons proposed. The statement was made here that "To accept EPA's proposal for further, indefinite, and virtually open-ended extension of the time for compliance, without a more convincing demonstration of evident impossibility would be to, in effect, repeal the Congressional mandate. Further to substitute the Court for making scientific determinations which the Court has neither the investigative tools, nor expertise, and, further, to grant an extension such as required by the EPA, would involve the Court, rather than Congress, in changing, qualifying, or amending, if advisorily, the unqualified, mandatory provisions of Section 7412 [of the EPA act]. "The court ordered that the regulations be put forth within 180 days of its decision.

Eleven months later, the regulations still had not been forthcoming. The Sierra Club again went to Court, this time in the suit *Sierra Club v. Ruckelshaus*. The latter had by then replaced Gorsuch as head of the EPA. Meanwhile the court had found the agency in contempt for failing to obey its earlier mandate. This time the same court, but with an opinion written by a different judge, upheld the decision to find the EPA guilty of contempt of court. The court also specifically ordered the EPA to file final radionuclide emission standards for nuclear facilities and in addition within 120 days file standards for emissions from uranium mines. An alternative was also given, namely that the EPA "make a finding on the basis of the information presented at hearings during the rulemaking, that radionuclides are clearly not a hazardous pollutant." This finding of contempt and the order for action took place in 1984, and at the moment, if such regulations were made, they are not readily available.

Suits by citizens or citizens' groups against the EPA have a long history. Among the first such cases is *National Resources Defense Council v. Train*, before the United States Court of Appeals, D.C. Circuit in 1975. The suit was brought in order to have the agency promulgate regulations controlling pollution by limitation of effluents under the Federal Water

Pollution Act of 1972, and Train was at that time head of the EPA. The issue was whether this group had standing to sue, and whether the agency was dilatory in its issuance of the requested restrictions. The law, which is dealt with extensively, and also is included in the appendix to the case, had two categories of discharge to be regulated and had set time limits for their regulation. Although the judge found the agency derelict in its actions, and admitted the right of the Council to sue, he also pointed out that the EPA could appeal to the courts if it found that it had neither staff nor resources nor sufficient information to draft regulations. However, even so there had to be limits as to the time the EPA could take before issuing final restrictions upon emission of pollutants. This, incidentally, is among the first of many cases in which the Council, a national organization of concerned citizens, will play the leading role as plaintiffs suing the government for its failure to follow the Congressional Mandates set forth in the various acts referred to previously.

The EPA was again sued in 1984 in the case *National Resources Defense Fund v. U.S. EPA*. Again the issue was failure to act expeditiously in issuing regulations under the TSCA. In this case the plaintiff was joined by the AFL-CIO while the Chemical Manufacturers' Association and the American Petroleum Institute joined in the defense. The U.S. District Court, S.D. New York, had to deal with a multiplicity of legal issues here, including the right of the original plaintiff to bring suit, and certain actions of the EPA which were taken to exempt itself from the suit. The court determined that citizens' groups had the right to

intervene as plaintiffs, and the claim by the EPA that
they had withdrawn the regulations objected to by the
other defendants was held not to be legally permissible.
In all there were sixteen claims to be adjudicated, and
the court found sufficient violation of regulations in
some of them so as to deny the summary judgment sought
by the defendants and to order compliance with the
mandate of the court so that the matter could be brought
to final judgment. Of great interest is the appendix to
this case, as it lists no less than twenty chemicals,
some common to many manufacturing processes, all coming
under EPA regulation. The hazards from these chemicals
are also listed and include such possible results from
overexposure to them as being cancer, genetic damage,
birth defects, and general environmental effects.

Another facet of radiation danger to the
environment arose in *State of Wisconsin v. Weinberger*,
tried before the United States District Court, W.D.
Wisconsin in 1984. The U.S. Department of the Navy had
sought to install a special facility in Wisconsin which
would use very low frequency radio signals to
communicate with submarines. The case is of long
standing as the Navy had begun its studies in 1969,
originally proposing a very large grid (6300 square
miles) of buried antenna. Later this proposal was
modified to planning to build two much smaller
facilities in Wisconsin and Michigan. In 1978, President
Carter decided to terminate the project for an
indefinite period. Subsequently in 1981 the project was
reinstated. As far back as 1972 the Navy had conducted
biological and other tests in order to prepare a
preliminary EIP. The study was updated in 1975, and a
final EPI was prepared in 1977.

The Navy continued to conduct studies of the effects of this type of radiation using monkey behavior, bird migration, animal population studies in the area where the transmitters were to be based and a general ecological monitoring program. While some implications of possible harm might have appeared in the primate studies, they were deemed relatively insignificant. However, other research, not sponsored by the military, began to appear, suggesting that there could be damage from this project. Among other suggested harms were leukemia and possible birth defects, and epidemiological studies suggesting a higher rate of suicides in the areas where overhead power lines were located. Subsequently new information concerning these radio waves became available, suggesting that low frequency radiation might not be harmless after all.

Another legal issue was whether an EIP was needed at all. If the court were to find that the order to build the facilities was a "discretionary act" by the President of the United States, rather than a decision by the Navy, the courts would have no jurisdiction over the case as such acts are not subject to suit. The judge did not deem this to be a "discretionary act, "and found that it was the Navy who had ordered the surveys and it was the Navy, not the Presiden,t whom the plaintiffs were suing, and the Navy would have to follow the requirements of the EPA.

The court also decided that the Navy must file a supplementary EPA statement. The original one dealt with the first proposed facility, and it was deemed that the change in the plans for the facility did not require another EIP statement. The reason for this decision was that the change in the plans *lowered* the chances of

damage and therefore did not essentially change the
first EIP. However, the judge ordered that the Navy deal
in a meaningful way with the new information concerning
the potential hazards of low frequency radiation. "In
summary, I find and conclude that in proceeding with the
reactivation of project ELF [The Navy's acronym for this
proposal] without undertaking a thorough and
comprehensive review of the significant new information
on biological effects of electromagnetic radiation that
has been generated since 1977 the Navy has abused its
discretion. In so proceeding, the Navy acted in
violation of the National Environmental Policy Act." In
so stating, the judge granted a permanent injunction
against further construction of the facilities until
such time as a supplemental EIP was prepared.

This opinion, like so many others dealing with
environmental cases, is lengthy and deals with many
technicalities of both law and low frequency radiation.
One of the problems seems to be a dichotomy among
scientists investigating the entire field of
electromagnetic radiations, with some finding evidence,
as claimed above, for definite effects, and others
finding either no effects, or no statistically
demonstrable effects of these waves. The judge,
attempting to strike a balance, took time in his opinion
to look at both claims and counterclaims, and apparently
felt that there was enough evidence to require the Navy
itself to do the evaluation which the court could not in
reality carry out.

Conflict between state and federal laws was the
basis of another environmental case dealing with atomic
power. In 1983, the United States Supreme Court decided
Pacific Gas and Electric v. State Energy Resources

Conservation and Development Commission. The issue here
was a section of the California Public Resources Code
providing that before a nuclear power plant could be
built there must be demonstration of adequate capacity
for interim storage of spent fuel. The Power Company
held that this was a state preemption of the Atomic
Energy Act of 1954. Obviously, state laws must bend
before federal laws. The issue was complicated as it
is clear that Congress has "preserved the dual
regulation of nuclear-powered electricity generation;
the Federal Government maintains complete control of the
safety and 'nuclear' aspects of energy generation,
whereas the States exercise their traditional authority
over economic issues such as the need for additional
generation capacity, the type of generating facilities
to be licensed, land use and rate making."

The majority of the Supreme Court, in an opinion
written by Justice White, found that the California
statute was directed solely at the economic issue of the
need for the plant, and therefore did not come into
conflict with federal laws. The issue of safety was not
considered to be the heart of the matter. Another
lengthy opinion was written in this case, with a
concurrence in the decision, but for other reasons. The
concurring judge (Justice Blackmun) in this instance
found that California would have had adequate
constitutional grounds to ban the building of the plant
on *safety* grounds alone, and that the 1954 Atomic Energy
Act did not apply to this kind of case. "The Atomic
Energy Act's twin goals were to promote the development
of a technology and to ensure the safety of that
technology. Although that Act reserves to the NRC
decisions about how to build and operate nuclear plants,

the Court reads too much into the Act in suggesting that
it also limits the States' traditional power to decide
what types of electric power to utilize. Congress has
simply made the nuclear option available, and a State
may decline that option for any reason. Rather than rest
on the elusive test of legislative motive, therefore, I
would conclude that the decision whether to build
nuclear plants remains with the States. In my view, a
ban on construction of nuclear power plants would be
valid even if its authors were motivated by fear of a
core meltdown or other nuclear catastrophe."

The complex issue of protection of industrial
secrets as opposed to the "right to know" acts arose in
1985 when the suit, *New Jersey State Chamber of Commerce
v. Hughy* was heard by the United States District Court,
D. New Jersey (Hughy is the Commissioner of
Environmental Protections in that State). In addition,
the suit was joined by a dozen chemical companies for
the plaintiff, and by the Commissioner of Health and the
Commissioner of Labor as well as the State of New Jersey
for the defendant. To complicate further the reading of
the opinion this case was joined to another in which
more than a dozen perfume manufacturers brought suit
against the same defendants.

New Jersey passed the *Worker and Community Right to
Know Act* in 1983, to become effective in 1984, which
would require manufacturers to disclose the chemicals
used and to make them available to workers in the plant
as well as to medical personnel and fire fighters who
need to know what chemicals are being utilized in a
manufacturing plant both for personal and
public safety. Furthermore, the New Jersey Act is most
implicit in requiring labelling not only of chemicals,

but of "pipelines, including valves, drains, and sample
connections" through which the chemicals may travel,
and, in fact a label is required for almost every part
of a chemical plant. The label could, however, simply
bear the number of the trade secret assigned to it by
the registry of trade secrets. Reading the opinion one
cannot but be impressed with the details of the act, and
the onerous burden of complying with it as written. The
act specifically deals with mutagenic or carcinogenic
chemicals as well. Several basic issues are involved.
First, does the act preempt the federal OSHA standards
for workers in a plant; second, does the act preempt
standards for those not directly involved in
manufacturing; third, does the revelation of trade
secrets constitute an unlawful taking under the Fifth
Amendment of the Constitution.

The manufacturers complained not only of the
difficulties of compliance, but also of the fact that
the labels will in themselves result in the disclosure
of trade secrets to anyone reading them. Trade
secret provisions are not applicable to known hazardous
chemicals, of which some 2051 items including 335 known
"carcinogens, mutagens or teratogens and are considered
special health hazards in a pure form or in a mixture at
very high concentrations, e.g. 80%, 90%, 95%." In
addition to this state provision, there are also
provisions dealing with hazardous chemicals under the
federal OSHA act. The manufacturer under OSHA
regulations may claim "trade secrets" but must file a
full material safety sheet concerning the hazardous
chemicals, and this must be available to health
professionals, but is to be kept private so that others
cannot learn of the exact chemicals or processes being

used. Such revelation *per se* might be sufficient to allow competing companies to determine the method for production of the end product, and as this would be public knowledge they might then be able to copy the procedures and thereby capitalize on other companies's research, development as well as manufacturing methods.

The court dealt with each of these issues and concluded that the requirements for the in-plant labelling was preempted by OSHA, but that as there were then no regulations regarding the safety of those not directly involved in the manufacturing processes, the state law could prevail. There was no legislation in New Jersey dealing with the protection of trade secrets and the court stated, citing an earlier case not described here, "As in the case of New Jersey's Right to Know Act, there was no pre-existing legislation protecting trade secrets submitted by registrants. In such a situation the entity submitting data cannot have a 'reasonable investment-backed expectation' that the agency receiving the data will maintain it in confidence. Consequently disclosure of the data is not a taking for which the state must pay compensation under the Fifth Amendment or the Fourteenth Amendment.

"Employers may face the unpleasant choice of disclosing trade secrets or limiting or shutting down operations in New Jersey....as long as the employer is aware of the conditions under which the data are submitted and as long as the conditions are rationally related to a legitimate government interest, a submission under the Right to Know Act does not constitute a taking." The decision then granted summary judgment for some grounds, namely preemption of federal standards, but did not grant dismissal for those employers, such as the Chamber of Commerce, who were not directly involved in manufacturing. Summary judgment on the issue of the taking of trade secrets was also not granted. However, the perfume manufacturers were granted summary judgment on the first two issues, and the court therefore felt it unnecessary to deal with the trade secrets issue as the summary judgment removed this as a part of the case.

This case seems somewhat difficult to understand. It is not precisely clear why the two cases were joined and then the decision reached different results in the two cases. Also, although OSHA does not deal with health hazards in the environment, it would seem that EPA and other agencies, also issuing federal regulations, could enter the case for those not directly involved in the plants. The opinion is, as usual for this type of case, and while certainly legally sound, perplexing as to the reasoning to reach the final conclusions. As it dealt only with whether summary judgments and preliminary injunctions could be brought, it is obvious that this case, or at least the first part of it, will be retried, while the whole case certainly will be appealed for a clarification of the legal status of the Right to Know Act, and a decision here will probably set a standard for other states to follow.

An example of confusion in the minds of a court as to Congressional intent in their drafting of some parts of the Clean Water Act and the revisions of it is found in *Chemical Manufacturers' Association v. Natural Resources Defense Fund*, decided by the United States Supreme Court early in 1985. This case, joined to another, resulted from two quite opposite decisions by lower federal courts, and the Supreme Court granted

certiorari in order to resolve the issue. The crux of
the case was whether EPA had the power to grant
"fundamentally different factor" (FDF) exemptions for
discharge of various toxic pollutants listed under the
Act when requested by manufacturers. In a typical 5:4
decision, the Court upheld the EPA. Basically the entire
opinion is a discussion of Congressional intent, with
each side reading into it its own interpretation of the
meaning of the Congressional Act. The majority felt
that there were two grounds for issuance of a FDF,
namely the economic effect if an FDF were not granted
such that a company might face bankruptcy in attempting
to comply, and secondly if it could be shown that
variations would not result in an environmental impact.
The dissenters felt strongly that Congress had intended
no such FDF exceptions whatsoever when it came to toxic
pollutants. The case is replete with references to other
cases, not cited here, and is in itself a good
background for those interested in pursuing the history
of the FDF controversy, as well as the history of the
enactment and modifications of the original Clean Water
Act. The majority also felt that the purpose of an FDF
was to "remedy categories which were not accurately
drawn because information was either unavailable to or
not considered by the Administrator in setting the
original categories." The dissent, drawing from the same
cases used by the majority disagreed on every point.

Particularly to their point they stated, drawing from
the remarks of Senator Muskie, the major drafter of the
Act, "The seriousness of the toxics problem is just
beginning to be understood. New cases are reported each
day of unacceptable concentrations of materials in the
aquatic environment, in fish and shellfish, and even in

mother's milk." He went on to cite a statistical relationship between cancer mortality due to drinking water in New Orleans containing pollutants, elimination of life in the James River by Kepone, PCB's in the Hudson River, and other cases. The dissenters took this, and other such statements as an intent by Congress to grant no exceptions to the act for toxic pollutants.

The dissent also disagreed completely with the majority on the intent of Congress in drafting and amending the CWA, implying that the majority interpreted those intentions incorrectly. Nonetheless, the majority opinion is now the law of the land, and EPA can, under strict conditions issue FDF exemptions even for toxic pollutants.

The Appellate Division of New York also found itself involved in the environmental case *Askey v. Occidental Chemical* in 1984. As might be suspected from the title, this case dealt with the Love Canal site. There was a new facet to this case: "The novel issue presented is whether those persons who have an increased risk of cancer, genetic damage, and other illnesses by reason of their exposure to the toxic chemicals emanating from the landfill, but whose physical injuries are not evident, should be certified as a class for the purpose of determining their right to recover the costs of future medical monitoring services to diagnose warning signs of disease." Some 2,855 persons residing in the area might be affected. The fact that the results of exposure, if any, may not be manifested until many years have passed is a major part of the case. An expert testifying for the plaintiffs stated, "any remedy 'should provide that when the invisible genetic damage becomes visible, the persons involved in the class

action should have immediate, free access to any such
medical care as they wish to have', and that 'there
should be surveillance of the population at risk in the
public health sense of the word' so that care may be
provided 'at an early enough stage [to] minimize the
impact of the disease.'"

The lower court had found two issues in the suit:
first the claim for a class action for immediate
damages, and second the class action for future damages.
It had denied both. The Appellate Division agreed
with the denial of the former on the basis that the
class was so large and so indefinite that it could not
be adequately described. Although, as the Appellate
Court pointed, out there is no basis in tort law for
compensation for future damages, "there is a basis in
law to sustain a claim for medical monitoring as an
element of consequential damages."

An additional concern in the case was the
applicability of the statute of limitations and "whether
plaintiffs have a claim for injuries not yet present ...
whether plaintiffs have a claim for 'potential' injuries
which they may be afflicted with in the future. The
apparent answer given by the Court of Appeals is 'Yes'
provided that proper medical proof is adduced." The
court admits to the difficulty of proof of such future
claims, but does not deny their possibility. In
addition, however, the court felt that it could not
admit a class action suit for future damages as there
was no possible way to determine how to delineate the
class. "The only fact common to the proposed class, as
established by the record, is that all of its members
live or have lived in an area adjacent to the landfill
at some time during the last thirty years. There is no

proof whatsoever of the nature and extent of the contamination which resulted from the various chemicals deposited over the years at the landfill. Consequently there is no way to determine as a threshold matter the identity of those persons who may need medical monitoring." In this manner the decision of the lower court to deny class action status for future injury was also upheld, but it would seem possible that individual suits will be permitted under this opinion.

A novel twist to the environmental cases is found in *Commonwealth of Pennsylvania v. Capitolo*, before the Superior Court of Pennsylvania in 1984. Ms. Capitolo was one of a number of protesters against the Shippingport Nuclear Plant who had carried out their protest by crawling underneath a fence surrounding the facility and then refused to vacate the premises when warned they would be arrested for criminal trespass. They were then arrested and found guilty of that crime. Their defense was that they were exempt from the Crimes Code under the defense of justification and that "they reasonably believed that the conduct was necessary to avoid a harm or evil and that the harm or evil sought to be avoided by the conduct was greater than that sought to be prevented by the law defining the conduct as criminal trespass." (The quotation is from the summary prepared by others, and not the court's words.) The evil they claimed was the possible damage to the environment and to the genetic constitution of those exposed, as well as the possibility of causing cancer and the risks entailed in shipping radioactive materials in and out of the plant. The basis for their appeal was that they were not given the opportunity through testimony of experts to prove their point which would have justified their actions. The Superior Court in a split decision ruled in favor of the plaintiffs, reversing their sentences and remanding the case for retrial. Two dissents were entered, holding that the facts of the case were so clear-cut that no defense against the charges would stand up in a reasonable manner. The dissenters feared that the remander of the case for retrial would now set a precedent since, if an the offender could claim exemption due to the "lesser evil" theory for any act, criminal justice would be set back considerably. The case, however, is one in which persons devoted, even if mistakenly so, to preservation of the environment, have sought interpretation by courts of the realities and meaning of environmental protection.

To illustrate another type of attempt to use environmental law, the case *Illiois Pure Water Committee v. Director of Health* before the Supreme Court of Illinois in 1984, may be cited. An anti-fluoridation group in Illinois attempted to use the state's EPA to prohibit the addition of this compound to the drinking water by a water company. A specific statute in Illinois permits this action in an effort to promote dental health, and the plaintiff also claimed a constitutional privilege denying freedom of religion, and requested validation of a class action suit against the state. These two complaints did not survive judicial scrutiny, but the third complaint against enforcement of the statute by the Department of Public Health and the state EPA was considered. The usual claims pro and con fluoridation were made and "experts" for both sides testified voluminously. The experts for the plaintiff were, however, found somewhat wanting in their fields, and one admitted to prejudice against fluoridation while

the other admitted that he was in the employ of a group
which had hired him specifically to testify against this
procedure. The defendant's experts, on the other hand,
were found by the court to be well-qualified, and the
court readily decided that no violation of either the
Police Power of the State, nor of the EPA had occurred;
they therefore dismissed the claim. While the case may
seem somewhat trivial, it illustrates some of the ways
in which the EPA acts may enter our lives without our
knowing it. Perhaps also the earlier case dealing with
transportation of spent fuel through New York City,
mentioned in the chapter on radiation, could fall in
this category of environmental cases and it should be
referred to in this context.

Other similar suits involving environmental safety
will continue to arise. For one example, according to
the New York Times (6/9/84), a number of major oil
companies are being sued on the basis that their
"reckless handling of toxic wastes" has caused specific
injuries including 9 cases of cancer and 5 children born
with birth defects alleged to have been caused by what
the plaintiffs call "a witches' brew" of toxic wastes
including more than a score of various compounds. The
suit, to be tried in a Federal Court, is interesting in
that it does not ask for specific damages, as those who
could be included would amount to "hundreds or
thousands" of persons living near the toxic dump sites.
Instead the plaintiffs requested that there be a halt to
further contamination, the buying of homes in the areas
and long term studies of the persons supposedly exposed.
There seems to be a contradiction in the article,
however, as the suit seems also to ask for both punitive
as well as actual damages. The case is obviously being

vigorously defended by the oil companies involved, and will probably not come to trial until 1985.

Perhaps it is just to avoid such future cases that Congress had before it a new "tougher" toxic waste law. The bill, passed 93-0 by the Senate, is not as restrictive as the bill passed in the House, and will have to go to conference, but even the Senate action seems a step forward. The Senate provides exceptions for small businesses generating less than 1,200 pounds of toxic waste a month, but also restricts landfills and the recycling of hazardous wastes for such things as oiling roads, etc. (Details are given in the New York Times, 7/26/84.) As of this writing, the final form of the bill and the action of the President cannot be predicted.

This has been a highly selected list of cases, chosen as each seems to throw some new light on Genetics in the Courts. As mentioned many times, these are difficult cases, and they lead to long opinions. In fact what has been done here is to summarize briefly several hundreds of pages of opinion. Within many of these cases cited are references to scores of other similar cases and to deal with all of them would require several volumes of text.

The environmental cases are certainly not going to cease and the activities of the federal bureaus continue to occupy newspaper headlines. A few such headlines, indicating the strong public interest, are as follows:"Gypsy Moth Chemicals Banned by U.S. Judge;" "E.P.A Proposes Allowing Emissions of Cancer-causing Substances;" "Court Hears Debate on Test Lab for Germ Warfare;" "Filing-Period Extension Expected in Toxic Suits." (the latter deals with an attempt to pass a bill

in New York to change the statute of limitations to
permit suits such as those in the New York case
discussed above in the Love Canal cases); "Illinois
Town's Battle over Radioactive Waste Disposal Tests U.S.
Policy." These headlines are from the New York Times
during the past two years. Furthermore, articles from local
papers are entitled "Lab Veils Test that Show Benzene's
Harm to Animals;" "32,000 Scouts Exposed to Toxic Dioxin;"
etc. Similarly the Journal of the American Association
for the Advancement of Science, which includes each week
a section dealing with science and social issues includes
such headings as "EPA Dumps Chemical Data System; How
Safe are Engineered Organisms?" or "Mill Tailings a $4-
Billion Problem." All these are only a small sample of
the attention given to environmental matters by the
press and it is obvious that there is a strong public
interest in these stories or else they would not be
found so frequently. Also, of course, it seems obvious
that behind each such headline there is a potential
legal issue and the basis for a future suit.

What seems most necessary is an urgent need for
clarification, simplification, and perhaps unification
of the myriad of federal and state regulations
concerning the environment and the effects of pollutants
upon it. Meanwhile, the cases cannot help but continue
to arise in increasing frequency, as each new regulation
is put forth and challenged in the courts. Perhaps
judges, in addition to the course in genetics suggested
earlier, may now need to take time to educate themselves
in basic ecological principles as well.

ALPHABETICAL LIST OF CASES CITED

CASE	CITATION	YEAR
Askey v. Occidental Chemical	102 AD2d 130	1984
Balelo v. Baldridge	724 F.2d 753	1984
Bokum Res.v.N.M.Water	603 P.2d 285	
Quality Control	(N.M.)	1979
Cabinet Mountains Wilderness		
v. Peterson	685 F.2d 679	1982
Chem Mfrs Ass'n v.Natural		
Res.Defense Coun.	105 S.Ct. 1102	1985
Commonwealth of Mass. v. Clark	594 F.Supp. 1373	1984
Commonwealth of Pa. v. Capitolo	471 A.2d 462 (Pa)	1984
Conner v. Andrus	453 F.Supp. 1027	1978
Foundation on Economic Trends		
v. Heckler	587 F.Supp. 753	1984
Foundation on Economic Trends		
v. Heckler	756 F.2d 143	1985
Ill.Pure Water Commission	470 N.E.2d 988	
v. Dir.Pub. Health	(Ill)	1984
N.J.State Chamber of Commerce		
v. Hughey	600 F.Supp.606	1985
Nat.Res.Def.Council v. Train	510 F.2d 692	1975
Nat.Res.Def.Council v. U.S.E.P.A	595 F.Supp. 1255	1984
Nat.Res.Def.Council v. U.S.E.P.A.	595 F.Supp. 1255	1984
Pac.Gas & Elec. v.St.Energy		
Resources Conserv.	103 S.Ct.1713	1983
Save our Ecosytems v. Watt	U.S.D.Ct D.Or.	
	(Slip)	1983
Sierra Club v. Ruckelshaus	602 F.Supp. 892	1984
State of Wis. v. Weinberger	578 F.Supp. 1327	1984

CHRONOLOGICAL LIST OF CASES CITED

YEAR	CASE	CITATION
1975	Nat.Res.Def.Council v. Train	510 F.2d 692
1978	Conner v. Andrus	453 F.Supp. 1037
1979	Bokum Res.v.N.M.Water Quality Control	603 P.2d 285 (N.M.)
1982	Cabinet Mountains Wilderness v. Peterson	685 F.2d 679
1983	Pac.Gas & Elec. v.St.Energy Resources Conserv.	103 S.Ct.1713
1983	Save our Ecosytems v. Watt	Dist.Ct D.Or. (Slip)
1984	Askey v. Occidental Chemical	102 AD2d 130
1984	Balelo v. Baldridge	724 F.2d 753
1984	Commonwealth of Mass. v. Clark	594 F.Supp. 1373
1984	Commonwealth of Pa. v. Capitolo	471 A.2d 462 (Pa)
1984	Foundation on Economic Trends v. Heckler	587 F.Supp. 753
1984	Ill.Pure Water Commission v. Dir.Pub. Health	470 N.E.2d 988 (Ill)
1984	Nat.Res.Def.Council v. U.S.E.P.A	595 F.Supp. 1255
1984	Nat.Res.Def.Council v. U.S.E.P.A.	595 F.Supp. 1255
1984	Sierra Club v. Ruckelshaus	602 F.Supp. 892
1984	State of Wis. v. Weinberger	578 F.Supp. 1327
1985	Chem Mfrs' Ass'n v.Natural Res.Defense Coun.	105 S.Ct. 1102
1985	Foundation on Economic Trends v. Heckler	756 F.2d 143
1985	N.J.State Chamber of Commerce v. Hughey	600 F.Supp.606

This chapter deals with a wide variety of cases in which genetics plays a role, decisions based on sexual discrimination, on racial discrimination, on I. Q. tests, on real property law and other laws, etc. None of these as yet has sufficient numbers of decisions to entitle them to a chapter of their own. Perhaps in the future many of these aspects of the law will expand into a larger number of cases, and a book similar to this will need separate chapters devoted to each of the type of cases discussed here. Indeed, this is the case with the environmental chapter; originally it was to have been a few paragraphs under miscellaneous cases.

SEXUAL DISCRIMINATION

Turning to the genetic differences between the sexes, one is of course aware of the chromosomal determination of sex. All normal women inherit one X chromosome from both their mother and father; males inherit the X chromosome from their mother and the Y from their father. All other chromosomes are identical in the two sexes. The inheritance of two X chromosomes sets into motion early embryological patterns which will normally lead to females; the presence of a Y normally leads to a male. The Y chromosome is relatively inert genetically; its main role is to establish a male developmental pattern. The females would normally carry more genetic information because of twice the number of X chromosomes. However, it is now known that in females it is usual for one of the two X chromosomes to become inactive so that the genetic balance between the sexes is maintained with each sex having but one active X chromosome.

Although, as stated earlier, a text dealing with cases decided upon genetic differences between the sexes or between races would take an entire book of its own, a few such cases will be cited here to show the type and features of these genetic differences. In these days, when great attention is focussed on women's equal rights, it is interesting to know that there have been many legal cases involving equal rights for men. Perhaps among the first of these is the classical case of *Stanley v. Illinois*, decided by the United States Supreme Court in 1972. "Joan Stanley lived with Peter Stanley intermittently for 18 years, during which time they had three children. When Joan Stanley died, Peter Stanley lost not only her but also his children. Under Illinois law, the children of unwed fathers become wards of the state upon the death of the mother. Accordingly, upon Joan Stanley's death, in a dependency proceeding instituted by the State of Illinois, Stanley's children were declared wards of the State and placed with court-appointed guardians. Stanley appealed, claiming he had never been shown to be an unfit parent and as married fathers and unwed mothers could not be deprived of their children without such a showing, he had been deprived of the equal protection of the laws guaranteed him by the Fourteenth Amendment." This forms the factual basis of the plea.

The Supreme Court in an opinion written by Justice White in which three other judges joined, agreed that the Illinois statute did indeed violate the Fourteenth Amendment and was as such unconstitutional and that Stanley was entitled to a hearing on his fitness for custody of the children. It might be noted that again

the opinion was not unanimous; two justices filed a
dissent, and two did not participate in the case, but
the right of an unwed father to attempt to claim his
children was upheld. Citing other cases, the majority
opinion notes that "These authorities make it clear
that, at the least, Stanley's interest in retaining
custody of his children is cognizable and substantial."
The claims by the State of Illinois that Stanley could
attempt to adopt or apply for custody of the children
did not invalidate the fact that the law did not grant
him equal protection.

Similarly, in 1976 the United States Supreme Court
again held an Illinois law to be unconstitutional. In
this case, *Trimble v. Gordon*, the issue was whether
illegitimate children could inherit by intestate
succession only from their mothers. Again in a five to
four decision, the Court found that the statute denying
the rights of illegitimate children violated the XIV
Amendment. The majority opinion stated in part, quoting
from an earlier opinion, "The status of illegitimacy has
expressed through the ages society's condemnation of
irresponsible liasons beyond the bonds of marriage. But
visiting this condemnation on the head of an infant is
illogical and unjust. Moreover, imposing disabilities on
the illegitimate child is contrary to the basic concept
of our system that legal burdens should bear some
relationship to individual responsibility or wrongdoing.
Obviously, no child is responsible for his birth and
penalizing the illegitimate child is an ineffectual--as
well as an unjust--way of deterring the parent."

Although the deceased could have left a will or
legitimized the child by marrying the mother, these were
not the issues at hand. What was of importance was the

statute whereby illegitimates could inherit only from
their mother, and legitimates could inherit from both
parents. Finding, as Illinois had, that an illegitimate
could not inherit from an intestate father was in the
opinion of the majority in itself a violation of the
rights of the illegitimate.

On the other hand, in 1979, The Supreme Court dealt
with a different facet of legitimacy and decided against
the right of the acknowledged father. In this case,
Parham v. Hughes, the issue was whether the father of an
illegitimate child could sue for the wrongful death of
the child caused by an accident. Here the Court ruled
that the Georgia statute which forbade such suits was
constitutional and upheld the Georgia Supreme Court on
the basis that the three-pronged reasoning forbidding
such suits in that state did not violate the Fourteenth
Amendment: "(1) the interest in avoiding difficult
problems of proving paternity in wrongful death actions;
(2) the interest in promoting a family unit; and (3) the
interest in setting a standard of morality by not
according to the father of an illegitimate child the
statutory right to sue for the child's death."

The Supreme Court carefully differentiated this case
from the previous one, noting that in that case there
was an issue of benefit to the child, but here, as the
child was deceased, the only benefit could be to the
illegitimate's father. In this instance he could, by
marrying the mother, have legitimized the child prior to
its death. The summary of the case (not written by the
Court) noted, "...does not reflect any overbroad
generalizations about men as a class, but rather the
reality that in Georgia only a father can by unilateral
action legitimate an illegitimate child." Again,

reflecting the makeup of the court, this was a 5:4 decision, with the usual dissents filed.

Another sexual bias found earlier was the right of a husband to bring suit for loss of his wife's services, but not vice versa. An early challenge to that condition was made in 1968 in the New York case, *Millington v. Southeastern Elevator*, decided by the Court of Appeals. The facts were simple; Mr. Millington was 37 years old when he became paralyzed from the waist down as the result of an elevator accident. His wife brought suit for "loss of consortium basing her claim upon the fact that her husband will spend the rest of his life as an invalid. She further alleges that this has caused a radical change in their marriage..." The Court quoted lengthily from a Michigan Supreme Court decision. "We come then, as we must ultimately in every case, unless we are to continue to utilize fictions, or unless we are to dispose of the case on a narrow point of procedure or pleading, to a balancing of interests. On the one hand we have a wife deprived of the affection of her husband, his companionship, his society, possibly deprived even of the opportunity to bear sons and daughters. On the other, we have a defendant whose liability because of his act must involve the violation of a duty of care with respect to it, and, furthermore, whose liabilities as a result of his negligent act must have some reasonable limitation. So analyzed we see the problem not as a unique and peculiar historical anomaly but as part of a much larger pattern, as a part of a clearly discernible movement in the law. * * *

"The gist of the matter is that in today's society the wife's position is analogous to that of a partner, neither kitchen slattern nor upstairs maid. Her duties and responsibilities in respect of the family unit complement those of the husband, extending only to another sphere. In the good times she lights the hearth with her inimitable glow. But when tragedy strikes it is part of her unique glory that, forsaking the shelter, the comfort, the warmth of the home, she puts her arm and shoulder to the plow. We are now at the heart of the issue. In such circumstances, when her husband's love is denied her, his strength sapped, and his protection destroyed, in short, when she has been forced by the defendant to exchange a heart for a husk, we are urged to rule that she has suffered no loss compensable at the law. But let some scoundrel dent a dishpan in the family kitchen and the law, in all its majesty, will convene the court, will march with measured tread to the halls of justice, and there suffer a jury of her peers to assess the damages. Why are we asked then, in the case before us, to look the other way? Is this what is meant when it is said that justice is blind?" Needless to say, after this citation, the court decided in favor of Mrs. Millington. The court, unusual for the most part in the decisions quoted earlier, stated, "Finally we turn to the argument that the change should come from the legislature. No recitation of authority is needed to indicate that this court has not been backward in overturning unsound precedent in the area of tort law... Abiding by that principle, ...and to recognize a cause of action for consortium in the wife, thereby terminating an unjust discrimination under the New York law."

A similar case arose in 1981 in the United States District Court, N. D. Georgia, Rome Division. This case, *Timms v. Verson Allsteel Press Company*, also involved a

serious accident to the husband. Apparently a steel
press descended upon his hands due to the failure of a
safety device. He sought damages for the loss of several
fingers, and his wife brought a suit for loss of
consortium. The court found this latter claim to be a
case of first instance in Georgia, and reviewed the
pertinent cases dealing with such a claim. Although it
found some similar ones, it is surprising that the case
just cited above was not discovered, and the present
case was decided without that precedent. The court found
that if her husband suffered some tortious personal
injury and she could establish her own personal injury
thereby, her suit for loss of consortium would be
proper. Again using an earlier decision the court
stated, "No appellate court in Georgia has ever denied
the wife a right of recovery in such an action and, as
Georgia courts make their own interpretations of common
law, this court is not acting beyond its powers in
recognizing the right at this time. Such decisions as
this do not involve a disregard of statutes, or sound
rules of conduct or any constitutional provision. ...It
is appropriate in this day, when human rights are on the
tongues and in the hearts and minds of men, women, and
children everywhere, and when the very existence of
civilization depends on whether fundamental human rights
shall survive, for this court to recognize and enforce
this right of a wife, a right based on the sacred
relationship of marriage and home." Based on these two
decisions, it now appears that women now also have
consortium rights, rights previously restricted by many
laws to men only.

The question whether a husband can obtain injunctive
relief to prevent his wife from having an abortion
arose in the case, *Coleman v. Coleman*, before

Appeals in Maryland in 1984, dealt with the right of a husband to seek injunctive relief to prevent his wife from having an abortion. The husband sought relief under the Ninth Amendment and for a number of other causes. Originally a judge had issued a temporary restraining order against the abortion but upon hearing further testimony he dissolved the order and denied injunctive relief. The woman then had the abortion. As the court pointed out, normally this would have rendered the case moot, but it felt the issue might arise again and sought to prevent further such appeals by husbands. The opinion was detailed, citing the *Roe v. Wade* decision of the U.S. Supreme Court as well as Maryland State Laws which were found in conflict with that opinion. It also took note of a special appeal made by the husband that the U.S. Supreme Court's opinion was no longer valid as viability of a foetus, according to a French expert he introduced for this case, supposedly began earlier than the end of the first trimester of pregnancy, the period during which the U.S. Supreme Court had ruled that abortions were permitted with no restrictions except for the advice of a physician. The expert testimony was found unnecessary as the court here felt that it should not deal with that point. Therefore, the court issued a *per Curiam* order that the original injunction had indeed been illegal and the subsequent lifting of the injunction was completely valid. Although this was a case of non-therapeutic abortion, it is cited here as cases of abortion during the first trimester or early in the second trimester for sound genetic reasons might subsequently involve other attempts by a husband to block this procedure.

A much earlier case, *Byrn v. New York City Health*

and *Hospitals Corporation*, deals mainly with the abortion issue, but there are also both genetic considerations as well as the rights of women to voluntary abortion. The case was heard by the Appellate Division of New York in 1972. Byrn had himself appointed *guardian ad litem* for all infants of less than twenty-four-weeks gestation and sought an injunction to prevent the carrying out of abortions on all such foetuses then scheduled for abortion. The opinion of the Appellate Division is of great value to those who wish to study the history of abortion in New York State. The principle reason for citing it here is the fact that the court took notice of the issue of abortion for eugenic purposes, for example Down Syndrome. The court was not unanimous, but ruled that with their decision, the preliminary injunction sought by Byrn against the hospitals (which would have prevented the carrying out of any abortions except those necessary to preserve the health of the mother) was dismissed as was his request to inspect the records of scheduled abortions. It should be stressed that this was a constitutional challenge against the abortion laws, but that cognizance of the legality of abortion for genetic reasons was also involved, as were women's rights in this issue.

A U.S. Court of Appeals in 1984 was faced with a claim of violation of the Pregnancy Discrimination Act of 1978 which stated that "discrimination on the basis of pregnancy is discrimination on the basis of sex." The case *Hayes v. Shelby Memorial Hospital* involved the dismissal by the hospital of an Xray technician following her informing her supervisor that she was pregnant. The supervisor in turn informed the medical director, who was also the radiation safety officer, of the radiology department who recommended that the plaintiff be moved to another area of the hospital to avoid any possibility of Xray damage to her unborn child. The hospital claimed that she was fired because there was no alternative employment available for her.

A major factor in the court's decision in her favor was that the hospital could not show that the exposure to radiation she might receive while performing her work was an "*unreasonable* risk of harm to the fetus." The opinion referred to testimony by the radiation officer that "*any* dose of radiation is excessive and therefore potentially harmful to the fetus. Dr. Eland [the radiation-safety officer] even stated that it would be dangerous for a pregnant woman to sunbathe in a bikini, which would expose her fetus to radiation from the sun." It must be added that danger to a foetus from this activity is most unlikely; any irradiation from the sun is from ultraviolet rays and these cannot penetrate more than a very small distance through tissues. The real argument, however, was whether there is a danger to the foetus from radiation possibly received and what is meant by the "maximum safe dose" during pregnancy while the mother worked as an Xray technician. Included in the court's deliberations was the interpretation of dosages as indicated by film badges, and of the fact that the plaintiff worked the night when there was much less use of the radiological facilities. The court then decided that as she would receive considerably less irradiation than the permissible dosage there would be no potential of harm to the foetus arising from Haye's normal duties. Taking these facts into consideration, the court found that the hospital had shown clear sexual discrimination even though its intent to protect a foetus might have been admirable. Additionally, because of its failure to make a serious attempt to find an alternative position for her, the hospital had further violated the law and therefore was liable for the awarding of the modest amounts of damage claimed by Hayes.

Other cases involving sex discrimination and equal rights for men or women might be cited here. Some, such as the removal of women from the workplace where there may be teratogenic compounds will be considered with the compensation cases, to be dealt with later, as that seems the more appropriate place for them. However, it must be noted that many cases involving equal pay, equal worth, etc. will be a standard type of genetic case for the courts in the future.

RACIAL DISCRIMINATION

Cases involving race distinctions are examples of genetic differences and must therefore be considered part of *Genetics in the Courts*. The basic problem is that the term "race" is so emotionally loaded. To the geneticists, race is a useful term, describing a real situation. The term refers specifically to populations which are primarily inbreeding, and which differ by the relative frequency of certain genes within the population. Note the plural used here; races are not recognized due to differences in a single trait, but due to differences in many traits. Note also that it is not the existence or non-existence of any particular single gene which differentiates racial groups. For example, in man, groups may be distinguished on the basis of the relative frequency of such traits as blood type, color, hair form, body structure, body hair, biochemical differences in proteins such as hemoglobin or enzymes, ability to taste various chemical compounds, etc. No one group of man is without some individuals who carry a gene mainly found in other groups; it is the concatenation of different frequencies of these and other traits which define to a geneticist the various races. An example of this type of difference has already been given in the paternity cases where the frequencies of the genes controlling the HLA antigen system are known to differ from race to race.

It is also noteworthy that in many non-human groups races are recognized without prejudice or value judgment. For example, although all dogs are of a single species, no one can help but distinguish a great Dane from a chihuahua. But except for the fancier of such breeds, no one would claim that one is more primitive or inferior to another. Yet, there are those who would take equal or lesser differences found in man and attempt to attribute superiority of one group over the other.

One must further consider the uniqueness of man in the biological world. All races of man, regardless of how classified, are interfertile; in genetic terms man is a single species. Contrast this, for example, to insects where it is believed there may well be close to a million species of beetles alone, most of which are infertile in crosses. This may have come about as each of these insect species is adapted to a specific environment over which the species has no control. This adaption is delicate, and obviously has great survival value to the larger class to which all beetles belong. Elimination of one species cannot eliminate the class.

Man is different for perhaps the single reason that man does not so much adapt to the environment as he adapts the environment to himself, and there is no biological need for speciation in order to survive. An Eskimo in his igloo is able to create a climate in which he survives; an African tribesman living in the tropical heat is equally able to find means of survival by moving into shade or limiting activities during the hottest part of the day. And of course in this day of air conditioning, climatic differences are of little concern to industrial societies.

In addition, man is an omnivore and his diet is not restricted to either meat or vegetable matter so that he can obtain his needed dietary requirements from either animals or plants. Most other animals, however, are specially adapted for one kind of feeding, either by adaptation of their digestive organs or by special mechanisms of predation or grazing which restrict them to a particular diet. The ability of man to survive with a proper diet of either type of food can be seen in the many persons who practice vegetarianism today; they would not be considered different species or races by any thinking person.

Still another factor which has kept man a single species is the fact that genetic speciation requires both isolation and time. Man has never been completely isolated for sufficient periods of time for this to occur. There has always been sufficient means and potential for even the most distant groups of humans to meet, mate, and interchange genes. No group of humans today, or in the past, has been totally isolated; mankind remains a single species. No group of humans can be thought of as primitive; all are coequal in time and none is ancestral in any way to another. Races as defined exist, but in man they replace species in other forms. Racism is not based upon genetic principles, and is both scientifically and therefore morally incorrect.

Another incorrect idea is that religions are in some way genetically determined. Modern religions are recent developments in terms of overall time. One is no more genetically a member of a religion than he is a member of a country. The accident of one's birth to a particular nationality or to a particular religious faith has absolutely nothing to do with genetics. Cultural differences do, of course, exist between groups, but these are imposed, not inherited. The confusion by all too many of religious or cultural differences with genetics differences is a tragic misunderstanding by modern society. There is simply *no* known genetic principle to which bigots might point which is supported in any way by contemporary genetic knowledge; those claiming the opposite simply overtly display their ignorance of modern genetic knowledge.

A number of cases involving sickle cell anemia have strong racial overtones. This disease, caused by a recessive gene, affects primarily members of the black population. Persons carrying two recessive genes are severely ill much of their lives. The effect of the disease is to cause the red blood cells to become misshapen; instead of having a normal spherical shape these cells become shrunken into the sickle shape. As a result the red blood cells are much less able to carry oxygen and the person may suffer from anemia, and from other acute symptoms, always painful, and often fatal. Persons who have only one of the alleles (a pair of genes at the same genetic site) of the recessive type and the other normal are known as carriers of the trait. To all intents and purposes, they display none of the symptoms of the disease, although there is, of course, a

25% chance that any offspring of two carriers will suffer from the disease and 50% of the children will be carriers also. At present there is no known cure for the condition, although a possible cure has been recently reported. It should be noted, however, as is the case in any genetically caused disease that the cure, to date, is for the symptoms only, and the genetic risk remains the same whether the person with the disease shows the symptoms or not.

The cases involving this disease come under a number of different legal categories, but all of them involve racial discrimination one way or another. A few sample cases follow.

In 1976 the case *Smith v. Olin Chemical Corporation* was tried before the United States Court of Appeals, Fifth Circuit. The case centered around the discharge from employment of the plaintiff due to the fact that he suffered from the sickle cell disease. Smith was a laborer who was hired after a preliminary physical examination. A subsequent second examination, routinely given by Olin some ninety days after employment begins, showed that Smith had "bone degeneration with a prognosis of possible aseptic necrosis or further bone degeneration in his spinal region." Smith then informed the doctor that this might be due to his history of sickle cell disease and the disease was confirmed by his own physician. Olin then discharged him as unfit to do manual labor. Smith charged this was racial discrimination under the Civil Rights Act of 1964. The fact that some 10% of the black population, and none of the white population, suffered from this disease led him to claim that blacks would be discriminated against in employment because of the disease. The lower court dismissed the claim, but the Court of Appeals took a closer look at the case and remanded it on the basis that there was a "possible liability of Olin based on employment practices that are fair in form but discriminatory in action." The appellant stated his case cogently as follows: "Because sickle cell anemia affects Blacks almost exclusively, probability dictates that bone degeneration, a common condition resulting therefrom, would be present among proportionately more Black workers in a labor force than among workers of any other race. If the presence of bone degeneration is used to automatically disqualify a worker from a position which involves manual labor, proportionately more Black workers would be disqualified than workers of any other race. This is unquestionably a discriminatory effect." The court here made no statement as to their own belief whether this practice was discriminatory, but remanded the case for further trial to determine this fact.

Riley v. State Employee's Retirement Commission arose before the Supreme Court of Connecticut in 1979. Mae Riley had been dismissed from her job and sought a service-connected disability retirement benefit. Examination showed that she suffered from "congenital sickle cell trait." The Commission refused her request for the disability retirement several times and she sought relief in the courts. The Supreme Court upheld the State Employees' Commission, finding that "The plaintiff is both obese and suffers from a congenital sickle cell trait, which conditions are not service connected. The Commission concludes that '[a] person with this illness can obviously have successive recurrences of bad reactions through external contacts, but these are due to the inherent nature of the disease

itself which is permanent and not through the results of
a single episode.'" A further factor was that she had
injured her leg while at work, but the Commission
pointed out this was temporary, and it appeared that the
real reason for her not being accepted back at work was
her obesity. Indeed it was stated, "...if she lost
weight she would be eligible for reconsideration for
reemployment in her previous position." Thus the
critical issue would seem to have been obesity, not
sickle cell disease. It is also perhaps noteworthy that
no claim of civil rights violation was made here, a
claim which might have been examined more thoroughly
than the simple claim of service-connected injury.

A much more serious case arose in North Carolina
and was tried before the United States Court of Appeal,
Fourth Circuit, in 1981. This case, *Avery v. County of
Burke*, clearly involved civil rights. The facts are
worth noting. Avery became pregnant at the age of 15 and
sought prenatal care at a clinic operated by the
defendant-county. A blood test was administered, and she
was told that she had sickle cell trait; the nurses then
urged her to consider sterilization. She was told she
would be unable to use birth control pills due to her
condition and she was also informed that pregnant women
with sickle cell trait are at high risk. A social worker
trainee was assigned to help her and accompanied her to
court where, based on the recommendation of the clinic
doctor's report, the sterilization was approved.
Subsequent tests showed that she did not have sickle
cell trait. Avery then brought suit against the Health
Department, nurses, the doctor, the social worker
trainee and individual members and director of the board
for deprivation of her rights of privacy and procreation

in violation of the Fourteenth Amendment. She claimed to
have agreed to the sterilization only because of the
"exhortations" and misrepresentations of individuals
associated with or employed by the boards.

The lower court (a United States District Court)
had entered summary judgment for the defendants, and the
appeal to this court resulted. The court, after first
asserting that the suit was properly brought, then went
on to deal with the legal validity of the claims. This
court dealt most harshly with the defendants. First it
determined that Avery "need not prove, however, that
members of the board personally participated in or
authorized her sterilization. Official policy may be
established by the omissions of supervisory officials as
well as from their affirmative acts. ...It is not
essential, however, that Avery show that all persons
suspected of having the sickle cell trait have been
mistreated. It is enough that an identifiable group of
people, of whom Avery is a part, is subject to
constitutional deprivations through the inaction of the
boards. ...Read in the light most favorable to Avery,
the evidence discloses that a trier of fact could
reasonably find that the boards did not adequately
discharge their statutory duties. Burke County has a
substantial black population, and the sickle cell trait
affects approximately 10% of the black people in the
United States. One member of the Board of Health
testified that as a practicing physician he was aware
that the board was involved in the sterilization of
women. Nevertheless, prior to Avery's sterilization,
neither board adopted policies, rules, or regulations
for counseling and sterilizing persons believed to have
sickle cell trait. ....In addition both Health

Department nurses who recommended Avery's sterilization testified that they had no special training in handling sickle cell cases. The social work trainee testified that she was assigned to counsel Avery about sterilization only eight days after she was hired." In plain English, North Carolina was found to "run a sloppy ship."

Further the court then went on to give a ringing affirmation of the right of procreation and examined the action of the board which deprived her of that right. The court determined that the decision of the lower court to grant summary judgment in favor of the defendants was wrongly reached. "Therefore the issues that divide the parties should be resolved by plenary rather than summary proceedings. The judgment of the district court is vacated, and this case is remanded for further proceedings." There can be no doubt that the issue of deprivation of the right to procreate was a major issue here; the racial issue also seems to be of importance; one cannot help but wonder, given similar genetic considerations, whether a white woman would have received the same treatment. Also of course the third issue, the incorrect results of the blood tests, will form an important issue in the retrial when it occurs.

A class action suit against the United States Air Force Academy arose in 1981 when a black cadet was expelled for being a carrier of the sickle cell trait. Stephen Pullens had been admitted to the Academy and was forced to resign when it was found that he was a carrier. His previous record showed him to be a stellar athlete. His dismissal was based on the grounds that he "might die" if he stayed for the rigorous training at 7000 feet where the academy is located. He had, previous to his acceptance by the Academy, been a mountain climber and had always felt that he performed better in high altitudes. When he was dismissed, he was escorted to assure that he would not discuss the case further. At the same time five other blacks were also dismissed for the same reason. The Air Force here was acting, it claimed, for his benefit. However, other opinions previously had stated that except for pilots and copilots persons suffering from sickle cell disease should not be prohibited from flight duty. The Air Force based its policy upon the fact that in 1972 two black sickle cell carriers had collapsed after strenuous exercise. The evidence suggests that both were suffering from viral infections at the time, and that the collapse was in no way related to their being carriers. Further it was stated that black athletes in the 1968 Olympics, held at high altitudes in Mexico that year, had failed to show any effects of their being carriers. Thus the Air Force case may have been based on an intent to restrict the opportunities of blacks.

Subsequent to the filing of the class action suit, Pullens was readmitted to the Academy, but the suit, as of this time, has not been withdrawn, nor has it come to trial. It would appear that with the threat of the suit the Air Force has changed its policy concerning sickle cell carriers, a policy which was certainly medically unsound on any basis.

Another case dealing with sickle cell anemia from a quite different aspect arose in 1980 in *State of New Jersey v. Meighan*. Here the issue before the Superior Court of New Jersey, Appellate Division, was the placement of a convicted felon in a jail. The issue of guilt in the manslaughter case was not involved as

Meighan pleaded guilty to this charge. Also involved was
an issue of search under warrant of his house, of a
bundle carried by his wife from the house after his
arrest, and of search of his wife in the home of another
person, the latter two searches without warrant. As
twenty-five grams of marijuana were found in his house,
the court ruled that the police had no need for a
warrant to conduct the latter two searches, and that is
not the primary issue. The interesting issue in terms of
genetics is the plea by the convict that he would not be
able to receive proper treatment for his sickle cell
condition in the prison to which he was sentenced for
five to ten years. Since his incarceration, he has
suffered one crisis requiring hospitalization, and he
sought a medical release, claiming that he could not be
properly treated in prison. The court refused to grant
such release and obviously felt that due to the serious
nature of the crime, and the fact that hospital
treatment had been given, there were no grounds to
overrule the sentence imposed by the trial court.

These cases involving genetic disease indeed
indicate the type of interaction between genetics and
the law. Again they involve both a knowledge of genetic
facts, and in most cases discrimination against a
genetically distinct group. The courts must deal with
medical genetics particularly, as pointed out
previously, as more knowledge of human genetic disease,
inherited and mostly incurable, is obtained. Defenses
based on genetic grounds will continue to increase, and
again the knowledge of genetic facts must play an
increasing role in court decisions.

Two new issues have come to the fore, the genetic
issue of what constitutes a person's race, and the issue

of the use of race in settling custody suits. The first
issue arose in Louisiana when an obviously white woman
applied for a birth certificate and found herself listed
as black. The laws of that state held that any person
who was considered to have 1/32 Negro Blood be
considered as belonging to that race. The law, dating
from the eighteenth century and found again in a 1970
legislative enactment, is believed to have been
established originally to prevent illegitimate children
of a slave owner, or their descendants, from claiming a
share in his estate as at that time blacks had no legal
or civil rights. The genetic basis for this is of course
non-existent. Each individual receives half of his/her
genes from each parent and the various genes for racial
traits are passed on randomly if parents have different
traits. Thus the first generation from a cross between a
black and a white will, of course, have half the genes
of each parent. But the second generation, should this
person have issue with a white partner, would again on
the average halve the number of genes coming from the
black parent, and also from the original white parent.
After 5 generations (1/32) of mating to whites, there is
little chance of many of the original genes from the
black parent still being present. Thus the arbitrary
setting of 1/32 is simply that-totally arbitrary-and
genetically meaningless. The ridiculousness of such a
law was emphasized by the case at hand where an
obviously white woman went to court to have the racial
characteristic listed on her birth certificate changed.
The issue came before the Louisiana Court system which
decided that she was 3/32 black and therefore the
listing on the certificate was legally correct. There
was national ridicule of this decision and in 1983 the

legislature of that state repealed the law but the
change is still far from adequate as it simply
substitutes the wording of "1/32" to read "traceable
amount," whatever that may mean. (Louisiana has dropped
the designation of race from its birth certificates, but
retains this information for statistical purposes in a
confidential file.) The use of the term "preponderance
of evidence" in the new state law is probably
intentional in its vagueness. "If the state says you are
colored, then your blue eyes, blond hair, white skin and
sister's color don't matter, because we don't know what
makes a person colored." (Quotation from NYT 7/6/83,
A10.) Indeed, within the past year, a Louisiana
Appellate Court has upheld the constitutionality of the
original law and prohibited the woman from having her
racial designation changed on her birth certificate. The
Louisiana court decided that changes on a birth
certificate could only be made if there had clearly been
an error. But the Court's decision was not their own as
they based it on an earlier Louisiana Supreme Court
decision dealing with the same case. This Appellate
Court further underscored their unhappiness with the
decision it felt bound to make by stating, "In reaching
this conclusion, we have not lightly dismissed
appellant's well-founded argument that [the law] was
constitutionally defective.

"To the contrary, we have recognized, in the past,
as we do today, that the statute was based upon wholly
irrational and scientifically insupportable
foundations." This was the same court which had
previously declared the law to be unconstitutional, only
to have the higher court overthrow that decision, and
the court here had no choice but to follow the rule of

the higher court. Perhaps the Supreme Court of Louisiana needs a lesson in genetics!

The second case involved a custody suit after a divorce. In this case the man was white and the woman black. The court, supposedly acting in the best interest of the child, awarded custody to the white husband, who took the child to another state. The woman appealed on the basis that the decision was made purely on racial grounds, although other reasons were purportedly given. The U.S. Supreme Court remanded the case to the original jurisdiction on the basis that racial characteristics were illegally considered. However the woman has still many problems ahead as there is now a question of which state court, Florida, where the original action occurred, or Texas, where the man and child now reside, has jurisdiction of the custody suit. It is also by no means clear that once this issue is settled the woman will regain custody as the courts may be able to find other, supposedly non-discriminatory, reasons to deny her claim.

RACE AND INTELLIGENCE

The use of I.Q. tests to determine educational placement of children troubled the California Courts for a number of years. The cases began in 1972 when *Larry P. v. Riles* first arose in the United States District Court, N.D. California. The facts were simple: I. Q. tests were being employed by the State of California to place children in classes for the educable mentally retarded (EMR). A child once placed in this category had that fact documented permanently upon the educational record, even in the rare event of the child being returned to normal classes. Furthermore it was clearly shown that while black students in the San Francisco School District made up only 28.5% of the school population, 66% of all students in the EMR group were black. The claim was made that the tests were being used to carry out racial segregation. Interestingly, the defendant Riles who was Superintendent of Public Instruction in California was himself black and found himself attempting to defend a suspected segregation of black students. The plaintiffs claimed irreparable injury due to the minimal nature of the curriculum in the EMR classes, the low expectation of the teachers of these classes for their pupils, the ridicule students in the classes underwent, and the subsequent feelings of inferiority. The permanence of the placement record revealed on the students' transcripts which would be available to employers, college admissions offices and the armed services was also claimed to be of consequent great harm to the students.

The defense claimed that the curriculum and special attention would be beneficial for such students. However the defense did not dispute the fact that students who were placed in the EMR program but who did not belong there would suffer the harms claimed above. The Court, examining other tests or qualifications which discriminated against blacks, held that the tests, as used (only I. Q.) were discriminatory, and harmed the children. In part to avoid this charge, the school system had, beginning in 1971, changed its operational method for placement by administering the I. Q. tests only after the student had been referred to a counsellor because of an apparent failure to adapt or to achieve adequate academic performance in the normal school system. The I. Q. test even then could only be administered with parental consent.

The Court made a careful study of the laws and the literature concerning the use of I. Q. tests, particularly in their cultural bias against minority students, and of the fact that despite all other factors presented by the defense, these tests were being used as the primary agent for placement. The court ruled that an injunction would not be granted against keeping black students in the EMR who were already in these classes. However their yearly performance must be evaluated by means which do not deprive them of equal protection. Further, and most important, "No black student may be placed in an EMR class on the basis of criteria which rely primarily upon the results of I. Q. tests as they are currently administered if the consequence of use of such criteria is racial imbalance in the composition of EMR classes."

The first appeal came before the United States Court of Appeals, Ninth Circuit, in 1974. The decision of the lower court was upheld, and the injunctive relief against the use of the I. Q. tests as a primary factor was upheld.

The issue arose once again in 1979 before the United States District Court for the Northern District of California. The court must have set a model of patience as the final 85-page opinion, it is pointed out, is based on over 10,000 pages of testimony! After tracing the history of the use of I. Q. tests and their use in California, the court then went into a lengthy, detailed, and careful analysis of the tests themselves. It dealt with the issue of the impossibility of measuring intelligence, including reasons for disparity in black and white scores. The court examined the

genetic arguments for heritability of intelligence, the
socio-economic arguments concerning the disparity, and
particularly the cultural bias of the tests. Following
this a detailed legal analysis was made dealing with the
issues of civil rights, equal protection under the
Fourteenth Amendment, state claims, etc. The court found
that there appeared to be no question that the use of
the I. Q. tests was clearly racially discriminatory. The
Court now granted permanent injunctive relief, stating,
"Defendants are enjoined from utilizing, permitting the
use of, or approving the use of any standardized
intelligence tests including those now approved pursuant
to Cal. Admin. Code 3401 for the identification of black
EMR children or their placement into EMR classes without
securing prior approval of this court." The court also
spelled out the conditions of possible approval which
would include a wide variety of statistical and other
data supporting the validity and non-discriminatory
results of any such tests. The defendants were also
required to monitor and to eliminate disproportionate
placement of black children in California's EMR classes.
A detailed report of how this was done, for example,
"When a disparity of one standard deviation in the rate
of EMR class placements of blacks above the district
rate for whites continues or occurs after the three-year
period, the state defendants must bring the imbalanced
district to the attention of this court, and the court
after hearing may order such further relief as it deems
appropriate." In addition, the court ordered the
mandatory review of all children in the EMR classes.
Essentially, the Court has outlawed the use of not only
I. Q. but other uniform tests which result in
disproportionately low scores for blacks, and made the

use of such tests illegal.

One might think that with this careful decision by the Court just discussed the issue would be closed. Unfortunately it is not. Later in the same year, 1980, in the case *Parents in Action v. Hannon*, another United States District Court, N.D. Illinois, came to just the opposite opinion. The reasoning was quite different. This court examined in detail three tests, the WISC, WISC-R and the Stanford-Binet test (the I. Q. test used in California).

The judge, apparently taking upon himself the role of a psychologist, concluded that he had found only a few questions in the three tests which might be culturally biased. Despite the evidence that of the children in the district involved 62% were black and in the special classes 82% were black (almost certainly more than the one standard deviation cited previously) he found no evidence of racial discrimination, and went out of his way to criticize the *Larry P. v. Riles* decision. "Judge Peckham's lengthy and scholarly opinion is largely devoted to the question of what legal consequences flow from a finding of racial bias in the tests. There is relatively little analysis of the threshold question of whether test bias in fact exists, and Judge Peckham even remarked that the cultural bias of the tests '...is hardly disputed in this litigation...' I find reference to specific test items on only one page of the opinion. ... As is by now obvious, the witnesses and the arguments which persuaded Judge Peckham have not persuaded me." It is difficult, having read Judge Peckham's opinion carefully, to see how this other judge can make these statements as Judge Peckham, while admittedly not looking at every single question in the tests, had dealt with the issue of basic unfairness of the tests in detail. Some of the questions he had found biased dealt with the identification of the author of Romeo and Juliet, the discoverer of America, and the inventor of the light bulb. The opinion in this case also did not mention the effect of a single wrong answer, several points, in the I. Q. tests. The opinion here concludes, "Intelligent administration of the IQ tests by qualified psychologists, followed by the evaluation procedures defendants use, should rarely result in the misassessment of a child of normal intelligence as one who is mentally retarded. There is no evidence in the record that such misassessments as do occur are the result of racial bias in test items or in any part of the assessment process currently in use in the Chicago public school system." There is no explanation, other than perhaps an implied one, as to the imbalance in percentages of blacks in the normal and specialized classes. The judge here listened to the various educational psychologists testifying about the tests and decided upon his interpretation of them that they are not racially biased. It is hard to see how two judges on courts of equal rank could disagree so sharply. Perhaps it is because Judge Peckham undertook to examine the wider implications of social and legal injustice while the judge in this case limited himself to an interpretation only of test questions. Be that as it may, the issue was again open, and with these two opposite opinions, the decision should be made by a higher court, taking into account the facts and opinions represented by the two lower courts.

It was perhaps inevitable then that the California I. Q. case should be appealed. Thus *Larry P. v. Riles* was heard by the U.S. Circuit Court of Appeals, Ninth Circuit, in 1984. As usual, the opinion is quite lengthy, but the reasoning of the lower court was upheld fully by the Court of Appeals. Little new by way of argument was offered, and there seems no need to go into details again. However, it should also be noted that there was no mention of the second I. Q. case, *Parents in Action v. Hannon*, mentioned in the text, and there still remains the disparity in the *Riles* case and the Illinois decision.

Another case involving I. Q. and race will be discussed in the section of this chapter dealing with libel suits. The fact that this libel case also deals with racial discrimination, although in a different fashion, illustrates the difficulty of neatly catagorizing some cases, but it is felt that the case more properly belongs in the category of libel cases and therefore it will be dealt with later in this chapter.

There also is a pending suit in Georgia contesting that its State Regents Examination required for graduation from high school is racially biased. The Federal Government has threatened to cut off all Federal funds for education to that state on that basis. In addition, there is a pending suit against the State of Alabama for maintaining a "dual system" of public colleges for blacks and whites. Of course both these and other similar suits imply a genetic racial difference in intelligence which almost all geneticists believe doed not exist. The cases should either be settled or come to trial soon.

That some specific learning disabilities may be genetically caused is suggested by an article in *Science* in 1983. Here it was found that a specific reading

disability (not named in the article) has a strong
familial pattern. The study was based on white families
only, and carries no racial implications, but is given
as an example of how single gene differences may account
for various personality traits. However, a later letter
to the editor of that journal claims to refute the
original findings.

WORKMEN'S COMPENSATION

Another legal issue where one might not expect
genetics to be involved is that of workmen's
compensation cases. Nevertheless, genetic issues are
arising in these cases with increasing frequency. The
basic genetic issue is whether the claims arose from
job-related causes or whether the injury was in reality
caused by inherited genetic defects to which the
claimant was subject. The issues may be intertwined in
that the workman may then argue that an injury, either
physical or mental, has aggravated an already present
genetic predeliction for the harm claimed, while the
defendant corporation seeks to absolve itself of all
responsibility on the grounds that the seeker of
compensation would have suffered the injury regardless
of any job-related events.

Among the earlier cases was *Simpson v. City of
Tulsa*, decided by the Court of Appeals of Oklahoma,
Division 2, in 1980. The plaintiff alleged that as a
result of cleaning a fire station he developed a
multitude of medical conditions, including high blood
pressure, dizziness, chronic nervous condition and
hypertension. There was some difficulty in the claim as
it was shown that despite this host of symptoms he was
working not only as a fireman, but also as a carpet

cleaner and a real estate salesman. It should be noted
that he continued the latter two occupations after he
brought suit for compensation for his injuries as a
fireman. His claim for 100% disability was even more
peculiar as it was found that after his claim had been
entered one of his physicians had advised him not to
pick up over 200 pounds. "In our opinion if this five
foot eight inch man weighing 177 pounds is able to lift
objects weighing 200 pounds without physical harm and is
otherwise able to carry out his duties required of him
as a fireman, carpet cleaner and real estate agent, he
is not 100 percent disabled within the meaning of the
Worker's Compensation Act...But it had more to support
its [the defendant's] conclusion. It had the testimony
of...an internal medicine specialist. ...He said he
examined Simpson at the request of the City of Tulsa and
found his blood pressure to be 'an entirely normal
130/86.' He was unable to find any 'remedial' cause for
the hypertension which he conceded must have been quite
high in view of the amount of antihypertensive
medication the patient was taking. After mentioning that
'present medical opinion holds that environmental
factors have little to do with the production of high
blood pressure, and [that] genetic factors are much more
important' causationally, he concluded Simpson had
nothing wrong with him that could be causally related to
his work as a fireman." For this reason, the court
concluded logically that the claim of 100% disability
was not allowable. It is also of interest again to note
that genetic factors entered the case, and that whatever
hypertension might exist was due to inheritance and not
to job-related conditions.

A more serious case is illustrated by *New Hampshire*

Supply Company, Inc. v. Steinberg et al. In this case
the defendant's late husband died from a heart attack
suffered while an employee of the plaintiff's. The lower
court had found that his widow was entitled to benefits
under the workmen's compensation act. The basic issue
was whether the stress of the deceased's job was a
causative agent in inducing the heart attack. "This
man's medical background pointed to some of these
[factors in evoking the heart attack]: hypertension,
obesity, cigarette smoking, positive family history of
coronary disease in both parents, indicating genetic
susceptibility. The question as to whether or not long-
term emotional stress and tension play any role in this
disease is still unanswered medically." It should be
noted that in two earlier decisions, the same court had
accepted that a causal connection "can exist between
work-related stress and a heart attack. In both
[earlier cases] the precipitating factor was physical
stress, whereas in Steinberg's case the precipitating
factor was work-related psychological stress and
overexertion. ...There is no valid reason to sustain a
claim based on physical exertion and to deny one based
on mental or emotional stress." There was conflicting
medical testimony; one physician testified that the
heart attack was not work-related, and another that it
was. The court, in answer to the plaintiff found that a
"choice" between these two opposite medical opinions
should be made either at the trial level or by some
legislative act. The court, however, pointed out, "If we
were to leave this decision to the trial court, the fate
of the litigants would then depend upon the identity of
the judge who happens to hear the case. It would be idle
to say that each litigant has a fresh chance to prevail

before a given judge, for the judge could hardly accept antithetical theories in successive cases merely on the basis of the relative persuasiveness or creditability of the particular experts who happen to appear. Thus cases would virtually be decided by the very act of assignment for trial." Again, the citation is from an earlier opinion. "...The claimant has a right to benefits if the deceased's injury met the requirements set forth in the law. Therefore we reject the position that work-connected overexertion and sustained psychological stress can *never* provide the medical causal nexus for compensable myocardial infarctions and a single precipitating event is not a legal requirement. ...The trial court, understandably, did not apply the rule of law we enunciate today: that protracted work-related psychological stress can cause a heart attack which can be compensable under our workman's compensation law. We, therefore, remand the case to the trial court to determine, from the record and in accordance with our opinion, whether Steinberg's heart attack was legally and medically caused by his work-related stress." It seems hard to understand how this will resolve the objections raised earlier about the variability of judicial decisions on a case by case basis. The lower court will now have to weigh all the evidence, including the possibility of genetic predilection to heart attacks, and it would seem that this in itself will yield the results which the earlier citation strives to avoid.

An opposite opinion was delivered in 1981 by the Supreme Court of Minnesota in the case *Lockwood v. Independent School District*. Here the claim for disability was made by a school principal who alleged

that the mental pressures of his job had caused him to
take a medical leave and to consult a psychiatrist.
While he was on leave, an audit of his purchases was
conducted by the school board and led to his indictment
by a grand jury; the criminal charges were subsequently
dismissed. In examining the case the court took note of
the fact that "... there is a genetic predisposition to
the disorder. He was of the opinion that the stress of
Lockwood's job caused his mental disorder. ...Employer-
insurer's psychiatrist testified that Lockwood's problem
was a schizophrenic-type reaction. It was his opinion
that Lockwood's illness was precipitated by the criminal
action in the fall of 1976 and was not work-related."
The court then tried to interpret the intentions of the
legislature when the compensation laws were enacted.
"Unquestionably, disablement resulting from a mental
illness caused by mental stimulus is as real as any
other kind of disablement. Nor do we disagree that there
can be no medically valid distinction made between
physical and nervous injuries. We are unwilling,
however, to construe our statute as affording worker's
compensation-coverage for mental disability caused by
work-related stress without physical trauma because we
are unable to determine that the legislature ever
intended to provide such coverage." Lockwood's claim for
disability was thereby denied. A dissenting opinion
disagreed strongly, citing cases both in Minnesota and
other states, and would have provided compensation for
Lockwood. The dissenter interpreted the Minnesota
statute in exactly the opposite of the majority-namely,
under the pertinent acts Lockwood should be compensated.
Again, the issue of a possible genetic predisposition to
schizophrenia was raised, en passant, but not fully

dealt with. The courts must sooner or later decide the limit of compensable mental injuries, and whether the "stress" caused by the occupational status is sufficient to induce mental breakdowns in individuals who may have a heritable tendency towards such behavior. The further issue whether persons with this genetic tendency should be placed in positions where stress may bring out the latent tendency will also be of legal importance when cases involving genetic screening, to be discussed later, arise.

Another compensation case involving different facts, but with the possibility of genetic predisposition to injury, arose in 1982 in the case *Grable v. Weyerhauser Company*. The case was decided by the Court of Appeals of Oregon. Here a worker had sustained a back injury while lifting heavy blocks of wood. The injury was compensable and the worker received due payment until his return to work some months later. After his return to work, he again injured his back; however this time he sustained his injury while lifting a steel pipe onto his roof at home. The basic legal issue was whether the on-the-job injury was a major contributing cause to the second injury. The laws are contradictory in this instance regarding the off-the-job injury. The genetic issue arose when it was testified by a medical expert that there was a congenital weakness predisposing the worker to disc injuries. There was further testimony by the physician that he felt no doubt that the first injury predisposed the claimant to further injuries; that statement was unchallenged. The court therefore concluded that Grable was entitled to have the compensation case remanded and that the lower court should take into account the factors cited by the higher court. Again, the "predisposition" which can be interpreted only as a genetic cause, is part of a workmen's compensation case.

Among other genetic issues is the case, *Beaty v. Thiokol* before the Court of Appeal of Louisiana, Second Circuit, in 1982. The claim for injury was ruptured abdominal aortic aneurysm which occurred while the claimant was on the job. Despite the defendant's claim that the tendency for this type of injury was genetic, the court awarded compensation to the claimant.

The issue in *Champion v. S&M Trayler Brothers* before the U. S. Court of Appeals, District of Columbia Circuit, was whether there was a genetic cause of persisting asthma and obesity induced by his job as a laborer in the tunnel system for the Washington, D. C. subway. Here the court decided that the claimant's genetic tendency had been sufficiently aggravated by the work conditions as to validate his claims for disability.

The same Court, in *Liberty v. District of Columbia Police & Fireman's Retirement & Relief Board* found that the claimant's genetic suffering from ectasia of coronary arteries was neither caused by nor due to his duties as a patrolman and denied benefits. They cited his familial history of the disease and thereby denied him the benefits which he sought on the basis of the claim that his work conditions aggravated his genetic predisposition to such a disease.

The Court of Appeal, Third District of California, in the case *Davis v. City of Sacramento*, also held that the claimant's cause for disability benefits was not valid as he had a genetic predilection for this mental disease and it was not job-related. There was also a strongly questioned issue whether the mental illness was feigned or real. The Court cited the instance when several sane people feigned mental illness and all were admitted to mental institutions, pointing out that "...of the twelve admissions, eleven were diagnosed as schizophrenic, and one as manic depressive."

Cornelius v. Sunset Golf Course decided by the District Court of Appeal of Florida, First District, dealt with the issue whether plaintiff's accidental exposure to the herbicide Diquat caused his subsequent suffering from schizophrenia. The issue was clouded by various technicalities, primarily that Cornelius did not have proper legal counsel and that he served as his own attorney even after being declared incompetent by the Deputy Commissioner. The defense claim that his injury was inherited and not induced by his exposure was not accepted in this case which was remanded for retrial with an appropriate attorney to serve for Cornelius.

New cases involving genetics and compensation arise each year. Several 1984 decisions will be used to demonstrate both sides of the issue of the relevance of genetic predilection and injury while on the job. The first, *Dean v. Cone Mills Corporation*, before the Court of Appeals of North Carolina, dealt with exposure of a worker to cotton dust and his subsequent respiratory difficulties which led him to choose early retirement. There was conflicting testimony concerning the hazards of his job, with physicians testifying either that the dust did, or did not, contribute to his 35% pulmonary impairment and total disability from lung and heart disease. The fact that the plaintiff had been a heavy smoker did not help his case; the court decided, despite the fact that the exposure to cotton dust may have contributed to a possible genetic predilection to such disease, that he had not adequately demonstrated the claimed dangers of his exposure and compensation was denied.

On the other hand, in the case *IML Freight v. Industrial Commissioner of the State of Colorado*, before the Colorado Court of Appeals, Division II, in 1984, permanent disability benefits were awarded to a truck driver who suffered a stroke while working. There was much argument whether the event was an "accident" or an "occupational disease," but the court cut its way through this legal tangle by deciding that this was not the true issue. What was at issue was whether the stress of being an over-the-road truck driver contributed to a genetic predilection for having such a stroke, and the court decided that this was indeed the case. As a result the driver was awarded full compensation under the Workmen's Compensation Act.

Further examples are found in 1985. Two additional cases will serve to illustrate the problems faced by the courts with genetic issues. First, in *Sandusky v. W.C.A.B. (Chicago Bridge and Iron)*, the question arose whether a welder, exposed to work-related gases could recover on the claim that this "aggravated the claimant's preexisting lung disease to the point of total disability." The disease claimed, alpha-1 anti-trypsin deficiency, is known to be of genetic origin, and the defense here claimed that there was no proof that his occupation had in any way caused his disability. The Commonwealth Court of Pennsylvania decided, however, that there was sufficient causal relationship and remanded the case to the Workmen's Compensation Appeal Board for rehearing. The court based

its opinion in part on a 1984 case, *Pawlosky v.
Workmen's Compensation Appeal Board*, in which the
claimant's exposure to caustic substances had aggravated
his genetic susceptibility to asthma.

Christensen v. Saif Corporation dealt with the
claim that ulcerative colitis was a stress-induced
result of being president of an automobile insurance
company during a time of difficult litigations. The
opinion is replete with medical testimony, on both sides
of the issue, some claiming this disease to be induced
by stress in those prone to suffer from it and others
claiming that there was no relationship between
ulcerative colitis and stress. The latter apparently
were more convincing as the court found for the
defendant and denied Christensen's claim for compensation.

A case in which women's rights and workmen's
compensation are both involved was found in *Matter of
Claim of Sanna Pond v. Workmen's Compensation Board*,
again in 1985. In this instance the claimant who was
employed as a secretary for a farm implement business
became pregnant and was advised by her doctor to stop
work due to the presence of paint fumes which it was
felt might harm the foetus. The employer ceased all
painting operations when she informed him of this
possibility, but the smell of paint could still be
detected in her work area; as a result she left the job.
After the delivery of a normal child, she brought claim
for compensation for the time lost during the pregnancy
and her subsequent recovery from a Caesarean section.

It was apparent that there was no other position
which the claimant could have filled without still being
subject to the paint fumes. This, plus the advice of the
physician to leave her position was held by the courts

to be sufficient cause for her to be awarded compensation for the time she was forced to be away from her position.

One is faced with conflicting decisions concerning what may be contributory to the aggravation of genetic tendency for a particular disease. If it can be proven that the claimant would have suffered the disease *regardless* of the occupation involved, then obviously no compensation will be awarded; if the work conditions are strongly contributory to the inherited tendency towards a physical condition, then compensation may be granted. An example of this is found in the earlier case, *Adams v. New Orleans Public Service*, before the Supreme Court of Louisiana in 1982. The issue here was whether a recently hired worker was entitled to Workmen's Compensation when he suffered an attack of angina. The plaintiff entered a hospital where tests indicated that as of the time he went to the hospital he was found to have normal responses to all tests for heart disease. Testimony was brought forward to show that the patient suffered from arteriosclerosis, a genetic condition which lead to his angina attack. Adams requested upon his return to work that he be given light duties so as to prevent a recurrence of the angina. The company refused, citing its policy of not giving such a transfer to an employee still considered to be temporary after three months with the company, whereupon Adams sought compensation.

The court was divided in its decision. The majority upheld the claim for compensation, while the dissenters found that the condition was not work-related, but was due to his previous genetic condition which made him susceptible to angina attacks. The majority held that as the event occurred while he was on the job, it made no difference whether a previous condition was the cause of the accident. The four dissenters felt that the angina attack was not in any way job-related and he should not receive compensation because of it.

An example of how rapidly the understanding of human genetic diseases develops can be found by consulting Victor A. McKusick, *Mendelian Inheritance in 6ManS* (6th ed. Johns Hopkins Press, 1983). This encyclopedic publication which appears at 3-4 year intervals, lists all of the known mutations in man, both proven and suspect. The total of the known or suspected mutations has risen between 1978 and 1983 from 2,811 to 3,386, and many of them are the causes of genetic diseases such as those given in these cases dealing with compensation. Perhaps both lawyers and judges will need the latest edition of *McKusick* on their reference shelves as often as such standard legal texts as *Prosser on Torts* or *McCormick on Evidence*. Yet, each case must be decided upon its own merits, and the judges must be fully educated as to the meaning of genetic tendencies for individual diseases.

INSURANCE CLAIMS AND GENETIC DISEASE

In *Burke v. Occidental Life Insurance Company of California*, the issue was whether a known genetic defect, Kleinfelter's Syndrome, was a mental or a physical disease. This was an insurance, rather than a compensation case. The plaintiff was shown to suffer from the disease, which is caused by an abnormal number of sex chromosomes, namely XXY. Men with this chromosomal abnormality are sterile, and frequently suffer from reduced mental capacity. The issue here was

whether this was a physical or mental disease under the
terms of the policy holder's insurance clause; Burke
claimed that he suffered from a physical disease, namely
the presence of an extra X chromosome and should not
have been denied the higher benefits from physical,
rather than mental disease. Unfortunately for Burke, his
attorney did not introduce the evidence of Kleinfelter's
disease in the original trial, and the Louisiana Court
of Appeals had no choice but to deny the introduction of
the evidence upon appeal. Therefore the issue whether
this was a physical or a mental disease was not decided
in this case.

An involved case seeking payment of a health claim
under a major medical expense policy arose in
Connecticut General Life Insurance v. Shelton. A woman
had undergone tubal ligation during her first marriage
in order to avoid having further children after several
had been born in ill health due to Rh incompatibility.
She was advised then that any further children would
die and underwent sterilization so as to avoid any
possibility of such an event. Subsequently, she married
Shelton, and it was determined that her second husband
was Rh compatible, and were it possible for them to have
children there would be no risk involved due to this
factor. She also learned that in rare cases it might be
possible to undergo surgery to have the Fallopian tubes
reunited, making it possible for her to have children by
her second husband. She underwent the surgery and it was
successful; she and her husband now had a "reasonable
expectation" that a healthy child might be born.

She then entered a claim against the insurance
company for recompense for the operation and surrounding
medical costs. The lower court in Texas affirmed the

claim, and the insurance company appealed to the Court
of Civil Appeals of Texas, Fort Worth. In 1974 this
court decided in favor of the Connecticut General Life
Insurance Company, setting aside her claim on the basis
that the surgery was not covered as there was neither
injury (defined in the policy as "all injuries received
by an individual in any one accident") nor sickness
(defined as "... physical sickness, mental illness,
functional nervous disorder and covered pregnancy"). The
court here reviewed the history of insurance and found
that "...the beginning theory was that risks or the
chance of loss, to be insurable, must be pure risks and
should not be based on moral hazard. In addition to be
insurable the risk must be measurable in quantitative
terms for which purpose the law of large numbers and the
theories of probability and chance are employed. The law
of large numbers is based on the observation that the
larger the number of instances taken, the closer the
result approaches the theoretical probability..." The
claim was therefore denied on the basis that the
operation did not fall under the protection of the
policy. A strong dissent was entered in which the judge
gave an excellent review of the problems with Rh
incompatibility. He dealt with this issue of chance
equally strongly. As so often a dissenting opinion in
one case becomes a majority opinion sometime later, it
is worth citing from this dissent. "It would be far more
accurate to describe Mrs. Shelton's situation when she
'chose' to have a tubal ligation sixteen years ago as
one in which she was *forced* to *trade* one illness for
another. ...Mrs. Shelton did not choose to be sick; she
merely chose from limited possibilities what form that
sickness would take. It was chance that made her sick,

chance that created the Rh negative and chance that
exposed her to the perilous environment of conceiving
children with a male who was Rh positive.

"To argue that surgery to avoid the complications
arising from her Rh negative factor was elective is
preposterous. To argue further that the inability to
produce healthy, live born children does not somehow
demonstrate a serious impairment in a female body
borders on the far reaches of common sense. A primary
ability of a healthy, normally functioning female is the
ability to successfully bear children. The fact that Mae
Shelton, as previously married and without a tubal
ligation, would have birthed still-born children or
miscarried indicates a deficiency in an otherwise normal
female.

"...Are we to presume that ability to give birth to
healthy, alive children is not the function of a healthy
female? Somehow, because there temporarily and frailly
exists a child to accept, unwillingly, the onus of this
failure, we are expected to conclude that such a
condition affects only the health of the child. As these
frail souls depart into the next world we are expected
to sit here content in the knowledge that the mother is
a healthy and unimpaired woman. In fact it is her body
that is possessed of the disease. The miscarried child
exists as sad testament to this disease. Only by curing
the mother can the tragedy to the child be avoided. Mae
Shelton did not choose to become sterile, not in the
true sense of the word. It was a decision forced upon
her as she sought the lesser of two evils. Such a
Hobson's choice is a choice in name only. And now she
seeks to have removed a deficiency the yoke of which she
had no choice but to struggle under." In other words,

the inability to have children *is* a disease, and should
come under the insurance policy.

Another interesting point is raised by the majority
opinion in this case, namely that of the use of
probability. If one recalls the difficulty in paternity
cases in accepting the HLA antigen tests based on
probability, it seems strange that an insurance company
in 1974 can be held not liable because a "disease" must
come under probability considerations while
establishment of paternity could not admit probability
evidence tending to establish paternity. It should be
remembered, however, that at that time the HLA system
was not developed, and therefore the use of probability
only for exclusion did not have to deal with the high
index of paternity found in cases after the legal
admission of the HLA tests. The case cited here,
however, shows that courts did, even in 1974, deal with
the admissibility of probability evidence. Of course the
need to prove that a disease is genetic is obvious and
again *McKusick* may become a standard legal as well as a
medical reference.

LIBEL CASES

It would seem likely that genetics would not play
any part in libel cases; as can be seen this is not
true. A case was recently tried in Georgia involving an
alleged libel of Dr. Shockley, a Nobel Prize winning
physicist. Dr. Shockley has put forth a theory holding
that most blacks are genetically inferior in
intelligence and suggests that there be a voluntary paid
sterilization program for such persons. This suit was
against the Cox Enterprises and specifically against
their newspaper, the Atlanta Constitution, for comparing

his views to those of Hitler and particularly a
paragraph which, as quoted by the New York Times,
September 6, 1984, states "The Shockley program was
tried out in Germany during World War II, when
scientists under the direction of the government
experimented on Jews and defectives in an effort to
study genetic developments." The Times further quotes
the alleged libelous article as stating, "Mr. Shockley
was a man with an idea that there are too many black
people around, and he is asking them to eliminate
themselves. The logic is absurd. Compliance is hardly
likely." Dr. Shockley planned to call as one of his
witnesses a Dr. Jensen who is a psychologist and whose
views concerning the lesser intelligence of blacks have
been widely published and equally criticized by many
human and evolutionary geneticists. He also expected to
have an expert witness to testify on libel laws. The
defense called upon a noted anthropologist and a human
geneticist. A six-member jury was selected, including
one black; the plaintiff exhausted his preemptory
challenges to eliminate several other blacks from the
panel. The main argument was over the accusation of
libel, but it is clear that the case must deal with the
accuracy of the genetic claims made by Shockley. In
view of the nature of the suit, libel, little further
comment will be made considering how the genetic
community views Shockley's ideas. Perhaps it is not
libelous to point out that his witness, Jensen, based
many of his original data upon research done by the
English psychologist, Sir Cyril Burt. Later examination
showed that many of that authority's data were not
valid, and there has been widespread discussion over
whether Burt "made up" the existence of his co-authors

and whether in the capacity of editor of the journal in which his articles appeared he passed upon his own work. (Jensen has since modified his views to eliminate the data by Burt.)

The jury decided that the defendants were guilty of libel and fined them the magnificent sum of one dollar! To paraphrase one of the defendants, "not bad, fifty-cents apiece!" However, both sides have appealed the decision, and it will be interesting to see how an appellate court handles this as most courts are loathe to interfere with jury decisions unless a real outrage has been perpetrated.

In an earlier case in 1978, also dealing with genetic issues, the J.P. Lippincott Company published the book *In His Image: The Cloning of a Man*. The book written by David M. Rorvik purported to be the true story of the genetic cloning of an egocentric millionaire and was published as a non-fiction work. An immediate furor arose and in the same year a Philadelphia court declared it to be "a fraud and a hoax." Subsequently, Dr. Derek Bromhall, one of the scientists mentioned in the book, brought suit against the author for misuse of his data, and for damages to his reputation on the basis of two charges. First he claimed that the author had invaded his privacy by "quoting his cloning techniques" and mentioning his name in a footnote as a "personal communication" without Bromhall's permission. Second, Dr. Bromhall claimed that Rorvik had fraudulently "acquired a copy of an abstract of Dr. Bromhall's doctoral thesis by representing himself as a scientific researcher preparing a project on mammalian cloning." The original suit for defamation of character asked for $7,000,000. Later the plaintiff's

claims were restricted only to invasion of privacy. What
might have been a most interesting, and certainly
controversial case dealing with genetics in the courts
was, however, avoided. There had been plans to have many
scientists testify, and the weight of their evidence
would have had to be determined by a judge or a jury.
Prior to trial in 1982, the case was settled out of
court by payment to Bromhall of $100,000 and a letter of
apology to him. Almost all geneticists felt that the
book was an interesting work of fiction, and as
Rorvik refused to name the specific persons involved
with the purported cloning, presumed, in agreement with
the court, the book was to be a fictional work. These
cases are presented here to point to still another part
of the law where genetics is involved, unexpectedlyG:;
i.e. libel law.

As an aside, the cloning of frogs is well known,
using a technique of separation of the products of early
cell divisions. Some work has been done with mammals,
but there is no evidence to date that this technique is
as yet possible with humans. However, the likelihood of
such a technique being used on humans in the future is
pointed out in the earlier mention of the cloning of a
cow. It should be noted that in that instance the
cloning was also done by separating cells from the first
cell division of a fertilized egg. The sole claim for
human cloning is the work by Rorvik, a non-scientist,
and in this case the source of the clone was supposedly
not from a developing egg, but from adult body cells.
The courts have already spoken on their disbelief of
this in the opinion cited earlier.

The question is posed, however, by this case as to
the legal issues which might surround cloning should it

be possible. Were it possible to clone a human (and
again, it should be stressed that there is no scientific
evidence this has been done) the legal status of the
cloned offspring who is genetically identical to the
source of the clone would be difficult to determine. In
the sense that the genetic identity would make the clone
an identical twin, problems of inheritance, self-
identity, etc. would arise. Until it is proven that this
can be done in humans, there is no way to interpret how
the courts will deal with such an issue.

ZONING LAWS

Although one would scarcely expect that cases
involving zoning ordinances have a genetic element in
them, nevertheless they do appear. The basis of these
cases is usually quite specific, namely does the term
"family" refer only to a genetically related group of
persons, or does it have broader legal meanings? This
difficulty will be illustrated by the cases which
follow. *Freeport v. Help of Children*, before the
Supreme Court, Special Term, Nassau County, in New York
State in 1977 is one of the earlier cases which dealt
with zoning. *The Association for the Help of Retarded
Children* purchased a home with the intent of making it
into a family-like residence for eight mentally retarded
young women, with "house parents," a married young
couple living with and supervising the young women. The
Town of Freeport claimed that this was in violation of
their zoning ordinance restricting the area to single
family dwellings. The ordinance in question defined a
family as "one or more persons related by blood,
marriage or adoption, living and cooking * * * as a
single housekeeping unit." The Supreme Court held this

to be unduly restrictive and cited an earlier case that
"zoning is intended to control types of housing and
living and not the genetic or intimate internal family
relations of human beings... The community residence
concept as envisioned by the Department of Mental
Hygiene, and as here attempted by the defendant, AHRC,
clearly intends the creation of a family unit. It is not
an 'institution' but rather is designed to replace the
usual institutional setting. Such a 'community
residence' bears the generic character of a family unit
as a relatively stable household and is consonant with
the life style intended for a family-oriented
neighborhood, and thus conforms to the purpose of the
village zoning ordinance." In these days of modern
mental health care this decision may become critical.
Most neighborhoods fiercely oppose the establishment of
such dwelling units on the basis that their
establishment may lower property values and they often
go to court to prohibit them. In light of this decision,
there may be increasing difficulty in maintaining that
such units can be excluded from single family
neighborhoods by zoning restrictions.

This issue of what constitutes a legal "family" in
terms of zoning ordinances continues to come before the
courts. In addition to cases in New York, many other
states now have court decisions dealing with this; only
two recent ones will be mentioned. The Missouri Court of
Appeals, Eastern District, Division Two, dealt with the
case, *City of Vinita Park v. Girls Shelter Care, Inc.* in
early 1984. Here a proposed shelter for girls to be
operated by the organization under the auspices of the
St. Louis County Juvenile Court and leased for that body
by the home, with eight unrelated young women and three

unrelated adult house-parents, was proposed in an area
zoned for single family occupancy only. The plaintiff
appealed to the court on the basis of the violation of
the definition in the zoning ordinances which stated,
"FAMILY: One or more persons related by blood or
marriage occupying a premises and living as a single
housekeeping unit." The court dismissed the action
against the home in part by stating: "We recognize that
the occupancy by eleven unrelated persons under the
group home principle does not conform to the letter of
either of the ordinances which defines family, but we do
believe that it conforms to the spirit of the
ordinances. ... It was concluded in a recent commentary
that most group homes, including the one in the instant
case, are family style, community based and function as
a single housekeeping unit, sharing responsibilities,
means and recreational facilities. In order for the
'family' to be as normal as possible it is essential
that the home be integrated into a residential district,
for that is the very type of atmosphere which it seeks
to emulate." The court determined that it was in the
public interest for such homes to be established, and
after citation of other similar cases, decided that the
home could be placed in this area. Clearly the *genetic*
issue of what is a "family" is not a consideration here,
but rather the *social* meaning of the word "family" takes
precedence.

Two months later in 1984, an almost identical issue
arose in the case *Knudtson v. Trainor*, in the Supreme
Court of Nebraska. The main difference here was that the
dwelling in question was to be used to house a group
home for five mentally retarded persons in an area zoned
for single family dwellings. The Nebraska court surveyed

similar cases in Michigan, North Carolina, Iowa,
Minnesota, New Jersey, and Wisconsin where those state
courts all had decided that the definition of single
family housing was broad enough to permit homes of this
type. They quote from each of the courts; a citation
from the Michigan Court may be most appropriate here.
"The residents are more than a group of unrelated
individuals sharing a common roof. They do not have
natural families on which to rely, and due to their
unique circumstances, it is unlikely that these women
will ever rejoin their parents or marry and form
independent families. The substitute family provided by
the group home allows the residents to lead more normal
and meaningful lives within the community than would
be feasible were they institutionalized." (The quote is
from *Malcolm v. Shamie*, 290 N.W.2d. 101 (1980.) It would
now appear, with many different states agreeing that
various homes of this type are not in violation of
zoning ordinances, that the issue may be settled, but it
is not totally unlikely that similar cases will arise in
states where such issues have not as yet been raised.
Judging by the unanimity of the courts which have so far
ruled on this type of case, there should be no legal
problem in establishing these homes in areas zoned for
single family occupancy provided other provisions of
safety and order are maintained.

There is also a somewhat amusing possible abuse of
such a defense concerning the meaning of family. At my
own college some years ago a fraternity occupied a house
in a wealthy, single-family district much to the
distaste of some of the neighbors. While the fraternity
was seriously looking for a more suitable dwelling, the
members tried, semi-seriously to maintain that, as

they were all "brothers" they constituted a single
family. The claim was never seriously pursued in the
courts, however. Indeed, the motto of Union College is,
"We are all brothers under the laws of Minerva." (There
is some difficulty with the motto since the college
subsequently became coeducational, but this is beside
the point.) If the motto were literally interpreted, all
members of the college could have lived together in a
single large dwelling and claimed to be a "family"!

SOCIAL SECURITY BENEFITS FOR ILLEGITIMATE CHILDREN

Whether insurance benefits under the Social
Security Act can accrue to children born out of wedlock
is the basis for two 1983 decisions. The first,
Donaldson v. Secretary of Health and Human Services, was
heard by the U.S. District Court, W.D. New York, in
1983. The facts were obvious: although a wage earner
had not legally established paternity before his death
there was no question of his relationship with the
mother. The man was killed in an automobile accident
some 8 1/2 months prior to the birth of the child, and
obviously he could not establish paternity before the
birth of the child. No one, including the government,
disputed the paternity, however. The usual regulations
governing insurance for children apply only if a court
of competent jurisdiction enters a decree of paternity
during the lifetime of the father. *After* the man's
death, a Family Court entered an order of filiation
determining that the decedent was the father. However,
another section of the law specifically states, "(c)
Your mother has not married the insured but the insured
is your father and he has either acknowledged in writing
that you are his child, been decreed by a court to be

your father, or been ordered by a court to contribute to your support because you are his child." It is important to note that this section does not state that the court must decree filiation *before* the death of the father, and on that basis the Secretary of Health and Human Services was ordered to award child benefits to the infant.

The second case, *Charles v. Schweiker*, also in 1983, was before the U.S. District Court, E.D. New York. The facts were much the same, the father died before the birth of the infant, and an order of filiation had also been entered from a Family Court. Nonetheless, the Secretary once more sought to deny insurance benefits. Here, however, the issue was the content of the New York State laws, and those laws state that the order of filiation must be made during the lifetime of the father, an impossibility as he had died suddenly at the age of 31, approximately two months before the child was born. Thus the order of filiation was obviously entered after his death and after the birth of the child. However, the New York law deals specifically with cases concerning inheritance by illegitimates of intestate fathers if "paternity has been established by clear and convincing evidence and the father of the child openly and notoriously acknowledged the child as his own." Both conditions were met, first by the filiation order, and second by the fact that the father had repeatedly told the woman, his mother and his sister that the woman was carrying his child, thereby leading this court to decree that the Secretary must make insurance payments to the child. In these two cases, at least, it is apparent that in the U.S. District Courts in New York east meets west with no difficulty.

However, there are, possible problems. What would be the case if the male had denied paternity before his death? Obviously no blood tests could be obtained, and the possibility of a woman naming a decedent as father in order to receive insurance benefits is certainly a real one. It might well be that a pregnant woman, who believed she might make a case, would try to get a filiation order which would name a dead man as the father, whether he was or not, simply to obtain insurance money from Social Security for her child. Certainly not all cases will be as clear-cut as the two presented, and the problem of illegitimate children receiving insurance from Social Security when the alleged father died during the time of a woman's pregnancy is not going to be an easy one to solve.

CASES INVOLVING DNA

Further examples of technical genetics involved in court cases can be made by citing a few opinions briefly, so that one can see the need for some technical background in genetics.

In 1978 the opinion by the United States District Court for the District of Columbia contains the following statement (the case *Mack v. Califano* dealt with an attempt to enjoin the government from carrying out certain genetic experiments): "Plaintiff seeks a preliminary injunction to prevent an experiment testing the biological properties of polyoma DNA (deoxyribonucleic acid) cloned in bacterial cells. The experiment is to be conducted in building 550, Frederick Cancer Research Center at Fort Detrick, Maryland." The plaintiff was obviously trying to block recombinant DNA work on the then felt danger of such experiments. He

lost the case, but the court had to investigate the
dangers of such genetic experimentation.

In *United States v. The Progressive* (the famous
case involving publication of data concerning the
manufacturing of an H-Bomb) part of the recorded opinion
reads, "The group notes that it is possible for example
that a technology such as recombinant DNA could someday
surface [as a] means of destruction that ought not be
published..." (United States District Court, Western
Division, Wisconsin, 1979).

A third opinion of the same ilk is found in *Phillips
v. United States* before the United States District
Court, South Carolina, Charleston Division, in 1980. A
note reads, "For example, one could hypothesize a
technological break-through in genetic engineering,
focusing perhaps on the transduction or transformation
of chromosome material through recombinant DNA (gene-
splicing) techniques, controlled mutagenesis, or
microsurgery, or in euphrenics, which would allow a
particular genetic effect to be treated in *utero* during
the early stages of pregnancy." The court is quoting
from a standard textbook of genetics, Strickberger,
Genetics, page 822. The text is used mainly in two
semester genetics courses in many colleges.

In the case *Industrial Union Department, AFL-CIO
v. American Petroleum Institute*, heard before the
Supreme Court in 1980, the following appears in the
opinion: "The so-called 'one hit' theory is based on
laboratory studies indicating that one molecule of a
carcinogen may react in the test tube with one molecule
of DNA to produce a mutation. The theory is that, if
this occurred in the human body, the mutated molecule
could replicate over a period of years and eventually

develop into a cancerous tumor." The case here involved OSHA regulations dealing with the toxicity of benzene and its possible effect upon workers in the manufacture or use of this compound.

Also in 1980, the opinion in the case *State of Washington v. Smith* found the Supreme Court of Washington stating in part, "...informed the trial judge that there are other frequently recognized and often debated effects of marijuana. These include effects on: (1) chromosomes, (2) the endocrine system, (3) testosterone (a hormone), and (4) the formation of deoxyribonucleic acid (DNA) (the substance of which genes are composed). There is evidence that marijuana may cause deleterious effects not caused by alcohol. Thymidine is a chemical which is the building block of DNA..." Here the court had to deal, as noted, with conflicting claims of the effect of marijuana, but in order to do so had to understand the molecular basis of DNA and of possible mutations.

Another instance of the courts dealing directly with the understanding of genetics involving effects on DNA can be found in the 1981 case *Marshall Minerals v. Food and Drug Administration*. The opinion is by the United States Court of Appeals, Fifth Circuit. It dealt with regulations concerning the use of the dye, gentian violet, in chicken feed; the plaintiff sought to use this as a fungicide and the federal regulations forbade this. The court stated in part, "FDA stated in the 'safety' section of its order that gentian violet is also known to produce various chromosomal anomalies in cultured mammalian cells, 'that it is capable of inducing damage to DNA... the ability to produce chromosome anomalies and induce DNA repair are two phenomena inducible by gentian violet treatment for which positive correlations with carcinogenicity have also been claimed.' FDA acknowledged, however, that 'gentian violet does not produce apparent genetic damage to chick embryos.' ...Other studies cited by FDA have implicated gentian violet with genetic damage to organisms other than chickens, although at higher dosage rates than those recommended for Marshall's premix."

Indeed, the legal problems which DNA work may lead to are numerous. They are well summed up by a review in the journal *Gene* in 1981 (15 Gene 1). The author of this brief review is himself a lawyer and he deals with the commercial aspects and legal consequences of applications of work with DNA.

The validity of patents by Stanford University for recombinant DNA work which have been extremely lucrative are being challenged for being insufficient and this case will probably soon reach federal courts and the judges will be forced to study and understand the highly technical aspects of this genetic process. There can be no doubt that genetics will have to be understood by the judiciary, lawyers, lay juries and all others who may take part in future cases. Molecular genetics has indeed "gone public."

GENETIC CAUSES OF INJURIES

Courts continue to be forced to examine human genetics in order to reach decisions. For example, it is sometimes imperative to determine whether a given event has occurred because of an accident or because of a genetically caused condition. For example, in 1982, *Juge v. Cunningham* was before the Court of Appeals of

Louisiana, First Circuit. The plaintiff here claimed that an injury sustained in an automobile accident caused his unborn son to suffer sufficient harm so as to be born with tuberous sclerosis, a disease causing permanent mental retardation. The lower court had awarded damages to the mother for physical injuries and mental anguish "associated with receiving an injury while pregnant." Mr. Juge had been awarded special damages also, but the court had denied damages to the defective child on the basis that the automobile accident did not cause his condition, but that it was a genetic defect. Of particular interest was the fact that there was no other member of either parents' families showing the defect. The defense expert testified that 80% of all cases of this disease are similar and that it is due to a new mutation affecting only the child. The Appellate Court accepted this genetic fact and agreed with the lower court that no damages should be paid for the infant's defect as it could not be related to the accident.

Another example of the need for courts to understand genetics is shown by the case *Mose v. Brewer.* Mose sued for damages incurred in an automobile collision and claimed that the accident was the cause of the onset of his diabetes, to which he was genetically susceptible. The Court of Appeal of Louisiana, Third Circuit, in 1983 dealt with this issue. It awarded him full damages, stating in part, "There is no question that Mr. Mose, who was 60 years of age at the time of trial, now suffers from the condition of diabetes mellitus. There is likewise no question that Mr. Mose did not suffer from this condition until shortly after the accident. The medical testimony concerning the diabetes consisted of the testimony of Drs. Vidrine, Seabery and Prosser. All three physicians agreed that trauma is not a direct cause of diabetes, but that trauma and other factors can aggravate or precipitate the onset of diabetes in a genetically predisposed person. They further agreed that Mr. Mose had a genetic predisposition to develop diabetes." Although one of the three physicians was of the opinion that the accident did not induce sufficient trauma to cause the condition, the court found that the testimony of the three regarding the possibility of this having occurred was sufficient to uphold the lower court's verdict in favor of the plaintiff. The complete testimony is not available from the opinion; it would be of interest to a geneticist to know how the establishment of the genetic predisposition was made; it would seem that evidence from parental health records, as well as evidence of diabetes in siblings or offspring, would be needed to establish this genetic cause. But the main point to be stressed is the need for judges to be able to evaluate genetic evidence and to allow its proper weight in court.

EDUCATION OF HANDICAPPED CHILDREN

The right of handicapped children to an appropriate education is now becoming the subject of law suits. This will probably become more prevalent as more parents attempt to secure adequate training for their learning-disabled children, many of whom suffer from an inherited condition. Two cases illustrating this type of suit arose in 1982. The first, *Colin v. Schmidt*, before the U. S. District Court, D. Rhode Island, concerned the attempt by a school committee to "mainstream" two children into a classroom where the pupil/teacher ratio might be as high as 10:1, rather than to pay the costs of a private residential school. The Court agreed with the plaintiff parents, finding that the learning disability and psychological condition of the children would be exacerbated in the public facility and that it was to the best interests of the children that they be placed in the private school. The childrens' right to a free education therefore required the requested placement and the payment by the school board of the much higher costs.

In *The Matter of Board of Education v. Ambach*, also in 1982, an Appellate Judge in New York dealt with the rights of learning handicapped children to a high school diploma. Ambach, the New York State Commissioner of Education, had upheld a school board's refusal to grant high school diplomas to two children who failed to pass the required competency tests necessary for the degree. The *guardian ad litem* for the children brought suit to compel the granting of their diplomas. The court, however, upheld the rights of schools to maintain standards for the degree, and found no reason to interfere. They found no discrimination against the handicapped as the children had been given opportunity equal to non-handicapped children, and they were allowed a full right to education, equal to that of non-handicapped children.

These two cases indicate the difficulties courts will have determining how to deal with the education of learning- disabled children. It would appear that these children have full rights to an education to whatever level they may achieve, but cannot expect to be granted a high school degree if they do not meet the minimal standards for such a degree. Indeed, the State of New York is now considering granting a special "diploma" to such persons which, while not the equivalent of regular diploma, would attest to a certain level of attainment and certainly to the attempt by the handicapped to do as well as possible.

Cothern v. Mallory, decided by the U.S. District Court, W.D. Missouri, C.D., in 1983, dealt with the proper educational placement of a Downs Syndrome child. The issue of due process was also involved. The State sought to place the child in a special school for the handicapped which the parents felt would not allow the maximum development of their genetically defective child, while the State claimed that this placement was in the best interests of the child. The child had originally been placed in a different school, but was withdrawn and placed, at his parents' expense, in a private school where he remained; the parents sought funding from the State for this private education. The court felt that the placement by the State in the special school met all federal standards for education of the disadvantaged, and that the procedures during the detailed hearing granted to the parents by the state in no way violated due process. The decision by the state to place the child in the special school was upheld. Of course this does not mean that parents may not continue the child in the private school at their expense, but simply that the state has no obligation to pay for a private education when, in the opinion of the court, alternate and equally capable public educational facilities are available. It would appear that the courts are trying, as best they can, to balance the public interest in terms of financial expenditures for the handicapped children against the best interests of these children; when they can find no conflict, as in the above case, they will opt for the less expensive, but equally adequate and educational system.

INTERNATIONAL LAW

Genetic issues have arisen in cases dealing with international law. This can be illustrated by two court opinions, one early and one more recent. The first, *Petition of Risdal & Anderson. Inc*, came before the U. S. District Court, D. Massachusetts, in 1968. Here the issues were complex. A fishing boat manned by a Norwegian crew, but owned by an American company, sank with all hands aboard. Claims were made that the vessel was unseaworthy and also that its loss was due to the negligence of the captain. The question then arose as to whether the suit brought by the survivors should be tried in the United States or Norway. The genetic issue which arose was whether illegitimate children of one of the Norwegian crew members could claim insurance benefits under the "Death on the High Seas" act and other U.S. statutes. Obviously, as the father was drowned, no tests of paternity could be made.

The U. S. Court held that the suit was properly brought in this country as the ship's owners were U. S. citizens. The claim by the illegitimate Norwegian children was also allowed on the basis that, while the children had never been legitimized, they were fully recognized by both the decedent and his spouse, and all those who knew them, as his children. There was no claim here by the U. S. owners that the illegitimate children were not sired by the late Norwegian, although it would have been impossible to decide such a claim on any genetic evidence. Certainly, in several states in the

U.S. the failure to legitimize the children might have been taken as evidence that they could not claim benefits.

The second case, *In re Richardson-Merrell, Inc.* is more difficult. Although technically this is not as yet a genetic issue, it will serve as a precedent for cases which will certainly arise in the near future. The dilemma facing the United States District Court, S. D. Ohio, in 1982 dealt with whether a case involving birth defects supposedly induced by the drug Bendectin should be tried in the United States where the drug was manufactured, or in the United Kingdom where the user of the drug resided. Here the Court found the doctrine of *forum non conveniens* to apply. It found that the United Kingdom had adequate laws covering this type of case, that the needed evidence could easily be obtained there and that the substantive British tort laws would govern such a case; therefore the case should be tried in the United Kingdom. The Court also reviewed other cases, not essentially of this type, and found that if there were no adequate laws in the foreign country where the plaintiffs resided the United States should take jurisdiction. But here as the plaintiffs could seek relief under foreign laws the case should be transferred and brought there.

While this is not strictly a genetic case, if the drug used had been one shown to induce genetic effects such as chromosomal abnormalities, the issue would arise as to which country, that of the manufacturer of the drug or that of the persons suffering from the effects of the drug, would have jurisdiction. In this day of increasing litigation over birth defects, it is almost certain that the doctrine of *forum non conveniens* will have to deal with genetic facts.

Additional causes for suits involving genetic defects caused by the use of various drugs manufactured in the U. S. and employed by citizens of other countries will undoubtedly arise. There are strict rules concerning the introduction of new drugs into the U. S. markets, but these rules do not apply to drugs made here and sold abroad. It is quite possible that some such drugs, perhaps not sufficiently tested, might prove to be teratogenic, or to cause chromosomal defects and thereby render the manufacturer subject to suits brought by non-citizens. The doctrine of *forum non conveniens* will have to be examined in each case and it will also have to be determined which country has jurisdiction over such suits.

The case previously discussed in the chapter on radiation genetics, *Juda v. United States*, dealing with the rights of the inhabitants of Bikini, might also have been considered as a case in international law. As this case only dealt with the United States and its protectorate, and the real issue was the effect of radiation on the natives of that island, it was dealt with there.

GENETIC SCREENING

Another major genetic issue involving both law and ethics is the right of employers to use genetic screening as a condition of employment. There are as yet no-clear cut cases dealing with this issue, but they are certain to arise. Perhaps an older case may serve as a model for what will be expected if these cases do come to court. *Taylor Diving & Salvaging Company v. U.S. Department of Labor*, the United States Court of Appeals,

Fifth Circuit, in 1979, concerned the use of physical examinations of persons to be hired as divers and the medical examinations required for such personnel. The genetic issue is somewhat masked by a host of regulations. There is a requirement of several sequential physical examinations, and among these is the determination whether an applicant has sickle cell anemia. While the wording of the OSHA regulations concerning this occupation are not readily clear, it appears that this genetic trait would be sufficient to bar employment.

The basic issue is much more complicated, however. Some industrial firms are examining the genetic background of employees, and restricting employment in various work areas if the prospective employees possess certain genetic traits. As reported by the press, "59 Top U.S. Companies" plan to use genetic screening as a condition for employment, on the basis that "it can determine the genetic predisposition to serious illness that might be set in motion by materials used in a variety of workplaces." The legality of such actions is certain to be challenged. There are many reasons to question the procedure. As indicated earlier, this may be an easy way to exclude blacks due to sickle cell disease or as carriers of that disease. But it also would appear likely to have other serious consequences. For example, it might be possible to exclude women on the basis that exposure to compounds which might be either mutagenic or might cause miscarriages of pregnant women, thereby denying women of equal rights of employment. At present American Cyanamid, as reported by the Civil Liberties Journal, is banning women of childbearing capacity from production jobs which could endanger the health of their future children by exposure to lead. Only women who could prove they are sterile or beyond childbearing age could keep jobs in this area. It is claimed that at least five such employees underwent sterilization in order to keep their jobs. A suit is now being prepared as a class action suit on the basis that such regulations are simply a device to keep women from high-paying, traditionally male jobs. The American Civil Liberties Union is considering bringing this suit on the basis of discrimination against women.

There is also here the basic question whether, if such a risk exists for these women, the companies involved should take appropriate steps to remove the conditions which bring about such a risk of genetic damage. Also involved in the pending suit is the fact that genetic damage to offspring may not be only related to females. Some indications show that damage may be more harmful to men than to women. Such a possibility that radiation to sperm may be more likely to cause genetic damage, due to mutation, in the offspring is certainly a consideration. If this type of discrimination against women in the work place is eventually before the courts, the constitutionality of limiting only women from hazardous working conditions will certainly be a major issue.

In addition, no uniform standards dealing with procedures for genetic screening are present. Indeed many of the laboratories used by the various companies do not seem to have suitable competence to carry out some of of these tests. The adequacy and correctness of the testing seems questionable in some instances, adding to the difficult legal issues which arise. Certainly the question whether such testing is in

violation of many sections of the Bill of Rights will
have to be resolved by the courts.

In 1983 the U.S. Office of Technology issued a
detailed report entitled *The Role of Genetic Testing in
the Prevention of Occupational Disease*. It deals with
all aspects of genetic screening, including frequency,
current technology, and the legal and moral issues
involved. Perhaps this report will provoke further
discussion and study of these issues and lead to some
standards and even, eventually, to some Congressional
action on what seems a major genetic-legal issue today.
Indeed, in 1981 a Congressional Hearing was held by a
subcommitee of the Science and Technology Committee of
the House of Representatives resulting in the
publication of *Genetic Screening and the Handling of
High-Risk Groups in the Workplace*. The two reports cover
almost all aspects discussed here. However, despite the
fact that several years have elapsed since the hearings
were held and the two reports were published, there
appears to be no legislation forthcoming. For that
reason all cases involving screening continue to present
issues for which there are no solid legal decisions
other than those from the courts.

PRIVACY OF MEDICAL—GENETIC AND LEGAL RECORDS

In *Burgos v. Flower Hospital* before the Supreme
Court, Special Term, New York County, in 1980, the
plaintiff was a learning-disabled child, and through
his guardian brought suit against the Hospital on the
basis that his defect was caused by negligent prenatal
care. The defendant counterclaimed that the defect was
genetic, and not medically caused. To prove this point,
the defendant sought to have the birth records of the

siblings of the plaintiff as well as records concerning tubal ligations performed upon his mother who was not a party to the case. The court found reason to have the requested hospital records considered to be privileged on the basis of invasion of privacy of the siblings, although the court recognized the need of such records to defend the case. The Court based its opinion upon the New York State Laws which state "Upon objection by a party privileged matter shall not be obtainable." One interesting exception was allowed, however; the records of the mother were admissible for "the period during which the plaintiff was *in utero*". During such period there could be no severance of a mother from child. "Neither can we sever the infant's prenatal history from the mother's medical history during that period. As the infant's privilege has been waived we cannot allow the mother's privilege to be interposed to the defendant's right to all of the infant's medical history." Thus the suit could now be brought, but with only the records of the infant and of his mother's pregnancy to be allowed. Obviously, this will be a serious impediment to the hospital's defense as a pattern of familial learning defects, or other birth defects, would be a strong claim that the condition was indeed heritable. Without the complete information concerning the abilities of the siblings, no basis for the defense claim can readily be established.

A somewhat similar attempt by the defense to open medical records occurred in California in 1981. The consolidated case, *Jones v. Superior Court for County of Alameda*, decided by the Court of Appeal, First District, concerned suits against Eli Lilly & Company over the use of the compound diethylstilbesterol (DES) during

pregnancies. It has been shown that the use of this
compound may lead to development of specific cancers in
female offspring. This was therefore a suit brought for
negligence, product liability, breach of express and
implied warranties, and enterprise liability. The suit
against the court, however, was the main issue here and
involved constitutional rights of privacy.

Lilly had requested, as part of its defense, the
complete medical records of the mother, including *post
partum* records, which she refused to give. She freely
testified to the use of DES during pregnancy, but
refused to testify as to subsequent medical history.
Lilly claimed "it would be important for these purposes
to know, in addition to facts relating to the mother's
pregnancy with the plaintiff, whether the mother's
history reflects any genetic defects which could cause
obstetrical problems; all details surrounding her
miscarriages and/or abortions; any symptoms presented
during subsequent pregnancies and the pharmaceuticals
which were prescribed for her." (It should be pointed
out that one of a pair of twins died shortly after birth
following a later pregnancy.) However, the basic issue
was whether having testified freely about the medical
history during the pregnancy, she had thereby waived her
legal rights to withhold information concerning her
subsequent medical history. The Court upheld her right
of privacy. "It does not follow that petitioner, by
disclosing portions of communications relating to her
consumption of DES and her pregnancy with plaintiff, has
waived her privilege as to all otherwise protected
communications during her lifetime. Even the waiver
which follows upon a tender of a medical issue by a
party to the litigation does not extend that far. ...If

there is need for broader discovery in a case of this sort, the remedy lies with the Legislature, and not with the courts."

As to the constitutional rights of privacy, the court at one point noted, "The state of a person's gastro-intestinal tract is as much entitled to privacy from unauthorized public or bureaucratic snooping as is that person's bank account, the contents of his library or his membership in the NAACP. We conclude the specie of privacy here sought to be invaded falls squarely within the protected ambit." (This statement is from an earlier case concerning privacy of medical records.) The court further bolstered its opinion with many citations from federal cases involving rights of privacy, and decided that the mother did not have to disclose the record sought by Lilly as part of its defense. Again a history of subsequent difficulties during pregnancy might point towards a genetic cause of injury rather than a DES-induced injury.

Another suit involving DES brought against Eli Lilly was tried in 1979 before the Appellate Court of Illinois. This case, *Morrissy v. Eli Lilly*, was a class action suit in behalf of daughters of mothers who had been taking that drug during pregnancy. The court here dealt with this differently, mainly on the basis that at the time of suit no actual injury had been sustained and that for each case there would have to be a proven record of injury caused by the drug. The plaintiffs had sought not only damages, but also an order that all daughters of mothers taking this drug be located and notified of the possibility of the carcinogenic effect believed to be caused by it. A request for the sum of $41,250,000 to be set aside for this purpose was made.

The court disallowed this request as well as others and again emphasized that, among other things, "Included among them are whether the product was properly prescribed for the mother's then existing medical condition; the dosage of DES prescribed; the amount of DES actually ingested; the point during the pregnancy at which DES was started and ended; the hereditary and genetic history and background of the patient and incidences of cancer of the other disorders suspected; the patient's exposure to known carcinogenic agents in the environment; and the personal habits of the individual subjects." The defendants' claim that the alleged cancers might be due to multiple causes not related to DES and that only examination of each case could determine whether DES was involved was upheld. It is important to note that again the court found that the genetic background of the mother would have to be determined. By implication, there may be the suggestion that some cancers are genetically determined.

Several other cases, all involving DES, have been recently decided by the courts. While not strictly genetic, they all allege induced cancer in daughters of DES treated women. The irony here is that the drug was originally prescribed in order to maintain a pregnancy which might otherwise have ended in miscarriage had the drug not been used. However, should it be shown in the future that the induction of the specific cancer is indeed due to the drug's causing a genetic change which in turn produced the cancer, this would be of necessity included when dealing with *Genetics in the Courts*. The cases include *United States v. 2,116 Boxes of Boned Beef* before the U. S. District Court, D. Kansas (1981), dealing with addition of DES to beef cattle, *Renfroe v.*

Eli Lilly, decided by the U .S. Court of Appeals, Eighth Circuit, (1982), dealing with the effects of DES upon the daughter of a woman who used DES during pregnancy, and *Payton v. Abbott Labs*, decided by the Supreme Judicial Court of Massachusetts (also 1982), a class action suit, particularly based on emotional suffering by the daughters whose mother used DES during pregnancy. Various legal arguments were used in these cases: failure to prove that DES was the causation of the cancers, statutes of limitations, and the identity of the state in which the alleged injuries occurred. The court therefore found for the defendants in these cases without having to investigate thoroughly whether these were genetic defects.

A different type of case arose in *Mitchell v. Superior Court of City*, before the Court of Appeal, Fifth District, in 1984. Here in addition to the genetic facts there was the important issue of the privileged communication between lawyer and client. The plaintiffs in a class action suit involving over a hundred other plaintiffs sought recovery for damages from exposure to dibromochloropropane (DBCP) because there might have been a contamination of their water supply by DBCP and they further claimed this chemical had carcinogenic and mutagenic effects. In addition the plaintiffs sued for emotional suffering due to their worry about the effects of their exposure to such compounds. A difficulty here was that it may well have been their lawyer himself who suggested these possibilities, thereby causing the emotional damage. Before the court could decide upon the claim for damages it was felt necessary to divulge the conversations between the lawyer and the plaintiffs. The plaintiffs held that this type of communication is privileged, just as is a communication between a person and a doctor or clergyman. The court decided that before the merits of the case were discussed, it would be essential to reveal details of the discussion with the lawyer. As the summary of the case put it, "...Since plaintiffs had identified counsel as one of the sources of warning which underlay her claim for emotional distress based on knowledge of the chemical, fundamental fairness required that plaintiffs either waive the privilege and reveal that information transmitted by counsel and relevant to their claim or maintain the privilege and abandon the claim."

The opposite side of the coin was found in *Morrison v. Syntex Laboratories, Inc.*, decided in 1984 by the U.S. District Court, D.C. Here suit was brought to have the Laboratory release information concerning the use of one of their products, Neo-Mull-Soy, a formula for feeding infants. The plaintiff complained that the use of this for feeding her son resulted in his mental defects and sought to have access to case study data which might reveal a similar effect in other infants. Some 20,000 children had used this product and reports on 230 of these infants had been made to the Center for Disease Control and to the Food & Drug Administration; those were the records which were sought. Syntex claimed that the number of reports was so low as to be statistically insignificant and further that approximately 5% of the population suffer from some genetically or environmentally caused learning disabilities so that the reported cases would fall into this group. The judge ruled that it was better to err on the side of allowing release of the data so that plaintiffs could use the information in a jury trial for damages to the infant rather than have the information suppressed. The information was needed quickly as the main case would be coming to trial within a few months. The privacy of the infants involved was also to be protected as much as possible by using coded identities. Only the data reported to the Agencies and not the complete supporting documentation would be needed as the accuracy of the information supplied to the Agencies involved would be assumed. With the proper data it would then be up to the plaintiff in a suit against Syntex to attempt to establish a statistical case, and it would be up to a jury to determine the weight of such evidence.

Certainly, other cases similar to these will be before the courts. Defenses against these claims have need to ascertain that the claimed injury was drug-induced, and not genetically caused. On the other hand, the right to privacy concerning medical and genetic records makes this a genuine conflict in the law. Resolution seems most difficult: the right to privacy is basic in the Bill of Rights, while the need to defend onself against false claims requires just that privileged information, and the right to an adequate defense is also basic.

SWINE FLU VACCINE v. FERES

A set of cases involving genetic factors, not specifically by name, arose when the Federal Government rushed through the Swine Flu Act of 1976. Both genetic and legal issues abound in this case. First, the genetic issue is a rather simple one. The group of influenza viruses is known to be highly mutable, and a vaccine prepared against any one influenza virus is ineffective against mutated strains. As new strains such as the Swine Flu, the Hong Kong Flu and others, appear, a special vaccine must be prepared for each.

There is also a strong historical precedent for the quick action taken by the government in this instance. In 1918 a virus epidemic, beginning at Fort Dix, New Jersey, swept the country and caused more deaths (an estimated 480,000) than the total U.S. casualties in World War I. The 1976 Swine flu was again first discovered at Fort Dix, and it was believed to be the same virus which had caused the 1918 epidemic. The government was naturally concerned and set about a massive vaccination program whereby every person in the United States would receive immunization against this potential killer. They turned to the manufacturers of vaccines for a crash program to produce the vaccine. However, these manufacturers found that they could not obtain adequate insurance to compensate them for what they feared would be a host of claims, not necessarily valid but brought nonetheless, against them. At this point President Ford made a national address urging the program, and Congress responded by rushing through, without debate, the Swine Flu Immunization Act of 1976, whereby the government would stand as a substituted party to defend all claims of injury due to the vaccination program, thereby relieving the need for excessive insurance costs.

The major case, *Hunt v. the United States*, was heard by the U.S. Court of Appeals, District of Columbia Circuit, in 1980. Hunt and others were military personnel who brought suit under the FTCA for damages inflicted upon them by the vaccine, damages which were real and serious enough to cause some to be discharged from the military with permanent medical disabilities. There is now a further complication to this case as the parties suing were members of the armed forces at the time they received the vaccine, and the government claimed immunity on the basis of the *Feres* doctrine. The case might have been more properly named "The Swine Flu Act v. *Feres*." The long opinion described the history of the FTCA, the history of the *Feres* doctrine and the history of the Swine Flu Act. In this instance, the Court found that *Feres* did not apply as the suit was in reality against the manufacturers of the vaccine for whom the government had agreed to accept substitute liability and not against the United States *per se*. Therefore, in this instance the servicemen were allowed to bring suit.

The court was particularly critical of Congress for the haste with which it passed the act, and the opinion is laced with witty barbs and quotes from some of the Congressional testimony with obvious delight in showing the inadequacy of that body during its haste to rush through this piece of legislation. Incidentally, it should be pointed out that the dreaded epidemic did not occur, and the entire program was an example of excessive concern and hastily conceived action by all parties except the manufacturers who wisely passed their responsibilities on to the government. It is also an example of the not unusual thinking that if one dose is good, two are better; the members of the armed services were given much stronger amounts of vaccine than that given to civilians. The basis for this was that the security of the nation depended upon the strength of the armed forces and the stronger dose was assumed to afford better protection.

Civilians who also were innoculated fared differently. Depending upon where they brought suit, recovery for damages caused by the program was either

granted or refused. The main issue here was that of
informed consent. In the 1982 case, *In re Swine Flu
Immunization Program Products Liability Litigation*,
before the U.S. District Court, D. Utah, the plaintiff
established that she had contracted transverse myelitis
as a result of the vaccination, a hazard known to exist
in this instance. Although again the opinion is long,
the court summarized it well in its opening paragraphs.
"In our view the intent of Congress in passing the Swine
Flu Act and the expressed policy of the Department of
Health, Education and Welfare, support just compensation
for a plaintiff whose malady was predictable under the
state of medical science in 1976 and was proximately
caused by the swine flu vaccine. Our holding rests on
narrow grounds and applies only to those vaccinees who
meet the requisites of proof stated herein.

" A recommendation is also documented calling for
implementation of a national program to compensate
victims of other national immunization programs
predicated on no fault principles." The court then
approved the full amount of damages which had been
requested by the plaintiff.

Quite to the contrary, in the 1983 case, *Daniels v.
United States*, the U.S. Court of Appeals, Eleventh
Circuit, found that under Alabama law, where the alleged
tort occurred, the plaintiff had no standing as she had
signed an informed consent form for the vaccination. The
consent form, which is reproduced in full in this short
opinion, was held to absolve the government from the
induction of a permanent neurological injury resulting
from the use of the vaccine. They found that under
Alabama's laws the signing of the informed consent form
prior to receiving the vaccination was sufficient to

deny any claims from harm resulting from its use.

The government was again absolved of any blame for the use of swine flu vaccine in 1984. In this case, *Carter v. United States*, before the U.S. District Court, W.D. Michigan, S.D., the essential issue was whether the vaccine had caused serious injury to the plaintiff. After a barrage of medical testimony, pro and con, the court decided that there was no "causal nexus" between the use of the vaccine and alleged connective tissue disorders suffered by him.

Once again, these cases illustrate the thesis already stressed in this book that an all important factor in any law suit involving genetics or other matters may be not just the facts, but also the court in which it is brought!

PURPORTED GENETIC ISSUES IN NON—GENETIC CASES

The word "genetics" is becoming a new catch phrase for legal damage cases. An example is *Curtis v. City of New Haven*. The issue really was not genetic, but dealt with the misuse of the compound MACE, used to control dangerous persons who threaten the safety of police officers. Curtis and another person had received MACE in their eyes and face (this in itself was contrary to the proper use of the compound), and the police did not follow proper procedures set forth in the police manuals to help them after the use of the MACE. Curtis and others individually sued New Haven and won considerable damages. However in this instance the plaintiffs sought a permanent injunction against all use of MACE by the police force under any conditions. They claimed that not only was it dangerous to the recipient but also that innocent bystanders might be affected.

"Each plaintiff also alleged that the defendants knew or should have known that MACE is a highly toxic poison capable of causing serious and permanent injury to a human being's skin and eyes and that the active ingredient in MACE [chloroacetophenone] is believed to be a carcinogen and capable of causing genetic damage." The Court, the U.S. Court of Appeals, Second Circuit, in its 1984 opinion did not deal with the genetic issue at all, but simply found that the seekers of the injunction had no legal standing in this case for obtaining the injunction. The case illustrates, however, the use to which claimants and their attorneys now will try to bring *genetic* issues in order to strengthen their causes.

INFORMED CONSENT, BATTERY, WRONGFUL LIFE AND GENETIC DEFECTS

A case involving the above could well have been placed in any of several categories and serves to illustrate the complexity of the interaction between law and genetics. This unique case, *Grieves v. Superior Court (Fox)*, was decided by the Court of Appeal, Fourth District, Division 3, of California in 1984. Suit was brought against a physician and a hospital by a woman seeking both personal and penal damages against both the physician and the hospital due to the following circumstances. The woman had instructed the physician that after the birth she wished to have a tubal ligation, but *only* if the child were normal. The child was born with a chromosomal disease, and the tubal ligation was carried out despite the woman's instructions. The woman, stating that she had not given consent for the tubal ligation under these circumstances, brought suit on her own behalf for most of the above claims and for both wrongful life and wrongful death on behalf of the child. The lower court sustained demurrers by the defendants against all claims. Thus the suit was against the lower court for its action.

There were many issues, but the main one which the court decided was whether the tubal ligation constituted a "battery." After examining earlier cases, the court made an interesting differentiation between this case and others. In many instances, a surgeon will find it necessary to carry out further surgery resulting from the discovery of unexpected medical conditions during the surgery. Here it was decided that the reliance by the defendants upon a previous case dealing with that situation was incorrect... "We find their reliance on *Cobbs* [the previous case] to be misplaced. *Cobbs* refers to complications which *result* from a surgical procedure, not complications which occur *prior* to surgery" (emphasis in the opinion). The Appellate Court, dealing first with the issue of battery, found the allegations by the woman to be adequate to sustain such a claim. The court, however, agreed that all of the other claims were not legally valid, and also that punitive damages could not be sought from the hospital as it had in no way acted to cause these events. From reading the opinion, it is not even clear how the issue of wrongful life entered the case although such a suit was brought in the name of the deceased infant.

All these diverse miscellaneous cases illustrate the importance of the understanding of human genetics in many diverse fields. The courts must take cognizance of this field of science if they are to render adequate justice. It must be repeatedly stressed that knowledge of this field is essential to all parties to a case, the plaintiffs, the defendants, the judges, and the juries. No matter what the issue may be, litigants will attempt to bring genetic issues to the courts, and the latter must be adequately prepared to deal with them. In a litigious society each side will attempt to frame its case in such a way so that genetic evidence will support its case. Only an informed judiciary can deal with this; the real issue is how to inform the judiciary.

ALPHABETICAL LIST OF CASES CITED

Matter of Pond v. Oliver	(Slip Opinion)	
	No. 49331 3d Dept	1985
Millington v. Southeastern		
Elevator Co.	22 N.Y.2d 498	1968
Mitchell v. Superior		
Court (Fox)	203 Cal.Rptr. 556	1984
Mitchell v. Superior Court		
of Fresno	200 Cal.Rptr. 57	1984
Mitchell v. Superior Court		
of Fresno	208 Cal.Rptr. 886	1984
Morrison v. Syntex		
Laboratories, Inc.	101 F.R.D. 747	1984
Morrissy v. Eli Lilly	394 N.E.2d 1369	
& Company	(Ill)	1979
Mose v. Brewer	428 S.2d 1212 (La)	1983
Mtr. Board of Educ. v. Ambach	A.D.2d 227	1982
New Hampshire Supply Co.		
v. Steinberg	400 A.2d 1163(N.H.)	1979
Parents in Action v. Hannon	---F.Supp. 831	1980
Parham v. Hughes	441 U.S. 347	1979
Payton v. Abbott Labs	437 N.E.2d 171(Mass)	1982
Petition of Risdal &		
Anderson, Inc.	291 F.Supp. 353	1968
Phillips v. United States	508 F.Supp. 537	1980
Publisher settles suit;		
clone book a fake 216	Science 391	1982
Recombinant DNA and the law	15 Gene 1	1981
Renfroe v. Eli Lilly & Co.	686 F.2d 642	1982
Riley v. State Employees'		
Ret. Comm.	178 Conn. 438	1979
Rorvik v. Bromhall	296 Nature 383	1982
Sandusky v. W.C.A.(Chicago		
Bridge & Iron)	487 A.2d. 1019 (Pa)	1985

Scientists Settle Cell Line Dispute	220 Science 393	1983
Simpson v. City of Tulsa	620 P.2d 921(Okla)	1980
Smith v. Olin Chemical Corp.	535 F.2d 862	1976
Specific Reading Disability	219 Science 1345	1983
Stanley v. Illinois	405 U.S. 645	1972
State of New Jersey v. Meighan	414 A.2d 576(N.J.)	1980
State of Washington v. Smith	610 F.2d 869	1980
Taylor Diving & Salvage v. U.S. Dept. Labor	599 F.2d 622	1979
Timms v. Vernon Allsteel Press Company	520 F.Supp. 1147	1981
Top U.S.Co. Plan Genetic Screening	N.Y. Times 6/23	1982
Trimble v. Gordon	430 U.S. 762	1977
U.S. v. 2,116 Boxes of Boned Beef	516 F.Supp. 321	1981
United States v. The Progressive	467 F.Supp. 990	1979
Women Workers Sterilized or lose jobs	343 Civ. Lib. 6	1982
Wright v. Olin Corp.	585 F.Supp. 1447	1984

CHRONOLOGICAL LIST OF CASES CITED

1980	Burgos v. Flower Hospital	108 Misc 2d 225
1980	Daniels v. U.S.	704 F.2d 587
1980	Hunt v. U.S.	636 F.2d 580
1980	Ind. Union Dept.AFL-CIO v. Amer.Petr.Inst.	448 U.S. 607
1980	Parents in Action v. Hannon	---F.Supp. 831
1980	Phillips v. United States	508 F.Supp. 537
1980	Simpson v. City of Tulsa	620 P.2d 921(Okla)
1980	State of New Jersey v. Meighan	414 A.2d 576(N.J.)
1980	State of Washington v. Smith	610 F.2d 869
1981	Air Force Academy Sued over Sickle Cell	N.Y.Times 1/3
1981	Air Force Challenge on Sickle Trait	221 Science 257
1981	Avery v. County of Burke	660 F.2d 111
1981	Conn.Gen.Life Insurance v. Shelton	611 S.W.2d 928(Tex)
1981	Jones v. Super. Court, Cty of Alameda	174 Cal.Rptr. 148
1981	Lockwood v. Independent School District	312 N.W.2d 924(Minn)
1981	Marshall Minerals v. Food and Drug Adm.	661 F.2d 409
1981	Recombinant DNA and the law	15 Gene 1
1981	Timms v. Vernon Allsteel Press Company	520 F.Supp. 1147
1981	U.S. v. 2,116 Boxes of Boned Beef	516 F.Supp. 321
1982	Adams v. New Orleans Public Service, Inc.	418 S.2d 485 (La)
1982	Beaty v. Thiokol Corp.	414 So.2d 1292(La)

1982	Burke v. Occidental Life Ins. Co. of Cal.	416 So.2d 177(La)
1982	Champion v. S&M Traylor Bros.	690 F.2d 285
1982	Colin v. Schmidt	536 F.Supp. 1375
1982	Cornelius v. Sunset Golf Course	423 So.2d 567(Fla)
1982	Davis v. City of Sacramento	188 Ca. Rptr. 607
1982	Grable v. Weyerhauser Co.	639 P.2d 677(Ore)
1982	In re Richardson-Merrell, Inc.	545 F.Supp. 1130
1982	In re Swine Flu Immunization	533 F.Supp 703
1982	Juge v. Cunningham	422 So.2d 1253(La)
1982	Liberty v. District of Columbia Police	452 A.2d 1187(D.C.)
1982	Looking at Genes in the Workplace	217 Science 336
1982	Mtr. Board of Educ. v. Ambach	---A.D.2d 227
1982	Payton v. Abbott Labs	437 N.E.2d 171(Mass)
1982	Publisher settles suit; clone book a fake	216 Science 391
1982	Renfroe v. Eli Lilly & Co.	686 F.2d 642
1982	Rorvik v. Bromhall	296 Nature 383
1982	Top U.S.Co. Plan Genetic Screening	N.Y. Times 6/23
1982	Women Workers Sterilized or lose jobs	343 Civ. Lib. 6
1983	Charles for Charles v. Schweiker	569 F.Supp. 1341
1983	Cothern v. Mallory	565 F.Supp. 701

1983	Donaldson v. Sec.Health & Human Services	567 F.Supp 166
1983	IML Freight v. Indust. Com'r of State	676 P.2d 1207 (Co)
1983	Mose v. Brewer	428 S.2d 1212 (La)
1983	Scientists Settle Cell Line Dispute	220 Science 393
1983	Specific Reading Disability	219 Science 1345
1984	Carter v. U. S.	593 F.Supp. 505
1984	City Vinta Park v. Girls Sheltercare	654 S.W.2d 256 (Mo)
1984	Coleman v. Coleman	471 A.2d 1115 (Md)
1984	Dean v. Cone Mills Corp.	313 S.E.2d 11 (N.Car)
1984	Hayes v. Shelby Memorial Hospital	726 F.2d 1543
1984	Knutson v. Trainor	216 Neb. 653
1984	Larv. Riles (Slip Opinion)	U.S.Ct.Appeals 9th Cir
1984	Mitchell v. Superior Court (Fox)	203 Cal.Rptr. 556
1984	Mitchell v. Superior Court of Fresno	200 Cal.Rptr. 57
1984	Mitchell v. Superior Court of Fresno	208 Cal.Rptr. 886
1984	Morrison v. Syntex Laboratories, Inc.	101 F.R.D. 747
1984	Wright v. Olin Corp.	585 F.Supp. 1447
1985	Matter of Christensen v. Saif Corp.	Slip.Ct. Als. (Ore)
1985	Matter of Pond v. Oliver	(Slip Opinion) No. 49331 3d Dept
1985	Sandusky v. W.C.A.(Chicago Bridge & Iron)	487 A.2d. 1019 (Pa)

The previous chapters have dealt with cases involving genetics which have been decided under state or federal laws dealing with general issues such as malpractice, tort, contracts, paternity, etc. This chapter deals with statutes at various levels which have been enacted solely to deal with genetic issues.

GENETIC SCREENING OF NEWBORN

It may be somewhat of a surprise to learn that all fifty states have enacted laws regarding screening of newborn children. In each of these states every child born in a public or private hospital must be screened for several genetic defects before the child is released from the hospital. The tests are relatively simple and for the most part require only the drawing of a few drops of blood from the heel of a newborn. As there may be religious objections to such a procedure, in many states exceptions to this type of neonatal screening are allowed should there be parental objections to the procedure. However, it is not clear, as will be seen, whether the parents are fully informed of this type of action.

The laws vary somewhat from state to state, but all are designed to detect birth defects, some of which, if found early, can be alleviated or avoided by early detection and certain forms of medical care. For example, the genetic disease phenylketonuria (PKU) which normally results in extreme mental deficiency in infants possessing both recessive genes responsible for the disease, can be avoided if the child's diet is kept free from the amino acid, phenylalanine. This is, however, an extremely difficult task as this amino acid is common in most foods, and restricting the child's diet means rigorous control for at least the first decade of the child's life.

Another genetic disease which may be avoided by early recognition and treatment is hypothyroidism. If this genetic disease is detected early, the missing hormones may be given to the child so that normal growth occurs. Essentially the aim here, as in other genetic testing of the newborn, is to take preventive measures from the time of birth so that infants born with recognized genetic defects will be spared the inevitable effects of these defects.

Sadly, many of the traits being screened for even when detected cannot, at this time, be ameliorated by any known treatments. Examples of diseases for which there is no known treatment are sickle cell anemia (although recent progress has been reported using gene therapy) and many of the enzyme deficiencies referred to earlier. One state is screening for maple sugar urine disease, and statistics indicate that if all the new-born in this state are screened only one case in ten years would be found due to the rarity of the disease in the population at large. The question then arises as to which particular genetic disease should be screened for and the cost efficiency of such screening needs to be examined further. While one cannot and should not put a cost on a life, it must be recognized that when there are limited public funds for the screening of newborn these funds should be used for the greatest benefit of all, and legal and economic as well as genetic factors must also be examined. The manner in which the decision to impose screening is decided is an important part of these laws and deserves attention and consideration.

The state laws will be examined first, and some comments as to their enactment and their utility must be made. It is of importance to examine whomever the law designates to decide what is to be screened for in the neonates, and how this is done. This varies from state to state, but an example of two widely diverse methods will suffice to show how the differences occur.

The New York law demonstrates one extreme. Here the legislature in 1974 passed the Conklin bill, "An Act to amend the public health law, in relation to testing for newborn diseases and conditions and making an appropriation therefore." The bill specifically listed a series of diseases mandated for testing under the terms of the bill. These included specifically phenylketonuria, sickle cell disease, hypothryoidism, galactosemia, maple syrup urine disease, homocystinuria, adenosine deaminase deficiency, and histidinemia, and *such other diseases and conditions as may from time to time be designated by the commissioner [of health] in accordance with rules and regulations prescribed by the commissioner*" (emphasis added).

In a letter from the counsel to the State of New York Department of Health to the Governor's Counsel regarding the action to be taken by the Governor, it is stated, "The Department recommends approval of this bill because of the need for the early discovery of these diseases as well as the flexibility it gives the State Health Commissioner in authorizing him to designate additional diseases or conditions." It should be noted that only three of the diseases on this list can be ameliorated by early treatment of the neonate, while the others are simply informational. The bill failed to provide any provision whatsoever to insure

confidentiality of the records, as well as a lack of any
penalty for failure to provide the information required.
Additionally, the financial support stipulated was
inadequate in that the amounts allocated were far too
low. Nevertheless, the bill was enacted and tests and a
form, in triplicate, are now required to be filed for
all children born in private or public hospitals.

Another noteable factor is that the triplicate form
provides for copies to the state, to the physician, to
the hospital, but none to the parents! Thus vital
information might not be transmitted to the parents in
time to take steps to help their children. Some consider
this act, while well-meaning, to be deficient in many
ways, as are many other state laws which follow
approximately the same procedures.

The type of bill which many feel should be a
model one was adopted by the State of Maryland in 1973.
It is therefore worth examining this is some detail.
First, the act did not invest power in a single
individual as was the case in New York, but created a
Commission of Hereditary Disorders. The Commission, to
be appointed by the governor, consists of eleven
members. One member is to be from the state Senate,
appointed by the President of the Senate, and one from
the House of Delegates, appointed by the Speaker of that
body. Five members are directly appointed by the
Governor and may not be members of the health
professions, or of any related health care organization.
Four are licensed medical professionals "knowledgeable
in the diagnosis and treatment of hereditary disorders"
and one each must be appointed from a list submitted by
the various medical societies or schools in the state.
The Commission is thereby representative of the

electorate, of lay persons, as well as those in the
medical profession.

The law further specifies that the Commission must
meet at least twice a year, and that it determine its
own internal organization. The Commission is empowered
to establish rules for genetic screening, to gather and
publish information for the enlightenment of the public,
to establish rules for recording data (confidentiality),
to reevaluate continuously the value and need of the
screening procedures used, and, importantly, to
investigate "unjust discrimination resulting from
identification as a carrier of a hereditary disorder."

The act also established that no mandatory
provisions were to be established, specifically that
refusal to have one's newborn tested would not preclude
welfare benefits. Further, the act required that genetic
counseling be available to all persons in the screening
programs, with nondirective counseling used not to
discourage further reproduction, but to inform parents
of risks involved. Further, the results of all
diagnostic procedures are to be made available to all
individuals 18 years old or to parents or guardians if
the child is below this age. A specific provision for
confidentiality of records is included. Another clause
specifically prohibits insurance companies from
discriminating against carriers of the diseases.
Maryland presently tests for 5 diseases, PKU,
hypothyroidism, maple syrup urine disease,
homocystinuria and galactosemia.

The contrast between these laws, while not meant to
be unfair to New York State, is an obvious one. No one
person in Maryland can make unilateral decisions, both
medical and lay persons must be consulted, there is an

attempt at strict confidentiality of records, and the
person tested or the parents have direct access to the
results of the tests. In addition, instead of merely
informing the physician as is the case in New York,
genetic counseling must be available to parents upon
request. The Maryland statute would seem to be a model.
By looking at all factors- cost, seriousness of the
genetic defect as well as the frequency of the defect in
the population-the Commission can keep the list of
diseases to be screened not only flexible, but also cost
efficient.

Further, some states which are or were screening in
such a manner as to be discriminatory would not be able
to do so under this type of law. For example, one
state's law required screening of all non-whites for
sickle cell; another mandated screening the entire
prison population, primarily black and obviously not in
a position to carry out reproduction. It has been
estimated that possibly as high as 90% of all blacks in
Boston may have been screened for sickle cell anemia
under a Massachusetts law. While some other states give
lip service to the confidentiality of the record, which
probably would be covered by the federal rights of
privacy, others do not. The Maryland law also makes the
screening voluntary, and this too may be more in accord
with the right to privacy. The newborn obviously cannot
give legal consent, and in many cases the parents in
some states give consent unknowingly, or possibly due to
a threatened loss of welfare of other social benefits.
For these reasons, the Maryland laws would seem to be
the model to follow. It would well behoove other states,
particularly those which regard themselves as moderate
or liberal in their laws to reexamine their codes in

view of the enlightened one just described.

FEDERAL LAWS REGARDING GENETIC DISEASE

In addition to the state laws, there exist federal laws regarding genetic screening and its use. These too must be considered in light of their aims and accomplishments. More such laws will be forthcoming, and the way in which their enactment occurs might also be worth some scrutiny. No cases involving the application, or constitutionality, of these laws have come before the courts as yet, although some protests over their application have been made.

Attempts have been made from time to time to enact federal laws regarding genetic screening of the newborn. For example, in 1972 the "National Cooley's anemia act" was passed. The act dealt with screening, treatment, counseling, research and information and educational programs. However, as new genetic diseases were found, and new pressures brought to bear on the Congress, a sort of "disease of the month" handling of these could have resulted. Consequently, in 1976 a blanket act was passed. A subsection of Public law 94-278, namely Title IV, of the act was now entitled "Genetic Diseases." The section dealt with testing, voluntary counseling, etc. It instructed the Secretary of the then Department of Health, Education and Welfare to carry out "a program to develop information and educational materials relating to Genetic Diseases and to disseminate such information and materials to persons providing health care, to teachers and students, and to the public generally in order most rapidly to make available the latest advances in the testing, diagnosis, counseling, and treatment of individuals respecting genetic disease."

It made possible the granting of funds for these purposes, with special emphasis upon Cooley's anemia and sickle cell anemia, but did not exclude other genetic diseases as they were discovered. It also provided for voluntary participation and made clear that other social benefits were not to be withheld if a person chose not to participate in the programs, and provided up to $30,000,000 for each of three years for the programs. Obviously the terms of this act come close to those of the Maryland statute with the exception that funds were appropriated for grants studying genetic disease. A definitive part of the act required that if a state or agency applies for funds it must conform to the exact terms of the act. Importantly, there was a provision in the act that all such grants must show that personal information, unless consent was given, would be confidential and the information would be public only for the overall statistical information obtained. There was also a section requiring public participation "where appropriate in the development and operation of voluntary genetic testing or counseling programs funded by a grant or contract..." This bill, originating in the House, paralleled one in the Senate to a large degree, and after conference it was passed and signed into law.

It is not clear, however, whether the funds allocated were ever fully authorized, and the impact of the bill was therefore probably limited. The importance of the act thus may be more in its recognition that a federal role in studying genetic disease was a legitimate part of national health care.

LEGISLATION AND RECOMBINANT DNA

Recombinant DNA also has been the subject of state

legislation and originally attempts were made to enact
federal laws concerning this technique. Initially there
was a great fear of the effects of this new technique,
leading first to a self-imposed moratorium by scientists
working in the field, then the Asilomar Conference, and
eventually the adoption of strict guidelines by the
National Institutes of Health. These guidelines, it
should be pointed out, were only restrictive for those
scientists working under grants from that agency, but
they became more or less the basis for all such work.
They were extremely detailed, spelling out which
experiments could be done under which conditions, and
totally banning certain types of work, such as
introducing the ability to produce deadly toxins into a
bacterium which normally could not produce them.

The National Institutes of Health directive was
extremely detailed and restrictive; for example, the
exact way in which floors must be mopped was put forth
in great detail. Nonetheless, the aim was admirable in
that the NIH sought both to permit important work in
this field to continue, and also to be certain that
there was no abuse of it. By setting up various
parameters including laboratory precautions and the
types of organisms which could be used the NIH attempted
to reach a reasonable compromise between the advance of
scientific research and a possible abuse of that
research.

As may be recalled, there was at first a public
outcry against the use of any recombinant techniques,
including among the opponents of this technique many
geneticists and other biologists as well as groups
opposed on philosophical grounds. At the height of the
furor, bills were introduced into Congress which would

have banned all such research or been even more
restrictive of biological research. The comparison to
the development and hazards of atomic projects was
frequently made by some of the more vocal opponents. The
furor was such that many geneticists found themselves
reluctantly supporting some of the more moderate federal
proposals such as that introduced by Senator Bumpers or
by Senator Kennedy on the basis that if a moderate bill
were not passed public clamor might result in a bill so
restrictive that scientific research of all kinds might
come under its blanket. Fortunately, none of the bills
was ever passed, and with the passage of time pressure
for federal action has essentially ceased or been
strongly diminished. However, given the great amount of
attention now directed to problems concerned with
release into the environment of genetically designed
organisms, there may ensue a renewed interest in
Congress in some way to regulate such procedures.

Several states have rushed in to fill the gap
left by the lack of federal acts. In 1977, the State of
New York passed "AN ACT to amend the public health law,
in relation to the certification of recombinant DNA
experiments." Again, the responsibility for both
developing guidelines and their enforcement was given
to the Commissioner of Health, with power to "exclude
those types of recombinant DNA experiments conducted in
a cellular environment which he determines to be beyond
the scope of the legislative purposes of this article."
There was a suggestion that the N.I.H. guidelines
referred to previously be used as a frame of reference,
but like the laws concerning blood testing, there were
no references to participation by others in the
Commissioner's decisions, and no apparent limitations

upon his power to approve or outlaw this research.

A similar measure was introduced in 1977 to enact the California Biological Research Safety Act. A commission was to be formed made up of 17 members, 11 appointed by the Governor with the consent of the Senate, and 3 each appointed by the leader of each of the legislative houses. Perhaps, in part, the Maryland genetic testing laws may have provided some basis for this diversified body. A deficiency in the regulation of recombinant DNA may have been the basis for Califonia's concern in that the guidelines by the N.I.H. only regulated grants from that agency, and did not apply to industrial or private research interests. Unlike the New York act, there was a provision for a fine of up to $10,000 for each violation of the proposed law.

These legislative acts are an indication of the public concern about the use of recombinant DNA. There is, perhaps, a much more serious implication both to law and science. If some types of research can be banned, who is to determine which type it will be? The principle can be extended to all investigation which might be controversial, not just genetic science but any research in social science or the humanities. Such studies might offend enough legislators to lead them to try to ban them on the basis that it might be dangerous or not in the public interest to know the results.

Now that recombinant DNA has set a precedent for a state's banning of some types of research, it will take vigilant care to assure that this principle does not extend to other research in every field. It will also be of interest whether such state acts are in themselves constitutional under federal law. No specific cases of this kind have arisen, but the possibility of a

challenge of the right of states to interfere with any
type of research will probably depend upon whether there
is sufficient danger to the welfare of its citizens for
the state to intervene, as has been the case with the
environmentalists' challenges to the use of atomic power
or to the use of recombinant organisms in field trials.

It is possible that in the near future, the United
States Congress will take more interest in genetic laws,
judging by the hearings during the past few years. The
first, held in 1980 by the Committee on Science and
Technology, chaired by Representative Fuqua of Florida,
dealt with "Genetic Engineering, Human Genetics and Cell
Biology." The second, before the same Committee in 1981,
dealt with "Commercialization of Academic Biomedical
Research," while the third, held also in 1981 before
another subcommittee of the Committee of Science and
Technology, dealt specifically with the general problem
of "Genetic Screening and the Handling of High-Risk
Groups in the Workplace," a problem mentioned previously
in the chapter dealing with miscellaneous cases. While
it is clear that hearings do not always lead to
legislation, it is also clear that these genetic
problems are coming under closer scrutiny, and where
committees of Congress find abuses it is frequently the
case that correction of these may be attempted by the
introduction of new Congressional acts. It might be of
particular interest, in light of the previous
discussions, to examine the participants and some of the
evidence offered in the third of the above hearings.

Included in the participating witnesses in the
Genetic Screening hearings were several persons with
opposing views. Particularly pertinent were the
statements by two persons on the last day of the two-day

hearings. Ms. Joan Bertin appeared on behalf of the American Civil Liberties Union to decry the use of such screening, citing cases of women who underwent sterilization in order to be eligible for certain jobs where women who were considered to be under the risk of pregnancy were excluded. Dr. Bruce Karrh, of Du Pont de Nemours & Company, testified that the company had kept such females from various areas, including those who, while not planning pregnancies, might accidentally find themselves in that condition. The fact that this large company was using data concerning the genetic risks of various work environments indicates that awareness of genetic dangers played a role in the hiring or placement of personnel. Again, no legislation resulted, but the possibility of this happening seems enhanced by the data and policies presented in the hearings.

In addition to the hearings on genetic screening, a conference report from the American Society of Human Genetics, dealing with this problem, was sent to President Reagan in April of 1981, indicating the Society's interest and concern. In general, the Society expressed concern over the lack of labor's participation in a decision to use screening, and the use of screening to ban workers from certain jobs. Here, although not implicitly stated, is perhaps a strong position taken by the leading society in the field. This may also be an indication of the growing opposition to that practice, and a confirmation of other earlier statements that laws should focus attention on cleaning up the job environment and prohibiting discrimination against workers because of sex or a genetic predilection to some disease related to the working conditions. It would seem almost inevitable that pressures from various sides will call increasingly for federal laws dealing with the use and abuse of genetic screening of employees.

The issue of privacy of genetic records will also have to be legally resolved, and although mentioned earlier in some of the state codes, there may be a serious problem arising. The use of computerized records in genetic screening is becoming standard. One review article contains the statement, "There is going to come a point at which I am sure it will be necessary to code all of these (clinical) findings on patients, so that when a next case is encountered, a search of records may find somebody who will match up with this condition, and on the basis of this, allow a better prognosis and better advice." The aim of the use of computerized records is admirable, but nowhere is there any statement as to how the privacy of the medical records is to be assured. In these days when computer "hackers" seem to be able to obtain access to almost any computer data, the confidentiality of medical records is of obvious concern.

Another area in which one may expect legal action is the use of amniocentesis. For example, in Nebraska women over the age of 40 have been urged to undergo this process if they become pregnant and thereby run the many-fold increased risk of producing Down Syndrome children. An early estimate of the cost of rearing such children and maintaining them through a normal life span was made in the 1970's. The figure then, without any correction for inflation since that time, was about $5000 per year for each child. The total cost, considering the number of such children born annually and the now possible normal life span of such individuals, reaches into the billions. Most of this cost is for hospital or institutional care and much of it will have to come from public funds. The question whether society can afford this cost may in turn lead to proposals that all high-risk pregnancies be monitored. Indeed while some insurance companies will pay for the cost of amniocentesis, there has been a suggestion that some other companies might refuse child support and medical care if the high-risk mother *does not* submit to amniocentesis, implying that if genetic defects are detected the insurance companies will not pay for the health costs for such children. Of course, this implies a threat of making abortion mandatory for women carrying potentially defective children when these women are not in a position to assume the financial costs independent of their insurance.

The danger here is, of course, the possibility of state-enforced eugenics. The shadow of Nazi Germany is still very real; eugenic laws, no matter how well-meant originally, could be used to restrict the constitutional right of choice of a woman to procreate, and could lead to the sort of racial or religious prejudices so well known from that era. There is also the question of who is to determine what a "defect" is in terms of society. Certainly easily reparable defects such as cleft-palate would not appear to be serious enough to deny procreation, but where is the boundary between the need to "protect" the financial interests of society and the cases where a genetic defect is so minor as to warrant no interference? While serious genetic defects, many detectable *in utero*, amount to approximately 1-2% of all live births, is there any social or economic reason to mandate amniocentesis to avoid this expense and tragedy? And should such laws be passed, is there a constitutional basis for requiring abortions of all fetuses found to carry such defects? While these questions may, at present, seem naive, it should be remembered that many states made genetic screening mandatory for all new-born, and from there to requiring amniocentesis and abortion is not as great a step as it might seem.

In 1975 the Federal government endorsed amniocentesis after a study by the National Institute of Child Health and Human Development. This suggests that mandatory statutes might be attempted, and the constitutionality of these may have to be tested. One must be aware that such laws may be proposed and one can be almost certain that the courts will in the future have to deal with them.

As society changes, so does the law, albeit usually somewhat more slowly. It is therefore impossible to predict what new laws dealing with genetics will appear, particularly as that science is changing so rapidly. But it would be equally foolhardy not to suggest that new genetic laws will be proposed from time to time and the statutory law dealing with genetic issues will continue to develop. As pointed out in the chapter dealing with miscellaneous cases, the problem of educating the judiciary must now be extended to education of the public as well as legislators of both state and federal bodies. To a geneticist this poses both a formidable and a challenging task.

GENERAL REFERENCES

Commercialization of Academic Biomedical Research, Hearings before the Subcommittee on Investigations & Oversight and the Subcommittee on Science, Research &

Technology of the Committee on Science & Technology, U.S. House of Representatives, 97th Congress, First Session, 1981 [No. 46]. (U.S. Govt. Printing Office).

Genetic Engineering, Human Genetics & Cell Biology. Evolution of Technological Issues, Biotechnology. Report prepared for the Subcommittee on Science Research & Technology, U.S. House of Representatives, 96th Congress, 2d Session, 1980. Serial DDD (U.S. Govt. Publishing Office).

Genetic Screening and the Handling of High-Risk Groups in the Workplace. Hearings before the Subcommittee of Investigations & Oversight of the Committee on Science & Technology, U.S. House of Representatives, 97th Congress, First Session, 1981. [No.53] (U.S. Govt. Publishing Office).

State Laws & Regulations on Genetic Disorders. U.S. Dept. Health and Human Services, DHS Publication No. (HSA) 81-5243, 1980.

Chapter 14
EPILOGUE

There really can be no end to this book. Even while the various chapters were being written, new issues, new cases, new decisions and new statutes appeared. Additionally, the reader should be cautioned that many of the cases discussed in this book may yet appear before an appellate court and some of the decisions may be overruled. For example, within the past year, two more states have allowed wrongful life claims so that the costs for maintaining a birth-defective child beyond the time when the child reaches majority are recoverable. Interestingly, while one of the two states is West Virginia, the other is Illinois where the first wrongful life case appeared, namely *Zepeda*.

The first birth of a child by a surrogate sister of a sterile woman has been announced, and the *in vitro* fertilization of a donated egg with a husband's sperm has been apparently successfully implanted in a woman without ovaries. At the moment there is a case coming before the California courts in which the mother of an AID child has gone on welfare and the state is trying to assess the costs of raising the child to be paid by an apparently known doner to a sperm bank. In addition, the New Jersey courts are hearing a case involving the legality of surrogate contracts

As pointed out, The United States Supreme Court will hear the Louisiana Creation Science case referred to in the text. As a geneticist, I can only hope this will be the final word upon this issue involving separation of church and state.

It has been decided that a patent may be issued for a single gene, specifically for a gene controlling the production of an essential amino acid by corn.

The list of bitter court arguments over patent rights for other genetic recombinations or organisms is typified by the vigorously contested patent suit between American and French researchers referred to earlier, each group claiming the isolation of the virus believed to cause AIDS Syndrome as their own discovery. The "ownership" of this virus may be of commercial importance as it might be used to produce a vaccine against this disease.

Paternity cases, despite the original hopes that blood tests would allow these to be settled out of court, swell the dockets of all courts, many of them, as has been seen, simply repeatedly trying old issues over again. As each new type of evidence arises, the courts have to resign themselves to deciding the admissibility of the evidence under *Frye*, and must keep abreast of current developments in molecular genetics.

A very recent report in the journal *Nature* deals with guilt in criminal rape cases. It claims that modern DNA techniques will offer 100% proof of guilt or innocence even on blood or semen stains several years old. Further indicatations are that it is possible to identify differentially DNA from sperm residues and DNA from cells sloughed from the vaginal tract. Whether this report will suffice under the *Frye* standards remains to be seen. When the first case using this method appears in an American court one can be certain of long and detailed arguments over its validity.

Many new cases are brought by environmentalists dealing with potential abuses of the land, air, or water, and the possible genetic harm both to persons and to other organisms by industrial, governmental or other actions. In addition to the earlier cases referred to

there are now cases either coming to court or about the
be brought involving purported unauthorized release of
recombinant organisms into the environment, further
requests for release of other organisms, and cases
involving the issue of which particulal federal agency
can make the determination whether any release at all is
permissable. A specific case deals with whether the U.S.
Department of Agriculture may grant perfmission for the
commercial use of an genetically altered virus to
produce immunity to a disease of swine without any input
from the EPA and the appropriate DNA committees. In this
instance the issue is also whether a virus from which
genetic material has been deleted so that it cannot
reproduce is a recombinant or simply a modification of
an existing virus. After an initial banning of further
use of the swine virus, there has been a reversal of
opinion and it may now again be used.

In particular such release of various recombinant DNA
or altered organisms is the basis of bitter
controversies. Another dispute has arisen concerning the
testing of a bacterium to induce frost resistance by the
injection of trees growing on the roof of a greenhouse,
deemed by some to be in essence a release in the
atmosphere and not contained in a greenhouse as the
permit for testing had specified. The problem here
arises because the trees are too tall to be contained in
the greenhouse, and also because there may have been
some observed damage to the trees not fully reported to
the EPA.

At present legislation dealing with the release of
recombinant organisms is being moved forward in the U.S.
Congress. Included in a proposed study and bill will be
regulations determining which governmental agency should

be controlling this process, and suggestions have been made that it should be the responsibility of both the Department of Agriculture and of the EPA, depending upon the nature of the proposed release.

These are just a few of the many examples of new types of cases which the judiciary will have to decide in the near future. The study of the use of genetic data in legal decisions is obviously not one which will ever be concluded.

Further, each year the various legislative bodies at all levels pass new laws dealing with all of the issues raised in this book, and changes in the laws often result in appeals to the courts. For example, the Attorney General of New York State continues to attempt to convince the legislature there to introduce and pass a "toxic torts" law which would allow victims of toxic chemicals or of other agents which cause genetic damage to bring suit not at the *time* of the act, but at the time of *discovery* of the claimed injury, thereby avoiding the statutes of limitations for such injuries.

New interpretations of tort law, such as the wrongful life decisions and others, occur as the courts are faced with new scientific evidence and its validity and admissibility. New genetic discoveries which will possibly involve court actions occur with increasing frequency and this type of new evidence must be evaluated almost on a case by case basis. Indeed, scientific discoveries in themselves now seem to predict new litigation as the commercial values of such new genetic techniques as recombinant DNA, gene splicing, etc. are realized.

Some idea of the increased litigation involving genetics may be grasped from the fact that during the period between June 1984 and March 1986 over one hundred cases containing genetic issues were quickly found with a computerized search. These cases were only a few found in a rather cursory search, and each contains references to many more. As noted, given these conditions, a study of *Genetics in the Courts* can never be completed.

If there can be a single unifying theme in this study, then it must be the way in which these new discoveries become part of the legal systems, and the need for scientific awareness and expertise by the judiciary. This is exemplified in many of the cases discussed, particularly those in which each side presents its own "experts" and the judge or jury must decide which of the experts are correct.

Perhaps it is for this reason that many seemingly identical cases are decided in different ways depending upon the state in which the case is brought. Each judge, or jury, must decide the scientific issues as they arise. Also, as the federal courts must base their decisions upon the laws of the state in which the case arises, these courts may also reflect opposite opinions to the degree that state laws require. Additionally, as can best be realized by the citations chosen, judges like the rest of us are human, and they differ in background, training, and philosophy, bringing different opinions to the law. One only has to look at the many dissenting opinions to realize this.

This might best be summed up by saying that the results of any particular case may depend to a large degree upon the state in which it arises, the court before which it is tried, and which judge is presiding. Some would even claim that the results may even depend upon whether the judge has had a good breakfast!

GLOSSARY OF GENETIC, LEGAL AND MEDICAL TERMS

The definitions listed below are based in part on
Blakiston's Gould Medical Dictionary Third ed. McGraw-
Hill, N.Y. 1973, *Mendelian Inheritance in Man*, Sixth
ed., Victor A. McKusick, Johns Hopkins Univ. Press,
Baltimore, 1983, and *Black's Law Dictionary* Fifth ed.,
West Publishing Co., St. Paul, Minn. 1979.

Act of Proximate Cause: An act which provokes an injury
which would not have occurred in the absence of the act.
Adenosine Deaminase Deficiency: A disease of the immune
system causing a lowered cell immunity towards disease.
Dominant, but with some closely related forms recessive.
Admissibility of Evidence: Determination whether a
particular type of evidence is legally permitted.
Contrast to Weight of Evidence, q.v.
Agglutinogin: A factor in the blood which causes a
combination with a foreign, introduced substance.
AID: Artificial insemination donor (not to be confused
with the disease AIDS) and contrasted to insemination by
a husband.
Allele: One of a pair of contrasting genes carried by an
organism. An organism may be homozygous, i.e. carrying
two identical alleles, or heterozygous, carrying one
each of a pair of contrasting alleles.
Alpha-1 Anti-trypsin: Lack of a particular protein which
leads to degenerative lung disease and death in middle
life. Dominant.
Amicus Curiae: Friend of the Court; a person or group
not directly involved in a lawsuit, but allowed by the
court to file a brief in behalf of one of the persons or
organizations directly involved with the suit. The brief

represents the interests of the agency filing it.

Amniocentesis: The withdrawal of fluid by means of a needle from the amniotic sac surrounding a foetus for purposes of obtaining foetal cells and examining them for possible genetic defects. Usually not performed until after the twelfth week of pregnancy.

Anoxia: Complete deprivation of oxygen.

Aortic Aneurysm: A weakening of the wall of an artery causing the vessel to swell and possibly burst. Possibly dominant, but not sufficient information for a determination.

Arguendo: A statement by a judge as an argument or a hypothetical fact.

Arterio-venosus anomaly: Changes in the connections of arteries and veins, or involving capillaries. Suspected to be dominant pattern of inheritance, with some types possibly recessive.

Atrophic Rhinitis: A disease of swine of uncertain etiology, resulting in deformities of the anterior head.

Backcross: A genetic cross of an offspring to one of its parents, usually to the homozygous recessive.

Battery: In medicine, any act which exceeds the consent of a patient for a procedure.

Benefit Rule: Determination of damages by the difference between the damage done and the advantages received by an act.

Carrier of a Trait: A person or organism which, while not expressing a genetic characteristic, nonetheless contains the gene for that trait. Carriers have one dominant and one recessive gene and the expression of the recessive gene is masked by the effect of the dominant. The chances of passing on either gene are equal.

Case of First Impression: A case which is of a type not previously arising in the jurisdiction of the court hearing the case and with no legal precedence.

Case of First Instance: a case arising for the first time in legal history.

Catatonic: A state of benign stupor or unresponsiveness; usually associated with schizophrenia, q.v.

Certiorari: Most usually associated with the U.S. Supreme Court; permission to bring a case before the court. Only those cases granted certiorari will be heard by that Court which as a rule chooses only cases involving major legal or constitutional interpretation.

Chromosomal Abnormality: A change in the structure of the chromosomes of an organism. These may be of several types: deletions in which part of a chromosome is lost, duplications in which part of a chromosome is present more than once, inversions in which part of a chromosome is found in reversed order, and translocations where a part of a chromosome is found in association with another chromosome not normally carrying that piece.

Chromosomal Analysis: Examination of the chromosome complement of an individual.

Chromosomal Disease: A disease caused by an abnormality of the chromosome; also may be caused by the presence of an extra chromosome or part of one, or by the deletion of part of a chromosome. Examples: Down Syndrome, Cri-du-chat, etc.

Chromosome: The carrier of genetic traits. Each organism has a typical chromosome number, for example in man 22 pairs plus the sex chromomomes (XX in females, XY in males).

Chronic Glomerular Nephritis: Kidney defects leading to heart failure, hypertension, and fibrosis. Recessive.

Compensatory Damages: Damages paid to "make the individual whole" as a result of an act by the person or organization being sued.

Conservator: guardian.

Cri-du-chat: A chromosomal disease caused by the loss of one part of a specific chromosome; characterized by the sound of a new born resembling the meow of a cat. A fatal disease.

Curator ad hoc: A court appointed person to represent a party unable to represent him or herself in a single matter.

Death on the High Seas Act: Congressional Act which provides reimbursement to survivors of a loss of life beyond the U.S. territorial waters if the death resulted from a wrongful act.

Derivative Claim: A claim originating from another, preceding claim.

Dextrocardia: A displacement of the heart to the right side of the thoracic cavity, with the apex of the heart directed to the right. Possibly X-linked.

Diabetes melitis: A hereditary disease caused by a failure of particular cells in the pancreas to produce the hormone insulin. As a result normal metabolism of carbohydrates is deficient and, unless controlled by administration of insulin, the disease is fatal with such symptoms as elevated sugar levels in the urine and eventual coma. Recessive.

DNA: Desoxyribonucleic acid. The chemical basis of the genes. Also sometimes written Deoxyribonucleic acid. Unfortunately also an acronym for Defense Nuclear Agency, the organization in the Department of Defense responsible for evaluation of the effects of exposure to nuclear radiation by service personnel.

Dominant Gene or Trait: The expression of a gene in either homozygous or heterozygous condition. Contrasted to a recessive trait which must be homozygous for its expression.

Down Syndrome: (formerly Mongoloid idiocy) A syndrome of general and specific effects caused by the presence of an extra 21st chromosome or a part thereof. Symptoms usually include mental retardation, cardiac defects, and a change in the formation of the eyes to resemble that of the Mongolian Race, hence its name.

Duchenne Muscular Dystrophy: Probably due to changes in the myocardium, the muscles of the heart. Progressive disease, usually beginning at an early age and with the victims seldom living beyond the age of 20. X-linked.

Ectasia: The swelling of a hollow structure such as a blood vessel.

En ventre sa mere: Literally "in the stomach of his/her mother." Used as a legal term to describe an unborn child.

Enjoinment: Prohibition of a person from carrying out an act.

Exhaustion of Remedies: A person may not bring a suit until all other possible means of presenting a case have been used, such as hearing agencies, lower courts, etc.

Federal Torts Claim Act: The Congressional Act which allowed suits to be brought against the Federal Government without first obtaining governmental consent.

Felony: A major crime; under federal and many state laws any crime which can bring about a punishment more severe than one year's imprisonment.

Filiation: The judicial determination of paternity.

Forum non conveniens: Discretionary power of a court to determine that there is a more convenient place for a suit to be tried.

Gamma Globulin: A blood constituent frequently used in an attempt to overcome the possible deleterious effects of rubella during pregnancy.

Gene: A localized region of a chromosome controlling a trait. Estimates of the number of human genes are in the order of 10,000 or more. Each gene is characterized by a unique sequence of the components of DNA.

Genetic Screening: Examination of a population, or part of a population, for the presence of genetic traits which might make that group more susceptible to certain environmental conditions, or to determine in some instances whether specific members of the population are carriers for a trait and would have a risk of producing defective children.

Glomerulonephritis: See Chronic Glomerular Nephritis.

Guardian ad litem: Similar to Conservator; a court appointed representative in a particular issue, usually for a minor or a mentally incompetent person.

Heterozygous: carrying different alleles. Contrasted to homozygous, q.v.

HLA Antigens: A complex series of antigens controlled by many genes not in allelic condition. Each antigen is thereby expressed independently of all others. Many such genes are relatively rare and the expression of a series of types in both an accused male and the offspring can be used if there is no match, for exclusion of paternity. If there is a match, the data then can be used to determine a probability of paternity to a high degree of confidence.

Homocystinuria: Disease due to improper metabolism of vitamin B 12. Usually involves mental defects and may be fatal. Recessive.

Homozygous: Having both alleles the same for a particular gene.

Hornbook Law: the rudiments of law.

Hypothyroidism: Insufficient amounts of the thyroid hormone causing poor mental and physical development. Extreme cases lead to cretinism, a form of idiocy. There are a number of genes known to cause different types of thyroid deficiencies; all are Recessive.

Hypoxia: Less than the normal amount of oxygen present; a less severe condition than anoxia.

Immunosuppressive drugs: Drugs administered in an attempt to defeat or neutralize the normal rejection reaction of the body to foreign tissues, particularly in organ transplants.

Implied Warranty: The implication that an item sold serves the purpose represented by the seller and is free of defects which would make the item unsuitable for the use claimed.

In Camera: In private, hearings in the judge's chamber and not before a jury.

In Esse: In reality; the way things really are.

In Utero: An unborn in the uterus of the mother.

In Vitro Fertilization: Fertilization of an egg by sperm carried out "in glass," the resulting conceptus to be then implanted into the uterus of a woman who has been treated with hormones so that she will be receptive of the product of fertilization and not reject it.

Informed Consent: Permission to carry out an act after the granter of the permission is made fully aware of the consequences both beneficial and harmful of the act. Frequently used in hospital procedures where consent is granted for treatment beyond that originally sought when unexpected circumstances arise.

Isograph: A graft between two genetically identical persons.

IUD: Intrauterine device to prevent pregnancy by preventing the implantation of the fertilized egg in the uterus.

Kleinfelter's Syndrome: A chromosomal disease caused by the presence of an extra X chromosome, namely XXY. Males of this type are sterile and may not show such secondary sexual development as a change of voice at puberty, facial hair, etc.

Larsen's Syndrome: Dislocation of knees, wide eyes, flat nasal bridge, prominent forehead, misshapen fingers. Dominant.

Long Arm Statutes: Laws which permit a state to have jurisdiction over corporations or individuals not residents of that state.

Lord Mansfield's Rule: Presumption of legitimacy of a child born to a married couple.

Lupus Erythrematosus: Skin and viscera disorders. Maybe viral, and possibly due to a dominant gene or interaction between a virus and the gene.

Maple Syrup Urine Disease: Characterized by smell of urine like that of maple syrup. Sufferers have mental and physical retardation and feeding problems. Recessive.

Master-Servant Relationship: Employer-Employee relationship. In cases where the employee is acting under orders of the employer, the latter may be legally responsible for the acts of the employee.

Mongoloid Deficiency: See Down Syndrome.

Moot: The situation where a law suit cannot have any bearing on an existing circumstance as the issues involved are either now academic or "dead."

Mores: Common customs of society.

Mutation: Change in the genetic material; change in one or more of the components of DNA within a single gene. Mutations may be spontaneous, i.e. arise from unknown causes, or induced by agents such as chemicals or irradiation.

Myelocytic Leukemia: overproduction of white blood cells by bone marrow. Possibly due to a chromosomal translocation.

Neurofibromatosis: Skin tumors which may reach large size, also "liver spots." Dominant.

Oxytocin: A hormone used to induce labor.

Parens Patriae: The role of the government acting as a guardian for persons unable to act themselves.

Paroral Evidence Rule: The rule that written contracts may not be changed by later oral declarations.

Phenylketonuria: A genetic disease causing mental retardation unless diagnosed at, or shortly after, birth. Persons suffering from this disease must severely restrict their diets, avoiding any foods containing a particular amino acid, a most difficult restraint as the amino acid in question, phenylalanine, is so widespread in foods. If a successful diet can be established and maintained the individual may develop normal intelligence. Recessive.

Plasmid: A self-replicating DNA molecule found separately from the chromosomal DNA of an organism. Most usually plasmids are found in bacteria.

Plenary Jurisdiction: The determination that a designated court has complete jurisdiction over the matter before it.

Polycystic Kidney Disease: A fatal kidney disease due to the presence of numbers of cysts interfering with kidney function. Recessive.

Post partum: Following birth.

Presumption of Legitimacy: The presumption that a child born to a married couple is theirs. See also Lord Mansfield's Rule.

Public Fisc: Public funds or public treasury.

Punitive Damages: Damages awarded over and above compensatory damages. The purpose is to penalize the wrongdoer for what may appear to be a deliberate act, and to set an example so that others may not attempt the same act.

Qua Parents: Literally "as parents."

Recessive Gene: Gene which must be present in the homozygous condition in order to be expressed. Contrasted to Dominant Gene which is active in either a homozygous or heterozygous state.

Recission: The right to cancel a contract if the other party has not lived up to its terms.

Recombinant DNA: The ability to insert pieces of DNA from one organism into the DNA of another, different, type of organism. The usual technique is to employ bacterial plasmid DNA as the recipient of the DNA from any organism, and then reinsert the plasmid into the bacterium in the hopes that the latter will now make a product directed by the inserted genetic material.

Redhibition Law: Voiding of a sale if the object purchased is so inherently flawed that the buyer would not have purchased it were the defects known.

Remand: Return of a case to the original court for further action at that level.

Renal Dialysis: The use of a machine to replace kidney function. The machine serves to remove potential or actual compounds from the body; if these are not removed their presence will rapidly lead to a fatal condition. Used when either kidney transplants are not available or the transplant operation cannot be undertaken for various medical reasons.

Res Judicata: A matter already decided. Once a matter is decided the same issue cannot be brought to a second trial.

Rh Incompatibility: A condition whereby, due to the Rhesus factors in blood, a deformed or stillborn child will result.

Rogue Seeds: Seeds which do not give rise to the type of plants which would have been expected had the seeds been genetically pure.

Rubella: German Measles. A disease which, if incurred early in pregnancy, will invariably lead to a defective child.

Rubenstein-Taybies Syndrome: Small cranium, facial abnormalities, mental retardation, small stature, finger and toe anomalies. Probably recessive.

Schizophrenia: A mental disease characterized by pronounced alterations in concept formation, misinterpretation of realities and various other behavioral and mental disturbances. Although the hereditary pattern of this mental disease is not fully clear, it is acknowledged by most that there is a strong hereditary component to the acquisition of the disease.

Sensu Strictu: Literally in the strict sense.

Sex Chromosomes: The chromosomes which determine whether a person (or any mammal) is male or female. All normal females have two "X" chromosomes while normal males are "XY." The Y chromosome is considered to be relatively free from genetic traits; the X chromosome carries many genetic traits, such as color vision impairments,

Duchenne Muscular Disease, etc.

Sex-Linked: Genes carried on the X chromosome. As normal males have only a single X chromosome all genes will be expressed whether dominant or recessive. Females, heterozygous for a recessive trait, will not express the trait but are carriers for it. Anomalies in the number of sex chromosomes result in a type of chromosomal disease.

Sickle Cell Disease: Characterized by change in shape of red blood cells and subsequent inability to carry sufficient oxygen to the tissues. Carriers apparently normal. The heterozgyotes may be more resistant to malaria. Found mainly in blacks and other groups from the Mediterranean area. Recessive.

Social Darwinism: The extension of the Darwinian Theory of Evolution to socio-economic events.

Stare Decisis: Let the previous decision rule. If a suit is sufficiently similar to a previous suit, the decision in the earlier case will determine the present case.

Statute of Limitations: Laws governing the maximum time after an event when a suit dealing with that event may be brought. In the case of minors the laws may not be applicable until after the minor reaches legal majority.

Sua Sponte: Of his own will; refers to an action by a judge without apparently consulting extensive legal precedence.

Substituted Judgment: The approval by a court of a judgment in a case involving a minor or incompetent unable to act on his own behalf.

Surrogate Mother: A woman who carries a child for another couple when the wife is unable to carry out a successful pregnancy.

Syndactylism: A fusion of the fingers or toes. Dominant.

Taye-Sachs Disease: (Infantile amaurotic familial idiocy) An enzyme defect in the ability to store lipids. Fatal within a few years; newborns may seem normal. Recessive.

Teratogenic: An agent which will cause birth defects if a pregnant woman is exposed to it. For example, the drug thalidomide which causes children to be born without limbs.

Thrombocytopenia: Decrease in blood platelets leading to undue bleeding. Dominant.

Tort: A legal wrong. There are narrow legal restrictions as to what activities and relationships must be present before an act may be deemed tortious.

Tort-feasor: A person who commits a tort.

Transverse myelitis: A disease of unknown etiology due to inflammation of the spinal cord, resulting in back pain and loss of sensory functions below the point of inflammation.

Tubal Ligation: A surgical practice whereby a woman's Fallopian tubes are severed surgically. An egg ovulated from the ovary thus cannot be fertilized. When properly done this surgical procedure is a certain method of avoiding pregnancy.

Tuberous Sclerosis: A disease causing changes in the brain with symptoms resembling epilepsy and usually with mental deficiencies. Dominant.

Tutrix: A female guardian.

Type Culture Collection: A repository of organisms usually maintained by an institution in order to preserve certain genetic traits. Frequently the organisms such as bacteria are maintained in a frozen state in liquid nitrogen and can be brought from this state by appropriate defrosting techniques.

Uremic Syndrome: kidney failure, often preceded by anemia and acidosis. Uncertain genetic origin, thought to be dominant.

Vasectomy: The male equivalent of tubal ligation; surgical severing of the vas deferens so that sperm may not be released. Neither this process nor tubal ligation interferes with libido.

Von Recklinghausen's Disease: see Neurofibromatosis.

Weight of Evidence: The determination by a jury whether the evidence is of sufficient import as to render a decision in a certain manner. Contrast to Admissibility of Evidence.

Wrongful Birth: Suits brought by parents claiming that a child would have been aborted had proper advice been given. Recognized as a legal claim in many states.

Wrongful Conception: Suit brought by parent for having conceived, usually after a failed vasectomy or tubal ligation. Recognized as a tort equivalent to malpractice.

Wrongful Life: A suit brought by an infant claiming the infant would have been better off not to have been born. Recognized as a tort by only a few states.

X & Y Chromosomes: The chromosomes controlling sex.

XO Condition: Turner's disease. Women of this abnormal genetic condition are usually short in stature, show no secondary sexual changes such as breast development or menstruation at puberty. They are also sterile as the result of this chromosomal disease, caused by the lack of a second X chromosome.

XYY Condition: Individuals of this abnormal genetic condition were once thought to have violent or criminal dispositions and in some countries this was accepted as the equivalent of an insanity plea. Now thought not to

be valid as a claim against criminal acts as most
criminal acts committed by XYY males are now believed to
be against property, not persons. XYY males are sterile
as the result of this chromosomal disease.

Xray Damage to a Foetus: Use of Xrays during early
pregnancy is known to be teratogenic.

ALPHABETICAL LIST OF CASES CITED

The first number refers to the page where the case is discussed; the number after the colon refers to the exact citation reference. Where the only number given follows the colon, the case is included in the citations for completeness, but is not discussed in the text. If the number is only before the colon the case is described briefly in the text but not fully referenced.

STUDIES IN HEALTH AND HUMAN SERVICES